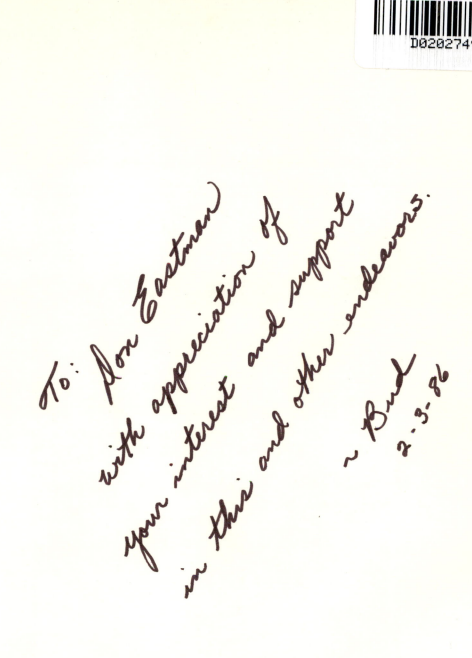

To: Don Eastman

with appreciation of

your interest and support

in this and other endeavors.

~Bud

2-3-86

LATIN AMERICA

LATIN AMERICA
Fifth Edition

Preston E. James
Syracuse University (Emeritus)

C. W. Minkel
University of Tennessee

with assistance by
Eileen W. James

JOHN WILEY & SONS

New York   *Chichester*   *Brisbane*   *Toronto*   *Singapore*

Cover and interior design by Kevin J. Murphy

Copyright © 1942, 1950, 1959, 1969, 1986, by John Wiley & Sons, Inc.

Library of Congress Cataloging in Publication Data:

James, Preston Everett, 1899–
    Latin America.

    Bibliography: p.
    Includes index.
    1. Latin America.   I.   Minkel, Clarence W.,
1928–   II James, Eileen W.   III. Title.
F1408.J28   1985   980   85-9519
ISBN 0-471-05934-X

Printed in the United States of America

10 9 8 7 6 5 4 3 2 1

To the geographers of
Latin America
whom we have known,
and with whom we have worked,
this book is respectfully
dedicated.

PREFACE

This book is a geography of Latin America. A fundamental objective of geography is the understanding of area, which in this case is translated to mean an understanding of Latin America from a spatial, or geographic, perspective. That objective is a large assignment for author and reader alike. The coauthors of this book have dedicated their professional careers to teaching and conducting research related to Latin America. They have spent much of their lives traveling and working within the region. Yet, Latin America is so vast and complex, and so rapidly changing, that no amount of exposure can provide an adequate degree of expertise. Latin America plays an increasingly important role in world affairs. We must know what that role is, and what to expect from it in the years ahead. This involves a knowledge of places, facts, and concepts. It also requires a dedicated investment of time and effort to achieve some degree of genuine understanding.

Latin America is one of the world's great culture regions and includes that part of the Americas lying southward, or more specifically southeastward, from the United States. Similarly, the northern part of the hemisphere, including Canada, the United States, and Greenland, is referred to as Anglo-America. Such terms are generalizations, of course, and should not obscure the fact that each of the two major culture regions includes some areas that are neither Latin nor Anglo in historical background or cultural identification.

South America and North America are continents, hence physical regions, divided arbitrarily at the border between Panama and Colombia. Less clearly defined are the terms "Middle America" and "Central America." Middle America as a geographical term probably owes its origin as much to pedagogical convenience as to any reality, but is most commonly considered to be that part of North America lying between the United States and mainland South America, including Mexico, Central America, and the islands of the West Indies. Central America may be considered to include all of the isthmian countries between Mexico and Colombia. However, due largely to historical reasons, Belize and Panama are often excluded.

Within Latin America there are nearly 40 independent countries and numerous dependent territories. Most of the latter are islands, and they constitute only a tiny fragment of what were formerly vast overseas empires. Probably the most outstanding characteristic of the entire area is its diversity — physical, cultural, economic, and political. Even within the individual countries conditions are so diverse that it has been difficult, in many cases, to achieve national unity and sustained progress. In recent decades, nevertheless, substantial progress has been made, at least in material welfare for most segments of the populace. This, in turn, has led to rising expectations, especially since the advent of modern electronic systems of communication. Growing demands for a share of "the good things in life" for all segments of society have placed increased strain on fragile political, economic, and social structures in the developing countries throughout the world. In Latin America, as elsewhere, this condition is reflected in political instability. Unfortunately, it is the acts of terrorism, the *coups d'état*, and the revolutions which capture the attention of the news media in Anglo-America and Western Europe.

• • •

It must now be recognized, in the United States, that successful dealings with Latin America can be effected only on the basis of mutual respect and understanding. Countries of the region are no longer willing to be treated as political and economic dependents of Western Europe or the United States. External aid for development will be accepted from the technologically advanced nations and from transnational corporations, but only on terms that properly recognize local sovereignty and domestic interests.

A heady nationalism will prevail for some years to come, reflecting the newfound sense of self-determination. This already has been illustrated in actions such as the expropriation of foreign investments, membership in international cartels to control raw material prices, territorial disputes, and the expansion of national sovereignty to include extensive areas of adjacent seas. There is a growing sensitivity to any foreign pressure, especially if it touches on conditions considered to be of purely domestic concern. Yet, while feelings of nationalism may be strong, it has been shown that no people can in isolation carve out their own destiny in the modern world. All are interdependent, and so will remain.

Despite substantial economic growth in recent decades within the region as a whole, many problems and opportunities remain. These form the major themes of the present edition. The problems, in most cases, are not new. Rather, their origins can be traced to conditions that have prevailed for extended periods of time. Their solutions are therefore not likely to be found within any given decade. Development, or modernization, must be recognized as a continuing endeavor — one that has no specific point of termination. Moreover, while both foreign and domestic writers have tended to dwell on the problems of Latin America, it would be unwise to shroud oneself with pessimism when studying the region and contemplating its future. The population of Latin America is young, resourceful, and energetic; natural resources remain to be developed, and there are frontiers yet to be conquered. The region, and each country thereof, is worth knowing and understanding.

In writing this volume on Latin America, the authors have received valuable assistance from many sources. The maps, more than 100 in total, were prepared by the Center for Cartographic Research and Spatial Analysis of Michigan State University, under the direction of Richard M. Smith and Michael Lipsey. Statistical reports were gathered and forwarded from many countries by the local project directors of the Inter-American Geodetic Survey, whose collaboration in any worthy cause is an institutional tradition. The photographs were supplied largely by the Inter-American Development Bank and by the

World Bank, as a generous contribution. The manuscript was typed at the University of Tennessee by Jean Lester and Frances Houser, and Hugo Bodini contributed substantially to the development of the Bibliography. Throughout all aspects of the writing and publication, the contribution of Eileen W. James has been well beyond measure. To these and the many others who have assisted in this endeavor, the authors express their sincere gratitude.

Preston E. James
C. W. Minkel

CONTENTS

INTRODUCTION

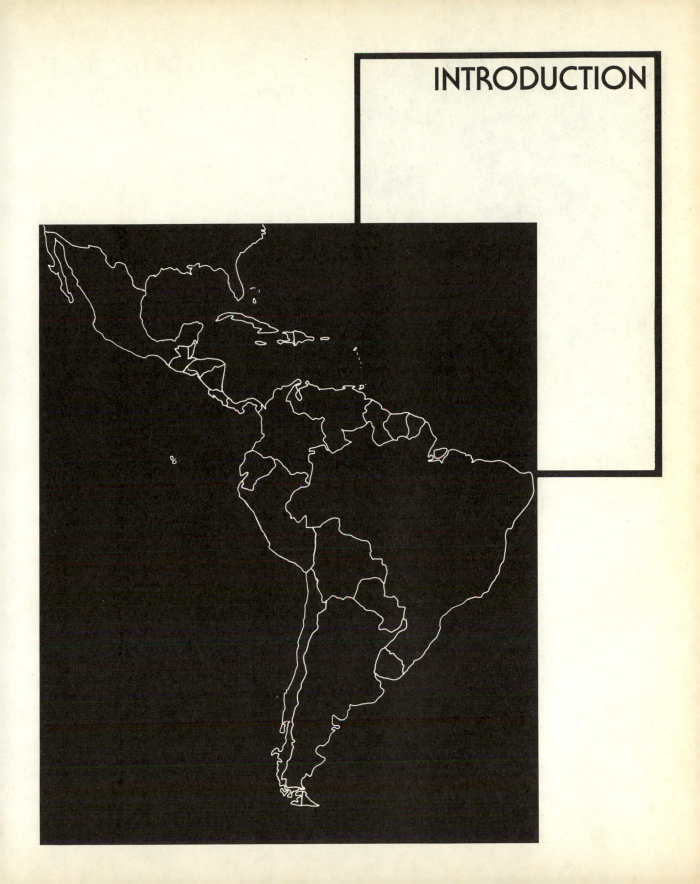

PRINCIPAL
CHARACTERISTICS

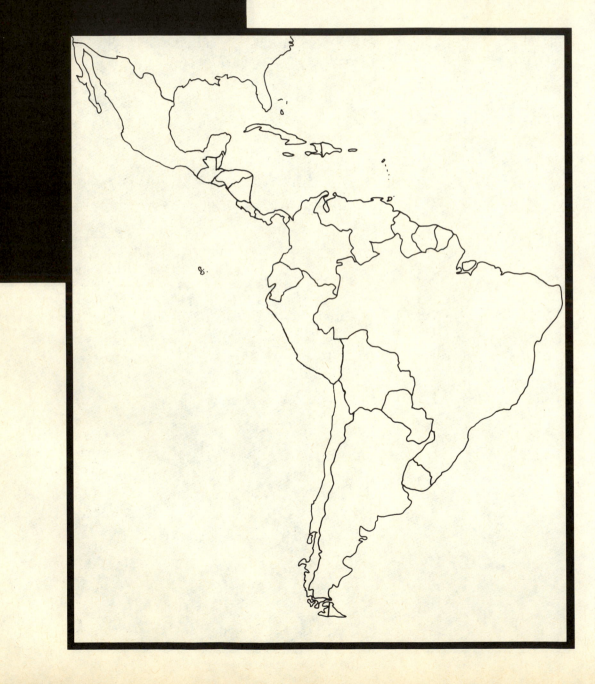

No one can have followed the news from the Americas in recent years without a realization that Latin America is an area of worldwide concern. The Revolution in Cuba in 1958, and that country's subsequent support of international communism, caused major political upheaval throughout the hemisphere. The discovery of vast reserves of petroleum in Mexico, along the Gulf Coast, has altered that nation's economic prospects and its potential role in world affairs. Central America has become a battleground with international overtones, and in the Lesser Antilles many small islands have become independent nations, with full participation in the United Nations and other international organizations. Brazil has demonstrated the potential to become one of the world's major powers of the twenty-first century, and Argentina has challenged the United Kingdom, by open warfare, over control of the Falkland or Malvinas Islands. To citizens of the United States, whose attention is usually focused on domestic concerns or events in Europe, a flood of illegal aliens from lands to the south has created conditions that can no longer be ignored.

Few parts of the earth have experienced change more rapidly in recent decades than has Latin America. This change has been mostly stimulated by striking improvements in the technical aspects of transportation and communication since World War II. As a result, millions of people in Latin America are joining in a demand for better living conditions and for an end to the inequities that have traditionally characterized their society.

Both the people and the natural resources of Latin America have long been exploited. Some lands were worked and abandoned by Indians long before the arrival of Columbus. In the centuries that followed Columbus, the so-called New World was ransacked by Spaniards, Portuguese, French, British, Dutch, and other people of European origin. Latin America is not, then, a virgin land. It is an old land in which many of its sources of accumulated treasure have been depleted and deserted, and in which many of its landscapes have been profoundly altered by human action.

What, then, are the principal characteristics of Latin America about which Americans of all nationalities should be aware? There is, of course, an almost infinite number of answers that could be given. Yet, certain general similarities seem to prevail, each with its own subset of features and factors. Six characteristics especially worthy of consideration are: (1) physical diversity of the land, (2) cultural diversity of the people, (3) inequalities in the human condition, (4) rapid population growth, (5) nationalism, and (6) emphasis on development.

Organization of American States

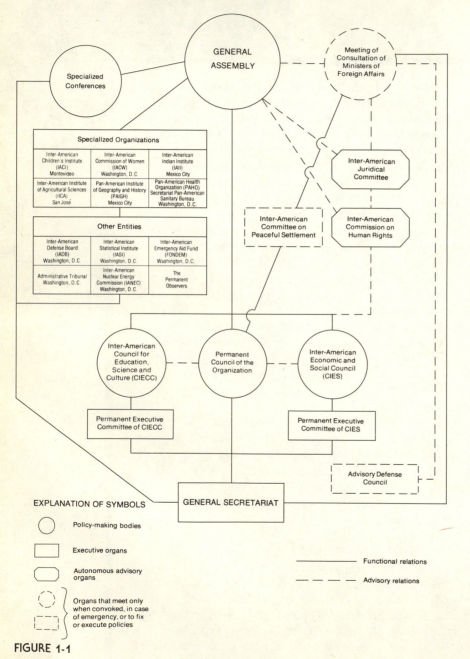

FIGURE 1-1

Diversity of the Land

Physical diversity is characteristic of almost any major world region and is one of those features that make the earth such an interesting planet on which to live. Yet, that diversity is particularly evident in Latin America and must be understood to appreciate the variety of circumstances in which human societies within the region have evolved through time.

In Latin America tectonic forces are responsible for such features as deep oceanic trenches off the Pacific Coast from Mexico to Chile; volcanism in central Mexico, Central America, the Lesser Antilles, and the High Andes; and extensive folding of the rock structure under, and emerging from, the Caribbean Sea. Earthquakes of devastating force have almost completely leveled major cities such as Antigua, Guatemala; Managua, Nicaragua; Port Royal, Jamaica; and Popayán, Colombia. Least affected are ancient crystalline rock areas, such as the Brazilian Highlands, which comprise the core of the South American continent.

Surface Features

The geologic structures and surface forms of western Canada and the United States continue southward into Mexico. Conspicuous features are the elongated peninsula of Baja California and the Mexican Plateau, separated by a rift valley occupied by the Gulf of California and adjacent lowlands. The Plateau rises in elevation as it extends southward from the U.S. border to about latitude 20°N. There, it is abruptly terminated by a northwest-southeast chain of towering volcanoes. Southern Mexico, Guatemala, Honduras, Belize, and much of Nicaragua, belong to a structural region that extends under the Caribbean eastward to Jamaica, southeastern Cuba, Hispaniola, Puerto Rico, and the Virgin Islands — an area of folded and faulted rocks with a generally east-west trend. Between this "Central American-Antillean" region and connected to South America lie two chains of volcanic mountains: the Lesser Antilles and the highlands of El Salvador, southwestern Nicaragua, Costa Rica, Panama, and western Colombia.

There is as well a pronounced eastward orientation in the entire southern portion of the North American continent. Thus, anyone traveling straight southward from any point in the United States west of Pittsburgh, Pennsylvania, would miss the South American continent entirely. This eastward thrust in Middle America and eastward location of South America help, of course, to account for the significance of Miami, Florida, as a center for travel and trade with Latin America.

Three main surface divisions form the major lineaments of the continent of South America. On the west are the relatively young Andes Mountains, paralleling the Pacific Coast and flanking the Caribbean. On the east are the Guiana and Brazilian highlands, which are geologically much older than the Andes. In the central portion of the continent lie the plains of the Orinoco, the Amazon, and the Paraguay-Paraná-Plata.

The massive ramparts of the Andes stand unbroken from Trinidad to Tierra del Fuego. They are barely 200 miles wide, except in Bolivia where the width is doubled. Many peaks exceed 18,000 feet in altitude, and Mount Aconcagua (22,834 feet) is the highest mountain in the Western Hemisphere. The Andes are formed mostly of folded and faulted structures, but in three distinct areas there are groups of active volcanoes. These are in southern Colombia and Ecuador, in middle and southern Peru and along the border of Bolivia and Chile, and in the southern part of middle Chile.

South America east of the Andes is composed mostly of highlands, which extend with few interruptions from southern Colombia and Venezuela across Brazil to the northern bank of the Río de la Plata and reappear in Patagonia. Throughout this vast territory three main surface elements predominate. There is a base of ancient crystalline rocks that forms a hilly upland; above this in a few places are the stumps of old, worndown mountains; and covering the crystalline base, especially in the interior, is a residual mantle of stratified rock, forming tabular plateaus with steeply scarped margins.

Between the sandstone strata in southern Brazil, and in small patches throughout eastern South America, are sheets of dark-colored lava known as diabase. The diabase is especially resistant, and the edges of the lava sheets stand out prominently as cuestas. Some of the great waterfalls of South America occur where rivers plunge over the edge of the diabase formations. The Paraná Plateau of southern Brazil is one of the world's largest accumulations built by successive lava flows, similar in origin to the Columbia Plateau in the northwestern United States or the Deccan Plateau of India.

The plains of South America are arranged quite differently from those of North America and occupy a much smaller proportion of the continent. The Orinoco Plain is separated from the Amazon Plain by a belt of highlands. The Amazon Plain, which is wide along the eastern base of the Andes, narrows to only a ribbon of floodplain along the great river east of Manaus. In contrast to North America, there is no substantial area of coastal plain along the Atlantic or Pacific.

There is another significant difference between the patterns of the two continents. Because the highlands of Brazil reach their greatest elevation in southeastern Brazil, near the coastal city of Rio de Janeiro, where the highest summits are just under 10,000 feet above sea level, the principal rivers flow inland from this region. The tributaries of the Paraná rise within a few miles of the coast in São Paulo state, flowing northwestward and then south. Those of the São Francisco, and other rivers that eventually reach the Amazon, also flow northward, away from the southeastern coast.

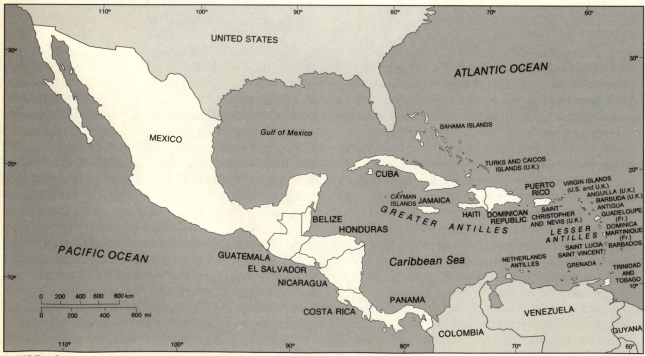

FIGURE 1-2

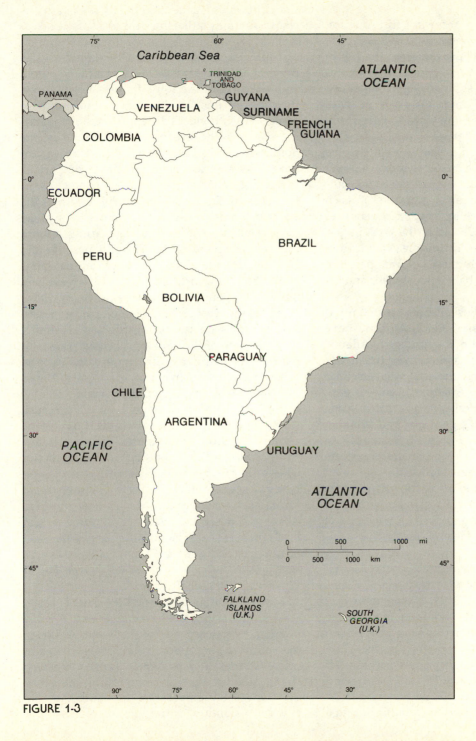

FIGURE 1-3

Climate

Many different climates occur in Latin America. Great extremes of temperature, however, are rare. Low latitude and proximity of the sea in most parts of Middle America help to make this one of the world's most genuinely "temperate" regions. Very hot summers do occur in small areas of Mexico, such as along the lower Colorado River, the northern part of the Mexican Plateau, and in the lower Río Balsas Valley to the south. Low temperatures are even more unusual. During winter in the Northern Hemisphere, outbursts of cold Arctic air ("northers") reach northern Mexico and may lower temperatures even in Central America and in the West Indies. Snow is extremely rare in Middle America but can be seen atop the highest volcanic peaks of Mexico, such as Orizaba, Popocatépetl, and Ixtaccíhuatl.

The great wind and ocean currents that influence the climate of most of Middle America move in a clockwise direction in the North Atlantic Ocean. Particularly important are the Trade Winds which flow across the equatorial latitudes and approach Middle America from the east. These winds are remarkably steady, moving at a speed of 10 to 15 miles per hour and bringing copious rainfall to east-facing slopes that lie in their path. A similar clockwise circulation occurs in the North Pacific Ocean. Hence, the prevailing wind and ocean currents off northwestern Mexico are from the northwest, or more poleward areas, and have cooler and drier climatic effects than do those on the eastern side of the continent.

Hurricanes, called "typhoons" in the Eastern Hemisphere, are a special feature of the Caribbean area. These are violent, rotating storms that originate over the warm waters of the Atlantic Ocean and gather momentum as they move westward, usually during the period July through October. These air masses may exceed several hundred miles in width and attain wind velocities well in excess of 100 miles per hour. As many as 10 or 12 hurricanes may occur during a

single year over the Caribbean, while an occasional hurricane generated over the eastern Pacific Ocean may strike the west coast of Mexico as well.

South America, poleward of latitude 40°S, projects a relatively narrow area of land into a wide expanse of the southern ocean. Therefore, the types of middle and high latitude climates associated with places distant from the moderating effect of the oceans are not found in the Southern Hemisphere. Nowhere, even in southern South America, are the winters comparable to those of Canada or the northern United States. In Tierra del Fuego, more than 50° south of the equator, temperatures average above 32°F in the coldest month, although in the warmest month they average below 50°F. The only part of Latin America, outside of Mexico, where temperatures above 110°F occur at any time of the year is in northern Argentina.

Along the west coast of both North and South America the sequence of climatic types is similar on either side of the equator. The cool, rainy climates of the higher middle latitudes on this coast occur in southern Chile and in southern Alaska and British Columbia. The coast of middle Chile, like coastal California, enjoys a Mediterranean type of climate, characterized by mild, rainy winters and cool, dry summers. Slightly equatorward of 30° on both continents, rainfall diminishes until only drought-resistant, or *xerophytic*, types of vegetation can survive. Southern California and northwestern Mexico, like northern Chile, are deserts. Equatorward from the dry lands the west coast is moist, especially that part of the coast in Panama, Colombia, and Ecuador which is bathed by warm water of the Pacific Equatorial Counter Current.

Between Bahía Blanca and the Strait of Magellan, the desert extends to the east coast — a rare phenomenon in any part of the world. Although this southern part of South America is crossed by many storms and is noted for its blustery, changeable weather, it receives little moisture. The ranges of temperature in South America reach a maximum in the interior of Argentina slightly north of the latitude of Buenos Aires, where the difference between the average of the warmest month and that of the coldest month is about 30°F.

Three-fourths of South America and most of Middle America lie within the tropics. This fact may be misleading, however, because of the common tendency to associate ill effects with a tropical climate and to think of all tropical climates as being more or less alike. Important distinctions are to be made between the cool, cloudy desert of the Peruvian coast, for example, and the hot, rainy conditions of Guyana or the hot, dry conditions along the Caribbean Coast of Venezuela. The highest temperatures in South America, as on other continents, do not occur along the equator, as is commonly assumed, but on the border between middle and low latitudes during the summer months. The parts of South America with temperatures averaging above 80°F in the warmest month are along the Caribbean and Guiana coasts, throughout the Amazon Basin, and across the plains of northern Argentina.

Climatic diversity is especially great in the Andes. In any mountain region the variations of exposure to sunlight and to rain-bearing winds produce intricate patterns of local climate, but there are also general altitudinal zones based chiefly on the decrease of temperature with increasing elevation. High-altitude climates are not at all similar to climates of the middle latitudes, for with increasing elevation above sea level the seasonal range of temperature decreases until it practically disappears.

Vegetation and Soils

Closely reflecting the conditions of climate, water, and soil are the various associations of plants that distinguish one region from another. The vegetation, or cover of noncultivated plants, is predominantly forest where water is abundant. Where water is less plentiful there may be woodlands composed of smaller and more widely spaced trees. In addition, in the transition zone between abundant and insufficient water there are areas of pure grassland or a mixture of grassland and scrub woodland. Where water is deficient, the plant cover is less dense, with bare ground between the plants. This is a distinguishing feature of dry-land vegetation. Temperature also has its effect on the nature of the plant cover. Tropical plants cannot survive frosts, whereas mid-latitude plants exist in climates where frost comes often enough to eliminate those species that are not resistant to it.

Wherever human beings have settled during the thousands of years they have occupied the Americas, the original vegetation has been profoundly altered. The fires of prehistoric people, started as an aid in hunting animals, pushed back the woodlands, increased the area of grasslands, and changed the nature of forests. Yet, the major types of vegetation that occur today have been in existence since prehistoric time, hence long enough to permit the development of distinctive soils and distinctive animal populations.

Vegetation can be classified into four broad groups. The *forests* consist of tall, straight-stemmed trees growing so close that their branches interlace. *Woodlands* are composed of smaller trees, gnarled rather than straight-stemmed, and usually spaced so that their branches may touch but are not interlaced. The *grasslands* may be without trees, or with a mixture of trees and grasses. *Deserts* include areas where there is bare ground between the plants, or where there are no plants at all.

The nature and quality of soils depend on the native rock material from which they are produced, and the influence of climate and vegetation. For example, volcanic soils are often thought to be fertile, but fertility depends on the mineral content of the material erupted and the length of time it has been weathered. Dry-land soils, including those of deserts, tend to be high in mineral content, but their productivity may depend on the availability of water for irrigation. Where temperatures are high throughout the year and rainfall is heavy, soils commonly are poor due to leaching of their minerals and the lack of any season in which they are frozen or otherwise in a dormant state. Yet another factor in soil fertility is erosion.

Diverse Racial and Cultural Elements

Another principal characteristic of Latin America is that the population is composed of diverse racial and cultural elements. In 1982 the total population of about 378 million was divided among 36 independent states, plus various other units with political ties outside the region. There were 231 million people in 18 countries who descended from Spain, nearly 128 million in Portuguese-speaking Brazil, 7 million in French-speaking countries, and more than 5 million in English-speaking countries that until recently belonged to the United Kingdom. There were also 3.3 million in the self-governing Commonwealth of Puerto Rico, about 25,000 in the Netherlands Antilles, and 100,000 in the American Virgin Islands. There are few native Indians in the Caribbean area, but many remain in Mexico, Central America, and the Andean countries of South America.

The Indians

Long before the arrival of European explorers, the Americas were occupied by people from Asia. These first native Americans came into the American Hemisphere by way of the Bering Strait, possibly at a time when the sea level was lower than it is today. Some tribes wandered off to the east, but the main current of repeated migration led southward, some groups even pushing through Central America into South America. There is little probability that any important numbers came by boat across the Pacific Ocean, and even less probability that any came from Africa across the Atlantic.

Certain native peoples of America were able to take a step that only a few groups in the history of humankind have equaled — they lifted themselves "from barbarism to civilization." These more elaborate cultures were developed by the

Mayas of Guatemala and Yucatan, the *Aztecs* and a few other groups in central and northwestern Mexico, the *Chibchas* of highland Colombia, and the *Incas* of highland Peru, Ecuador, Bolivia, and northern Chile. The cultural advance made by these people was reflected in a great increase of population that could be supported by the land. About three-fourths of all the native people in America at the time of the European conquest were located in the territories of these four advanced cultures. The explanation is that a sedentary agricultural economy supports many more persons within a given area than does an economy based on shifting cultivation or migratory hunting and fishing. Each of these cultures was based on that distinctly American food grain, maize.[1] In addition, the natives of Middle America made use of manioc,[2] beans, squash, tomatoes, tobacco, and cacao.

Both the Maya and the Inca civilizations passed the zenith of their development before the arrival of Europeans. The Mayan was the first great Indian civilization to evolve in pre-Columbian times, and it was already decadent by 1492. The Aztec and Inca empires were formed by the conquest and assimilation of formerly separate and distinct Indian groups. The nucleus of the Aztec state was in the Basin of Mexico, and from there political control was extended over a wide area. *Nahua*, a division of the linguistic family of *Uto-Aztecan*, is the language of the Aztecs, but their empire did not include all the Nahua-

---

[1] *Maize* is a word that refers to the grain which in the United States is called *corn*. According to general English usage, however, corn refers to any common grain. In England *corn* refers to wheat, and in Scotland to oats. In this book the word *maize* is used to designate Indian corn, the predecessor of modern hybrid corn. It is believed that maize was first domesticated in Mexico, a little south of present-day Mexico City.

[2] Manioc is also known as manihot, mandioca, cassava, and yuca. The last-named is not to be confused with yucca, a genus of the family *Lillaceae*. Manioc is a plant with an edible root which furnishes a starchy food widely used throughout the tropics, but formerly known only in the Americas. It is now produced commercially as the source of *tapioca*.

speaking tribes of Mexico. The Toltecs, also a Nahua-speaking tribe, dominated the center of Mexico in the tenth – twelfth centuries. The Incas extended their conquest from an original nucleus in the Basin of Cuzco and brought together in one great empire the various tribes included in the two linguistic families, the *Quechua* and *Aymara.* By the time of the Spanish conquest, however, the Inca state was already torn by civil strife. Another group, the Chibchas, belonged to a politically complex society, with an economy that stressed intensive agriculture and active trade with neighboring tribes.

The parts of America outside the territories of these relatively advanced Indian cultures were only thinly populated except for Hispaniola. Most of the hemisphere was occupied by many separate Indian groups, only vaguely related in certain broad linguistic families. The tribes of tropical America ranged from seminomadic hunters, fishers, and primitive farmers, such as the Caribs and Arawaks of northern South America and the Antilles, to shifting cultivators whose basic crop was manioc, including the Tupi and Guaraní of Brazil and Paraguay. In southern South America there were seminomadic hunters and fishers who practiced some incidental farming, such as the Araucanians of Chile, and warlike nomadic hunters, such as the Abipones and Puelches of the Argentine Plains, whose chief food supply was derived from the wild guanaco. These diverse tribes collectively comprised only about a fourth of the total population of the American Hemisphere at the time of Columbus.

The Europeans

Diversity of race and culture in present-day Latin America can be attributed not only to the original native inhabitants but also to the European conquerors. One can scarcely understand the Spanish or Portuguese conquest of the New World without considering the centuries of conflict between Christians and Moslems which immediately preceded the discovery of America. Iberian society was grouped in small, semi-independent units, each under the control of a lord. In each group, the people were sharply divided: on the one hand were the aristocratic landowners and the *caballeros* who supported them; on the other were the serfs, or *peones,* bound to the land and dependent on the lords and fighting men for protection. Ownership of an estate, service in the army, or the priesthood were the only real roads to prestige. Commerce and industry were left to the Jews and Moslems and in the cities entire districts were set aside for these people.

Two important events occurred in 1492. The Moors were defeated in battle and were forced to give up Granada, their last territorial possession in Spain; and Columbus discovered America. The energies of a united Spain were thus released for a conquest of the New World, but gold was needed to repair the war-wrecked economy of the country, and there prevailed a strong desire to kill infidels or to convert them to the service of God.

The relatively dense populations of the Aztecs and Mayas in Mexico, and those of the Incas and Chibchas in South America, exerted a special attraction. These peoples had already accumulated stores of treasures and, after the conquest, were ready to accept Christianity and begin working for their new masters. Since the Indians had no concept of private landownership, and no perception of the value of gold and silver in a commercial sense, the conquest was at first no more than a change of rulers.

Newcomers, after landing on the shores of the Caribbean in northern Colombia, Panama, and Vera Cruz, pushed their exploring parties far to the north and south. In the process of conquering people and seeking material goods, great sources of precious metals were discovered and partly exploited. The Spaniards combed the land for new sources of wealth and were motivated always by the prospect of sudden riches. They were not stopped by physical obstacles but were attracted to areas where there was a readily available laboring class.

Meanwhile, the Portuguese were settling Brazil. As a result of the Treaty of Tordesillas between Portugal and Spain in 1494, the Portuguese were entitled to all lands that might be discovered east of a line drawn north and south 370 leagues west of the Cape Verde Islands — approximately at 50°W longitude. The Portuguese in Europe had already achieved a much greater degree of national unity than was the case in Spain. Their interest in the New World centered less on the spread of Christianity and the opportunity to implant their institutions than on the discovery of new and needed sources of wealth with which to bolster the fortunes of the homeland.

The Africans

The use of African slaves to produce sugarcane was first developed on the Portuguese island of Madeira, but soon thereafter the Spaniards started their own sugar plantations on the Canary Islands. As early as 1505 some Africans were brought to Hispaniola to dig for gold in the stream gravels, but the first sugarcane plantation in the New World was established in 1515 near Santo Domingo. The Spaniards imported African slaves and planted sugarcane not only in Hispaniola but elsewhere in the Greater Antilles and in lowland areas of the mainland.

During the seventeenth century the Dutch attacked and occupied the prosperous Portuguese sugar colonies on the northeast coast of Brazil. They remained there for about 30 years, during which time they learned how to grow the cane, extract the juice, and prepare the sugar for export to the European market. When the Dutch were expelled by the Portuguese they established colonies in the Antilles, where their advanced technology was quickly passed on to the Spaniards, French, and British. The British seized Jamaica from the Spaniards, and the French gained possession of what is now Haiti. Soon British, French, Dutch, and Danish colonies appeared in the Lesser Antilles and in Dutch Guiana (Suriname), mostly in areas where Spanish settlement had been ineffective or non-existent. Wherever sugarcane was cultivated on a commercial scale in Latin America, African slaves soon outnumbered their European masters by a wide margin. Throughout the Caribbean area most of the people are of African origin, as are those of adjacent mainland areas.

The mixture of Africans and Europeans produced a racial type known as *mulatto,* whereas the term *zambo* refers to a person of mixed African and Indian origin. The zambo is less common in Latin America than either the mestizo or mulatto, however, since African slaves were not imported in large numbers to places where Indians provided an adequate supply of labor.

Recent Immigration

More recently, another element has contributed to the racial and cultural diversity of Latin America. Beginning in the nineteenth century, new immigrants from Europe, the Middle East, and Japan have come to the New World. The overwhelming majority of these immigrants have entered the United States, yet a considerable number have also gone to certain parts of Latin America. Consequently, these parts are quite different from the rest of the continent in their racial and ethnic composition. In most cases, this new colonization has been in regions that were previously little developed. From São Paulo across the southern states of Brazil, across Uruguay, and over the Humid Pampa of Argentina there was an expanse of territory almost devoid of precious metals and only thinly occupied by native peoples. Most of it was unsuited to sugarcane. Except for its strategic importance to the rival colonial empires of Spain and Portugal, this part of South America was of little use to the early conquerors. Because Spain had established a center of settlement in Paraguay and had used the Río de la Plata as a "back door" to Peru, and because Portuguese colonists had threatened to establish themselves permanently on the shores of the Plata, the Spaniards paid some attention to this strategic route. Yet, land bordering the Plata was only a remote part of the Spanish colonial empire and was used only for the grazing of cattle and mules.

This stretch of territory is composed mostly of

the descendants of people who came to South America during the past hundred years—Italians, Spaniards, Portuguese, Germans, Poles, and people of many other nationalities, including the Japanese.

In Latin America racial composition differs notably from one region to another. Six principal areas can be identified in terms of present-day racial characteristics: (1) areas of predominantly European population; (2) areas of predominantly Indian population belonging to the Quechua and Aymara linguistic groups; (3) areas of predominantly Maya Indians; (4) areas of predominantly Indian population descended from other than the Quechua, Aymara, or Maya groups; (5) areas with a mixed population, largely African; and (6) areas with a predominantly mestizo population.

During the past several decades, migration to Latin America has been limited, largely because the region is already quite densely populated in relation to its natural resources and is, itself, a source of migrants to other areas. However, some Mennonites from Canada and the United States have moved to northern Mexico and Belize; Salvadorans have gone to Honduras and Guatemala; Syrians and Lebanese have migrated to northern Honduras; and Jamaicans and Chinese are conspicuous elements in the population of urban centers, such as Panama City. Almost 17 percent of the Uruguayan population has emigrated, especially to Argentina and Brazil,[3] while a tide of Brazilian immigrants has populated the northern and eastern border areas of Paraguay. Much earlier, Hindus from India migrated to Trinidad as indentured servants, and their descendants still have an important influence on the island's economy and culture.

Human Inequities

The conquest of what is now Latin America occurred well in advance of the French Revolution,

with its radical notions of "liberty, equality, and fraternity." The native Indians were defeated, subjugated, and converted. Great numbers were also killed, either in battle or by diseases introduced to the New World by the Europeans. Land, which constituted the principal source of wealth, other than mines, was divided into large estates.

Wherever Spaniards settled in significant numbers in the New World, they established the *encomienda*, out of which the *hacienda* system evolved. This involved large areas of land that were given to a relatively few select individuals and especially to officers of the conquering armies. Indians already on the land were permitted to remain and were allowed to cultivate small plots of ground, often on hillsides or other less productive sites, to produce food for their own use. At the same time they were required to contribute their labor to the landowner, who normally devoted the best land to one or two high-value commercial crops. Thus, a class system was established, similar to that of Spain, in which the landowners became relatively wealthy and the peasants remained landless and poor.

In the late eighteenth century, new concepts of freedom began to spread around the world. During the first half of the nineteenth century, Mexico and most Spanish colonial areas of Central and South America established themselves as independent nations. Among the patriots who fought for independence were some who were genuinely devoted to the cause of individual liberty and equality before the law. However, most of the revolutionary leaders came from small circles of politically conscious people whose concept of freedom centered around a desire to be free from outside interference. In each country there were political leaders who demanded, and in some cases achieved, the right of local self-government.

First among the Latin American nations to achieve independence was Haiti, in 1804. This was in large part the result of a rebellion against French control, which included the massacre of most white inhabitants of the country. The great plantations of sugarcane, with their mills and irrigation works, were abandoned and fell into

---

[3] R. H. McDonald (1982): "The Struggle for Normalcy in Uruguay," *Current History* 81:472 69–73 ff.

ruin. The new leaders who acquired positions of power and privilege were not greatly concerned with the welfare of the general public. This was *political* independence, and most other countries of Latin America soon gained similar independence.

It was not until the Mexican Revolution, beginning in 1910, that problems of economic and social inequality were addressed in any major way. As late as 1930, almost 90 percent of the agricultural land in Mexico was under the control of haciendas. Then, many of these great estates were expropriated, and the land was distributed to the underprivileged rural population. This revolution, often described as a "revolution spelled with a capital R," brought about fundamental changes that benefited directly the great majority of Mexican people.

With the rapid advance in technology related to transportation and communication following World War II, the horizons of most Latin Americans were no longer limited to the boundaries of the hacienda or plantation on which they lived. The national road network was expanded, and trucks and buses reached every hamlet. The inexpensive transistor radio brought almost everyone into contact with the rest of the world.

The Cuban Revolution of 1958, the Nicaraguan Revolution of 1979, and current upheavals in Guatemala, El Salvador, and elsewhere reflect an increased desire of people to improve their immediate circumstances and to have a greater voice in their own destiny. Personal freedom, historically subject to rigid constraints, appears likely to be an increasingly critical issue in the years ahead, not only in Latin America but also throughout the world.

Rapid Population Growth

Compared with its land area and resource potential, Latin America has been only thinly occupied during most of its history. This condition changed rapidly with the advent of modern medicine, which resulted in a sharp decline in mortal-

ity rates and consequently lengthened greatly the average human life span. With population increase there has been a corresponding increase in pressure on the land. More land has been cleared of its original vegetation, more land has been placed under cultivation, and more land has been permanently damaged by erosion and the depletion of its fertility.

The population of Latin America more than doubled between 1940 and 1970, from 126 million to 278 million, and increased by more than a third to an estimated 406 million by 1985. It is estimated that by the year 2000 it will have attained a total population of 554 million, which will exceed that of Europe (492 million) or Anglo-America (264 million) by a wide margin. As a consequence of such rapid growth, more than 38 percent of the population is under 15 years of age. More than half of the age groups 5 to 14 is not enrolled in school, and adult unemployment is a major problem throughout the region. Yet, in 1985 Latin America included less than 10 percent of the total world population, and most countries included empty frontier areas remaining to be settled.

With the increase of population in rural areas has come an intensified demand for employment, land reform, and for colonization of the frontiers. Landholdings have been divided in some areas to the extent that individual properties are too small to provide a decent standard of living to their inhabitants. Hence, there has been a rapid increase in migration to the cities, mostly by young people and especially to the national capitals. Urbanization has proceeded at an uncontrolled pace, accompanied by all the symptoms of excessive growth—including substandard housing, unemployment, traffic congestion, pollution, lack of public services, crime, and political instability.

In the larger countries of Latin America, most migrants move from farms to nearby towns, and then to state or department capitals before reaching the great metropolitan centers. This process, known as *stage migration*, may or may not be completed within a single generation.

With a rapid increase in population, migration, and urbanization, there has been a corresponding increase in the role of the primate cities.[4] These cities are in most cases the national capital, the principal industrial center, and the center of artistic and cultural activities which best convey the unique characteristics of the nation.

In Latin America generally, governments are faced with a dilemma: whether to give priority to rural development and decentralization or to seek to alleviate the economic and social problems of the cities. Financial resources are seldom adequate to do both or either. Most countries have implemented a land reform program. Land reform is a politically sensitive issue, however, since the large landowners are also influential in government. Unless the land is taken by force, through conflict and violence, the owners of property must be remunerated. Few governments have the means to pay for such land reform on a scale adequate to meet local demands. More attractive is the colonization of empty lands in the interior, but this requires heavy expenditures for infrastructure, such as roads, extension service, grain storage, credit, housing, schools, and public utilities. Moreover, unoccupied frontier lands are fast disappearing throughout Latin America; they no longer exist in El Salvador, Uruguay, and most countries of the Caribbean area.

Another option to relieve population pressure is emigration, both seasonal and permanent. Currently, a steady flow of migrants from Mexico enters the United States, almost as if legal restrictions against such movements were nonexistent. Central Americans also enter in considerable numbers, largely to escape the problems of unemployment and civil disorder in their home countries. Cubans flock to the United States whenever permitted by the Castro regime, and from 1971 through 1981 about 44,000 Haitians entered the United States as illegal aliens. A heavy influx of immigrants from Puerto Rico occurs when economic conditions are favorable in the United States, and a reverse flow prevails when Puerto Rico experiences a period of relative prosperity. Oil-exporting Venezuela has one of the largest populations of illegal aliens in the Western Hemisphere. It is estimated that 2 million people from Bolivia, Peru, Ecuador, and Colombia are immigrants who supply labor for the economic sector, many of whom are subject to deportation if for any reason they are detained by local authorities. In some countries human resources appear to be expanding more rapidly than natural resources, and some elements of the population are restricted in their aspirations by social stratification. Hence, the rich tend to become richer, the poor become poorer, and many seek radical solutions to redress their grievances.

Nationalism

Each nation needs a set of values, a heritage, and certain traditions to remain viable as an independent state. In other words, most people must have some sort of "state-idea" that helps hold them together in the form of a modern nation. This is *nationalism*, which in a positive sense helps to provide people with a sense of identity and belonging, as is true of family, friendship groups, race, language, and religious affiliation. In a negative sense, nationalism is used to generate distrust and animosity toward those who look or think differently or who occupy a different segment of the earth's territory, especially adjacent territory. This type of nationalism is based particularly on different interpretations of history and different sets of political and economic objectives.

Nationalism has been facilitated by modern mass communication, with the effect that increased attention is given to frontiers and boundaries. Most recently, this has been reflected in the

---

[4] The "law of the primate city" was expressed by the geographer Mark Jefferson in 1939. This generalization states that each nation tends to have one city that is at least double the size of any other, and once having gained such primacy tends to maintain or enhance its position by attracting the best talent and major enterprises from throughout its tributary area.

establishment of territorial claims over the continental shelves for mineral exploration and a "200-mile limit" in the high seas for exclusive fishing rights and for purposes of national defense. Among the principal conflicts centered around nationalism in Latin America are an ideological dispute between Communist Cuba and non-Communist United States, a territorial claim by Guatemala over all of Belize and that of Venezuela over much of Guyana, disagreement between El Salvador and Honduras over their mutual boundary, a territorial dispute between Ecuador and Peru, and the Argentine-British conflict over control of the Falkland or Malvinas Islands. All of Latin America is considered critical to the defense of the United States, especially strategic features such as the naval base at Guantanamo, the Panama Canal, and such a critical resource as Mexican and Venezuelan oilfields.

Most nations of Latin America lack adequate domestic investment capital, infrastructure, and internal political stability, with the result that economic independence is difficult to attain. To provide a broader base of operations and an expanded market for locally manufactured goods, various international agreements have been signed. Most notable is a series of treaties under which the Central American Common Market (CACM) and Latin American Free Trade Association (LAFTA) were organized, beginning in 1960. The CACM, composed of Guatemala, El Salvador, Honduras, Nicaragua, and Costa Rica, was a major success in its early years, stimulating intraregional trade, industrial development, and international cooperation. LAFTA was a more ambitious enterprise, involving Mexico and all of the South American countries except the Guianas. It was designed to eliminate trade barriers among the member nations and eventually led to a common market such as had proved so successful for the European Economic Community. Yet, national interests have prevailed over regional solidarity to the detriment of both. The same can be said of other regional organizations, such as the Federation of the West Indies, which sought to unite politically the British islands of the West Indies, and the Caribbean Free Trade Association (CARIFTA), which has sought economic unity among nations of the Caribbean area. Viewed in the short run, local and national interests have consistently prevailed over international understanding and cooperation. Yet, human experience in Latin America, as elsewhere, must be viewed with a long-range perspective. A sense of regional unity, as opposed to narrow nationalism, may yet prevail within Latin America in the years ahead.

Emphasis on Development

The term *development* has not been well defined. At first, it was used primarily in an economic sense, and achievements were measured in financial terms such as gross national product, income per capita, and balance of trade. Such improvements as occurred were felt mostly in the capital cities and primarily by the privileged class. More recently, there has been an increasing awareness that significant national development can occur only with modernization of the component parts of a country; that regional development is best attained through full utilization of all available human resources.

Around the general theme of development has been a strong emphasis on planning: economic planning, national and regional planning, family planning, and planning in a variety of other forms. The implementation of plans has also been stressed, with emphasis on industrial development, mining, forestry, fisheries, agriculture, highways, electrification, transportation, communication, and modernization in general.

With development, if uncontrolled, comes destruction of a region's natural ecological conditions. Forests are cleared, soil is eroded and fertility depleted, mineral wealth is exhausted and the land permanently scarred. Hence, new voices are heard, urging caution and restraint.

Latin America, like the Middle East, is one of the world's strategically important areas. Peace and stability within the region are essential, yet are difficult to attain unless some form of development occurs that benefits the great mass of the population. For the immediate future, Latin America will merit special attention from the outside world and require a reasonable degree of understanding.

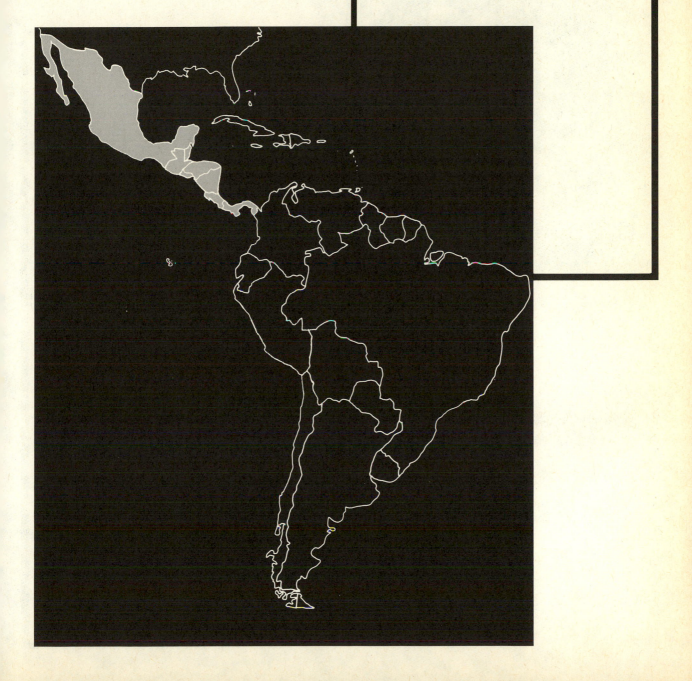

MEXICO

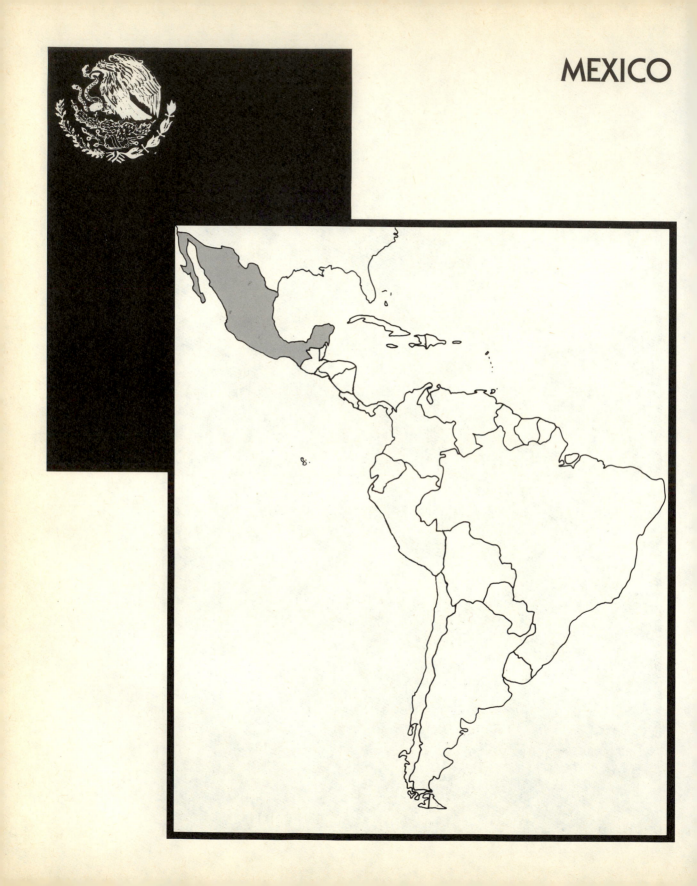

ESTADOS MEXICANOS UNIDOS

Land area   756,198 square miles

Population   Estimate (1985) 79,700,000
Latest census (1980) 67,382,581

Capital city   Mexico City 16,000,000

Percent urban   70

Birth rate per 1000   32

Death rate per 1000   6

Infant mortality rate   53

Percent of population under 15 years of age   42

Annual percent of increase   2.6

Percent literate   83

Percent labor force in agriculture   41

Gross national product per capita $2240

Unit of currency   Mexican Peso

Physical Quality of Life Index (PQLI)   76*

COMMERCE (expressed in percentage of value)

Exports

crude oil   67
food   8

Exports to		Imports from	
United States	54	United States	63
Spain	9	Japan	6
Japan	7	West Germany	6
France	4	Italy	3
United Kingdom	4		

---

* A nonincome measurement that summarizes three indicators: infant mortality, life expectancy at age one, and literacy.

Data mainly from the 1985 World Population Data Sheet of the Population Reference Bureau, the 1985 Britannica Book of the Year, and the 1985 World Almanac.

Mexico has discovered oil recently in vast quantities. Hence, new attention has been focused on that country, especially by its northern neighbor and other industrialized nations of the world. In an era of energy crises the discovery of great new petroleum reserves is a major event. Many will want to share in the benefits, but for Mexico there are two major hopes. One is to gain greater political and economic strength in its dealings with the United States; the other is to achieve more rapid internal development. To understand Mexican attitudes and objectives, it is essential to consider some historical events.

More than any other Latin American country, Mexico has felt the impact of an expanding and industrializing United States. In 1836 the English-speaking settlers who had been granted land in the northern part of Mexico declared their independence and defeated Mexican troops sent to put down the revolt. Texas was an independent nation from 1836 to 1845, and then became part of the United States. Between 1846 and 1848 U.S. armed forces invaded and occupied the entire northern part of Mexican territory from Texas to the Pacific and in 1847 captured Mexico City in the famous battle of Chapultepec. About 53 percent of the total national territory was lost. The history books read by Mexican school children did not describe this war in the same terms as did the books read north of the border. Yet, the Mexican government subsequently profited from friendly relations with the United States.

Mexico was the first country in Latin America to take effective measures to relieve the poverty of vast numbers of people and to address the problem of social inequality resulting from the traditional status of the landowning minority. The Mexican Revolution, which began in 1910, led to the expropriation of large estates and redistribution of land among the peasants. In 1938 the Mexican government expropriated all foreign-owned oil properties. Meanwhile, contacts with people of the United States became closer as many Mexicans emigrated to the United States —some permanently, others only temporarily to seek employment in agriculture—and as floods of tourists from the United States supplied much-needed dollars to the Mexican economy. However, among the Latin American countries, Mexico was most likely to reject policies announced in Washington. In the 1960s, for example, Mexico continued to maintain relations with Cuba. With its prominent position in the world of oil, Mexico has become increasingly assertive in its relationships with other counies.

The Appearance of Modern Mexico

In 1821, when the Spanish colony of New Spain became the Empire of Mexico, leaders of the independence movement were seeking to establish equal rights between those born in Mexico and those born in Spain. The great majority of people who lived in the new national territory were Indians or mestizos. Almost all were illiterate farmers, for whom the question of equality with people born in Spain had no meaning. Mexico began its independent existence with a negative state-idea that was of concern to not more than 5 percent of the people. From 1821 to 1876 the country was poverty-stricken and turbulent. There were revolts and counter-revolts, as younger army officers who were not in profitable government positions sought to replace those who were. Occasionally, a sincere reformer appeared, such as the Indian Benito Juárez, but reform movements aimed at improving the lot of ordinary people were soon diverted into traditional channels.

Porfirio Díaz was one of Latin America's most successful dictators. With a small group of trusted supporters and a powerful police force, he ruled Mexico by decree. Only 15 percent of the total population could read or write, and not more than 1 or 2 percent were wealthy and enjoyed political power. In central Mexico 95 percent of the heads of rural families owned no land. A striking contrast existed between the educated, cosmopolitan, and wealthy society surrounding the dictator and the great majority of Mexicans who were illiterate, sick, and hopeless in their poverty. Yet, under Díaz there was order and security, and foreign commercial interests thought of Mexico as a safe place in which to invest.

Díaz was 80 years old when he was elected President of Mexico for the eighth time in 1910. By then, almost all of the wealth of Mexico was concentrated in the hands of less than 3 percent of the people, and the regime was undermined largely because few people had a stake in it. Suddenly violence, disorder, and revolution occurred. Even the army and police deserted the dictator, who was forced to flee for his life.

Is Mexico a rich land or a poor land? This question has been debated since Cortés first reported to his king that he had found a land of superlative resources. Mexico is a rich land for the mining engineer who encounters little difficulty in locating ores where the rivers have done the work of excavation and where the cover of vegetation is sparse. Mexico is incomparable for the tourist who seeks spectacular scenery. Mexico could become a productive meat or dairy land, since it possesses ample physical advantages for these forms of economy. For agriculture Mexico is a poor land, because much of its area is too steep or too dry to be classified as arable. Yet, a large number of Mexicans are farmers.

The sea which has an abundant supply of shrimp, has become an important source of wealth for Mexico. However, with new techniques and constant overfishing in its waters, the country has found it necessary to protect itself from poachers. As a result of the First United Nations Conference on the Law of the Sea, coastal states are permitted to regulate fisheries around their shores beyond the 12-mile territorial limit and to explore and exploit the continental shelf. Both Mexico and the United States have continental margins that extend over 200 miles; it is here that a primary source of food, medicines, petroleum, and other minerals is found.

The People and Cultures of Mexico

The people and cultures that comprise the Mexican nation result largely from a blending of Indian and Spanish components. Millions of Indians were already firmly entrenched on the land in central and southern Mexico at the time of the Spanish conquest, including the sedentary agricultural people of the Aztec Empire. These were the Indian communities whose distribution most strongly influenced the course and pattern of Spanish settlement.

The Spanish contribution to the racial composition of the present population is relatively small numerically. The number of registered emigrants who left Spain during the colonial period to enter the territory administered from Mexico City, in New Spain, was only 300,000. Although each registration might include members of the family and servants, the total number of Spanish immigrants to New Spain was small compared with the number of Indians already there. Spaniards had the advantage of technical knowledge, however, which assured their political and economic conquest of the more numerous native peoples.

Racial Composition of the Mexican People

The Indians of Mexico were themselves of diverse origin and culture. While today Spanish is the common language throughout Mexico, it is significant that conduct of the national census in 1950 required the use of 50 different languages. Pure-blooded Indians are estimated to comprise only 29 percent, whereas Indians with some white ancestors form at least 55 percent of the total. People of unmixed European descent constitute only about 10 percent.

These proportions have been changing gradually through the decrease in number of people of unmixed ancestry and the increase of the mestizo population. During the colonial period more than 30,000 Africans were imported to work on sugar plantations and in other enterprises, but by 1805 the number of pure Africans was only about 21,000, or 0.2 percent of the total. The proportion of blacks is now less than 1 percent.

The Indian contribution to the Mexican mestizo is relatively large, since the typical mestizo has more Indian than white ancestors. Díaz attempted to stimulate European immigration, and about 11,000 Italians actually came to Mexico shortly after 1878, but by 1890 not more than 5,000 remained. Since then, the number of Europeans who have come to Mexico has been negligible. From a racial perspective Mexico is overwhelmingly Indian rather than Latin.

Growth of the Mexican Population

Until recently, growth of the Mexican population has been slow. After decimation of the Indians by diseases introduced along with the Spanish conquest, the number of people began slowly to increase again. It is estimated that Mexico had a population of about 5.8 million in 1805, probably less than the Indian population of the same territory three centuries earlier. After 1824 the Mexican population required 80 years to double itself. The Mexican population in 1956 was estimated at 30.5 million. In 1985 Mexico's population had grown to more than 79 million, while that of the United States was estimated at about 239 million.

Two factors help to account for the traditionally slow rate of population increase and lack of expanded settlement. First is the death rate, which until 1950 was among the highest in Latin America. Children under one year of age accounted for about a fourth of all deaths. Diseases were mainly those caused by malnutrition and poor hygiene—diarrhea, enteritis, and dysentery—but respiratory diseases were also widespread, especially on the high plateau around Mexico City. In 1985 the Mexican death rate was down to 6 per thousand, as compared with 27 per thousand in 1950.

The second reason for the slow rate of population growth in past years is that Mexico, during the entire 450 years of European dominance, was a country of emigration. During the colonial period Mexico supplied most of the people who occupied the Philippine Islands. From the Pacific port of Acapulco many Mexicans sailed to this still more distant colony of Spain. In the modern era large numbers of Mexicans have emigrated to the United States. Mexicans today form an important minority group in all of the border states from Texas to California, as well as in large industrial cities of the North, such as Chicago and Detroit.

Attitude Toward the Land

The traditional attitude toward land in the central area of Mexico was based on a mixture of Spanish and Indian ideas. Despite parallel institutions, the contrast between Spanish and Indian attitudes toward the land was enormous. For the Spaniard, the sure road to prestige and economic security was by the private estate, while only a small group of Aztec nobles thought of land + ownership as bringing prestige. Most Indian farmers who actually used the land thought of their small plots as belonging to the community, and they cared only to produce enough for their own needs. Commercial farming was almost unknown. The few items taken to local markets provided, then as now, more an excuse for the producer to take part in the social pleasures of the market than an element of economic support.

The Spanish-Indian Impact

The Spanish system of the *encomienda* did not differ greatly from the Aztec system of tribute. Ruling Aztecs exploited the labor of people they had conquered, and the Spaniards simply continued where the Aztecs left off. In many instances the same units were taken over, a Spanish officer replacing an Aztec lord without further dislocation of the system. Cortés himself received grants of *encomiendas* from the Crown in various parts of Mexico, including 22 villages with a total population of about 23,000 and an area of 25,000 square miles.

The *encomienda* system did not survive the first century of the Spanish conquest. In its place the grant of large tracts by the Crown gave the owner actual title to the land, not just the right to collect tribute. Gradually, the *encomienda* system was abandoned, and more of the land was placed in private hands. Some grants were less than 100 acres in size, but many included thousands of acres.

The Spaniards also brought parts of Mexico

under control through the establishment of missions. Especially on the remote northern frontier, the Jesuit, Franciscan, and Dominican orders founded new centers of settlement. Around each center they brought together Indians from many small scattered communities and reestablished them as farmers, teaching new agricultural techniques, importing new crops, and, incidentally, exposing the Indians to the ravages of epidemics.

The impact of the Spaniards on the Indians produced a struggle that lasted more than 400 years. This was for the right to own land. It involved two contrasted forms of tenure: the *ejido*, or landholding agrarian community occupied chiefly by persons of Indian descent; and the hacienda, or large privately owned feudal estate, usually in the hands of persons of unmixed or nearly unmixed Spanish descent. By the end of the Díaz regime in 1910, almost all of the public domain had been shifted to private ownership. In all but five states, more than 95 percent of the heads of rural families owned no land. Eleven million rural people (of a total population exceeding 15 million) lived in small, isolated communities, raising subsistence crops on land rented for that purpose and gaining a miserable additional wage by working for the owners. The great majority of Mexicans lived monotonously, in isolation, ignorance, and poverty, plagued by poor diet and disease. Such conditions formed the background of the brilliant aristocratic society of the capital in the days of Porfirio Díaz.

The Mexican hacienda was more than a large property; it was a way of life. Ownership of a hacienda provided two things that every Mexican desired but few could hope to achieve: social prestige and economic security. Because the owners were relatively free from land and labor costs, they could sell their products profitably even when they were inefficiently produced and when transportation costs were high. In contrast to the almost complete self-sufficiency of the rural workers, the landowners were closely tied to the world of commerce. Their standard of living was high, their diet was varied and hearty, their children were educated in Europe, and the entire family had frequent opportunities to travel and to develop a cosmopolitan familiarity with the outside world.

In 1857, Benito Juárez came to power with a similar movement, but like the others it was soon turned away from its basic objective — to do something about the land problem. Mexican laborers who had been in the United States and returned were largely responsible for increasing the widespread discontent of the people with a system in which they participated so little. The Mexican Revolution, which lasted from 1910 to 1917, was a conflict that set the stage for truly fundamental reforms.

Present System of Land Tenure

Land redistribution based on the Constitution of 1917 transformed rural life in Mexico. Between 1916 and 1934, some 25 million acres of hacienda lands were expropriated and assigned to peasant communities, by which a total of 939,000 farmers actually received property. During the presidency of Lázaro Cárdenas from 1934 to 1940, there was a great acceleration of the program. In this period almost 50 million acres changed hands, and more than 7.7 million individuals received land. By 1950 about 90 million acres had been taken from the haciendas and given to farmers in peasant communities. The traditional hacienda as an economic and social institution was destroyed.

In Mexico, the word *ejido* refers to a farming community that has received land under procedures established by the Constitution of 1917. The *ejido* is a rural peasant community, a farm village. In Mexico, as a whole, the average number of families per *ejido* is less than 100. Each family works its own land and in most ways treats it as private property. However, land thus parceled cannot be sold, and if the family moves away, title to the property remains with the *ejido*. The *ejido* is organized politically, with a general assembly, an executive committee, and a vigilance committee to watch the executives. Technical assistance, education, and credit are furnished by the federal government. The amount of land granted to each *ejidatario* (head of an *ejido* family) varies according to the potential productivity of the land.

Since 1947, another kind of rural holding has appeared. This is the small private farm. Land suitable for agriculture in the central area was already redistributed under the *ejido* system, but elsewhere there was still much land in private holdings, used chiefly for cattle ranching. In 1947 the government passed a new colonization law permitting the expropriation of range land and the establishment of farm colonies. To make this colonization possible, the government also undertook a major plan of irrigation and reclamation.

Land cultivated under the *ejido* system still supplies the greater part of Mexico's crops. *Ejidos* produce well over half of the wheat, rice, sesame, henequen, cotton, and tobacco. Yet, yields on *ejido* lands are lower than on private holdings, due in part to a lack of technical skills. The great cotton-producing area around Torreón, in the north of Mexico, has suffered seriously because low rainfall in the mountains to the west often leaves reservoirs almost empty.

It is commonly believed that agrarian reform has reached its full potential in Mexico and that further subdivision of large holdings can yield only marginal benefits. Thus, it seems likely that land reform will not be a major element in further modernization of Mexico, despite the fact that 4 million peasants are still without land and another 5 million have only minimal lots. The land redistribution program cannot be properly evaluated without consideration of the underlying qualities of the land itself and of differences in land use that distinguish one part of Mexico from another.

The Land

When the Spaniards first arrived in America, they were attracted by areas already densely settled by sedentary agricultural Indians, while outlying regions with their sparse populations were much less attractive. Concentration in the central area was so great that during the Díaz regime an attempt was made to promote the colonization of other parts of the national territory, but with discouraging results.

The Mexican land is one of extraordinary diversity. A large part of the national territory is mountainous, and the mountains include some terrain produced by the erosion of streams in areas of contorted rock structures and some produced by outbursts of volcanic ash and lava. Well over half of Mexico is more than 3000 feet in elevation, and only about a third of the country can be classed as level. Over all these surface features there are diverse types of climate, controlled partly by differences of altitude and partly by relationship to the sources of moisture.

Surface Features

The major element of surface configuration in Mexico is the great highland area that extends from the border with the United States southward to the Isthmus of Tehuantepec and occupies most of the width of the country. Although the highland is exceedingly complex in geologic structure and surface form, it is convenient to think of it as composed of two main parts: a *central plateau* and a *dissected border*. The surface of the central plateau is cut by few deep canyons, yet it is by no means flat, for above the moderate slopes of its bolsons and intermont basins stand block ranges and volcanoes. In the north the bolsons are mostly 3000 to 4000 feet in elevation, and the block ranges rise about 3000 feet above them. South of the Bolsón de Mayrán the general level of the plateau rises: the intermont basins are mostly between 7000 and 8000 feet, although some are as low as 5000 feet, and above these basins great volcanoes reach elevations from 12,000 to more than 18,000 feet. The dissected borders of the highland, unlike the central plateau, have been deeply cut by streams. The relief of the western and eastern dissected borders is made even more rugged by deep accumulations

FIGURE 2-1

of volcanic material, so that on these two sides the rim of the highlands is higher than the central part. On the southern dissected border, south of Mexico City, the general highland level between 6000 and 8000 feet is preserved, not in the basins but on the ridge crests, and streams have cut deep valleys below what was a continuous surface.

Beyond the great highland region with its dissected borders are three other surface divisions of Mexico: the block mountains and basins in the northwest; the lowlands of the Gulf Coast and Yucatán on the east; and the highlands of Chiapas on the border with Guatemala.

In the northwest, the surface features of southern California and Arizona continue into Mexico. The peninsula of Baja California is composed of tablelands and terraces surmounted by a few isolated block ranges with structures similar to those of the mountains east of San Diego. The Sonora Desert between the Gulf of California and the Sierra Madre Occidental is a mountain-and-bolson country, similar to the Mojave of southeastern California. Even the structural depression that forms the Imperial Valley of California continues southward to form the Gulf of California. Throughout the Mexican northwest rocky surfaces predominate, separated at wide intervals by steep-sided, flat-bottomed valleys typical of arid lands.

On the eastern side of the highlands, the Gulf Coastal Plain of Texas continues southward into Mexico as far as Tampico, where it is pinched out by outliers of the Sierra Madre Oriental and by isolated volcanic features. South of Tampico the coastal lowland is relatively narrow and in many places is broken by promontories where the highlands extend to the sea. The lowlands bordering the Gulf widen again at the northern end of the Isthmus of Tehuantepec, and the entire Yucatán resembles Florida in that it is composed of horizontal limestone formations of relatively recent age.

The Isthmus of Tehuantepec separates the southern dissected border of the great Sierra Madre del Sur from the highlands of Chiapas. The latter is the northwestern end of the mountainous region that extends through Central America to the lowlands of Nicaragua. In Mexico it is composed of parallel ranges of block mountains, enclosing a high rift valley. Along the Pacific is the crystalline range known as the Sierra Madre de Chiapas. Inland from this, and parallel to the coast, is the rift valley of Chiapas, drained by a tributary of the Río Grijalva. On the northeastern side of the valley are several other ranges of block mountains, composed of folded and faulted strata and capped with volcanic ash and lava. These mountains are much dissected by streams.

Climate and Natural Vegetation

In a land of rugged surfaces and contrasts of altitude within short distances, climatic conditions and the cover of natural vegetation have extremely irregular patterns. There is, however, a general vertical zonation which becomes apparent when one disregards the many irregularities of details.

Vertical zones in Mexico result from the general decrease of temperature with increasing altitude. The hot country, which Mexicans call the *tierra caliente,* extends to about 2100 feet above sea level on the slopes of Mount Orizaba near the east coast. Here are tropical forests and crops such as sugarcane that cannot survive frosts. The land above this altitude is called the *tierra templada,* or temperate country, rising to about 6000 feet. Oaks are the common trees, and the major crop is coffee. Above this is the *tierra fría,* or cold country. It includes the zone of conifers above 11,400 feet and extends to the upper limit of trees at about 13,100 feet. Between the tree line and the snow line is a zone of mountain grasslands suited for the grazing of cattle. The snow line on Orizaba is about 14,600 feet above sea level.

Most of the Mexican territory is deficient in moisture at least during part of the year. The entire northern boundary from the Pacific to the mouth of the Rio Grande passes through regions of arid or semiarid climate, but the driest sections are in the northwest and northcentral parts. A belt of aridity extends southward from western Texas almost to San Luis Potosí. Semiarid country includes all of the central plateau except the southern and southwestern part, bounded by a line drawn roughly from Aguascalientes to Mexico City.

Rainfall is adequate throughout the year in only two sections of Mexico. One belt of dependable rainfall extends southward from Tampico along the lower slopes of the Sierra Madre Oriental and crosses the Isthmus of Tehuantepec into the state of Tabasco. The other is along the Pacific Coast of Chiapas.

To appreciate the importance of surface and climate in terms of the distribution of people in Mexico, it is necessary to examine the various parts of the country in greater detail. For this purpose the general divisions used by the Mexican government are employed, in which the individual states are combined into five regions. These are not natural divisions of Mexico but follow arbitrary state boundaries. Their chief value is that they permit the use of statistics gathered and averaged by states. The five divisions, with the states included in each, are as follows:

1. The North Pacific region: Baja California, Baja California Sur, Sonora, Sinaloa, Nayarit.

2. The North: Chihuahua, Coahuila, Nuevo León, Tamaulipas, Durango, Zacatecas, San Luis Potosí.

3. The Gulf Coast and Yucatán: Veracruz, Tabasco, Campeche, Yucatán, Quintana Roo.

4. The South Pacific region: Colima, Guerrero, Oaxaca, Chiapas.

5. The Central region: Aguascalientes, Jalisco, Guanajuato, Querétaro, Hidalgo, Michoa-

cán, México, Distrito Federal, Morelos, Tlaxcala, Puebla.

The North Pacific Region

The North Pacific region extends roughly from Cape Corrientes to the border of the United States. Although it includes 21 percent of the total area of Mexico, this region is occupied by only 8 percent of the Mexican people. Like the west coasts of all continents between 20° and 30° of latitude, this is a desert, and human settlement is closely attached to oases. Its eastern side is bounded by the exceptionally rugged Sierra Madre Occidental. This mountain range, 150 miles wide and 750 miles long, is crossed by only three lines of communication. The first to be developed was the old colonial route from Guadalajara to Tepic. Originally just a road for wagons, it later became a railroad route and is now followed by one of the paved highways that connect Mexico City with the United States. During the 1950s another road was built between Durango and Mazatlán. In 1961 a railroad was completed from the border of Texas, through Chihuahua, and across some of the most spectacular terrain in North America to the Pacific port of Topolobampo.

The Sierra Madre Occidental

The Sierra Madre Occidental remains one of the most thinly populated parts of Mexico. On its eastern side the mountains rise gradually from the bolsons of Chihuahua and Durango, but on the western side there is a bold escarpment, notched in only a few places by deeply incised canyons. Within this mountain area the surface is extremely rugged: steep-sided longitudinal ranges with conspicuously even summits rise to elevations of 10,000 feet. Between them, deep valleys connected by short transverse gorges have been cut along the lines of weaker rock by seasonally torrential streams. A luxuriant cover of forest and grass offers shelter and food for a variety of game animals, and from these a nomadic hunting people gain a living.

Sonora, Sinaloa, and Nayarit

The states of Sonora, Sinaloa, and Nayarit, west of the Sierra Madre, descend toward the Pacific Ocean and Gulf of California through country broken by ranges and basins. A series of terraces and lava flows have been dissected by streams descending from the Sierra Madre into isolated mesas and plateaus interspersed with flat valleys. Along the coast is a lowland that varies in width from less than 10 to more than 50 miles. Summers in this area are hot and rainy; winters are mild and dry. The length of growing season, amount of rainfall, and number of streams bringing water from the Sierra Madre all decrease toward the north.

When the first Spanish explorers descended to the Pacific Coast from the primary settlement center of Guadalajara, they encountered a variety of Indian cultures. A few of the tribes practiced intensive agriculture with irrigation. Especially in the south, in what is now Nayarit and Sinaloa, a sedentary farming people grew maize, beans, and squash on the river floodplains. The Indian population, however, was neither rich enough in accumulated gold and silver, nor numerous enough to attract any considerable Spanish settlement. The sedentary Indians of Nayarit and Sinaloa were divided into *encomiendas,* and within the first few decades of the conquest these Indians were all but eliminated by the combination of slave raids and the ravages of disease. Exploring parties that marched far to the north, seeking the fabulous cities of Cíbola, failed to discover any dense Indian populations comparable to those of the highlands between Guadalajara and Puebla, or rich mines of precious metals. Eventually, the more accessible southern part of the North Pacific region was divided into large haciendas, and the Indians who survived were forced to work for their new masters.

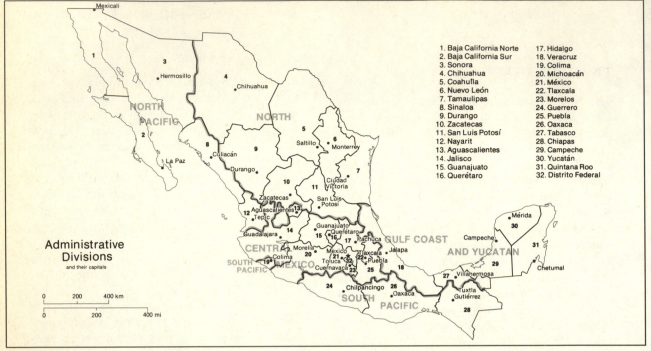

Administrative Divisions and their capitals

1. Baja California Norte
2. Baja California Sur
3. Sonora
4. Chihuahua
5. Coahuila
6. Nuevo León
7. Tamaulipas
8. Sinaloa
9. Durango
10. Zacatecas
11. San Luis Potosí
12. Nayarit
13. Aguascalientes
14. Jalisco
15. Guanajuato
16. Querétaro
17. Hidalgo
18. Veracruz
19. Colima
20. Michoacán
21. México
22. Tlaxcala
23. Morelos
24. Guerrero
25. Puebla
26. Oaxaca
27. Tabasco
28. Chiapas
29. Campeche
30. Yucatán
31. Quintana Roo
32. Distrito Federal

FIGURE 2-2

The more remote country of northern Sinaloa, Sonora, Baja California, and California were settled by the mission system. The usual practice was to establish a mission at a location carefully selected in advance, and then urge the widely scattered Indians to settle nearby. Before the arrival of missionaries, the native people lived in small groups. Some were seminomadic hunters and fishers; others practiced sedentary farming with maize, beans, and squash as their leading crops. The Fathers introduced new crops, taught methods of irrigation or adopted Indian methods, and grouped the Indians into compact communities where they could be instructed in the Christian faith.

The results were disastrous. The crowded settlements were exposed to contagious diseases from Europe, and soon the mission settlements were decimated by epidemics of smallpox, measles, and other forms of pestilence. After an epidemic came famine. Entire villages, stricken at the time of planting or harvesting, were unable to carry on the agricultural work on which they depended. When a settlement was so reduced that it could no longer exist alone, it was abandoned and its members were transferred to other mission communities. The abandoned land passed into the hands of a Spanish grantee, who used it for the grazing of cattle, horses, and mules. Thus, step by step, lands on which the Indians had depended for their subsistence became private property of the conquerors.

The geographer, Carl Sauer, estimated that the native population of what is now roughly the territory of the three states under discussion was at the time of the conquest about 540,000. The population of the same area did not again reach this figure until about 1920, almost 400 years later.

The hacienda system was firmly established in

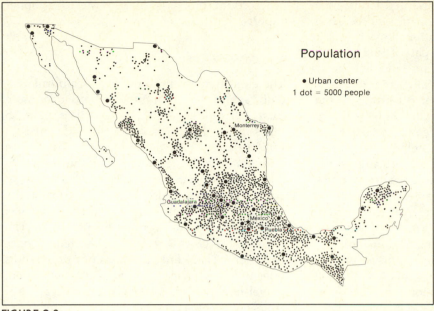

Population

• Urban center
1 dot = 5000 people

FIGURE 2-3

this part of the northwest until it was broken up during the 1930s. Only in remote and isolated valleys of the Sierra Madre Occidental did small Indian communities survive with their traditional communal holdings. The landowners used the fertile irrigated valley bottoms for wheat (which the Indian workers did not eat), cotton, oranges, and such Mediterranean crops as dates, figs, olives, and grapes. The basic food of the workers, maize, was grown mostly on the poorer lands; on these same steep or arid farms the peasants raised beans, squash, and chili. When mines in this region were active, workers were recruited from the neighboring territory; but when the mines closed, as they did periodically for reasons incomprehensible to the Indian workers, the latter drifted back to the haciendas, adding more hungry mouths to be fed. The gradual increase of population meant a steady decline in the standard of living, for there could be no extension of arable land without expensive irrigation, and there could be no increase in yields per acre without a fundamental change in farm practices.

Fundamental changes have now been made. First came an agrarian reform that eliminated the hacienda as an economic unit. In the southern part of the North Pacific region haciendas were replaced by *ejidos* and small private farms. In the north, where irrigation is essential for the support of agriculture, settlement since 1957 has been in government-sponsored colonies where the colonists own their own farms. Eight large dams were built, all after World War II, on rivers descending from the Sierra Madre Occidental. Millions of acres of new agricultural land were developed, and farms were provided for pioneer settlers who came mostly from the central area of Mexico.

The government has established three objectives in its agricultural policy: (1) to achieve self-sufficiency in basic foodstuffs, (2) to increase the production of export crops, and (3) to obtain higher incomes along with improved living standards and greater employment opportunities for the rural poor. With increased income through petroleum exports Mexico may be able to attain these goals. New areas of rainfed lands, especially in the humid tropical zone, will be developed and large-scale irrigation projects will be undertaken in the northern part of Sinaloa. More than 148,000 acres of saline land will be reclaimed in the Río Fuerte Valley and 188,000 in that of the Río Sinaloa. When the project is complete, rural underemployment will be reduced.

This part of Mexico is a leading source of cotton and wheat. Tomatoes, melons, peas, beans, cucumbers, peppers, and eggplant also grow rapidly and are shipped northward in direct competition with the winter vegetables and fruit of California, Florida, and Texas. Nayarit produces about 80 percent of Mexico's tobacco, with Veracruz and Oaxaca contributing the remainder. On the northern dry lands cattle-raising continues to predominate, serving an expanding domestic market and providing exports to the United States. Chihuahua state alone is able to export 170,000 animals annually.

Mazatlán, once a pirate refuge, now has seagoing ferryboat service to La Paz on the Baja California peninsula. Large ships from Los Angeles en route to South America stop at Mazatlán so tourist passengers can admire this picturesque part of Mexico. The city's economy centers on fishing, shrimping, leather-making, and the manufacture of fertilizer, shoes, and beverages.

Nor has the government been lax in the development of commercial fishing. The Mexican fishing industry has increased its fleet, modernized its ports, improved its marketing system, and trained its personnel. All this provides thousands of jobs, better diets for those with low incomes, and expanded exports of frozen and canned food. The variety of fish includes tuna and sardines from Baja California and shrimp from Mazatlán and Guaymas.

Numerous small communities in northwestern Mexico depend heavily on the mining industry. Government-owned copper refineries at Cananea in Sonora and near Santa Rosalía in Baja California are major employers in their areas. Mines at Cananea and La Caridad, in Son-

ora, account for a large share of Mexico's copper exports.

Baja California

The elongated peninsula of Baja California extends southward from the U.S. border for more than 750 miles, while averaging less than 75 miles in width, and includes the states of Baja California and Baja California Sur. The surface and climate of the northern part of this peninsula are similar to those of southern California. In Baja California there is, except for the Colorado, only one permanent stream, and this is a short one near the southern end of the peninsula. The terraces, therefore, are not as dissected as those of Sonora. There are vast areas of dry uplands, interrupted at wide intervals by steep-sided, gravel-filled valleys in which surface water appears only after a rain, but where settlement can be supported by wells. All but the southern tip of the peninsula is deficient in moisture. The northern half receives most of its low annual rainfall during winter, as does San Diego; the southern half receives only summer rains. Winters are mild, and summers are generally cool.

The northern part of Baja California has experienced several periods of boom development. Tijuana was a favorite attraction for thirsty North Americans during the prohibition era. In recent times it has redeveloped the recreation business and is visited each year by millions of visitors from the United States. The race track, gambling houses, and other tourist attractions bring wealth that helps to support the Mexican economy.

Since 1970 Tijuana has doubled its population, which now totals 600,000. Because peso salaries are higher here than in other Mexican cities, and because Tijuana is a free trade zone, people are naturally attracted to the area. Migrants en route to the United States stop here, and the city welcomes more than 50,000 tourists from the United States daily. South of Tijuana is the oldest community in Baja California, the renowned fishing port of Ensenada. Fishing is also significant in the Gulf of California, where about 600 species abound.

The Colorado Delta and Imperial Valley

The state of Baja California includes an irrigated area along the lower Colorado River which is distinctive in both physical character and human settlement. The great rift depression which forms the Gulf of California continues across the border into the United States in southeastern California. Downstream from Yuma, Arizona, the Colorado River enters this depression from the east and has built a huge delta of coarse alluvium at the head of the Gulf. The part of the depression north of the delta lies more than 200 feet below sea level, and water escapes from it only by evaporation. This northern end of the rift, lying partly in Mexico and partly in California, is known as the Imperial Valley.

Development of the Mexican part of the Imperial Valley around Mexicali has occurred since World War II. For a long time this potentially productive area was handicapped by isolation. Meanwhile, water from the Colorado was being used to irrigate that part of the Imperial Valley lying north of the border. Far upstream the state of Colorado was diverting water from the river by tunneling through the mountains to expand irrigation on the eastern plains. It was at this time that settlers from the United States secured land on the Mexican side of the border and claimed a share of Colorado River water.

Between 1936 and 1938, in the period of rapid expropriation, land in the Mexicali district was taken from the foreign owners and distributed to Mexican peasants. An agreement was reached between Mexico and the United States regarding the use of water along the border, including not only Colorado River water but also water from the Rio Grande. The Mexican government provided for the use of its share of Colorado River water to irrigate some 500,000 acres on the Mexican side of the border. Ginning equipment was installed and credit provided to farmers. There

was an enormous increase in cotton production, so this is now one of the major cotton-producing districts of Mexico.

Despite numerous opportunities to develop thermal power, the American nations have been slow to take advantage of this form of natural energy. Mexico, however, is taking steps to supplement its huge oil resources. In northern Baja, the Cerro Prieto plant has been in operation since 1973. It now produces electricity valued at $1.4 million per year. This advanced program is sponsored by the Federal Electric Commission. It seems probable that geothermal energy is a practical solution to power needs in those areas with little or no other energy sources.

Mining and fishing also offer significant prospects for development in Baja California. Waters offshore for a distance of 200 miles, now clearly under Mexican jurisdiction, abound with fish that are exploited for both sport and commercial purposes. Copper is mined near Santa Rosalía, and some manganese is also produced. Marine salts are recovered by solar evaporation on the desert west coast of the peninsula, in the world's largest such operation, and are shipped to chemical companies on the Pacific Coast of the United States. Agriculture is little developed in the southern part of the peninsula, except where groundwater is sufficient to support small-scale irrigation. The largest city in Baja California Sur is La Paz, which for many years was a pearl-gathering center. La Paz is now a free port stocked with merchandise from many parts of the world.

The North

The region known as the North includes 41 percent of the national territory of Mexico, but only 18 percent of its population. Like the North Pacific region, it suffers from deficient rainfall, and any significant expansion of agriculture or population will depend largely on new irrigation works.

As a result of the Mexican War, all territory

north of the Rio Grande and from El Paso westward along the course of the Gila River, was ceded to the United States in 1848. In 1853 the strip of land along the southern border of Arizona and New Mexico known as the Gadsden Purchase was added to provide the best route for building a railroad to the Pacific. A minor revision of the boundary within the city of El Paso, resulting from a shift of the Rio Grande, was settled in favor of Mexico in 1967.

During the twentieth century, the flow of border crossings has been mostly northward. With nearly 200,000 Mexicans joining the ranks of the unemployed each year since the early 1970s, there has been an intensive movement across the border by Mexican migrants who become illegal aliens in the United States. They have come by any means available to seek any kind of employment, but especially in agriculture. For the Mexican government this emigration is seen as a necessity, since it helps resolve the plight of thousands of underemployed and jobless young people. Mexicans have remembered the earlier long-standing U.S. policy of attracting migrant labor to meet its manpower shortage in the field of agriculture. Hence, they have come by the tens of thousands to supply cheap labor, while relieving demographic pressure in a country that doubles its population every 20 years.

San Luis Potosí, Zacatecas, and Durango

The states of San Luis Potosí, Zacatecas, and Durango are composed of basins and plateaus that stand at about the same altitude as those of the central area in the state of México. The Basin of Zacatecas is just over 8100 feet above sea level. The cities of San Luis Potosí and Durango are both just over 6000 feet. The surface between them is largely high, semiarid plateau country, with ranges of block mountains standing a few thousand feet above the general level.

The main interest of the Spanish conquerors who invaded this area, after gaining control of central Mexico between Puebla and Guadalajara,

was to search for gold and silver. The first places established, after Mexico City and Guadalajara, were mining communities such as the silver mining centers of Zacatecas, Guanajuato, Pachuca, Querétaro, and Aguascalientes in the Central region, and San Luis Potosí and Durango in the North.

These sixteenth-century mining towns also became political centers, dominating the rural districts around them. After Mexico gained independence from Spain, they became provincial capitals around which the new states were organized. During the modern period a wider variety of minerals has been exploited, and smelting and refining facilities have been expanded. Zacatecas today produces not only silver but also gold and copper. Durango produces silver, gold, copper, lead, and more than half of Mexico's supply of iron ore.

Mountain-and-Bolson Country

The northern parts of San Luis Potosí, Zacatecas, and Durango, together with most of Chihuahua and Coahuila, are included in the sparsely settled mountain-and-bolson country of the North. The basins, or bolsons, have typical desert landforms. The lower places are occupied by shallow, salty lakes with fluctuating shorelines, or by salt-encrusted flats. The bolsons are bordered by the gentle slopes of alluvial fans, which only thinly mantle the rock pediments around the base of each range. The ranges themselves, with steep, rocky slopes, stand abruptly above the bolsons.

Almost all of this part of the North is arid or semiarid. Most of western Chihuahua receives from 15 to 20 inches of rainfall per year, with a marked summer maximum. There is a belt of arid climate which extends southeastward from El Paso across the Bolsón de Mapimí and Bolsón de Mayrán into the northern part of San Luis Potosí state. Near Torreón, in the Bolsón de Mayrán, the average annual rainfall is only 10 inches. This vast region was not attractive to the colonial Spaniards. Its sparse nomadic, warlike Indian population could not compare with the sedentary agricultural Indians of the Central region as producers of wealth. The first settlements were mining communities, but they were fewer and not as prosperous as those farther south. The greater part of the region was, and continues to be, used for grazing.

In modern times mining activities of the North have been more productive than those in any other part of Mexico. In the state of Chihuahua there are mines of zinc, lead, and gold, and the first uranium processing plant in Chihuahua will be built at Las Margaritas. In 1971 the government-owned zinc corporation, ZincaMex, authorized construction of a large new plant near its Saltillo refinery for the production of zinc-based products, and a magnesium plant was already in operation at Laguna del Rey in Coahuila. The largest zinc plant in Latin America was built at Torreón where Mexico gained about $160,000 per day in foreign exchange. The mining communities are mostly small, fluctuating in population and by no means permanent features of the map.

The Laguna District

The Laguna district, about 200 miles west of Monterrey, occupies part of the Bolsón de Mayrán on the border of Coahuila and Durango. Torreón is the urban center for an irrigation district which at one time was the showpiece of the Revolution. Visitors came to Torreón from around the world to observe the most modern methods of land redistribution. In the mid-1950s this district produced about 60 percent of Mexico's cotton; by the mid-1960s it was producing only 13 percent. The Laguna district had become a disaster area, from which farm families were being evacuated to other parts of the country.

The land redistribution program was not applied to the Laguna district during the early years of the Revolution. The hacienda owners were permitted to seek their own solution to the problem of poverty, but the workers became increasingly dissatisfied with the lack of attention to their difficulties. Strikes against hacienda owners were met by the introduction of strike breakers from other parts of Mexico. In 1936, the federal government intervened, and within a brief period of 45 days an area exceeding one million acres was expropriated. To the Indian communities went 31 percent of the entire area and nearly 78 percent of its irrigable land.

So hurried a division of the land could not be accomplished without error. The surveying was in some places so inaccurate that confusion and litigation occurred over the new boundaries. The lack of a careful land survey in advance resulted in a pattern of *ejido* properties that bears only chance relationship to the sources of water — the irrigation canals and the wells. Most seriously, a land area of 470,000 acres was estimated to be irrigable. Actually, not more than 312,000 acres proved suitable for irrigation, which left some of the new property divisions without water. In an effort to provide more water, the Lázaro Cárdenas Dam on the Río Nazas was built, but by the mid-1960s 10 years of drought in the headwaters of the Nazas and the Aguanaval resulted in failure of the reservoir behind the new dam to fill up. Unfortunately, the lack of rainfall observations, which should have been gathered systematically before engineering works were undertaken, made it impossible to be certain whether a less-than-normal rainfall was really responsible for the drought.

When *ejidos* of the Laguna district were granted lands expropriated from the haciendas, the land was to be cultivated on a communal basis. A change in the original collective system was made in the 1950s. The cultivation and planning operations were still done on a collective basis, but thereafter each family was allotted a certain area of cotton. On this area the family was responsible for bringing the crop to maturity. Harvesting was done collectively, but the individual farmer was credited with the production from specific rows of cotton. A considerable increase of production resulted.

Then came a disastrous decrease in water supply. On the average, in each year between 1958 and 1962 there were 235,000 acres of cotton under irrigation in this district, but by 1963 this figure was reduced to 114,000 acres, and there was further decline in subsequent years. In 1978 the crop suffered from a fungus outbreak. About 70 percent of the irrigated area is still used for cotton, but total production is only a small part of the Mexican total. There are increasing acreages of alfalfa fields that supply fodder for dairy cattle, and some areas are now even used to grow grapes for wine.

Irrigation Along the Río Bravo del Norte

Along the Río Bravo del Norte (the Mexican name for the Rio Grande) are seven irrigation districts, totaling about 476,000 acres, on which the chief commercial crop is cotton. However, in contrast with conditions along the lower Colorado River, where the Mexicans are dependent on water flowing from the United States, irrigation along the lower Rio Grande on both sides of the border is supplied largely by water flowing out of Mexico. It was this fortunate balance along the U.S.-Mexican border that led to an international agreement regarding water rights.

The first of the seven Mexican irrigation districts is along the main stream of the Río Bravo, just southeast of Ciudad Júarez, with 37,000 acres. The second is along the valley of the Río Conchos in Chihuahua, just east and southeast of the city of Chihuahua. La Boquilla reservoir regulates the flow of water in this area and permits the irrigation of 173,000 acres. Third is the small irrigated area near the Río Bravo opposite Eagle Pass, with about 25,000 acres. Southwest of this, along one of the headwater tributaries of the Río Salado, is a small area of about 3000 acres developed by the former hacienda owner before the land was expropriated. The fifth is along the Río Salado, where the Don Martín reservoir provides enough water to irrigate 37,000 acres. The sixth area is along the Río San Juan, where there are 141,000 acres under irrigation downstream from El Azucar reservoir. Finally, along the Río Bravo, opposite the much larger and more famous lower Rio Grande irrigation district in

Texas, the Mexicans have some 50,000 acres supplied with water near Matamoros.

Border Development Program

Beginning in the 1960s, Mexico initiated what has become a highly successful border development program. The objectives are to attract industry, stimulate the economy, and provide employment in cities all along the 1900-mile northern frontier. Foreign companies, particularly from the United States, quickly established assembly plants under an "in-bond," or free zone, scheme. This permits manufacturers to send raw materials or parts to the plants, use Mexican labor in the manufacturing or assembly process, and ship the finished product to the United States or elsewhere, paying tariff charges only on the value added to the product. The advantage in labor cost is reflected in the minimum wage scales in 1980: $3.10 per hour in the United States and about 80 cents per hour in the northern border region of Mexico.

The number of border plants reached 520 in 1980, almost one-fourth of them located in Ciudad Júarez (population 800,000), across the Rio Grande from El Paso, Texas. Employment by these plants totaled nearly 110,000, and it was estimated that for every job in industry, two were created in other sectors of the economy. With such apparent employment opportunities, great numbers of job seekers have migrated northward to the border and beyond. The border cities have grown accordingly, with the provision of urban facilities and services becoming increasingly complex. The extension of a natural gas pipeline to these cities from the Gulf Coast, and completion of a major coal-fired electrical plant at Piedras Negras, Coahuila, provides further incentive for industrial employment and urbanization.

The Northern Sierra Madre
Oriental and the Gulf Coastal Plain

The remaining part of the North includes the northern end of the Sierra Madre Oriental and the Gulf Coastal Plain north of Tampico. The Sierra Madre Oriental is almost as great a barrier to communication as the Sierra Madre Occidental. Uplifted, folded, and faulted geologic structures have been eroded by torrential streams to form a bold mountain system through which routes of travel are not easy. South of Monterrey, the Sierra is composed of a series of great north-south ranges, rising to elevations between 6,000 and 12,000 feet above sea level. Where these ranges are unbroken, passage over them is almost impossible; but in a few places there are breaches in the mountain ramparts. Inland from Tampico the three rivers that unite to form the Río Pánuco have cut headward into the highlands. In this section there is a confusion of steep-sided valleys and ridges, but openings have been cut in the great ranges and several routes are offered between the piedmont and the highlands. The most important pass routes are in the North. Near Monterrey, the main fold axes turn from a north-south to an east-west alignment, and from there northward for a short distance are several easy passes through relatively broad valleys and basins that connect the highlands with the eastern piedmont. The ranges turn northward again in Coahuila before crossing the border into Texas. When Monterrey was founded in 1596, however, it was by no means certain whether the chief route of travel in the future would descend from Saltillo or would continue northward through Monclova to cross the Rio Grande at Eagle Pass. In fact, the latter route was used more commonly than the route through Monterrey during the period of Spanish and Mexican control of Texas.

North of Monterrey the Sierra Madre Oriental is too dry to maintain any important communities of farming people. The small settlements around Sabinas and Lampazos in northern Coahuila are supported by the mining of coal. In these districts Mexico is endowed with small deposits of fairly good bituminous coal, a natural resource that is rare in Latin America. Coal is supplied to the Mexican railroads and to the

INDUSTRIAL WORKERS IN A MEXICAN RAILROAD FREIGHT CAR PLANT IN SAHAGUN.

heavy industries at Monterrey. In recent years oil and magnesium have been produced in Tamaulipas. In a narrow zone along the mountain piedmont and extending south to Linares, sugarcane, oranges, cotton, maize, wheat, and beans are grown to supply the city market. Sorghum, the most important feed grain, is also raised.

In the North as a whole, only about 5 percent of the area can be used for crops. Cotton is grown extensively on the irrigated lands, with the surplus exported to Japan and China. There are also large acreages of wheat, including new drought-resistant varieties, and still larger areas devoted to open range useful only for the grazing of cattle or goats.

Monterrey

Monterrey has evolved from a pass city into one of Mexico's leading industrial centers. In 1880

Monterrey had no more than 30,000 inhabitants, its first period of rapid growth coming with construction of the railroad. Further rapid growth resulted from construction of the Inter-American Highway from Laredo through Monterrey. By 1980 its Metropolitan population was estimated to be 2 million.

Today Monterrey is the hub of an extensive system of railroads. The main line from the United States to Mexico City passes through Monterrey, tapping the coal fields near Lampazos on the way southward from Laredo and ascending to the highlands at Saltillo. A railroad also reaches the lower Rio Grande, another line runs southward to Tampico, and yet another crosses the desert country westward to Torreón. These railroads and their connections reach the most productive mineral and agricultural centers throughout the north of Mexico.

During the 1930s major change came to Monterrey and the district south of it with completion of the first section of the Inter-American Highway from the border at Laredo. Instead of climbing to Saltillo, the highway continues along the piedmont southward through Linares to the little Indian community of Tamazunchale. Then it makes a spectacular climb to the highland, over rugged terrain cut by the Río Moctezuma and its tributaries. It passes not far from Pachuca en route to Mexico City. Inland from Tampico this highway passes through country formerly inhabited only by scattered tribes of Indians who supported themselves on wild game from the forests. Now tourists in great numbers from the United States drive this route and encounter an abundance of modern air-conditioned motels, gasoline stations, restaurants, and gift shops. The tourist industry has developed rapidly during the past 20 years, and Monterrey has benefited accordingly.

The Steel Industry

The manufacture of pig iron and steel began in Monterrey in 1903, and the plant in Monterrey was the only integrated steel producer in Mexico until World War II. As a result of the war, with restricted imports of steel, the government entered the business by constructing a large new plant at Monclova, northwest of Monterrey, in 1944. Together, these two plants account for more than a third of total Mexican steel production. Another plant, using scrap iron and natural gas imported from the United States, is located just across the Rio Grande from Eagle Pass, and a $307 million sponge-iron and semi-basic steel plant was completed near Tampico in 1983. Largest of all is a new steel complex being developed in the Rio Balsas Valley.

The plants at Monterrey and Monclova use coking coal from nearby Sabinas, iron ore from Durango, local limestone, and manganese from northwestern Chihuahua. Adequate water supplies are available from nearby rivers. This part of Mexico is one of the few places in Latin America where raw materials for steel making can be brought together at low cost, and where the steel can be transported easily to its major market. Sabinas coal is among the best in Latin America. Similarly, the Durango iron deposits are high in quality and easily mined by open-pit methods.

Monterrey has the second largest industrial concentration in Mexico and is the headquarters for 15,000 manufacturing concerns. Among the largest are two steel mills, three cement plants, a brewery, and three glass factories. In total, the factories of Monterrey employ about 450,000 people, 19 percent of whom are women. The industrial sector in this sooty, but vigorous, city is characterized by diversified groups with affiliates throughout the nation. Outstanding is the ALFA concern, organized in three different fields: steel, paper, and food packing industries. It has also diversified into electronics, petrochemicals, and real estate.

Monterrey is Mexico's chief center of lead smelting and is a minor producer of silver, gold, copper, arsenic, bismuth, and antimony. Other manufactures include tile, furniture, beer, cigarettes, plastic, nylon, and television sets. The government has urged Mexican industrialists to build plants in remote areas to attract workers

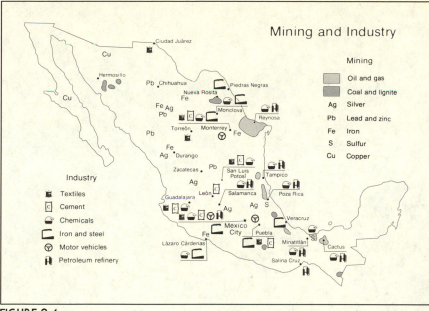

FIGURE 2-4

from the overcrowded cities. Many Monterrey companies have established plants in distant places such as Tlaxcala and Coatzacoalcos in an effort to conform with government policy.

The Gulf Coast and Yucatán

In striking contrast to all other parts of Mexico is the Gulf region, including the states of Veracruz, Tabasco, Campeche, Yucatán, and Quintana Roo. This region comprises 12 percent of the total area of the country and is occupied by about 12 percent of the population. Unlike most of Mexico, the Gulf region is abundantly supplied with moisture — 73 percent of it receives sufficient rain in all months. Also unlike most parts of Mexico, 53 percent of it is classified as level. In terms of physical quality of the land, this is the best maize-growing area of Mexico. Here are to be found the largest per acre yields and the fewest crop failures.

During recent decades Mexico has concentrated on the development of its wet tropical lowlands, especially in the states of this region. Malaria, the major obstacle to occupance of the land, has been overcome. Fortunately, there are few diseases that destroy crops; hence, rubber, pepper, and cloves flourish, as do cacao and bananas.

Veracruz and Tabasco

Only south of the Río Tamesí do those plants survive that cannot endure occasional freezing weather. Cold air masses from the north that bring frosts to the coastal plain of Tamaulipas continue southward across the Yucatán Peninsula as cool waves accompanied by heavy rains. South of Tampico, however, frosts are rare or entirely absent. Throughout the *tierra caliente* of the Gulf Coast humidity is high at all times of the year, averaging close to 80 percent at the port of Veracruz. Annual rainfall is abundant. Vegetation of the *tierra caliente* includes both forest and grasslands.

The first port that Cortés established on the Gulf Coast proved to be a poor one because of

FIGURE 2-5

insect pests and the difficulty of defense. In 1609 the old port was abandoned, and the new city of Veracruz was established a little farther south. From there, a road was built into the highlands by way of Jalapa, a route that functioned in Aztec times and is also closely paralleled by the modern railroad. Veracruz, now a city of more than 255,000 people, remains the chief port of Mexico and is equipped with breakwaters, modern docks, and other facilities for handling cargo.

The state of Veracruz has an excellent transportation system, ranking high in miles of paved highways and railroad lines. Veracruz city has an international airport, and other major cities in the state have air terminals and landing strips for local flights. The many ports handle tons of cargo, especially petroleum and petrochemicals. The primary ports for both ocean-going and coastal traffic are Veracruz, Coatzacoalcos, Tuxpán, and Pajaritos.

Veracruz is also a leading agricultural state, employing more than a half million people in farm activities. Yields of the major crops are high, and livestock raising is profitable. Fishing is important, too, with adjacent waters providing more than 40,000 tons of fish per year.

The chief concentration of people is in the coffee zone, focusing on such towns as Jalapa, the political center of the state, and Orizaba, an old center of cotton textile industries. Both of these cities are between 4000 and 5000 feet above sea level in the *tierra templada*. Lower, the plantations are fewer; yet, there is an important production of such tropical crops as rice, tobacco, bananas, vanilla, sugar cane, rubber, and chicle (the chief ingredient in chewing gum). From the rain forest, collectors gather valuable cabinet woods, gums, and various other products of small bulk but high value.

The number of plantations and settlers decreases toward the south in Veracruz, and the southern part of the state, or northern side of the Isthmus of Tehuantepec, is sparsely inhabited. The discovery of vast reserves of sulphur, petroleum, and natural gas in recent years may give the Isthmus added significance. Salina Cruz, on the Pacific, is already connected by pipeline with a refinery on the Gulf Coast, and offers dry dock and boatbuilding facilities, as well as highway and rail connections.

In 1947 the Mexican government started a project to establish colonies in the Gulf region. The basin of the Río Papaloapan, just south of Veracruz, was selected as the first colony. Five large dams were built to control floods and provide electric power. The lowland part of the basin, an area twice as large as Puerto Rico, was drained so that malaria was practically eliminated. Farm colonies were surveyed, and colonists were brought from the highlands, each family located on a 30-acre farm and provided with a house, farm equipment, and seed. Modern farm methods on fertile soils produced bumper harvests of cotton, sugarcane, and maize. Ciudad Miguel Alemán, an entirely new city, grew rapidly as the chief commercial center of the lower valley. By 1960 the success of the project was proved, and other farm colonies were being laid out all along the Gulf Coast. New all-weather roads made these colonies accessible, modern health measures and model homes made tropical living healthful and pleasant, and modern techniques of farming made possible continued high yields from what are now recognized as among Mexico's best soils. Subsequently, agricultural colonies were established in Tabasco, Campeche, and the eastern side of the Yucatán Peninsula in Quintana Roo.

For many years hot, humid Tabasco was underdeveloped and poor. The sparse population ate well on bananas, seafood, and vegetables, but there were no schools, railroads, or roads, and the occupants were dependent on the many rivers for communication with the rest of the country. Now, Tabasco has come into its own, for beneath the green, lush, swampy vegetation is a huge reserve of petroleum and natural gas. The discovery of oil has been a mixed blessing, however, for natives complain that the government petroleum corporation, *Petroleos Mexicanos*

(PEMEX), is siphoning off a natural resource and failing to develop the state. On the other hand, as the largest company in Mexico, PEMEX employs thousands of workers.

Tabasco is Mexico's leading producer of cacao, half of which is exported to more than 20 different countries. Copra and sugar are also produced commercially. A wide range of tropical fruits, such as papaya, mangoes, citrus, and avocados thrive in this climate, and it appears that if Mexico is to feed its future population it must take advantage of products such as these that grow well in the tropics.

The Gulf region has become a vital source of fish for the Mexican diet, especially snapper, bass, tuna, and sardines. The fishing industry of Tabasco and Compeche has especially great potential, although present production is limited primarily to oysters and shrimp.

Yucatán

At one time Yucatán was the well-populated habitat of the Maya Indians, who preceded the Toltecs and Aztecs. At the time of the Spanish conquest, however, the Maya culture had already declined. The small number of survivors carried on a shifting agriculture, whereas the cities had been abandoned and were almost lost in the forest cover. Yet, Yucatán was not entirely neglected. The town of Mérida was formed in 1542, and the few remaining Indians were allotted to the Spanish invaders in *encomiendas*. The land was partitioned into large estates on which cattle grazing was the chief economic activity.

There are many handicaps to successful utilization of the Yucatán. In the northwest, around Mérida, rainfall is low, although it increases toward the south and east until it is abundant on the borders of Belize and Guatemala. A scrub woodland with patches of savanna occupies the drier northwest of the peninsula, but as rainfall increases toward the southeast the woodland is replaced by tall trees.

Surface water is not abundant in Yucatán. The peninsula is underlain by horizontal beds of

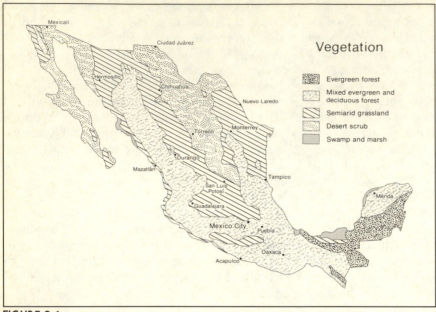

FIGURE 2-6

limestone in which rainwater forms underground solution caverns. There are no surface streams, and only where the cavern roofs have collapsed are there pitlike *cenotes* in which the plentiful groundwater can be reached. The native Indians, having no iron tools, were unable to dig wells and had to locate their permanent settlements near the natural sources of surface water.

After 1880, the northwestern part of Yucatán became the world's exclusive source of fiber from henequen, a type of maguey. Maguey plants are of the agave family, which supplies not only henequen and sisal fiber for export, but also tequila and mescal for local consumption. In ancient Mexico the maguey plant was the source of fiber for house roofs, shoes, and clothing. Its pulp was used for making soap, its spines were used as nails, and its sap was made into the beverage modern Mexicans call *pulque*. Crushed and despiked, it can be used as cattle fodder.

Henequen production was highly profitable after a world market was developed. The major use of this coarse, strong fiber was in the automatic, self-binding reaper used to harvest grain throughout the grasslands of the middle latitudes, a machine first marketed after 1880. Until 1911, Mexico was the only place in the world where henequen was produced, but, despite efforts to maintain the monopoly, agave plants were taken out and cultivated in Cuba, Puerto Rico, the Bahamas, and Florida, as well as parts of East Africa and Central America. In 1911 the product of these other plantations began to appear on the market. Although Mexican henequen production increased until the peak year of 1916 (217,300 long tons), Mexico's share of the combined world production of sisal and henequen decreased steadily.

Today, the Yucatán peninsula is experiencing a tourist boom, resulting in part from a Mexican law that allows foreign investors to buy land and to hold it for 30 years. A major project of the Echeverría administration was development of the resort city of Cancún. Instant wealth became possible as foreigners rushed to invest and formerly worthless land skyrocketed in price. Cancún has since become a leading tourist center and

MAYAN RUINS AT CHICHÉN ITZÁ, MEXICO.

provides a significant source of employment for local labor.

Quintana Roo, once part of Yucatán, became a territory in 1902 and a state in 1974. Its extensive forests include mahogany, rosewood, and oak, each a valuable export product. The principal crops of sorghum, rice, beans, and oranges are for the most part consumed locally. Industrial development is based largely on the manufacture of wood products.

Campeche abounds with Mayan ruins, some of which are in almost inaccessible areas. These attract tourists in modest numbers. By 1976, Campeche was involved in a major oil exploration boom, and discoveries within the state and offshore in the Gulf of Campeche have been extensive. The number of rigs and wells has been increased, and a pipeline has been laid from

Campeche to Dos Bocas, a terminal that handles exports from the Reforma and Campeche fields. Political and ecological considerations influence the total volume of production, but PEMEX has increased oil exports in an effort to improve Mexico's negative balance of trade.

The Oilfields

Oil production from the Gulf region is of extreme importance economically and politically in Mexico's foreign relations. The country's first successful oil well was drilled about 50 miles west of Tampico in 1901. The significance of the new fields was quickly appreciated by North American and British interests, and with the aid of foreign capital, Mexico's production increased rapidly. By 1902 oil was discovered on the Gulf side

A LUXURY HOTEL AND ITS GROUNDS AT THE COMPUTER-PLANNED RESORT OF CANCÚN IN YUCUTAN, MEXICO.

of the Isthmus of Tehuantepec, and in 1908 another major field was discovered inland from Tuxpán, just south of Tampico. Production peaked in 1921, when Mexico supplied 21 percent of the world's exports. However, the Mexican government then placed restrictions on oil company operations until 1938, when most foreign-owned properties were expropriated. The North American and British companies withdrew and, after a period of negotiations, were partially reimbursed for their losses.

Mexico's oil production and marketing were placed under control of the new government agency, PEMEX. By making good use of the Poza Rica field, PEMEX produced an average of about 40 million barrels a year from 1938 to 1945. In 1965 extensive geological exploration considerably increased Mexico's known oil reserves. New

fields have been discovered around Villahermosa, Matamoros, Torreón, and Chihuahua, as well as in Veracruz and Tabasco, and oil and gas pipelines have been built to the major cities of the Central region. Ciudad Pemex became a thriving community in what previously had been uninhabited tropical forest. In southern Veracruz and Tabasco, oil is associated with sulphur, which is exported through the port of Coatzacoalcos. Oil and sulphur also provide the basis for a local chemical industry, including the manufacture of fertilizers, insecticides, and plastic products.

By 1980 it appeared that Mexico might surpass even Saudi Arabia in oil reserves. Potential reserves exceeded 200 billion barrels, and Mexico seemed likely to become the world's largest single source of oil. However, the government has stated that the country should not become a "nation of oil derricks dedicated to feeding an oil-hungry world at the expense of internal progress." Recognizing that oil will run out some day, Mexico has taken a realistic attitude in its plans for exporting this precious commodity. New oil discoveries are commonplace and widespread. PEMEX revenues run as high as $17 billion a year, and further exploratory plans indicate expansion of the industry.

Dominating the news from Mexico in 1979 were reports of the world's largest oil spill, which occurred in the Gulf of Campeche. A well blowout caused the loss of 126 million gallons of precious crude oil, at a cost of more than $200 million. Oil slicks spread as far north as Texas, where $20 million more was lost as tourists stayed away from coastal hotels and polluted beaches. The nine-month gusher was finally brought under control in 1980. PEMEX was stunned, as were environmentalists in both countries, at the damage done to property, fish, and wildlife. Hence, PEMEX has adopted extreme caution in its drilling program both on land and offshore.

The South Pacific Region

The South Pacific region includes the states of Colima, Guerrero, Oaxaca, and Chiapas. In contrast to the Gulf Coast, less than 20 percent can be considered level. This is a region of towering mountains, green valleys, and hidden archaeological treasures. It includes about 12 percent of the nation's total area and an equal percentage of its population. The region is divided into eastern and western parts, separated by the Isthmus of Tehuantepec. In both parts the rural population is predominantly Indian, and Indian languages are still spoken extensively. In certain areas, particularly along the coast, a rapid transformation from the traditional rural economy is taking place.

Most of the southern dissected border of the highlands is drained by the Río Balsas and its tributaries. The main stream has opened a deep gulf well into the *tierra caliente* along the north-

ern border of Guerrero, and the tributaries extend in dendritic fashion back into the higher country on either side. In the course of stream erosion, however, certain more resistant structures in the Balsas Valley were exhumed and left standing as prominent and more or less isolated blocks. An example is the small block mountain in northern Guerrero not far north of the Balsas itself, in which the old mining town and present tourist center of Taxco is located. Another is the larger block bordering the Pacific in southern Michoacán, Guerrero, and southern Oaxaca, which is known as the Sierra Madre del Sur.

Flat places are remarkably few and widely scattered in this region. It is on these flat places that the people are concentrated, mostly Indian farmers growing subsistence crops. However, there are too many people in this area to be supported on a subsistence basis by the small areas of good land. Food must be raised on slopes so steep that it can be said with considerable truth that farmers could easily slip and fall out of their fields. The patches of maize on the mountain sides are temporary, for new clearings must be made each year for this type of agriculture.

The coast west of the Isthmus of Tehuantepec is extremely rugged. The Sierra Madre del Sur descends with steep slopes almost to the water's edge, leaving only a narrow fringe of sandy and generally hot and dry lowland. The ports all suffer from one handicap or another; either the water is too shallow if they are situated, like Salina Cruz and Manzanillo, near the outlet of one of the silt-laden rivers; or they are isolated from the interior by steep mountain slopes, as in the case of Acapulco. Because of its excellent harbor, the Spaniards selected Acapulco as the chief port of departure for the Philippines. However, the country behind Acapulco is so difficult to cross that no railroad has ever been built to connect it with the interior, and only in 1940 was the paved highway extended to it from Taxco. Despite its isolation, Acapulco has become a seaside resort of considerable popularity. Completion of the road opened it to an increasing flood of North

American tourists, a stream that converted Taxco, the sixteenth-century silver and tin mining center, into a profitable tourist center. Because a railroad connects Manzanillo, in Colima, with the interior in Jalisco, this city has become Mexico's chief seaport on the Pacific. Manzanillo has also become a significant center for tourism.

One of the largest industrial enterprises in Mexican history is being developed in the state of Michoacán near the mouth of the Río Balsas. The project is located on the Pacific Coast about 160 miles northwest of Acapulco, and the nucleus of the industrial complex is the Siderúrgica Lázaro Cárdenas-Las Truchas, S. A. (SICARTSA). The SICARTSA steel complex was conceived by General Lázaro Cárdenas, who as President of Mexico nationalized the Las Truchas iron ore deposits in 1936. His proposal for a major steel industry based on local resources and designed to function as a ''development pole'' was adopted before his death in 1970. Steel production is to evolve in two stages, with completion of the first

stage to yield an annual output of 1.3 million tons. The second stage will increase the installed capacity to 3.6 million tons annually, making SICARTSA the largest steel manufacturer in the nation. The bulk of the steel mill's iron ore requirements is supplied from Las Truchas, only 16 miles from the plant. A deep-water port will facilitate the importation of coal and other materials while encouraging the development of export-oriented activities. Electrical power for the steel mill, satellite industries, and the city is supplied by two hydroelectric plants, at La Villita and El Infiernillo dams, a short distance upstream on the Río Balsas. It is anticipated that the city of Lázaro Cárdenas will expand to a population of 250,000 by the turn of the century, and the industrial complex will stimulate employment and economic growth in the entire region.

Chiapas

The highland of Chiapas is the second of the two subdivisions of the South Pacific region. This

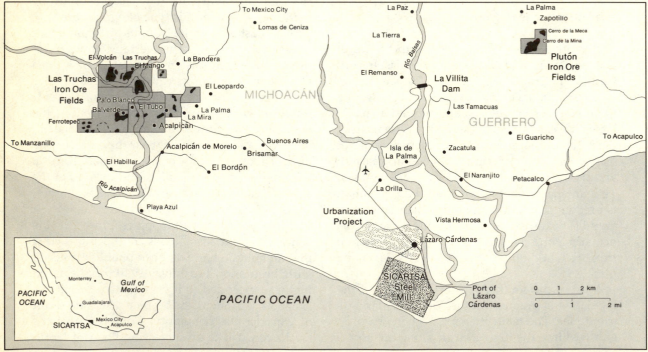

FIGURE 2-7

highland is composed of folded and faulted structures, partly covered by volcanic outpourings, with the main structural features running parallel to the Pacific Coast. Behind a narrow coastal lowland the first range of mountains, the Sierra Madre de Chiapas, rises to elevations of more than 9000 feet within 20 miles of the ocean. Immediately northeast of this range is a rift depression known as the valley of Chiapas. The floor of the depression is between 1500 and 3000 feet above sea level. It is drained by a tributary of the Río Grijalva which finds its outlet to the Gulf of Mexico in the state of Tabasco. To the northeast, again, is a succession of block mountains, each somewhat level-topped but deeply dissected by streams. From altitudes of 12,000 feet overlooking the valley of Chiapas, these mountains drop eastward to the limestone plains of Yucatán. Most of the people in this part of Mexico live in the valley of Chiapas, near the Pacific Coast.

Here, too, the natural vegetation and prevailing land use are arranged by vertical zones. The *tierra caliente* of Chiapas has greater natural potential for tropical crops than has the similar zone northwest of the Isthmus of Tehuantepec, because of increasing rainfall toward the southeast. Particularly productive is the Soconusco: the Pacific piedmont and coastal areas of Chiapas extending eastward to the Guatemalan border. On the lower slopes of the mountains in this district the Aztecs grew cacao, from which they made a ceremonial drink now known as cocoa. This is perhaps the native land of the cacao tree. Cacao is still produced here, although more important are the cotton plantations on the coastal plain. Since the mid-1950s the Soconusco has become one of Mexico's leading cotton-producing areas, and its mountain slopes have long yielded high-quality coffee. Cattle are pastured in the valley of Chiapas and on the high grasslands above the tree line, and higher parts of the valley are devoted to wheat.

Long characterized by isolation, the entire South Pacific region is now accessible to the rest of Mexico by rail and all-weather highways.

Thus, hundreds of scattered Indian villages have become part of the modern world.

Such are the outlying parts of Mexico. The clusters of people in these areas are highly diverse in origin and in their present relationship to the land. Yet, through all the diversity there are certain prevailing themes. All parts of the country have felt the effects of land reform, by the elimination of the haciendas and the landowning class. Now there is a mixture of *ejidos* and medium-sized private farms, operated by their owners. The crop that occupies the largest acreage is maize, grown near all the villages and often on slopes too steep to be used appropriately for this purpose.

The Central Region

The group of states that form the Central region comprises 14 percent of the area of Mexico and 52 percent of the Mexican people. In this region, too, are most of Mexico's industries. This core area of the Mexican nation accounts for nearly 20 percent of all agricultural land, nearly half of all the farmers, and more than half of all the area devoted to maize.

The Central region includes many areas that are only thinly populated and others that were well populated even before the arrival of the Spaniards. The land includes a number of well-defined basins, all drained by rivers that reach the sea, except for the Basin of Mexico in which the capital city is located. Between the basins much of the area is composed of gently rounded hills through which the rivers have cut deep, narrow valleys. The entire scene is given extraordinary beauty by a series of exceptional volcanic cones and by many lesser volcanic forms throughout the region. Among the most notable features is Parícutin, near Uruapan, in Michoacán state. One of the world's newest volcanoes, it erupted in a cornfield in 1943 and grew to a height of 1700 feet before it became dormant in 1952.

Indian groups migrating from the arid north

came upon this more humid country, with its many lakes and running streams fed by permanent snow fields, and saw in it a land of plenty. In about 1325 the Aztecs founded their capital city of Tenochtitlán on an island in Lake Texcoco, where it could be easily defended, and from this center they extended their conquest, or at least levied tribute, over much of the territory to the south. Each of the basins of the Central region, however, had its own distinct Indian culture which differed in language and customs from the cultures of other basins. The native peoples, even those brought under Aztec rule, were never molded into one culture.

The First Spanish Settlements

The conquest of Mexico by Cortés and his little band of men is one of the great epics of adventure in the New World. In 1519 Cortés arrived on the margins of the Basin of Mexico with about 350 Spaniards and more than a 1000 Indian allies. After occupying the palace of the Aztec ruler for several months, the Spaniards were forced into a disastrous retreat on the famous *noche triste* — June 30, 1520. Cortés returned the next year and destroyed Tenochtitlán. The objective of this first conquest, and the reasons that led Cortés to found the city of Mexico on the site of Tenochtitlán, were definitely *not* related to ease of living, coolness of climate, or productivity of soil for plantation crops. The objective was the securing of wealth: first, wealth that had already been accumulated and could be carried away, and then wealth in the form of minerals ready to be mined. To supply labor for the mines and provide the newcomers with an adequate supply of food, a large population of peaceful Indian farmers was needed. The dense Indian population also satisfied the other great purpose of the Spanish conquest: the conversion of infidels to the Christian faith. Sedentary Indians and precious metals were factors that guided the course of Spanish settlement, and both were abundant in the Central region.

The 50 years after the founding of Mexico City in 1521 witnessed the establishment of mining communities and the start of active mining operations at almost all the sources of precious metals in Mexico. Few new sources have been discovered since. Northeast of the capital was the silver town of Pachuca, still the richest in the world, and to the north was the silver mine of Zumpango. Northwest of Mexico City was the gold and silver center of El Oro; in the hilly country south of the Basin of Guanajuato was the silver center of Morelia; and in the mountainous country north of the Basin was another silver mining community, Guanajuato. To the west was Guadalajara, itself a center of spreading settlement, and locally enriched by silver mines and dense Indian populations. Farther to the north was Aguascalientes and the three other prominent silver mining towns — San Luis Potosí, Zacatecas, and Durango. South of the capital were the silver mines of Taxco. The restless search for silver, gold, and dense populations of peaceful Indians led the Spaniards to the extreme reaches of their vast territory, far beyond the present limits of Mexico; wherever Indians or precious metals were discovered, towns were founded and Spanish civilization established.

Pattern of Population in the Modern Period

The rural population of the Central region is still predominantly Indian, and despite the adaptation of Spanish customs, most of these people still live in their traditional manner. Therefore, concentration of the Indian grain, maize, in the Central region is a very different kind of agricultural localization from that of the Corn Belt in the United States. Actually, the Central region is poorly suited to the production of maize. There are none of the steaming hot summer days and nights which, in the Middle West of the United States, are known as "corn weather"; crop failures due to cold weather, even in the summer, are all too frequent, and even in good years the yields are low.

Some change appears underway. Important dietary changes have occurred in Mexico during the past 20 years, largely because of increased urbanization. The diet of city dwellers is based much more on wheat, meat, and milk and less on the traditional maize, rice, and beans. During the period 1960 to 1980, average maize consumption even in the rural areas decreased by 20 percent. At the same time the acreage planted to maize decreased sharply, being replaced in part by grain sorghum. Yet, maize remains dominant in the Mexican diet and on the agricultural landscape.

Population Clusters in the Central Region

Within the Central region the population is grouped in seven separate clusters. The principal Spanish colonial towns, which have since become the centers of political power, were originally located with reference to the silver and gold mines, while the concentrations of Indian farmers were closely related to the character of the land. The intermont basins were densely settled before the Spaniards arrived and have remained so since the European conquest. In most cases the centers of the basins are swampy. The people, therefore, usually occupy the borders of the basins and the lower slopes of surrounding hills and mountains.

The seven chief clusters of people can be described with reference to the basins around which they are grouped. That these clusters do not have a simple relation to the state boundaries, as is usually the case in Latin America, is a peculiarity of this area. The seven basins, and the states included in each area of concentration, are as follows:

Basins	States
1. Basin of Mexico	Federal District; part of México and Hidalgo states
2. Basin of Puebla	Part of Puebla and Tlaxcala
3. Basin of Toluca	Part of México
4. Basin of Guanajuato	Southern Guanajuato; bordering highlands of Guanajuato, Querétaro, and Michoacán
5. Basin of Jalisco	Part of Jalisco
6. Valley of Aguascalientes	Aguascalientes; nearby parts of Jalisco and Zacatecas
7. Valley of Morelos	Morelos; bordering edge of Guerrero

Basin of Mexico   The original Aztec center was the Basin of Mexico, an irregular-shaped depression some 30 miles east-west by 50 miles north-south, which includes the Federal District and parts of México and Hidalgo states. When the Aztecs knew this basin, its bottom was occupied by five separate lakes, each shallow and bordered by marshy zones. Tenochtitlán was built on an island in the midst of Lake Texcoco and was connected to the land by causeways. Indian communities, which the Spaniards eventually incorporated into their haciendas, were strung along the margins of the basin and the lower and gentler slopes of the bordering hills and mountains. The irregular shape of the basin produced, therefore, an irregular pattern of population.

The Basin of Mexico is the only one of the seven areas of dense population which is not drained naturally to the sea. In 1607–8, however, the engineer Enrico Martínez partly drained the basin by digging a ditch and tunnel northward across a low divide to the headwaters of the Río Pánuco. In the present century, additional works have further reduced the extent of Lake Texcoco. The results are not altogether fortunate, for the lake bed soils contain such a high percentage of salts that not even pasture grasses will grow on them without expensive chemical treatment. Therefore, instead of acquiring a huge area of exceptionally rich farmland, as was anticipated, the city of Mexico found itself bordered by a vast expanse of empty flats, only the edge of which could be utilized. During the dry winters the bare ground gives off great clouds of dust which contribute to the unpleasant and even unhealthful conditions of the capital. In addition, as the lake bed deposits dried out, the surface began to settle. The construction of modern tall buildings on such foundations presents a major challenge to architects.

The population of the Basin of Mexico, outside of the capital city, is distributed much as in preconquest days. On the outskirts of the capital, the settled area has crept well up the bordering mountain slopes, but generally only the lower slopes are occupied by villages. The cluster of people included in this area extends well beyond the immediate basin, however, both in the southeast and in the northeast. Along the western base of the great volcanoes Ixtaccíhuatl and Popocatépetl are several small valleys filled with agricultural communities, the largest of which is Amecameca. Toward the northeast, bordering the Inter-American Highway, are stretches of gently rolling upland used in scattered patches for crops. Sixty miles by road from the capital is the old colonial mining town of Pachuca, with crowded buildings along narrow, irregular streets wedged against the valley head where the world's richest silver mines are located.

Crops raised by farming communities in the Basin of Mexico are similar to those produced in the other parts of the Central region. Maize is still most important in terms of acreage, although yields per acre remain low, despite government efforts to provide assistance. Beans are sometimes planted in the same field with maize, and adjacent plots may be devoted to chili and alfalfa. The chief crop grown for sale is wheat, which is consumed largely by city people, while sorghum provides feed for livestock and poultry.

The Central region has long been the chief area of *maguey* planting. The maguey is a species of agave from which a sweet liquid is derived. When fermented, this liquid becomes *pulque*. The average Mexican once consumed about 15 gallons of pulque per year. This was the poor man's drink, and the *pulquería*, where the drink was sold, was the poor man's club. The maguey plant grows on thin stony soils too poor for any other use. The general rise in the economic well-being of Mexico has resulted in a notable drop in

pulque consumption, largely in favor of beer. In the state of Hidalgo alone, in 1934 there were 22 million maguey plants, while today more than 70 percent of the land under cultivation is in short-season crops, such as alfalfa, beans, tomatoes, barley, and wheat.

The basin of Mexico provides good conditions for wheat production and excellent conditions for the grazing of cattle. After the land-reform program was implemented, some increase in wheat cultivation occurred. The government also undertook to build up herds of dairy cattle, until 1946 when this program was temporarily interrupted by the discovery that some of the animals imported for this purpose had developed foot and mouth disease. Spread of the disease was halted, however, and since 1950, with the introduction of new strains, the number of cattle has again increased.

Basin of Puebla   The second cluster of people in the Central region is in the Basin of Puebla, which includes parts of Puebla and Tlaxcala states. This basin lies on the eastern side of the volcanoes, Ixtaccíhuatl and Popocatépetl, at an elevation just over 7000 feet. The basin, which is drained by a headwater tributary of the Río Balsas, receives more plentiful rainfall than does the Basin of Mexico and is also better favored for agriculture. At Mexico City the average annual rainfall is 23 inches; at Puebla it is nearly 35 inches.

The city of Puebla lies 80 miles by highway southeast of Mexico City and about 200 miles west of Veracruz. Founded in 1532, it now has a population exceeding 465,000. When the Spaniards arrived, they brought with them the industry of tile-making, and this city subsequently became the leading cotton textile manufacturing center of Mexico. Tile and textile industries continue, and others such as food processing, basic metals, chemicals, and automobile assembly have been added to make Puebla one of the nation's leading industrial centers. Meanwhile, the many little Indian communities that dot the floor of the Basin of Puebla and climb the lower slopes of the surrounding hills and mountains are similar in their economy and organization to communities in the Basin of Mexico. The predominant crops are maize, wheat, beans, barley, and apples.

The old center of the Indian communities in this basin was Cholula, the capital of the Toltecs. It is said that Cholula has a church for every day in the year and was the holiest of holy cities. In 1970 it became a college town when the University of the Americas moved its campus from Mexico City. Nearby Tlaxcala is noted for Indian handicrafts, including serapes, sweaters, rebozos, and rugs.

Basin of Toluca   West of the Basin of Mexico is a densely populated area that centers on the city of Toluca. The Basin of Toluca, highest in the Central region, is the first of a series of basins that drain to the Pacific via the Río Lerma. It stands more than 8600 feet above sea level, and to reach Toluca from Mexico City, a distance of only 40 miles by road, requires a climb more than 10,000 feet over the intervening mountain range. The center of the basin was once swampy, as in the Basin of Mexico, and for this reason settlements were strung along the lower slopes of the bordering hills. Crops are also similar to those of the Basin of Mexico, but with a somewhat larger proportion of land devoted to wheat. Unfortunately, poor farm practices on the slopes have resulted in serious soil erosion. Few parts of Mexico show more vividly the devastating effects of misuse of land than do the western and eastern margins of the Basin of Toluca.

Included in this district of concentrated settlement is the old mining community of El Oro. The mines of this district, unlike those of Pachuca, have not remained productive; El Oro itself was not selected as a political center, for the city of Toluca was made the capital of the state of México. Today there are about 250,000 inhabitants in the clean, comparatively uncrowded city of Toluca, and few highrise buildings mark its skyline.

Basin of Guanajuato   Largest of the basins, and the largest area of concentrated settlement in the Central region, is in the southern part of the state of Guanajuato. Leaving the Basin of Toluca, the Río Lerma plunges through a narrow gorge from which it emerges into the upper part of the Basin of Guanajuato only about 5900 feet above the sea. The Río Lerma forms a trench in the bottom of this basin at it proceeds westward. The floor of the basin is composed of drained lake beds and soils of exceptional fertility that have formed on accumulations of volcanic ash. The Basin of Guanajuato, the famed *Bajio,* has long been considered the breadbasket of central Mexico. The lush farmlands produce a variety of vegetables and fruits, in addition to maize, beans, peanuts, and livestock, and are especially noted for the cultivation of strawberries.

The first Spanish settlements in this general area were the cities of Guanajuato, Querétaro, and Morelia, all in mining districts of the bordering highlands. As mining decreased, however, the centers of population shifted. Guanajuato, once the wealthiest city in Mexico, has declined notably. Meanwhile, towns in the agricultural areas have all grown, especially Celaya, Irapuato, and León, and manufacturing has become increasingly important. Near Irapuato is a huge oil refinery, connected by pipeline with the Poza Rica oilfield. Querétaro has established an industrial park where many factories have been built to serve regional and national markets.

Basin of Jalisco   Still farther westward along the Río Lerma is the Basin of Jalisco, second in size to the Basin of Guanajuato. This basin stands about 5000 feet above sea level. Nestled against the base of the twin-peaked volcano Ceboruco, its surface is interrupted by several small volcanic cones, now inactive, and its soils have been produced by the weathering of lava flows and ash falls. In the lowest part of the basin is the Lago de Chapala, into which the Río Lerma empties.

The Basin of Jalisco is not far from the Pacific Coast, but reaching it from there is difficult. The Río Santiago enters a narrow gorge not far from the outlet of the lake, near the city of Guadalajara. In about 275 miles the river descends over falls and rapids 5000 feet to the Pacific Ocean. Despite rugged terrain, the route from Guadalajara to Tepic is the main line of travel from the central area to the whole northwest of Mexico.

Guadalajara, one of the primary settlement centers of colonial Mexico, has maintained its importance in the modern period. Today it is second only to the capital, having a population of about 2.5 million, and it is a major industrial center. It is the marketing and transportation hub for an extensive area, as well as an educational and cultural center. This city, home of the famous mariachi music, maintains much of its colonial tradition and charm. At the same time it has the largest and most modern shopping center in Latin America, the Patria Plaza.

The rural population around Guadalajara is fairly dense, and a larger area is devoted to subsistence crops than in the other basins of the Central region. The land is dedicated mostly to the usual combination of maize, beans, chili, and alfalfa. The maguey plant is cultivated extensively on the dry lands around the town of Tequila, where more than 20 distilleries process its juice into a popular national liquor of the same name.

Valley of Aguascalientes   A smaller, but important, cluster of people occupies the valley of the Río Verde, a tributary of the Santiago, where the mining town of Aguascalientes was founded in 1575. Its land area covers Aguascalientes state and parts of Jalisco and Zacatecas. Aguascalientes might have experienced a gradual decline similar to that of some other mining communities had it not been for the establishment there of maintenance shops for all the Mexican railroads. Aguascalientes is growing, although the production of its mines is of little importance. This town had 48,000 inhabitants in 1921; by 1980 it had grown to almost 300,000.

The land near Aguascalientes is relatively fertile and supplied with ample water for irrigation. Maize cultivation is extensive, as is the raising of livestock. This is also the leading center of viticulture in all of Mexico.

Valley of Morelos   The last of the seven clusters of people included in the Central region is in the valley of Morelos, mostly in Morelos state. Cuernavaca, the capital of Morelos, is 40 miles from Mexico City. It is only 4800 feet above sea level, however, a little more than half as high as the national capital. Cuernavaca has played the role of resort city for many years, because of its easy accessibility to the highland centers and to its climate, which is much more comfortable than at places of higher altitudes. Here is a region in the tropics in which people descend, not ascend, to seek more moderate temperatures. In the small valleys just below Cuernavaca, sugarcane was planted in colonial times, and Morelos became the chief source of sugar for cities in the highlands.

Industry in Morelos is centered around Cuernavaca and Cuautla, which together contain more than a third of the state's total population. Taxco, in adjacent Guerrero state is still a productive silver-mining center and tourist paradise. The most important agricultural products are sugarcane, tomatoes, rice, cotton, and fruit. By doubling the irrigated lands it is expected that flower and fruit cultivation will be increased, with Mexico City as a potential market.

Mexico City

About 20 cities in Latin America have passed the million mark in population, but Mexico City is growing faster than any of the others. Early in the nineteenth century it was the largest urban center in the American Hemisphere. It had more than 300,000 inhabitants in 1900 and more than a million by 1930. In 1980 there were more than 16 million people in the greater metropolitan area.

In many ways Mexico City is a typical Spanish-American capital. Its historical nucleus is the old cathedral and government buildings facing the *Zócalo,* as the central plaza is called. To the west of this plaza is the commercial core, built on a pattern of narrow, right-angle streets. The Mexican capital also has its baroque avenue, built in the nineteenth century and patterned after the Champs Elysées of Paris. This is the Paseo de la Reforma, a wide, tree-lined avenue designed to

A MONSTROUS TRAFFIC JAM IN MEXICO CITY.

provide a long, straight approach to Chapultepec Castle from the center of the city. The castle, long the residence of the Mexican presidents, stands atop a hill, some 200 feet high, which dominates the city on the west. Outside of the old central section of the capital, with its strictly rectangular pattern, suburbs sprawl without reference to any master plan. In the modern era a lack of wide thoroughfares leading to the center of the city produces insuperable traffic congestion.

Even the cutting of new avenues of approach and construction of a modern subway system have failed to relieve the problems of transportation. Traffic accidents cause 3000 deaths a year, and the National Chamber of Commerce reports that this metropolis is the second noisiest and the third most polluted city in the world. The prob-

lem of pollution relates partly to the concentration of industry but also to the city's 1.3 million vehicles, which spew out 560,000 tons of contaminants annually. The fact is that there are just too many people living within too small an area. Still, the lure of the city is great, and more than 1000 newcomers arrive each day. The chief district of factories and workers' homes is to the northwest in Tacubaya and Atzcapotzalco, yet new residents tend to settle in the northeast where rents are cheaper but public transportation is poor.

Near the international airport is the satellite city of Netzahualcoyotl, called "Netza" by its inhabitants. It is a slum of more than 1.8 million people who would rather be there than in poverty-stricken rural areas of the country. Some of the homes have electricity and running water, and the main streets have been paved. Most of the men work in low-paying positions, and some young women without families manage to obtain service jobs. Since education is compulsory only through the sixth grade, thousands of 12 year olds become eligible for employment each year.

Government officials are trying to modernize the city in many ways. New boulevards have been built, and a computerized traffic light system has been installed. A master plan for transportation within the city has been developed. This includes extension to the underground system, to occur at an average annual rate of 5 miles until the year 2010, at which time 20 new lines will have been added. The center of the city is changing, too, as teams of archaeologists and workmen discover and excavate relics of precolonial time. Nearby buildings are torn down while Mexicans watch with awe and excitement as remains of the Aztec Empire are unearthed.

The salty, dry bed of Lake Texcoco long delayed extension of the city eastward. As a fortunate result, this land was available for the airport, which is much closer to the center of the city than would have been possible if the city had grown equally in all directions. The bed of Lake Texcoco also has had an unhappy significance. The spongy lacustrine deposits offer no firm foundation for heavy buildings, and since the lake was drained the continued drying of the silt has caused a general drop in the surface level. The *Palacio de Bellas Artes*, begun in 1900 and completed in 1935, has settled more than five feet. The situation is further complicated by the fact that a fault line, a break in the rock crust of the earth, passes almost directly under Mexico City.

As a manufacturing center, Mexico City rivals São Paulo for leadership in Latin America. Before World War II the most important industries were textiles and food-processing plants. Then a rapid expansion of manufacturing occurred, in part financed by investments from the United States and other foreign countries. The factories of Mexico City now turn out a great variety of products. This metropolitan center, occupying just 1 percent of the total national territory, now has 47,000 industrial establishments and accounts for 52 percent of the nation's industrial activity. Important enterprises include automobile assembly, printing and bookbinding, and the manufacturing of pharmaceuticals, paper, tires, textiles, foods, and beverages.

Mexico as a Political Unit

The changes in Mexico City since 1940 are but a reflection of changes in the economy of Mexico as a whole. The Mexican Revolution uprooted established institutions such as the hacienda, eliminated the class of large landowners, and, at first, disrupted the economy. The internal warfare from 1910 to 1915, and the Constitution of 1917, set the stage for Mexico's transformation. It was during the presidency of Lázaro Cárdenas (1934–40), however, that the most important changes were made. By 1940 much that was traditional in Mexican life had been swept away. Since then a new Mexico has been built. It has already surpassed most other Latin American countries in economic productivity.

The Mexican Economy

Mexico has the largest population of any country in Spanish America and is second only to Brazil in all of Latin America. Its total number (79 million in 1984) is far greater than that of the second most populous Spanish American country, Argentina. The Mexican population, which grew at a rate of 3.5 percent in 1977, is now increasing at a rate of about 2.6 percent per year. Yet, even if the present decline in birth rate is maintained, it is estimated that by the year 2020, the 1985 figure will have more than doubled.

The net increase is due to an unprecedented decrease in the death rate, which is one of the first changes brought by modernization. Not only are professional services made available to more people, but also the increased number of people able to read and write facilitates the dissemination of knowledge about the causes of disease. In 1944 some 53 percent of the Mexican people were illiterate; by 1985 the illiteracy rate was less than 18 percent.

THE FUNDIDORA STEEL MILL AT MONTERREY, WITH ITS OPEN HEARTH FURNACE.

A country with a gross national product per capita of $2240 a year has little income that can be saved. Moreover, in the presence of a rapidly growing population, economic development must occur at a particularly high rate. Tremendous amounts of money must be invested to provide jobs for even a relatively small number of workers and to effect improvements in roads, railroads, docks, housing, and electric power. To partially solve this financial problem, Mexico made the decision to borrow heavily from many sources in an all-out effort to modernize.

Mexico is rich in mineral resources, but only a fraction of its territory has been explored and brought into production. Nevertheless, it is the world's largest producer of silver and fluorspar, and is among the world's "top ten" in graphite, lead, cadmium, sulphur, zinc, antimony, manganese, mercury, and gypsum. Altogether, 50 different minerals are mined commercially.

In the 1960s more than half the population of Mexico lived in houses with only one room. Twenty-four million Mexicans lived in houses with no running water. A government program hastened the construction of low-cost housing, but the 360,000 units provided were scarcely enough because the Mexican labor force increased more than 67 percent from 1960 to 1980. The labor force is growing by 600,000 to 800,000 annually, while unemployment exceeds 20 percent. The provision of adequate housing will long present a monumental challenge.

Many unemployed Mexicans now migrate across the northern border, causing problems for the U.S. government, yet benefiting farmers who employ them on a seasonal basis. Some are captured and returned, while others obtain illegal documents as they move northward in search of opportunity and employment. There is also a steady flow of drug traffic. In an attempt to halt the invasion of both people and drugs, the United States announced plans to build new fences and replace old ones in the heavily crossed sections of the border. The new fence was promptly dubbed the "Tortilla Curtain" and denounced by citizens of both countries. Mexico's petroleum bonanza created a new problem by 1980, if only on a minor scale. With "regular" gasoline selling at 45 cents a gallon in Mexico and $1.05 in Texas, a number of motorists saw prospects of gain by crossing the border with an "extra tank." Small-volume smugglers were subject only to small fines, but one entrepreneur was jailed when caught with his waterbed—containing 500 gallons of fuel!

The rate of Mexico's economic growth during the 1960s was 6.3 percent per year, but by 1980 it had dropped to about 5 percent. The government then drew up a plan that established regional priorities to encourage investment in medium-sized cities and away from the Mexico City area where much of the country's industry is now located. Borrowing overseas is expected to continue on the basis of oil revenues, with the government lengthening its maturities. An Industrial Development Plan is to reallocate oil income and promote alternative sources of energy, including nuclear, geothermal, hydroelectric, and coal-based thermal power.

Agriculture

After 1960 Mexican agriculture began to show spectacular growth, as a result of increased areas under irrigation. The use of insecticides, fertilizers, soil conditioners, and better seeds all led to increased yields per acre. An increase in the yield of maize came as a result of effective team work between the Mexican government and agricultural experts from the Rockefeller Foundation. Other improved crops were wheat, potatoes, sugarcane, and tobacco.

Better farming methods, all-weather highways giving improved access to markets, and the establishment of agricultural training schools seemed to indicate the transformation of Mexico's traditional culture. However, by 1980 growth in this sector had diminished. Wheat, maize, sugarcane, sorghum, and beans were still the chief crops, but yields were not sufficient for local consumption and Mexico was forced to im-

port more than 3 million tons of grain. Ten years earlier it had been self-sufficient.

Mexico sells to the United States about $200 million worth of winter vegetables a year. These include eggplant, squash, cucumbers, tomatoes, strawberries, and peppers. These vegetables, after petroleum, are among Mexico's largest export items and involve 200,000 Mexican agricultural workers. Their main competition is from Florida where growers likewise do a $200 million business.

Traditionally a major exporter of beef, Mexico has been plagued by feed and transportation problems. Frosts and drought have destroyed much good pastureland. Economic and political factors such as the government's embargo on exports of beef and live cattle add to the cattleman's woes. Government officials realize that population is outstripping the supply of beef in the entire domestic market, with shortages occurring mainly in Mexico City. In the exporting regions of northern Mexico, cattlemen have been experiencing a most difficult time. Government regulations periodically prohibit them from exporting

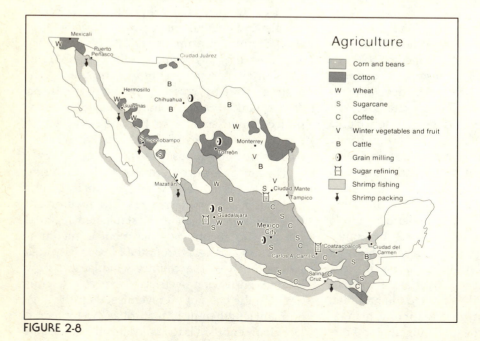

FIGURE 2-8

cattle, weather conditions often reduce pasture-lands, and transportation problems make it difficult to import feedgrains. Meanwhile, the poultry industry has expanded steadily, especially near the major centers of population.

A quickening of economic life is reflected in a rapidly expanding market for consumer goods and for services to the consuming public. Farmers who for the first time can raise a surplus to sell now wish to buy all kinds of possessions they have never sought to purchase before. Housewives want radios, televisions, sewing machines, washing machines, refrigerators, and new clothes that are shown in the shops in both small communities and large cities. Yet, most Mexicans cannot afford these products. Employees of the government and private companies, together with a privileged group of unionized industrial and urban workers, may enjoy such benefits, but they represent only 40 percent of the population.

Exports and Imports

Mexico's foreign trade has undergone major changes since the beginning of the Revolution in 1910. Before that date the principal exports were metallic minerals, and imports were mostly basic foods and railroad equipment. As late as the 1930s, the chief exports were still minerals (silver, gold, lead, zinc, copper, and antimony). Oil exports, which decreased in the 1920s, ceased altogether after 1938, when Mexico began using all of its production for domestic purposes. After World War II the makeup of exports changed radically. All agricultural exports in 1939 comprised 28 percent of the total, and by 1955 they exceeded 55 percent. In 1985 the leading export was crude oil (66 percent). The leading customer for Mexico's products was the United States (65.3 percent), and this may become increasingly true as oil exports grow and natural gas flows through a new 652-mile pipeline from Cactus (Chiapas) to the border at Reynosa (Tamaulipas). Of the imports, the United States supplied 62 percent, and Germany and Japan each 6 percent.

An important part of Mexico's foreign exchange earnings is derived from tourists, 85 percent coming from the United States. In 1934, tourists brought to Mexico only about $30 million; by 1980 the figure exceeded $4.4 billion. Until surpassed by the value of oil exports in 1978, the net income from tourism was, in fact, far greater than that from any other single source. By the end of the decade, Mexico expects to attract more than 7 million tourists annually. A boom has already taken place in the Yucatán Peninsula with the development of Cancún. This architecturally splendid city was not developed by city planners, nor was the site selected by geographers. Rather, it was chosen by a computer. The computer picked an L-shaped, 14-mile long island that promised 240 days of sunshine a year, cool evenings, and miles of beautiful beaches. Responsible for putting the computer to work was FONATUR, Mexico's National Trust for Tourist Development. Foreigners were urged to invest in this "new Miami Beach," and Mexicans who had acquired land before it was developed became wealthy. Others have benefited less dramatically by the many jobs created in this city which already exceeds 35,000 in population.

Tijuana, Mazatlán, Puerto Vallarta, and Zihuatanejo are favorite tourist resorts on the Pacific Coast, while inland places such as Cuernavaca, Mérida, Guadalajara, and nearby Lake Chapala are also popular. Acapulco still attracts hundreds of thousands of tourists from all over the world, despite the pollution of its bay and the presence of about 120,000 shanty dwellers on adjacent mountain slopes.

Economic Perspectives

For 25 years after World War II Mexico was thought of as a showcase, with its modernization and heavy capital investment in industrialization. However, during the 1970s the government found itself burdened with foreign debts and a mammoth trade deficit. Millions of people poured into the cities to escape from hunger or attempted to cross the border in search of jobs in the United States.

One problem, of course, is the exploding population; another is that Mexico is poor in arable land. Most of the peasants hold no land, or if they do own a small parcel, crop yields are minimal and for their own use. These "marginal Mexicans" supplement their income by outside work at low wages. Many become squatters in the cities, which are unable to provide essential services.

Mexicans hope that oil will transform the country, but it can do so only if the political system can meet the fundamental needs of the people and prevent the gap between rich and poor from widening.

Of vital necessity for integrating isolated communities and the decentralization of economic undertakings away from Mexico City is the modernization of transportation. At present there are 4 major north-south highways into the interior of Mexico from 12 border cities, and the government is planning 6 more. A fifth route traverses the Baja California Peninsula north-south from Tijuana to Cabo San Lucas. An ambitious program estimated to cost $144 million will construct 26 more roads throughout the country.

The Political Situation

Mexico has gone far since the Revolution of 1910–18 toward establishing order and coherence in the nation. It has formulated a state-idea powerful enough to command support from the majority of its people. The system is not democratic as that term is used in Anglo-America, but it is distinctly Mexican. Since adoption of the Constitution of 1917 and the election of General Carranza as President, Mexico has been governed by only one political party. This was called the *Partido Revolucionario Nacional.* In 1946, to indicate that the purposes of the Revolution were now accepted by everyone, the party was renamed *Partido Revolucionario Institucional.* (PRI). Although other political parties are allowed,

there has been no challenge to the authority of the PRI. When the President nears the end of a six-year term of office, and after consulting with other party leaders, he names a successor. The election, in which all citizens vote by secret ballot, has only one slate of officers. The vote is an expression of support for acts of the party leaders. The President backed by the entire PRI, names the state governors, candidates for Congress, and judges of the Supreme Court. A decline of popular support for the PRI has forced it to rely on local political machines, peasant leaders, and the Mexican Labor Confederation (CTM) for control.

The Mexicans are a deeply religious people. Mexican Catholics focus their attention on the Virgin of Guadalupe, whose shrine is located in Mexico City. The Virgin is supposed to confer special favor on Mexicans and is pictured as the ideal of Mexican motherhood. To a considerable extent, the Mexican state exists for the purpose of protecting and supporting the ideals symbolized by the Virgin of Guadalupe.

Delegates to the International Women's Year in Mexico City in 1975 made it quite clear that the family was all-important. They also focused on the status of women generally in society. They noted, for example, that the great bulk of the female labor force is in agriculture. Although women commonly do most of the farmwork, without compensation, it is the men who receive training in agricultural technology.

Finally, the Mexicans are very close to the United States. All school children study history, and their history books tell how the United States took the entire northern part of Mexico, and how U.S. forces stormed the heights of Chapultepec and defeated the cadets of the Mexican military academy who were defending it. Along with veneration of the Virgin of Guadalupe, homage to the heroes of Chapultepec is a part of the Mexican state-idea. Most Mexicans admire the technical expertise of the United States, but there is widespread concern that the United States might again use its power to force its weaker neighbors to adopt unwelcome policies. The 1900-mile border between the United States and Mexico is the longest in the world between a developing country and a wealthy country. This border im-plies a special relationship as each country seeks to resolve its problems involving oil, migration, trade, and international friendship.

GUATEMALA

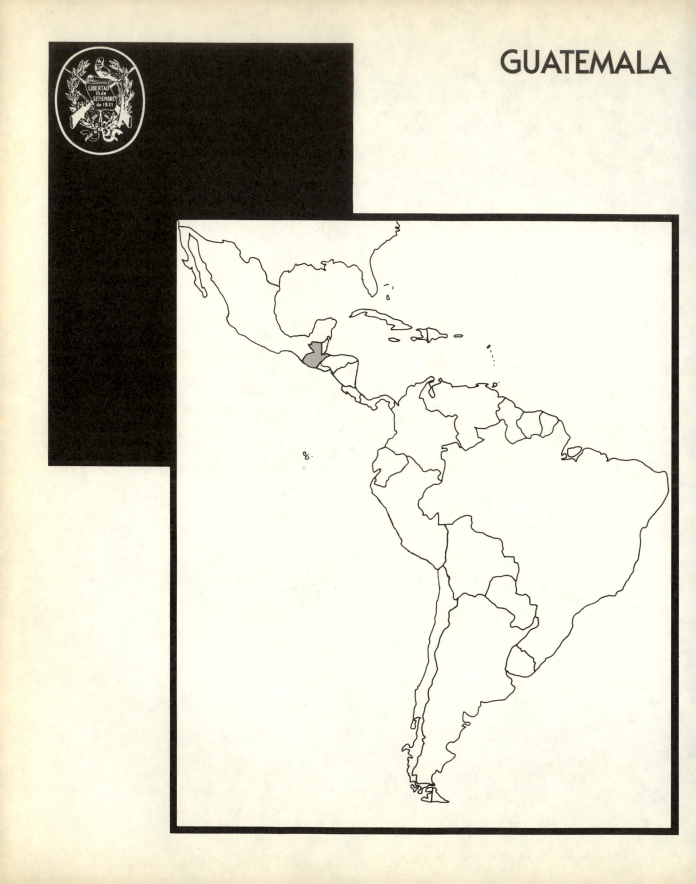

REPÚBLICA DE GUATEMALA

Land Area   42,042 square miles

Population   Estimate (1985) 8,000,000
             Latest census (1981) 6,043,559

Capital city   Guatemala City 1,307,000

Percent urban   39

Birth rate per 1000   43

Death rate per 1000   8

Infant mortality rate   62.4

Percent of population under 15 years of age   46

Annual percent of increase   3.5

Percent literate   51.1

Percent labor force in agriculture   53

Gross national product per capita $1120

Unit of currency   Quetzal

Leading crops in acreage   sugar, bananas, maize, coffee

Physical Quality of Life Index (PQLI)   59

COMMERCE (expressed in percentage values)

Exports

coffee	32
cotton	7
bananas	6
petroleum	4

Exports to		Imports from	
United States	27	United States	31
El Salvador	17	El Salvador	9
West Germany	7	Mexico	7
Japan	5	Netherlands Antilles	7
Costa Rica	5	Venezuela	6
		West Germany	6

Data mainly from the 1985 World Population Data Sheet of the Population Reference Bureau, the 1985 Britannica Book of the Year, and the 1985 World Almanac.

Guatemala, a country about the size of Ohio or Tennessee, has a long history of unrest and violence. In recent years terrorism has prevailed, with extremist groups, right and left, battling for control of the government. The variety of life and scenic grandeur of the country are astonishing to visitor and native alike, and it is difficult to reconcile the existing turbulence with the natural resources and beauty of this nation. There are reasons for dissatisfaction, however, and strong desires for fundamental change.

The People

There are about 8 million people in Guatemala, about half of whom are pure Indian descendants of the extraordinary Mayas. Most of the Indians live in the relatively isolated north and west, and many still do not speak or understand the official Spanish language. Now, highways, airplanes, radio, and television are forcing them into another culture, and they are adjusting rapidly to the best or the worst of two worlds. The Indians

are Catholic, but their religious ceremonies include mixtures of Catholic and pagan forms. Each community has its own rituals for appeasing the gods and ensuring good harvests. Religious practices enter into every aspect of life, from cultivation of the soil to methods of curing the sick, or to deciding on political policy.

The Indians traditionally have been separate from the rest of the Guatemalans. Their land was never included in the Spaniards' large private estates, as was the Indian land in Mexico. Yet, because their farms were too small to provide all the food they needed, and because they became dependent on other items they themselves did not produce—such as metal hoes, *machetes*,[1] adzes, needles, medicines, and even certain kinds of seeds—the Indians found it necessary

---

[1] The *machete* is a long-bladed knife with a short wooden handle used throughout tropical America for such work as cutting brush and harvesting crops, or as a weapon. For many decades almost all the machetes sold in Latin America were manufactured at Collingsville, near Hartford, Connecticut.

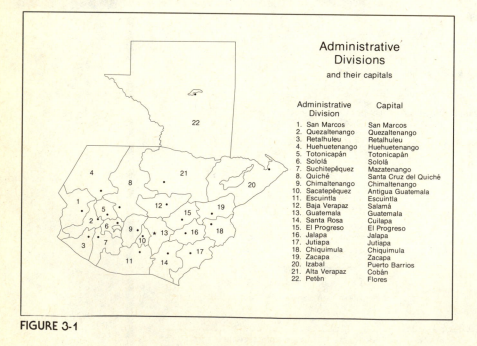

Administrative
Divisions
and their capitals

Administrative Division	Capital
1. San Marcos	San Marcos
2. Quezaltenango	Quezaltenango
3. Retalhuleu	Retalhuleu
4. Huehuetenango	Huehuetenango
5. Totonicapán	Totonicapán
6. Sololá	Sololá
7. Suchitepéquez	Mazatenango
8. Quiché	Santa Cruz del Quiché
9. Chimaltenango	Chimaltenango
10. Sacatepéquez	Antigua Guatemala
11. Escuintla	Escuintla
12. Baja Verapaz	Salamá
13. Guatemala	Guatemala
14. Santa Rosa	Cuilapa
15. El Progreso	El Progreso
16. Jalapa	Jalapa
17. Jutiapa	Jutiapa
18. Chiquimula	Chiquimula
19. Zacapa	Zacapa
20. Izabal	Puerto Barrios
21. Alta Verapaz	Cobán
22. Petén	Flores

FIGURE 3-1

to derive some income from this other society with which they were so closely associated geographically. Consequently, the Indians have been willing to accept temporary employment as farm laborers on the coffee or banana plantations, or on cattle estates. In fact, it would have been impossible to develop such enterprises without the availability of Indian workers.

The other culture group in Guatemala is known as *ladino*.[2] The ladinos include all people who no longer live as Indians. The Indian who learns to speak Spanish, wears shoes, gives up Indian dress, abandons the religious practices of the Indian communities, and perhaps moves away from the native community is classed as ladino. Ladinos also include the 5 percent of the total population that is of unmixed Spanish ancestry and the 42 percent of the people who are mestizos.

Guatemala has the largest population of any country in Central America, yet only the southern third is densely populated. Most of the people are concentrated in the highlands overlooking the Pacific Coast. The northern third of the country, known as Petén, has few inhabitants. The government has initiated a land-reform program in this northern area and has awarded 86,000 land titles, with farms averaging 25 acres. Cooperative associations have been organized as a means of reducing the rate of internal migration to the capital city, but rapid urbanization continues to cause much concern. Guatemala City, with more than 1.3 million inhabitants, is almost 20 times the size of Quezaltenango, the country's second largest city, and by the year 2000 it may be 30 times larger. The capital city now contains more than 75 percent of the nation's total urban population. It has been predicted that the Guatemalan labor force will increase from 6.2 million

FAMILIES ATTENDING MASS IN ANTIGUA, GUATEMALA.

to 14 million by the year 2000, and it is the young who will migrate to the cities.[3]

Guatemala's population growth has been impressive. When the country gained independence from Spain as a part of the Mexican Empire in 1821, the population was estimated at 500,000. In 1921, the figure exceeded 2 million and now has reached about 8 million. No country in Central America has spent more on public housing for its *campesinos*, or rural peasants, than has Guatemala. Thousands of household units also have been built in Guatemala City, and these have been quickly filled by families that pour into the city.

Despite rapid modernization, Guatemalans remain conscious of the greatness achieved by the Mayan Empire. For example, it is common to hear that the tallest man-made structures in the country are those built by the Maya at Tikal.

---

[2] Whetten points out that the word *ladino* is derived from *latino* and refers to persons who have adopted the way of living of the people of Latin origin. See Nathan L. Whetten, *Guatemala, the Land and the People* (New Haven, Conn., 1961).

[3] *Population and Urban Trends in Central America and Panama* (Washington, D.C.: Office of Information, Inter-American Development Bank, 1977).

A PANORAMIC VIEW OF GUATEMALA CITY.

Until 1970 this was true. Now, some buildings in Guatemala City reach a height of 260 feet, whereas the tallest known structure of the Mayan world, Temple IV at Tikal (built c. 741 A.D.), is slightly under that figure. The ancient Mayan city of Tikal is nevertheless awesome. Spread over more than 25 square miles are the ruins of a community that once sheltered 200,000 people. The mystery of why this advanced culture disappeared has not been unraveled, but one major puzzle has been solved. Scientists, for years, have wondered how the Mayas could produce enough food to feed millions of people. Now, airborne radar has revealed the secret, as 50,000 square miles of Guatemala and Belize display, in radar images, an extensive network of drainage canals hidden beneath dense rain forests. These canals, which probably date from between 250 B.C. and 900 A.D., enabled the Maya to create plots of dry land where maize and cacao could be grown intensively.

The Land

The national territory of Guatemala extends across the Isthmus of Central America from the

Pacific to the Caribbean. It includes four major divisions: the Pacific coastal lowland, the highlands, the deep valleys that drain to the Caribbean between steep-sided fingers of highland, and part of the densely forested Peninsula of Yucatán.

The Pacific coastal lowland is a southeastward continuation of the lowland of Chiapas. At the Mexico-Guatemala border it is about 25 miles wide, and it continues for about 150 miles across Guatemala into El Salvador. The coast is straight, with no harbors, and ships have had to load or unload while at anchor offshore. This lowland strip was covered, when the Spaniards came on it, by a wooded savanna. Along the base of the highlands, and on the steep slope facing the Pacific, the land was covered with forest. The lowland was long neglected by farmers and was used mostly for grazing cattle on the unfenced range. Only since 1960 have its agricultural possibilities been realized. The entire area is covered by productive, porous soils of volcanic origin, developed on material washed down from the highlands. All along the Pacific Coast, from Chiapas to Costa Rica, the Pacific coastal lowland is an area of recent settlement.

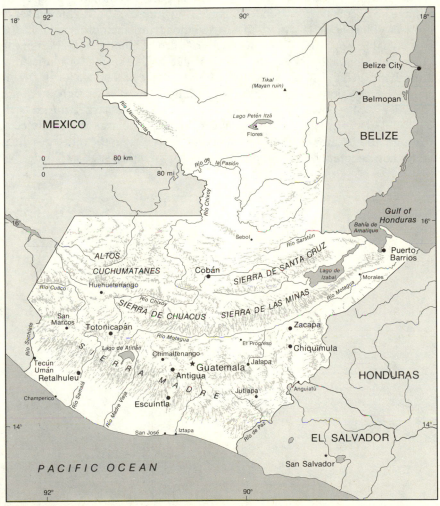

FIGURE 3-2

The highlands are similar to those of Chiapas in Mexico, with underlying geologic structures extending in a generally east-west direction. On the Pacific side they are deeply mantled with volcanic ash and lava, the accumulation being deepest in western Guatemala, where elevations vary from about 3500 feet to more than 10,000 feet above sea level. Standing above the upland surface are many volcanoes, some of which are still active. The highest mountain in Central America is Tajumulco, which rises 13,845 feet above sea level. Among the volcanic cones are several basins of irregular shape, one of which includes the magnificent Lake Atitlán. Headwaters of the Caribbean drainage have cut steep canyons into the easily eroded volcanic ash. Toward the east, where the volcanic cover is thinner, and where streams have cut their valleys down close to sea level, underlying geological structures are exposed, forming a succession of steep ridges, sharp divides, and deep valley lowlands.

The vegetation cover of the highlands exhibits two basic characteristics of mountain geography:

a general zonation by altitude, and an intricacy of detail that makes the vertical zones in some places difficult to identify. Generally, at higher altitudes the semideciduous forest of the Pacific slope and the thick rain forest of the Caribbean slope give way to oak, cypress, and pine. Above 10,000 feet pine grows only in patches, and there are wide areas of mountain grassland.

Three deep valley lowlands converge on the narrow head of the Gulf of Honduras. In the north, on the border of Belize, is the valley of the Río Sarstún; in the middle is the lowland in which Lake Izabal lies; and near the border of Honduras is the largest lowland, the valley of the Río Motagua. Long fingers of the highlands separate these lowlands from each other.

The northern third of Guatemala is part of the Peninsula of Yucatán. The surface is a limestone tableland, 500 to 700 feet above sea level. Drainage is largely underground, although there are numerous lakes and some large, northward-flowing rivers. Most of the area is densely covered with evergreen forest, in the midst of which lie the ruins of Mayan cities, such as Tikal.

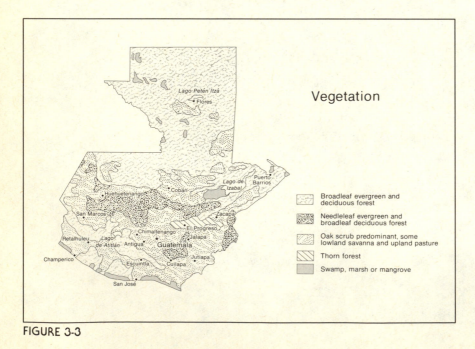

Vegetation

Broadleaf evergreen and deciduous forest

Needleleaf evergreen and broadleaf deciduous forest

Oak scrub predominant, some lowland savanna and upland pasture

Thorn forest

Swamp, marsh or mangrove

FIGURE 3-3

A HIGHWAY IN THE WESTERN HIGHLANDS OVERLOOKING LAKE ATITLAN.

The Pattern of Settlement

In Guatemala, as in Mexico, the distribution of Indians in the preconquest period is reflected in current patterns of population. The highland basins were rather densely occupied by sedentary agricultural people whose basic food crop was maize. The site of the present city of Quezaltenango was the focus of this pre-colonial settlement. Outside of the highland area, what is now Guatemala had a relatively small population of shifting cultivators or nomadic hunters and fishers.

Compared with Mexico, Guatemala yielded little in the way of quick wealth. There were silver mines near Huehuetenango, and these are still producing, but the ore bodies of this area proved disappointing. Many of the Spaniards therefore migrated southeastward along the isthmus. The few who remained were granted large estates which they developed in the traditional Spanish manner.

In time the highland area, which was the only part of Guatemala that was settled, was divided into two different regions. The line of division was roughly the 5000-foot contour, which in this area marks the division between *tierra templada* below and *tierra fría* above. The Spaniards preferred the *tierra templada*; the *tierra fría* was left for the Indians. Guatemala City, just under 5000 feet above sea level, on the divide between rivers draining to the Caribbean and those draining to the Pacific, was near the boundary between these contrasting regions of settlement. Land below 5000 feet is mostly in the southeastern part of the highlands, with drainage via the Río Motagua and along the escarpment where the highlands descend to the Pacific. That above 5000 feet is mostly in the southwestern part of the highlands. A third part of the country, the lowlands of the *tierra caliente* below 1600 feet, remained almost empty until the development of banana plantations in the twentieth century.

The department of Petén in northern Guatemala is covered with dense forest and was once laced by a maze of footpaths used by chicle gatherers going to and from the camps where chicle was prepared for shipment. This was the world's major source of chicle for chewing gum. Since 1948, however, synthetics have replaced chicle as the principal ingredient of this product, and the local industry has declined. The Petén area comprises nearly a third of the national territory but is occupied by less than 1 percent of the total population.

Indian Settlement of the High Country

The predominantly Indian part of Guatemala is in the highlands west of Guatemala City. Indian settlements are arranged around Quezaltenango, Huehuetenango, and the shores of Lake Atitlán. Indians are also concentrated farther eastward around Cobán. The Spanish conquerors never divided these areas into private estates. The indigenous communities range in population from

AN OUTDOOR MARKET IN CHICHICASTENANGO, GUATEMALA.

a few thousand to more than 50,000. In some districts the people live in villages or towns, and each day they go out to work the surrounding farms. In others, they live in family units on the land they cultivate.

Almost 60 percent of the working-age population is engaged in agricultural activities, and the Indians are particularly skilled farmers. They use hoes to build ridges and furrows that run like contours along the slopes. As a result of this ancient practice, there is little erosion even where steep mountainsides are cultivated. Maize is the principal crop, often grown with beans, chili, and other vegetables in the same field. Yields are relatively high, and maize provides more than 75 percent of the Indian diet by weight. Wheat is raised at the higher elevations and is sold as a cash crop to merchants for use in the cities, where people eat white bread. Wheat prices are subsidized by the government, and flour mills are required to use a fixed percentage of domestic grain as a support to the region's farmers. Above 9500 feet the only crop grown is the potato, which has limited commercial value. Land above 10,000 feet is used primarily for grazing sheep. More than 500,000 Indians are presumed to be migratory farm laborers, picking coffee and cotton, cutting sugarcane, or working on the banana plantations.

A high rate of investment in infrastructure will provide the country with abundant geothermal and hydroelectric power, open the extensive northern area with a paved highway system, and develop a port complex on the Pacific Coast. About 60,000 Guatemalans have already benefited from the free distribution of farms to landless peasants, with approximately 15,000 farm families now settled in the Northern Strip. An additional 40,000 families will follow in the next few years. Coffee, cardamom, cacao, and rubber will be the principal commercial crops.

An important petroleum reservoir has been discovered in the northcentral part of the country at Chinajá. Oil from this field and from nearby Rubelsanto is moved by pipeline to the port of Santo Tomás de Castilla (formerly Matías de Gálvez), near Puerto Barrios on the Caribbean.

Some 20 percent of the Guatemalan cattle

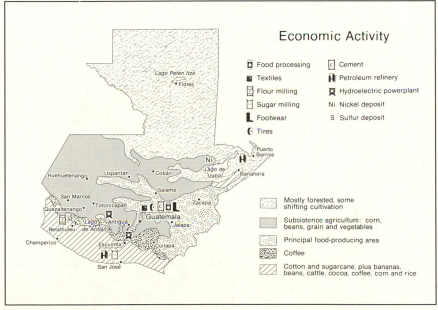

FIGURE 3-4

herd is now located in Petén, where ranchers formerly had serious transportation problems in reaching the national market. Conditions have improved rapidly with the paving of the main highway to Petén and construction of a bridge over the Río Dulce. Roads are being built into the new rural development areas of Ixcán and Quiché, and the entire communication system will be linked through construction of a national peripheral highway. This will bring modernization to an extensive area and, it is hoped, prevent thousands of people from leaving the rural areas for the city. Soon, maize, cereals, cardamom, and other crops will be moved from local cooperatives to urban markets by road, rather than being flown out by the air force.

Thousands of peasants who have been given land are expected to benefit from massive plantings of rubber and cacao trees that have begun in the Northern Transverse Strip. This program will supplement an already active forestry industry in the Department of Totonicapán. The latter area is the nation's principal source of inexpensive pine furniture and also provides lumber and firewood. Preservation of the forests is due largely to the institution of communal forest holdings, by which the Indian community is encouraged to use the forests wisely and to punish those who cut wood illegally.[4]

Commercial Agriculture in the Highlands

In striking contrast to the Indian subsistence farms are the plantations of commercial crops on large private estates. The people of Spanish origin were not interested in subsistence; they wanted a crop of sufficient value to cover the high cost of transportation. They introduced sugarcane, wheat, and domestic animals to Guatemala, but the highly stable society of the Maya has never made much of a place for these importations. The Spaniards did find two native Indian

crops from which they could derive a profit: cacao and indigo. In the second quarter of the nineteenth century, they imported cochineal insects which were fed on a variety of cactus known as *nopal*. From these insects a red dye was manufactured. Cochineal and indigo were of major importance around Antigua and Guatemala City until the manufacture of chemical dyes after 1857 put an end to the market for the more costly agricultural dyes.

After 1850 the large landowners of Guatemala turned to coffee production, following the lead of other Central American countries. For the past 90 years, coffee has been the chief export of the country and its principal source of revenue. There are 40,000 producers of coffee in Guatemala, of whom the top 1 percent account for 56 percent of production.

Before World War II there were two types of coffee-growers in Guatemala. Around Antigua and Guatemala City, and in the upper Motagua Valley, they were of Spanish descent or ladinos whose ancestors had lived in Guatemala for many generations. These landowners constituted the aristocracy of the country, enjoying political power and social prestige, while their plantations were worked by migratory labor. The other class of growers was of German extraction, usually descendants of emigrants from Germany during the 1860s who retained their German citizenship. These colonists settled as pioneers in two localities: on the Pacific slope of the southwestern highlands, overlooking Ocós and Champerico; and in the rugged highlands between Huehuetenango and Cobán. Although foreigners comprised only about 1 percent of the landowners of Guatemala in 1935, they held more than 30 percent of the cultivated land. About 48 percent of the large properties prior to World War II were owned by foreigners, and these accounted for almost two-thirds of all the coffee produced.

The best coffee lands of Guatemala are on the Pacific slope of the highlands, at altitudes between 1000 and 5000 feet. In this zone there is ample rainfall; there is also a dry season from

---

[4] T. T. Veblen, "Forest Preservation in the Western Highlands of Guatemala," *Geographical Review* 68:4 (1978): 417–434.

December to April in which the coffee is harvested. The deep volcanic soils are highly productive, and the coffee plants, grown in the shade of larger trees, protect the slopes from erosion. From this part of Guatemala comes some of the world's finest coffee.

Quezaltenango (population 65,526), the second largest city of Guatemala, is the regional trade center for the western highlands. During two brief periods between 1838 and 1849, it even served as the capital for a Central American nation known as Los Altos. Here, produce from the highlands is exchanged for that of the Pacific Coast, and a variety of manufacturing activities has developed. Chief among these are cotton and woolen textiles, food and beverages, wood products, and leather. As with most regional trade centers of Latin America, competitive advantages favor the nation's primate city in terms of market, credit, transport and communication, and electricity. Thus, while Quezaltenango is the urban focus for western Guatemala, it is also the leading source of internal migration to Guatemala City.

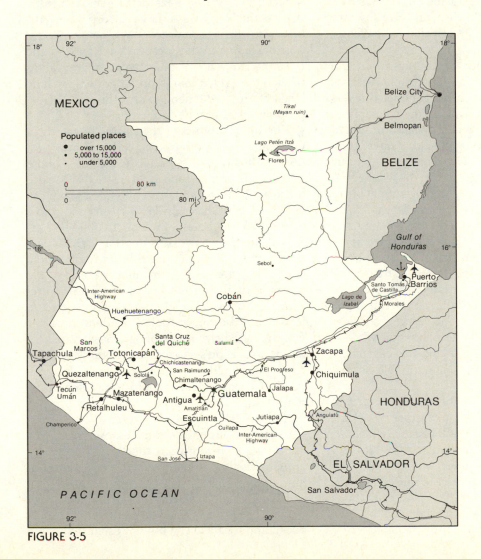

FIGURE 3-5

Commercial Agriculture in the Lowlands

The Pacific Coast is served by an Inter-Oceanic Highway and by the Inter-American Highway that parallels the shoreline. These highways intersect at Escuintla, a thriving city that is also developing an industrial base. A short distance south is the port of San José, and to the west is Champerico, where a lively shrimp industry exists. The shrimp are frozen and sent overland to Santo Tomás de Castilla, where they are shipped abroad. There is now sufficient export and import traffic on the Pacific Coast to justify a new deep-water port facility, and San José appears likely to be the site selected for development.

Historically, people of European origin avoided the lowlands of Guatemala because even short-term visitors often contracted fevers carried by mosquitoes. During the colonial period small areas of the Pacific lowlands produced cacao and sugarcane, with the aid of slave labor, but these plains were devoted primarily to cattle ranching. The animals grazed on the wooded savannas which were maintained by annual burning during the dry season. The wet lowlands on the Caribbean side remained mostly empty.

The United Fruit Company was formed in the United States in 1899 and soon began to make use of the Central American lowlands for banana plantations. In 1906 plantations were developed in the lower Motagua Valley, the fruit being shipped from special banana-loading docks at Puerto Barrios. Most of the plantation workers were highland Indians who were attracted by the prospects of higher wages, comfortable homes, and protection from disease.

Banana diseases, nevertheless, did occur and caused great damage. Panama disease, which attacks the roots, and Sigatoka disease, which attacks the leaves, first appeared in Costa Rica and reached the Motagua Valley in the 1930s. The United Fruit Company therefore shifted its plantations to the Pacific side. The new plantations on the Pacific side came into production by 1939 and produced 90 percent of the Guatemalan crop in 1956. The company developed a plantation at Tiquisate from which bananas were shipped by rail to Puerto Barrios for export. Despite the importance of the banana crop, it seems unlikely that it will increase greatly in the years ahead, because other crops, such as tomatoes, melons, tobacco, and strawberries, are assuming high levels of production.

Since the 1950s the Pacific coastal lowland has been an area of rapid settlement and agricultural productivity. Insecticides have eliminated malaria and yellow fever, and antibiotics have reduced greatly the hazards to general health. All-weather highways have incorporated the area into the national economy, and sufficient port development has occurred at San José and Champerico to handle the area's agricultural exports. However, there is no longer empty land available for settlement, except through the expropriation of private property.

The main product of the Pacific lowlands of Guatemala is cotton. In fact, cotton is now produced all along the Pacific Coast of Central America, from the Tapachula district of Mexico to Costa Rica. In 1950 this entire area produced fewer than 100 bales; in 1978–79 it produced a record 723,000 bales, of which 46,000 were assigned to domestic consumption. Yields are among the highest in the world for nonirrigated cotton, but heavy investments are made in the spraying of insecticides by airplane to control pests. Most of the cotton crop is exported to Japan, and the return trade is evidenced by the widespread distribution of Japanese transistor radios, television sets, cameras, motorcycles, and automobiles. Other economic activities along the Pacific Coast of Guatemala include the cultivation of sugarcane, tobacco, cacao, and pineapples; specialty crops such as cardamom, lemon grass, kenaf, citronella, and sesame are locally important. In recent years there also has been a growing fishing industry and the beginnings of tourism.

Cattle-raising is extensive. The animals are of modern breed and are fed on planted pastures. A packinghouse has been built at Escuintla where the animals are slaughtered and the meat processed. Refrigerated trucks haul the meat to Santo Tomás de Castilla, where the trailers are loaded on ocean-going ferryboats destined for Miami, Florida, via San Juan, Puerto Rico. These vessels carry shrimp, fruit, and vegetables, in addition to meat. This system of shipping food products to the United States has opened new markets for the produce of the Pacific lowlands of Guatemala, El Salvador, Honduras, Nicaragua, and Costa Rica.

Guatemala as a Political Unit

The rural population of Guatemala, largely Indian, is heavily concentrated in the southwestern highlands. With a continued high birth rate and a declining death rate, the inevitable result has been an increase of population pressure on the land. It is estimated that 7.5 acres should be adequate to provide the kind of living required by an Indian family, but in many areas the average Indian farm is considerably smaller. The farmland available to each community is too limited to provide for all the inhabitants, and even with high maize yields many people suffer from an inadequate diet.

Migration to Guatemala City has occurred on a grand scale, as evidenced by "urban sprawl" and by densely settled slum areas in the canyons, bordering the city. This movement has been largely by ladinos, however, and primarily from the secondary urban centers of the nation. Migration to the city holds little attraction for the average Indian, who has grown up experiencing the social cohesion of the traditional indigenous community. Yet, when the Indian farm family finds that it can no longer earn a livelihood from its small plot of ground in the highlands, survival takes precedence over tradition and supplemental income is sought from other sources. In recent decades additional income could be earned from seasonal employment on the coffee *fincas* of the intermediate slopes or the cotton plantations of the Pacific lowlands. These sources are no longer

expanding at a rate parallel to population growth, and other solutions must be found.

Owners of large estates, primarily on the Pacific lowlands, hold sufficient political power so that a major land-reform program has been successfully resisted. Therefore, agricultural colonization and settlement programs in that area have been limited largely to former United Fruit Company lands and the estates of exiled political leaders. Now, attention is focused on development of the northern lowlands as a means of relieving intensive political and social pressures that result from increasing population and a highly unequal sharing of the nation's wealth.

It is ironic that the lowlands of the Yucatán Peninsula, previously occupied by one of the great civilizations of all time, is now largely an uninhabited wilderness. Three nations — Mexico, Guatemala, and Belize — are today striving to occupy this land and to incorporate it into their effective national territory.

The Economic Situation

Guatemala remains a country with many poor farmers. Some 65 percent of the workers are employed as farmers or in related activities. In this high proportion of people supported by agriculture, Guatemala is surpassed in Latin America only by Haiti and Honduras. Nevertheless, Guatemala has ample resources. Beef, for example, can be sold to the United States for 50 percent less than the United States pays for it from several other countries. Since 1975, despite a major earthquake, a revolution, demonstrations, and terrorism, the economy has shown steady improvement, including an increased gross domestic product. Earnings have come from coffee and cotton, from tourism, and from expanded mineral production.

Hydroelectric projects and investment in nickel and petroleum production have especially improved the economic outlook. In 1980 about 70 percent of the electric power consumed in Guatemala was produced by thermal and diesel plants, while only 2 percent of the nation's abun-

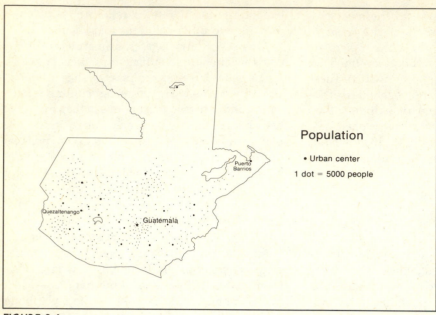

FIGURE 3-6

dant water power potential was utilized. This is changing dramatically with the installation of several major hydroelectric plants, the largest being the 300,000 kilowatt Chixoy project completed in 1982. A ferro-nickel complex on the north shore of Lake Izabal is the largest mining operation in Central America. Other mineral resources exist in Huehuetenango, where deposits of silver, copper, tin, antimony, and lead have been found. In 1979, 40 foreign petroleum companies were authorized to work in the northern Alta Verapez region near the Mexican border, and drilling in the Petén area continued. Meanwhile, it was estimated that Guatemala would no longer need petroleum byproducts as fuel for electric power, upon completion of a chain of hydroelectric plants.

Guatemala is far less dependent on the United States for its exports, or imports, than it used to be. Helping greatly to reduce this dependence on the part of Guatemalan commercial, labor, and manufacturing interests was the Central American Common Market (CACM), established in 1960 and composed of Guatemala, El Salvador, Honduras, Nicaragua, and Costa Rica. Nearly all

the country's light industrial products were sold within the CACM. Without access to it, Guatemalan industry would not have been able to reach the present stage of development, for the CACM provided a trade area of about 20 million people.[5] The leading industrial products marketed by Guatemala are beverages and foods, followed by textiles, clothing, and footwear.

The tourist industry has especially great potential for Guatemala. There are towering volcanoes, incomparably beautiful lakes, and the shorelines of two oceans. There are modern glass and steel buildings in the capital city, thatched Indian huts in the rural areas, and formidable ruins of the Mayan Empire throughout much of the nation. A favorite of tourists is Antigua, once the capital of all Central America and a city devastated by earthquakes.

The most advertised and photographed lake in all of Central America is Lake Atitlán, an important center of Indian culture. This clear, deep lake is bordered by majestic volcanoes and nu-

[5] "Guatemala: Economic and Social Position and Prospects," *A World Bank Country Study,* 1978.

merous Indian villages. East of Lake Atitlán is Chichicastenango, where the majority of the Indians live in the hills and woods but appear en masse on market and fiesta days. Yet, the main attraction for the tourist and native Guatemalan alike is the capital itself, Guatemala City. This city is clean, modern, and dynamic. Slums have been cleared, low-cost housing has been erected, and new suburbs are expanding in all directions. Guatemala City has an increasing number of hotels and good transportation facilities. Manufactures include textiles, pharmaceuticals, leather goods, foods and beverages, tires and batteries, metal and wood products, and cement.

The Political Situation

A basic fact about Guatemala is the existence of a sharply defined part of the national territory within which Indians comprise more than half of the population. The disintegration of a nation is more likely when people with one state-idea are geographically separate from people with a conflicting state-idea; and when one of these areas occupies a peripheral position the danger of disintegration is magnified. In Guatemala, however, the Maya Indians have not yet developed a group consciousness. Each community forms a miniature nationality, each with its own dialect, religious practices, and distinctive dress. The Maya until recently have remained aloof from the political life of the ladinos. Therefore, the peripheral position of the Indian area has not been a factor leading to disintegration.

Guatemala began its independent existence in 1839 with no positive state-idea. Large landowners and army officers had control of the new country, and no powerful voice suggested that the Indians should have any rights except to work. Political leaders tended to perpetuate their hold on the higher offices and could not be shaken loose until younger army officers, frustrated in their ambition to gain the financial rewards of high office, could develop the power to stage a revolution or *coup d'état*. When discontent

stirred too dangerously, governments in power found that they could command a kind of coherence and order by pressing claims for the territory of British Honduras (Belize) which was, and still is, claimed by Guatemala. The one thread tying all political groups together continued to be resentment toward foreign interference.

Matters were complicated when, in 1906, the United Fruit Company began operations in Guatemala. The company directed its attention to a part of the national territory which until then was almost unoccupied. It developed economic values in areas previously outside of the effective national territory and generated substantial tax revenue for the national treasury. Here was a foreign enterprise, operated on different principles from those common in Guatemala. Here, too, was part of the national territory in which even the laws were administered separately. In an economic sense it was good, but it was not Guatemalan. When the company used its influence to gain advantages with the Guatemalan government, resentment by the Guatemalans grew stronger. Nothing, not even a dispute with Great Britain over British Honduras, was as politically popular in Guatemala as an attack on the United Fruit Company.

Guatemala, like Mexico, was governed by a succession of military dictatorships. As in Mexico, some were liberal in policy, actually concerned with the increasing poverty of the farmworkers. All were strong when they started and, until overthrown, maintained peace and stability in the country. Most notable among the Guatemalan dictators was General Jorge Ubico, who ran the nation from 1931 to 1944.

Over the years an increasingly large group of people were influenced by the Mexican Revolution. With the usual few interested in revolution for its own sake, there was a core of genuine liberals, concerned with establishing democracy and with ending a system of power and privilege. When General Ubico's dictatorship was overthrown in 1944, the liberals, who had been in exile, took control of the government and demanded an election. In this election the liberal-minded civilian Juan José Arévalo was named President. A new constitution was adopted in 1945, in which numerous social reforms were presented. These included measures to give Indians the rights of citizenship and to provide workers with social security and the right to form labor unions. A program was implemented by which land not in use was to be expropriated and given to landless peasants. The law was applied vigorously to the property of the United Fruit Company and less vigorously to the property of other large landowners.

In 1951 Arévalo was succeeded by Jacobo Arbenz Guzmán. Arbenz has been described as an ambitious junior officer, long frustrated in his hope for advancement and in his desire for wealth. He was one of the three who plotted the revolution of 1944. When he became President in 1951, Arbenz relied heavily on support by the Communists.

In 1954, a group of exiled Guatemalan army officers under Colonel Carlos Castillo Armas organized a small force in neighboring Honduras. When this "army" marched across the border, the Guatemalan army refused to attack. Soon the capital was occupied, and Arbenz was forced to flee. Castillo Armas then assumed the presidency, which he held until his assassination in 1957.

The new government included supporters who were opposed not only to the Communists but also to the entire program of social reform. In many cases the land that had been expropriated was seized again by the landowners. Castillo Armas did, however, establish a new, non-Communist land-reform program. Idle land on large estates was expropriated but with proper compensation to the owners. On the Mexican model, government technicians attempted to turn the Indian subsistence farmers into producers of cash crops. Under the Communists landownership remained with the government, whereas the new law gave the colonist possession of the land and prohibited resale for 25 years.

Successive governments since 1954 have been struggling with the problems of social and economic change. The use of violence by one group has led to retaliation by the opposition. Continued violence and turmoil give evidence of a deep-seated unrest. It is not mainly the Indian group that provides the unrest, but rather the new urban class in the capital and smaller cities, and organized workers on the coffee and banana plantations. Basic concerns are to end the system of privilege, to secure equality of treatment for the Indian, to promote modern social legislation, and to achieve the opportunity to own land.

BELIZE

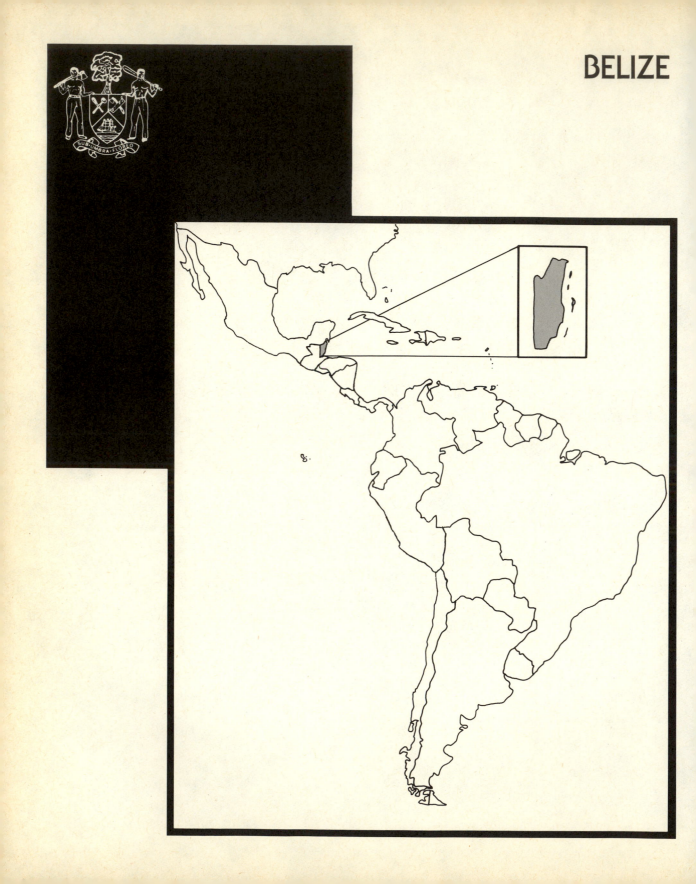

4

BELIZE

Land Area 8866 square miles

Population Estimate (1985) 200,000
Latest census (1980) 144,857

Capital city Belmopan 2940

Largest city Belize 39,887

Percent urban 50

Birthrate per 1000 32

Death rate per 1000 7

Infant mortality rate 27

Percent of population under 15 years of age 44

Annual percent of increase 2.5

Percent literate 90

Percent labor force in agriculture 29

Gross national product per capita $1140

Unit of currency Belizean dollar

Leading crops in acreage citrus, sugarcane and bananas

COMMERCE (expressed in percentage of values)

Exports

sugar	30		clothing	9
machinery	13		watches	7
dairy products	11			

Exports to		Imports from	
United States	47	United States	42
United Kingdom	44	United Kingdom	15
		Netherlands	9

Data mainly from the 1985 World Population Data Sheet of the Population Reference Bureau, the 1985 Britannica Book of the Year, and the 1985 World Almanac.

Belize, formerly British Honduras, is small in both size and population. Here, 200,000 people live in an area about the size of Massachusetts. Most of the inhabitants live in and around Belize City, the former capital, which is hot, crowded, and poorly drained. This tiny underdeveloped nation is often referred to as the "undiscovered country," "a land without people."

Belize is a mosaic of limestone plains, low coastal areas, swamps, and marshes. It has innumerable beaches and offshore cays, and broad rivers, forests, and jungles. The rugged Maya Mountains of the interior are topped by the 3680-foot Victoria Peak. Offshore is a great barrier reef, the second longest in the world after that of Australia. This low-latitude country is subject to temperatures ranging from 70° to 90°F and to humidity that is 80 percent most of the time. Belize has also been exposed to frequent droughts, floods, and hurricanes.

About 50 percent of the population are descendants of African slaves originally brought to this coast in the eighteenth and nineteenth centuries by the British to cut wood in the forest. About 20 percent of the population is the Indian-black combination known as "Black Carib." The remainder is European, East Indian, Chinese, Lebanese, and a religious group called Mennonite. There are even a few descendants of U.S. Southerners who came to the Toledo district following the Civil War. Despite a high birth rate, the net population growth is almost nil, a rare occurrence in Latin America. The reason is that many Belizeans emigrate. It is estimated that about 35,000 have gone to the United States, a number equal to the current Belizean labor force. Unfortunately, it is the more skilled who leave, and there is the distinct possibility that improving education might further increase the number of those who would leave. It is calculated that $10 million annually is added to the economy by Belizeans living in the United States who send their earnings home.

With the near destruction of Belize City by "Hurricane Hattie" in 1961, the Belizeans initiated construction of a new capital the following year. This city, named Belmopan, is located 50 miles inland on rolling terrain, near vegetable farms and citrus orchards. It now has a population of almost 3000. The name was chosen from some 300 entries submitted in a contest conducted by the Reconstruction and Development Corporation. *Mopan* is the name of an ancient Mayan tribe that never surrendered to the Spaniards; the prefix *Bel* refers to the New Belize. Two lucky contestants won 50' × 100' lots in Belmopan for submitting the imaginative name!

Economic Conditions

About 40 percent of the land in Belize is arable, yet only 5 percent is cultivated. This means there is ample space for growth and development, and with the help of international agencies the government intends to boost production in agriculture, fisheries, and forestry. Agriculture accounts for one-third of the gross national product and for more than a fourth of the country's employment. The agricultural structure is dominated by large plantations that produce sugar, citrus, and

HARVESTING SUGAR CANE IN ORANGE WALK, BELIZE.

FIGURE 4-1

bananas for export, while small-scale peasant farms produce crops for local consumption. There are many absentee landlords, and a certain bias favors plantations that produce export crops. Food comprises a fifth of total imports, although Belize is self-sufficient in maize, rice, and beans.

Citrus is exported from the Stann Creek area, but sugar is clearly predominant and is exported largely to the United States and the United Kingdom. Corozal and Orange Walk districts, in the north, are the main areas of sugarcane cultivation. Periodically, this commodity suffers from smut, froghopper damage, and excessive rains. Replanting always occurs, and farmers are now experimenting with smut-resistant varieties. A byproduct of the sugar is molasses, most of which is shipped to the United States. Honey is also produced in this area and in Toledo district to the south.

About 10,000 acres are devoted to citrus orchards, primarily in the Stann Creek Valley of southeastern Belize. All of the citrus crop is processed locally, and oranges are even imported from Mexico for processing. The country of Trinidad and Tobago is the major market for orange and grapefruit concentrate; the United States, the United Kingdom, and Canada are customers for juice, concentrate, and sections. Bananas now occupy about 4000 acres in the same general area.

In an effort to stimulate the beef cattle industry, Belize has begun a program of pasture development. The greatest potential for expansion is in the Belize River Valley of western Belize, an area well provided with roads, ferries, bridges, good irrigation, and drainage. The Cayo district is already the main cattle-producing district with large areas of improved pasture. Beef exports which began in 1973 are destined mostly for the United States and the Caribbean area.

Cacao plantations once flourished in the Cayo district but until recently had been abandoned. Now, the Hershey Foods Corporation has formed a local company and has invested $8 million in developing and rehabilitating large cacao plantations in the Sibun and Caves Branch areas along the Hummingbird Highway. The Hummingbird-Hershey Corporation intends to make Belize the leading producer in this part of the world. Some of the young cacao trees have been transplanted and interplanted with plantain suckers to provide shade for the young plants. Plantains will thus become a secondary crop, and rejected plantains will serve as feed for hog-raising.

The contribution of Mennonites to agriculture in Belize is substantial. Belize has long been recognized as a place for those seeking refuge, but the most significant immigration in relatively recent years was that of about 1000 Mennonites who entered between 1958 and 1961.[1] Coming from Chihuahua state in Mexico and previously from the Red River Valley near Winnipeg, Canada, they settled first in Spanish Lookout. This was the result of an agreement with the Belize government that gave them complete freedom to "practice their customs and beliefs, including the right to run their own schools." They are now located in four separate communities: Spanish Lookout and Barton Ramie in the Cayo district and Blue Creek and Shipyard in the Orange Walk district. The original group brought in about $1 million to invest in farming, and they immediately set to work clearing the dense forest. They built small, neat wooden houses and worked the land with horse and plow. Today the Mennonites own more than 145,000 acres of land, most of it bought with cash from private landowners. They are engaged in dairying, raise thousands of chickens, and produce rice, citrus, beans, sorghum, and maize. The Mennonites also manufacture gutters from sheet metal, make water tanks, and have developed a market in Canada for dehydrated papayas. Their migration has been a happy experience for all concerned.

Beginning in 1980, refugees from war-torn El Salvador arrived in Belize, often illegally, on foot, and without financial resources. Emergency aid

---

[1] T. A. Minkel, "Mennonite Colonization in British Honduras," *The Pennsylvania Geographer* 5:3 (1967): 1–5.

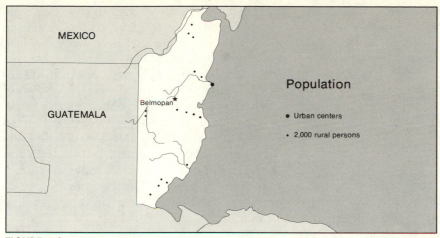

Population

• Urban centers

• 2,000 rural persons

MEXICO

GUATEMALA

Belmopan

FIGURE 4-2

in the form of clothing, medicine, and household needs was provided through the United Nations. Since the refugees were mainly small farmers, the government of Belize, in turn, provided land in the Belize River Valley near Belmopan. Initially, 140 Salvadoran families and 40 needy Belizean families were provided with 50-acre holdings, plus a village lot. The settlement itself became known as the "Valley of Peace."

More of a problem is the cultivation of marijuana, largely because of intensive efforts to exterminate the crop in countries such as Colombia and Mexico. Thousands of acres of marijuana have been planted illegally in small plots throughout the thinly populated and remote forest lands. The shipment of marijuana from Belize is part of the international drug traffic, with its market chiefly in the United States, and has proved difficult to control.[2]

For nearly two centuries forestry was the predominant industry of Belize, producing primarily logwood, mahogany, and chicle. Although forestry has declined, it still has a potential to support the manufacture of paper, veneer, and plywood. Uncontrolled cutting must be halted and reforestation take place, however, if this industry

is to survive and prosper. Another neglected sector is the fishing industry, although a three-year project has been developed through workshops and seminars that will teach the fishermen modern technology. With proper conservation and management, the adjacent Caribbean would provide sustained yields of spiney lobster, shrimp, turtles, scale fish, and conch. For the sports enthusiast, fishing along the reefs provides amberjack, kingfish, Spanish mackerel, bonefish, and many other varieties.

What manufacturing exists involves mainly the processing of agricultural products, such as sugar. There are also companies making clothing, furniture, cigarettes, matches, soft drinks, rum, and beer, and a local meat-processing industry has been initiated.

Belize is poor in mineral resources. Traces of gold, copper, and zinc were discovered years ago but have not been found in commercial quantities. Belize may have oil, if the Reforma and Campeche Sound fields of Mexico are extended into its territory. Extensive exploration is being conducted along the Mexican border, north of Belize City, and in the southern part of the country.

Tourism is little developed, but tourists from the United States are increasingly "discovering"

---

[2] "Another Kind of War," *The New Belize* 13:10 (1983): 5–8.

this quiet part of the world. There is a national forest reserve to visit and two major Mayan ruins, but three-fourths of the visitors spend vacations on the offshore islands. Most popular is primitive Ambergris Cay, which has no telephones, radio, or television.

The Political Situation

The colony of British Honduras was established in 1862, primarily as a source of forest products. Subsequently, both Mexico and Guatemala made claims on the territory now known as Belize. The colony gained full internal self-government in 1964, and the British government was prepared to have Belize become independent whenever it wished. Because of Guatemala's persistent claims, however, Belize did not rush toward independence. Complicating the matter was a dormant claim by Mexico to the northern part of Belize, a claim that was not pursued as long as the British did not seriously consider Guatemala'a demands. Guatemalans insisted that the British forfeited their claim to Belize because of failure to build a road to the Guatemalan area of Petén, as pledged in a treaty of 1859. Independence was finally declared in 1981, at which time Belize also became one of the Commonwealth nations.

There are three active political parties in Belize, the two main ones being the People's United Party (PUP) and the United Democratic Party (UDP). The PUP was established in 1950 and remained in power until 1985 under the leadership of Prime Minister George Price. Nearly 90 percent of the eligible voters do vote, including 18 year olds.

Although physically part of the Central American isthmus, Belize has stronger historical and cultural ties with the Caribbean area. It is a member of the Caribbean Community (CARICOM) but, at the insistence of Guatemala, was excluded from the former Central American Common Market (CACM). As an independent nation, Belize seeks to become a cultural and economic link between the Caribbean and Central American countries. It seems potentially well endowed to play that role.

REPÚBLICA DE EL SALVADOR

Land Area   8124 square miles

Population   Estimate (1985) 5,100,000
               Latest census (1971) 3,554,648

Capital city   San Salvador 445,100

Percent urban   39

Birth rate per 1000   28

Death rate per 1000   6

Infant mortality rate   42.2

Percent of population under 15 years of age   45

Annual percent of increase   2.1

Percent literate   57

Percent labor force in agriculture   47

Gross national product per capita $710

Unit of currency   Colón

Physical Quality of Life Index (PQLI)   66

COMMERCE (expressed in percentage of values)

Exports

coffee	58
textiles	9
food	9
cotton	8
chemicals	6

Exports to		Imports from	
West Germany	28	United States	27
United States	36	Guatemala	24
Guatemala	19	Mexico	9
Japan	3	Venezuela	9
Costa Rica	3	Costa Rica	4

Data mainly from the 1985 World Population Data Sheet of the Population Reference Bureau, the 1985 Britannica Book of the Year, and the 1985 World Almanac.

The Organization of Central American States (ODECA) includes the five countries of Guatemala, El Salvador, Honduras, Nicaragua, and Costa Rica. The smallest of these, and the most densely populated, is El Salvador. There are about 600 people per square mile in El Salvador, and the total population has reached 5 million. Moreover, there is a grave imbalance of labor and land, as more than 200,000 peasants are landless and there is no undeveloped area for agricultural laborers to occupy. Cultivation has reached its areal limit, but a growing population must be fed. Urgent economic and political problems are the result.

The population of El Salvador is predominantly of Indian ancestry. During the fifteenth century, shortly before the arrival of the Spaniards, the area that is now El Salvador was invaded by the Pipil Indians, a group that had adopted many culture-traits from the Aztecs. The Pipil settled among the more primitive tribes formerly occupying the area and intermarried with them. The density of Indian settlement was not as great as that of the Maya, however, and attachment to the land was not as close. When the Spaniards arrived, they quickly divided the land, and its Indians, into large private estates. Yet, few Spaniards ever came to settle. During the colonial period only about 1500 Spaniards remained permanently in El Salvador, an area with an Indian population estimated at 150,000. At the beginning of the nineteenth century, when the city of San Salvador had a population of 12,000, only 624 were listed as being of unmixed Spanish ancestry. Now scarcely a family can claim to be purely of European origin. Those of unmixed Indian ancestry comprise 10 percent of the population but, unlike the Maya of Guatemala, the people of Indian ancestry in El Salvador have abandoned traditional ways of living and the Indian language. They all speak Spanish. Mestizos comprise 89 percent of the population.

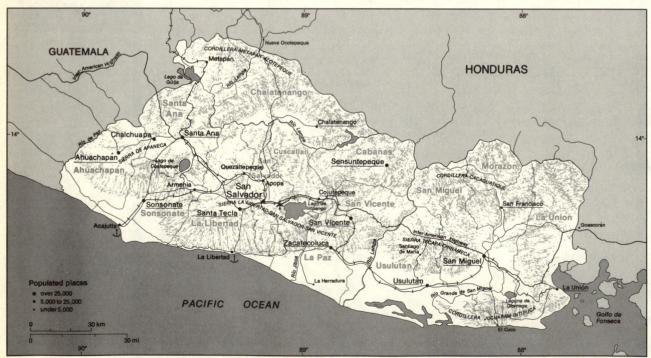

FIGURE 5-1

The Land

El Salvador is a country with no great elevations but with little flat land. Most of the country is deeply dissected by streams. The backbone of El Salvador is the volcanic highland which continues southeastward from Guatemala. In El Salvador the elevation of the highland decreases eastward to the Gulf of Fonseca. At the capital city, San Salvador, the elevation is only a little over 2000 feet. Surmounting the highland is a succession of volcanic cones that reach elevations between 7000 and 8000 feet above sea level. Most of them are dormant, yet the most active volcano in all of Central America over a period of almost two centuries is located at the edge of this highland. Volcán Izalco, the "Lighthouse of the Pacific," erupted from a flat plain in 1770 and continued to erupt with striking frequency and regularity until it reached a height of 6183 feet in 1957. Then, just as a hotel and other tourist facilities were being erected to capitalize on the view of this remarkable work of nature, the eruptions ceased. Among the volcanic peaks are numerous small intermont basins, deeply filled with volcanic ash and bordered by flows of lava. There are also numerous lakes, such as the large and beautiful Lake Ilopango near San Salvador, formed as the result of volcanic activity. As in Guatemala, the soils derived from lava and ash are highly productive. These soils erode quickly when the forest is cut and the surface allowed to dry. Erosion in El Salvador, especially on the steeper slopes, is a serious problem.

The gently sloping land is chiefly in the Pacific lowland, which continues southeastward from Guatemala as far as Acajutla. Just east of this port, the coastal lowland is pinched out by flows of lava that descend from a volcano to form rocky promontories along the Pacific. From the port of La Libertad a lowland plain extends beyond Usulután. This plain is 20 miles wide where it is crossed by the Río Lempa.

The Río Lempa is El Salvador's largest river. It rises in the highlands of Honduras and Guatemala and flows through rugged hill country toward the Gulf of Fonseca, until, without reaching the Gulf, it turns abruptly southward and crosses the coastal lowland to the sea. Along the Lempa are patches of floodplain, inundated dur-

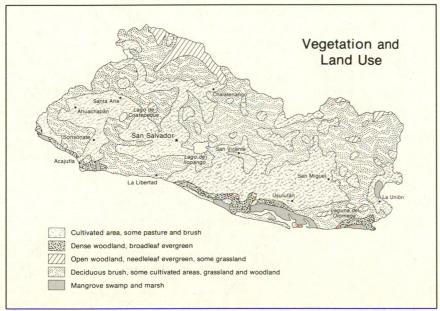

Vegetation and Land Use

☐ Cultivated area, some pasture and brush
▨ Dense woodland, broadleaf evergreen
▧ Open woodland, needleleaf evergreen, some grassland
⬚ Deciduous brush, some cultivated areas, grassland and woodland
▨ Mangrove swamp and marsh

FIGURE 5-2

ing the rainy season and dry and sandy during the remainder of the year.

The greater part of the country was originally covered with semideciduous forest. The density of population, as well as the fact that the major fuel is wood or charcoal, has resulted in extensive forest destruction. Today there is a serious shortage of fuel, and the removal of forest from the steep upper slopes has greatly increased soil erosion.

The Pattern of Settlement

The Spaniards who first entered El Salvador came from Guatemala. A road was built southeastward from the capital of the Captaincy-General to the settlements and ports of Nicaragua. This route is now followed closely by the paved highway that connects Guatemala City with San Salvador and continues around the Gulf of Fonseca into Honduras and Nicaragua. In the colonial period the Spanish settlers who decided to remain in this country made their homes in San Salvador. They laid out large properties along the road which gave access to Guatemala City, and also around Sonsonate, inland from the port of Acajutla. The Pipil Indians accepted Christianity and went to work for the new owners of the land. The chief crop raised by the Indians to supply their own food was maize, which remains the predominant crop in terms of acreage. The principal commercial crops, raised on the best land, were sugarcane and indigo. In the Lempa Valley cattle were pastured on wild grasses that grew there after the forests were burned. These cattle were driven to market in Guatemala City. In 1840 coffee was introduced from Brazil and soon became the chief commercial crop of the highlands.

Present-Day Agriculture

Coffee is still the main commercial crop of El Salvador. The best coffee is grown at elevations above 2000 feet, and most of the plantations are located on the gentler slopes around the margins of the intermont basins. Coffee normally provides up to 60 percent of the nation's export earnings and employs a similarly high proportion of its agricultural labor force, particularly during the harvest season. Because of guerrilla activity and the occurrence of coffee rust, export earnings from this crop were reduced to 45 percent in 1980 while rural unemployment rose dramatically. As civil war spread throughout the country, the future of coffee cultivation, and of all commercial agriculture, became increasingly uncertain.

Cotton-growing spread along the Pacific coastal lowlands of Central America for 20 years following World War II and employed the most modern techniques. Japan served as the leading export market. Sugarcane also expanded in acreage, mainly where irrigation water could be diverted from streams draining the mountainous interior. Like coffee-raising, the cultivation of cotton and sugarcane was drastically reduced as a result of internal warfare by 1980. In that same year, new land-reform measures became operative which expropriated all farms exceeding 500 acres, with those exceeding 250 acres to be next on the agenda.[1] Both acreage planted and crop production plummeted.

With more agricultural land in the hands of peasant farmers, the planting of so-called basic grains has been less adversely affected. Maize, rice, and beans are grown as food crops on many small farms, particularly on the coastal lowland. A grain sorghum called *maicillo*, introduced from Africa, has replaced maize in certain areas and is second only to maize in total acreage. Sisal plantations appear along the lower Lempa Valley and near La Unión. The fiber of this plant is used in the manufacture of bags for grain, coffee, and other produce.

Where the coastal plain is not under cultivation, there are extensive areas devoted to modern

---

[1] "El Salvador-Agricultural Situation," *USDA Attaché Report* (February 1981).

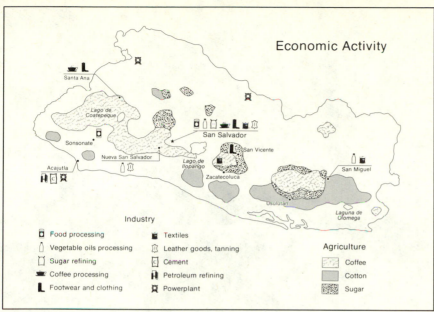

Economic Activity

Industry

🗆 Food processing	⬛ Textiles
🍶 Vegetable oils processing	🗆 Leather goods, tanning
🗆 Sugar refining	🄲 Cement
☕ Coffee processing	🄷 Petroleum refining
L Footwear and clothing	☒ Powerplant

Agriculture

░ Coffee
▨ Cotton
▦ Sugar

FIGURE 5-3

cattle ranches, and shrimp-fishing is conducted from each of the Pacific ports. Small-scale mixed agriculture, even subsistence farming, characterizes the more remote mountainous areas of the country.

Industry and Power

El Salvador has made important progress in manufacturing and power production. In San Salvador cotton-textile mills, food-processing plants, and factories produce appliances, electronics, pharmaceuticals, and plastics. Textile mills have also been built in the densely populated area near the Gulf of Fonseca. Subsistence farmers in this area have had great difficulty in providing a minimum living for their families. Each year hundreds of thousands of workers migrate to the highland coffee plantations for the harvest season from November to April. A start toward reducing rural poverty has been accomplished by the production of basic grains, the expansion of fisheries, and the construction of additional industries.

El Salvador once suffered from a shortage of electric power, but in 1956 a large dam was completed on the Río Lempa and hydroelectric power became abundant. Additional generating facilities have subsequently been installed at other places along the Lempa, notably at Cerrón Grande (270 megawatts) and San Lorenzo (180 megawatts), and the flow of the river has been regularized by a dam at the outlet of the Lago de Güija, on the border between El Salvador and Guatemala. El Salvador is also among the world's leading nations in the advancement of geothermal technology. Production at its Ahuachapán field, about 40 miles west of San Salvador, is being expanded from 45 megawatts to 95 megawatts; this site is the principal source of geothermal power in Central America.

Unfortunately, the territory of El Salvador is not only small, but it also lacks natural resources. There is some mining of gold and silver, and small quantities of mercury and lead have also been produced. El Salvador does have one other distinctive product. This is the so-called Peruvian balsam (*Myroxylon pereirae*), which, despite its name, has nothing to do with Peru. It is a lofty

THE OLD AND THE NEW: A CAMPESINO AND HIS OXEN PAUSE IN FRONT OF A POWER PLANT IN ACAJUTLA, EL SALVADOR.

tree that grows within a small area on the slopes of the highlands north of Sonsonate. From the bark comes a reddish-brown or black viscid oil that is used in the manufacture of perfumes.

El Salvador as a Political Unit

The problem of racial and cultural diversity that exists in Guatemala is not as apparent in El Salvador. Few countries exhibit a greater contrast of wealth and poverty, however, than exists between the coffee aristocracy and the great majority of the people. The landowners have not actually controlled the politics of El Salvador since 1932, but they do resist any infringement on their privileged position.

The Economic Situation

Economic development in El Salvador is taking place in the presence of an exploding population and is contributing to an acceleration of the explosion. The net rate of population increase, 2.6 percent per year, puts El Salvador in a class with Mexico, the Dominican Republic, and Costa Rica. This is occurring in a country that has no unused land for pioneer settlement and that has a population density that is already too great to be supported by a largely agricultural economy. With little land for new subsistence farming and with fewer jobs than potential workers, the poverty of a large proportion of the people has become critical.

During the 1960s El Salvador's economy enjoyed an unprecedented improvement. The gross national product per capita in 1955 was only $152, which placed this country among those states requiring massive inputs of foreign aid just to begin the process of development. If all the wealth of the original landowners, the so-called "14 families" had been confiscated, it would not have provided nearly enough money for the necessary rate of new capital formation. Foreign aid was granted in large amounts, and at the same time the tax base was expanded despite strong resistance by the wealthy minority. New capital investment went into basic facilities such as electric power plants, all-weather highways, docks and other port facilities at Acajutla, and manufacturing industries. By 1984 the gross national product per capita had risen to $700.

El Salvador gained enormously from the Central American Common Market during the 1960s and soon became the most industrialized country of the isthmus. The nation appeared prosperous and the future looked bright. There were basic flaws in the system, however. The conservative oligarchy that had dominated El Salvador for generations remained firmly in power, and the circumstances of the landless peasant became increasingly desperate. As domestic problems intensified, a steady increase of emigration from El Salvador to Honduras, Guatemala, and Nicaragua occurred. Of the 300,000 individuals who crossed into Honduras by 1969, about 250,000 settled in empty areas along the north coast and in the interior. By 1980 thousands also were emigrating to the United States and Europe.

Agriculture remains the cornerstone of El Salvador's economy, and until 1981 coffee alone generated 58 percent of the country's total export earnings. Cotton ranked second among exports, followed by sugar and a variety of lesser commodities. In the same year, the People's Republic of China replaced Japan as the country's principal market for raw cotton. The fishing industry was considered to have good potential for development, with emphasis on shrimp, lobsters, and tuna. Dependence on the United States for foreign trade continued, although there was a sharp decline after 1953.

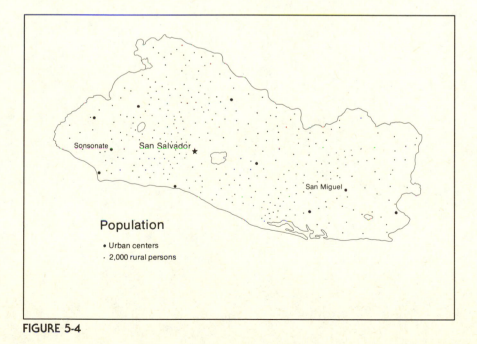

FIGURE 5-4

The Political Situation

Economic development in El Salvador has been closely tied to the distinctive way in which social change has been directed. There was a time when the "14 families" held veto power on matters of political policy. After 1930, political power shifted to the officers of the army. Beginning in 1931, General Maximiliano Hernández Ramírez ruled the country for 13 years, and for 50 years, except during one four-month period, every Salvadoran President was an army officer. Yet, stability eventually disappeared as terrorists of the political left and right created widespread violence. In 1979 a moderate military-civilian *junta* ousted right wing President Carlos Humberto Romero and then expropriated about 60 percent of the country's richest farmland.

To this point El Salvador had made rapid progress toward industrialization, which provided new employment opportunities. The literacy rate had increased, as had the general standard of living. Yet, a few thousand of the 5 million Salvadorans received about 50 percent of the total national income. This was the worst income inequality in all of Latin America.

The country had not yet recovered from the 1969 war with Honduras, and the subsequent collapse of the Central American Common Market, when the political violence brought about severe economic crisis and civil war. By 1981 the conflict threatened to spread internationally as the United States sent military advisors to aid government forces, while antigovernment guerrillas received support via Cuba and the Soviet Union. A nationwide presidential election was conducted in 1984, with international observers present to assure its validity. However, El Salvador had become a pressure cooker with population growth, limited resources, and social inequalities building irresistible forces inside. Explosion appears inevitable where corrective actions are rejected or too long delayed, and El Salvador has become a Central American battlefield.

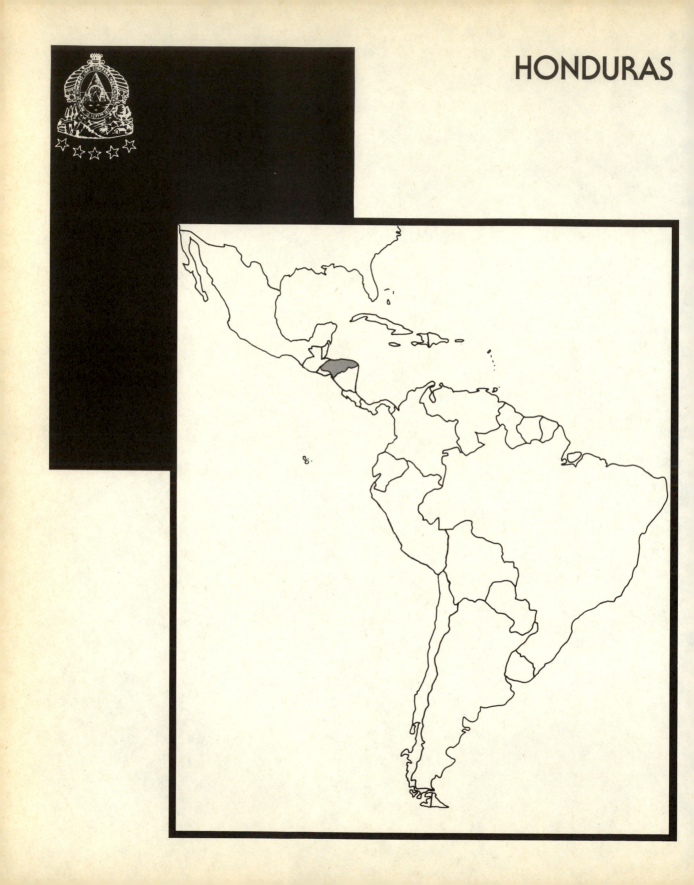

HONDURAS

6

REPÚBLICA DE HONDURAS

Land Area   43,277 square miles

Population   Estimate (1985) 4,400,000
Latest census (1974) 2,656,948

Capital city   Tegucigalpa 508,000

Percent urban   37

Birth rate per 1000   44

Death rate per 1000   10

Infant mortality rate   82

Percent of population under 15 years of age   47

Annual percent of increase   3.4

Percent literate   47

Percent labor force in agriculture   59

Gross national product per capita $670

Unit of currency   Lempira

Leading crops in acreage   maize, coffee, bananas

Physical Quality of Life Index (PQLI)   53

COMMERCE (expressed in percentages of values)

Exports

bananas	33	meat	5
coffee	23	metal ores	5
timber	7	shrimp and lobster	4
		sugar	4

Exports to		Imports from	
United States	53	United States	38
West Germany	9	Trinidad-Tobago	15
Belgium	6	Guatemala	7
Japan	6	Japan	6
Guatemala	4	Venezuela	4

Data mainly from the 1985 World Population Data Sheet of the Population Reference Bureau, the 1985 Britannica Book of the Year, and the 1985 World Almanac.

Honduras has experienced some handicaps in its struggle to achieve a national identity. Its population is grouped in several small and separate clusters, each with different economic interests, yet none of them large enough to provide a base for major economic development. The total population of the country, which was estimated in 1984 at 4.2 million, is less than that of tiny El Salvador. Furthermore, a large area of northeastern Honduras is still not part of the effective national territory. Within the area that is effectively occupied, rivalry between the separate areas of concentrated settlement has militated against national unity. However, the 1969 "Soccer War" with El Salvador, subsequent withdrawal from the Central American Common Market, and large-scale development projects have helped to give Honduras a new national consciousness.

The contrast between the dense population of what is today El Salvador and the thin population of the country to the north was a distinctive feature of the population map even before the arrival of the Spaniards. There was a great difference between the sedentary Pipil Indians and the migratory and more warlike Lencas of what is now Honduras. The Lencas, like the Pipil, have given up their Indian languages in favor of Spanish and no longer cling to Indian customs as do the Maya. According to the census of 1950, some 90 percent of the people of Honduras were mestizos. This remains true today. Another 5 percent are of African descent, 4 percent are Indian, and the remainder are Caucasian.

The Land

The main surface features of Guatemala continue eastward into El Salvador and Honduras. The volcanic region that forms the southern margin of the highlands of Guatemala extends into El Salvador, and only a little of southern Honduras includes areas of lava and ash. The geologic structures of the Central American-Antillean system, with their marked west-to-east or southwest-to-northeast trends, underlie most of Honduras. Long, steep-sided mountain fingers point eastward toward the Caribbean or run at a slight angle to the north coast on the Gulf of Honduras. The highest elevations are found in the westernmost part of the country, where there are peaks a little under 10,000 feet. In the vicinity of Tegucigalpa the peaks are about 6000 feet above sea level. To the north the elevations are much lower, but the slopes are even more rugged.

There is not much level land in Honduras. In the thinly populated eastern part of the country, Honduras shares with Nicaragua the extensive Miskito savannas, an area that begins east of the Río Plátano. Some of the valleys that have been cut westward into the highlands and drain to the Caribbean from the Río Patuca southward have ribbons of floodplain along their bottoms. The lowland of the Ulúa, on which San Pedro Sula is located, is about 60 miles north-south by 25 miles east-west. Smaller valley lowlands open to the north coast between the steep mountain ridges.

The highlands of Honduras are cut sharply across from north to south by a zone of faulted structures. Throughout this highland country with its steep-sided ranges, intermont basins, and rift valley, the arrangement of the vegetation is complex in its details. Altitude zones can be observed if the details of the foreground are disregarded. Most of the country receives an abundance of rain from May to November and very little during the rest of the year. Most of the *tierra caliente* was originally covered with a dense evergreen rain forest. Oak and pine, characteristic of the *tierra templada* and the *tierra fría*, may descend as low as 2000 feet above sea level. The Miskito Coast is covered with a savanna mixed with pine and palmetto. On some of the higher basins, east of Tegucigalpa, there are also extensive savannas. Forests cover about 45 percent of the total area of Honduras and may be the country's most valuable natural resource. The Olancho forest reserve of northeastern Honduras alone constitutes the largest undeveloped pine

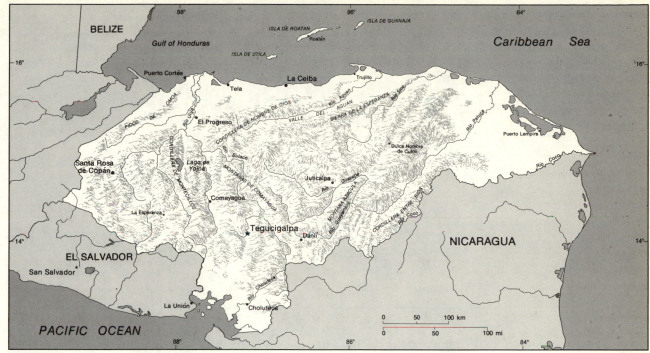

FIGURE 6-1

forest in Latin America, and there are also extensive stands of cedar and mahogany within the national territory.

The Pattern of Settlement

In pre-colonial times the land east of the Maya settlements and north of the Pipil settlements was thinly occupied by shifting cultivators of maize and squash, and by primitive hunters and fishers. Nevertheless, in the restless search for wealth and large Indian populations to Christianize and set to work, Spanish expeditions were sent out from Mexico and Guatemala. Cortés sent the first expedition into Honduras, but when its leader undertook to set up his colony as an independent unit, Cortés himself led a column of troops through the Gulf region of Mexico, across Yucatán, across the forested lower Motagua, and into what is now Honduras. In 1525 Cortés founded Puerto Cortés near the mouth of the Río Ulúa. The colony was brought back under the administration of Mexico and later placed under the Captaincy-General of Guatemala. San Pedro Sula was founded in 1536, and Comayagua in 1537. Comayagua became the administrative center for this part of the Captaincy-General of Guatemala. In 1578 silver ores were discovered in the highlands about 70 miles southeast of Comayagua, and the mining town of Tegucigalpa was established in a nearby basin at an elevation of about 3000 feet. In the rivalry that ensued among these miniature concentrations of settlement, Comayagua held leadership until 1880, when the capital was shifted to Tegucigalpa.

The Highland Settlements

The one important cluster of people in the highlands is located around Tegucigalpa, a city of almost 473,000. Other small areas of concen-

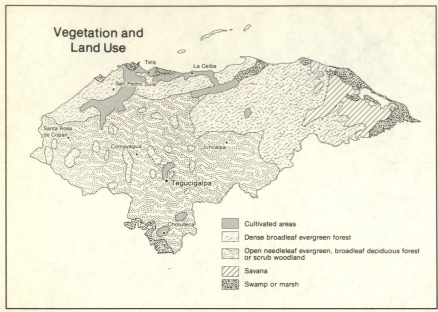

Vegetation and Land Use

Tela
La Ceiba
San Pedro Sula
Santa Rosa de Copán
Comayagua
Juticalpa
Tegucigalpa
Choluteca

■ Cultivated areas
▒ Dense broadleaf evergreen forest
▫ Open needleleaf evergreen, broadleaf deciduous forest or scrub woodland
▨ Savana
▤ Swamp or marsh

FIGURE 6-2

trated settlement are in the highlands, including Comayagua and Copán.

Except for some silver mines near Tegucigalpa, the economy of the highlands is based on lumbering, stock-raising, subsistence farming, and, in only a few spots, commercial agriculture. The pine forests offer an important resource that is used extensively only where roads make the timber accessible, as along the highway between Tegucigalpa and Comayagua. Trucks carry the logs to sawmills or to Caribbean ports for export. Pine logs are also used for the manufacture of paper. However, the form of land use that occupies the largest area in the highlands is cattle-raising. The large landowners have long been accustomed to burning the forest to increase the growth of pasture grass. In this way a considerable area of pine has been destroyed before the lumbermen could reach it.

When subsistence farmers occupy the land, they grow maize, beans, squash, and sometimes *maicillo*. Farmers are scattered throughout the highlands, but they tend to concentrate near towns or along the roads. A road from Tegucigalpa eastward has become an axis of pioneer settlement. Many settlers in this formerly unoccupied part of Honduras came as immigrants from crowded El Salvador or as refugees from the revolution in Nicaragua. Thousands of Palestinians, fleeing discrimination in their homeland, earlier settled in Honduras; although they show no inclination to become part of the farming community, they and numerous Syrians and Lebanese form an important part of the nation's industrial and commercial class.

The principal coffee plantations of Honduras are located around Tegucigalpa, both to the east and west along the highlands. There are also plantations on the slopes that descend toward the Gulf of Fonseca. The newest area of coffee planting is on the slopes on either side of Lake Yojoa. In addition to coffee, tobacco is grown commercially. A very old area of tobacco farming centers on Copán, and some new areas have been developed, in part by refugees from Cuba. These Cuban farmers now sell their tobacco in markets which used to import from Cuba.

Tegucigalpa, in addition to being the national capital, serves as the primary industrial and commercial center of the highlands. Its manufac-

A VIEW OF TEGUCIGALPA, HONDURAS.

turers specialize largely in consumer goods for the domestic market, such as processed foods, beverages, clothing, furniture, and appliances, but some products are shipped throughout Central America. Mining remains an important activity in the highlands and also focuses on Tegucigalpa. The largest existing mine, however, is El Mochito, west of Lake Yojoa. This mine, opened in 1948, produces silver, gold, lead, zinc, and cadmium.

The Transportation Problem

Much of the difference between El Salvador and Honduras results from the relative isolation of Honduras. The main colonial road between Guatemala City and Nicaragua went through San Salvador and skirted the Gulf of Fonseca, the route now followed by the Inter-American Highway. This route avoided the steep climb to the highland basins of Honduras. A branch of the Inter-American Highway reaches Tegucigalpa from Nacaome, near the Gulf of Fonseca, and a paved highway continues on through Comayagua to San Pedro Sula and Puerto Cortés. The Western Highway, from near San Pedro Sula to the border of El Salvador near Nueva Ocotepeque, is another important route. It was completed in 1963 and serves an area noted for the production of maize, coffee, tobacco, and beans. A road from Tegucigalpa northeastward to Catacamas is being extended to open up the almost unoccupied territory east of that town. Tegucigalpa is one of two Central American capitals with no rail connection.

Many rough trails are still used to move prod-

NEW SLUM SHACKS CROWD ON A HILLSIDE OF TEGUCIGALPA.

ucts to port from isolated farms, ranches, and mines, and it is clear that modernization is hindered by the lack of efficient, low-cost transportation. Honduras, in recent years, has contracted numerous international loans and has designated a high percentage of its national budget for improvement of the outdated transportation system.

Aviation suddenly lifted parts of Honduras out of the eighteenth century and transported them into the twentieth century. In a rugged mountain country, where other means of transportation are slow and expensive, airplanes can compete successfully for the shipment even of bulky products such as coffee. There are international airports at Tegucigalpa and San Pedro Sula which provide regular connections with the other countries of North and South America. In addition, within a radius of 100 miles of Tegucigalpa there are more than a hundred small landing fields offering passenger and freight service. Today it takes only 45 minutes to go from Copán to the capital.

The Caribbean Lowland Settlements

San Pedro Sula, the second largest city of Honduras, had about 200,000 inhabitants in 1980. Located 35 miles up the Ulúa Valley from Puerto Cortés, this town was on the main route of travel inland to Comayagua and Tegucigalpa. It became prosperous in colonial times on the basis of sugarcane grown on the rich alluvial terraces of the valley. However, the lower valley and the coast were avoided because of difficulty in controlling the rivers and clearing the tangle of wet

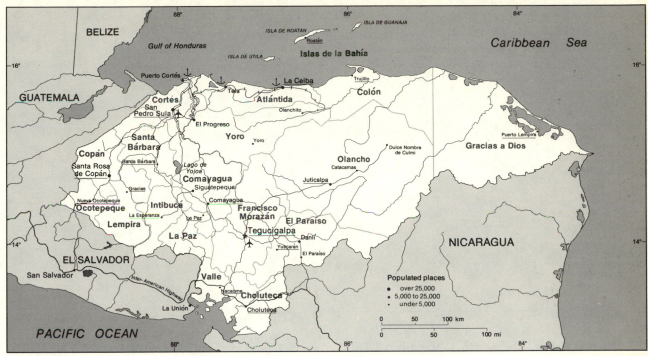

FIGURE 6-3

vegetation. Only in modern times has the Caribbean Coast become commercially the most important part of Honduras.

Before the soils and climate of coastal Honduras could be fully utilized, engineers had to devise methods of controlling the floods of the Ulúa and the Aguán. Engineers of the fruit companies found a way to divert the floodwaters, to reduce their rate of flow enough to permit the deposition of fine silt, and then to drain off the water through dredged channels to the Caribbean. Large acreages of new land have been built up high enough so that they can be protected with levees and are covered with 1 to more than 10 feet of rich, fine silt. Banana plantations have been developed along the lower river and on the fringe of plain along the Caribbean. Hundreds of miles of narrow-gauge railroad lines are used to convey the bananas to ports and to connect the ports with San Pedro Sula.

Two fruit companies, both with headquarters in the United States, have operated in this region.

The United Fruit Company built special banana-loading docks at Puerto Cortés and Tela. Aguán was the chief shipping port for the Standard Fruit Company. These companies cleared the forest, drained the swamps, controlled the river floods, eliminated disease-carrying insects, laid out towns, and built modern tropical houses, schools, and hospitals. The workers were recruited mostly from the highlands, and to get them to move to the lowlands the companies had to pay higher wages than those paid elsewhere in Honduras.

Shortly before 1940, the banana plantations of the Caribbean lowlands were threatened with extinction by the rapid spread of Panama disease and Sigatoka. Before methods of control could be applied in Honduras, a large part of the banana lands was abandoned, including all those in the Aguán Valley. By 1966 the plantations of the lower Ulúa Valley were being redeveloped, using spray delivered by airplane and practicing a regular flood fallow between plantings to control

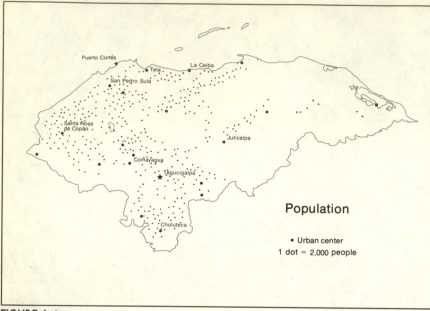

FIGURE 6-4

Panama disease. In addition, the United Fruit Company, a subsidiary of United Brands, developed new, disease-resistant varieties of bananas and began to diversify the production of these lowlands by introducing oil palms from Africa, abacá (Manila hemp), coconuts, and cacao.

The entire north coast of Honduras has experienced a period of rapid modernization during the past 25 years. Cattle-raising has expanded, based on improved breeds and the export of chilled beef to the United States. Sugarcane is grown extensively, and new mills have been constructed, again largely to serve the export market. Most important, major efforts have been expended to develop the entire northeast of Honduras through agricultural colonization and settlement. Thus, the Guayape, Catacamas, and Aguán valleys are again becoming productive, and trade is increasing through Puerto Cortés, Tela, La Ceiba, Trujillo, and Puerto Castilla. The primary focus of this regional development, however, is San Pedro Sula.

San Pedro Sula is the principal industrial center on the Caribbean Coast of Central America. Its location near Puerto Cortés, in the na-

tion's richest agricultural district and at the junction of transportation routes, has given the city a major advantage relative to other Honduran communities. The descendants of immigrants from Lebanon, Syria, and Palestine are prominent in local business and manufacturing which have experienced dynamic growth. This city is now noted as the center of Honduran industry, with factories producing a wide variety of products such as textiles, clothing, cosmetics, matches, sugar, metal goods, and cement.

A major obstacle to development of the Caribbean lowlands has been the frequency of hurricane damage. "Hurricane Francelia" caused extensive havoc in 1969, followed by the devastating "Hurricane Fifi" in 1974 which took more than 17,000 lives, drowned more than 50,000 cattle, and flooded more than 20,000 acres of bananas.

Honduras as a Political Unit

Honduras, unlike El Salvador, has a population that is small in relation to the resource base.

Moreover, Honduras has been pushed toward national coherence by the pressure of its neighbors, rather than by internal unity. During most of its history, the country has been composed of small, isolated clusters of people who had no sense of belonging together until circumstances and geography forced sovereignty on them. Except for the development of air transportation, the modern machine age seemed remote indeed.

The Economic Situation

Honduras has long been identified as the poorest, most underdeveloped country in Central America. Thus, when the Central American Common Market (CACM) was created in 1960, special provisions were made to support the nation's efforts toward modernization. Honduras responded well. New industries were established, roads and power plants were built, and trade with neighboring countries expanded rapidly. Yet, the other countries of the isthmus appeared to reap even greater benefits as the CACM was widely proclaimed to be a model of regional organization for economic integration.

The "Soccer War" of 1969 and subsequent withdrawal from the CACM, plus the ravages of several major hurricanes, had a disastrous impact on the nation's economy. Yet the war with El Salvador resulted in greater unity among Hondurans than they had ever known before, and this was followed by an era of stability while the neighboring countries of Nicaragua, El Salvador, and Guatemala experienced revolution and terrorism. Perhaps most important, the country's basic land resources were abundant in relation to the total population.

By 1980 Honduras was engaged in a variety of projects and programs aimed at modernization. These included the 292-megawatt El Cajón hydroelectric project on the Río Humuya to provide power nationwide and to control flooding in the San Pedro Sula area. Of major dimensions was the Olancho forestry and industrial development project to utilize the extensive pine forests for lumber, pulp, and paper. This project includes the construction of a road network, sawmills, and factories, and improvement of shipping facilities at Puerto Castilla. A free zone was established at Puerto Cortés, and on the Gulf of Fonseca a new

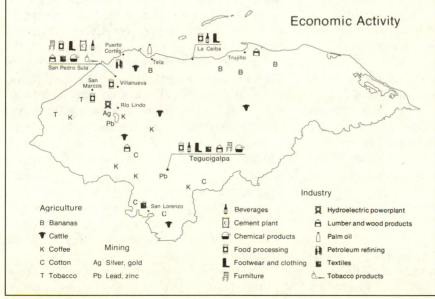

FIGURE 6-5

deep-water port was completed at San Lorenzo to eliminate the need for transshipment of goods by lighters through the port of Ampala on Tigre Island.

Plans continue to evolve for Honduras to become the primary supplier of basic iron and steel products for all of Central America. The raw materials are located near Agalteca, 65 miles north of Tegucigalpa, where the steel mill is to be built. An iron ore deposit of 10 to 15 million tons is supplemented by limestone available locally, and an extensive area has been planted with eucalyptus trees to provide charcoal for fuel. A steel plant at Agalteca would be expected to produce 100,000 tons annually, one-half of it available for export.

The main thrust of economic development is toward the virtually uninhabited territory of northeastern Honduras, with particular attention focused on the lower Río Aguán Valley. Agricultural colonization is proceeding, based on small farms and cooperatives, with emphasis on the cultivation of basic grains and of African oil palms, from which factories obtain palm kernels for the manufacture of soap, margarine, shortening, and cooking oils.

Honduras remains the poorest Central American nation, but change is evident. In 1984 the country's gross national product per capita was $660, as compared with Guatemala's $1130, El Salvador's $700, Nicaragua's $920, and Costa Rica'a $1280. However, its foreign commerce is increasingly diversified, and Honduras can no longer be characterized exclusively as a "banana republic." The economic future of the country will depend on continued large-scale investment in basic resources, successful occupance of the relatively empty northeast, and, above all, political stability.

The Political Situation

Traditionally, the political geography of Honduras has illustrated the problem of establishing a national identity. Historically, there has been a struggle for supremacy between Tegucigalpa and San Pedro Sula. When the administrative center was transferred from Comayagua to Tegucigalpa in 1880, San Pedro Sula suffered a setback. San Pedro Sula and the whole Caribbean lowland began to regain a competitive position in relation to the highland settlements when the banana companies began operations along the coast.

After World War II, the north coast was dominated by two banana firms that owned the land and controlled the economic, political, and social life of the region. The coastal swamps were transformed into pleasant, attractive landscapes where workers had access to recreation centers, modern homes, health facilities, and company stores that supplied goods of all kinds not available in the highlands. These conditions represented a foreign influence, however, and as such were resented by the more nationalistic elements of society.

Relationships with the banana companies continue to be of special concern in national politics. This was particularly evident in 1975, when a $1.25 million bribe to Honduran officials to reduce the banana export tax was revealed. The scandal, known locally as "Bananagate," resulted in the overthrow of General Oswaldo López Arellana as head of the government. In addition, a state corporation, COHBANA, was created to increase Honduran participation in the banana industry and to handle negotiations with the American-owned fruit companies. One effect was that the Tela Railroad Comapny, a subsidiary of United Brands, sold its railroad and port facilities to Honduras for the nominal sum of $1.

In 1980, the 11-year state of war between El Salvador and Honduras was ended officially by a peace treaty signed in Lima, Peru. The border dispute between the two countries was to be resolved within a period of five years, through the World Court, and it was hoped that normal relations would soon be established among the Common market countries.

Honduras has a long history of military rule, yet the country has been generally free of political terrorism and repression. Freedom of the press has been maintained, and laws have been enacted to provide social security, land reform, and benefits to labor. Two political parties, the Liberal Party and the National Party, have vied for control of the congress, while maintaining about equal strength and rejecting political extremism. Yet, tension persists between government and labor, between landless peasants and estate owners, and between the military and those who seek to establish representative democracy. In 1981 more than 80 percent of the eligible voters elected a civilian President, Dr. Robert Suazo Córdova, of the Liberal Party, thus ending 18 years of almost continuous military rule.

Of greatest continuing concern was a civil war in El Salvador and the threat of warfare with a Marxist government in Nicaragua. The United States provided increased military support to counterbalance Soviet and Cuban support to the Nicaraguan government and to guerrilla forces in El Salvador. By 1984, Honduras served as a shelter for 230,000 refugees from Nicaragua, including about 4000 Miskito Indians, and about 20,000 from El Salvador. Social needs and political security placed a heavy burden on an already impoverished nation.

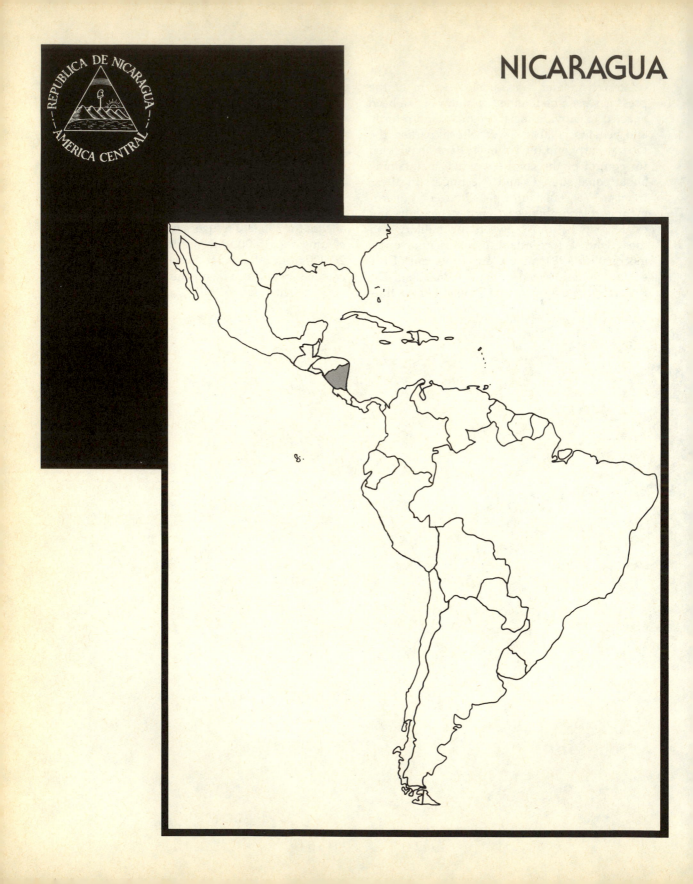

NICARAGUA

REPUBLICA DE NICARAGUA

AMERICA CENTRAL

7

REPÚBLICA DE NICARAGUA

Land Area   57,000 square miles

Population   Estimate (1985) 3,000,000
Latest census (1971) 1,877,972

Capital city   Managua 715,000

Percent urban   53

Birth rate per 1000   44

Death rate per 1000   10

Infant mortality rate   76

Percent of population under 15 years of age   47

Annual percent of increase   3.4

Percent literate   87

Percent labor force in agriculture   65

Gross national product per capita $900

Unit of currency   Cordoba

Physical Quality of Life Index (PQLI)   55

COMMERCE (expressed in percentage of values)

Exports

coffee	35
beef	9
sugar	8
chemicals	5

Exports to

United States	22
West Germany	14
Japan	12
Costa Rica	6

Imports from

Mexico	20
United States	19
U.S.S.R.	5
France	4
Cuba	4

Data mainly from the 1985 World Population Data Sheet of the Population Reference Bureau, the 1985 Britannica Book of the Year, and the 1985 World Almanac.

Nicaragua is the largest Central American nation, but much of its territory is thinly occupied. In total area it is roughly the size of Iowa. The population was estimated to be 3 million in 1985, considerably less than that of Honduras. The core of the country, its largest area of concentrated settlement, is located in the Nicaraguan lowland. It was in this lowland that the Spanish explorers found concentrations of Indians and here they, too, settled. Today it is estimated that about 75 percent of the people are mestizos. Pure Indians comprise only about 5 percent of the population, and these are found only in the more remote parts of the country. People of unmixed Spanish ancestry make up about 10 percent. The considerable concentrations of blacks along the east coast constitute another 10 percent of the total.

The Land

The land in Nicaragua is composed of four surface regions. The greater part of the country is a triangular wedge of highland that continues southward from Honduras. To the east is the Miskito Coast, a wide belt of swampy lowland extending from Punta Mico northward about as far as the Río Plátano in Honduras. The most important part of the country in terms of development is the third region, the Nicaraguan lowland. This conspicuous feature extends across the isthmus diagonally from the Gulf of Fonseca in the northwest to the valley of the Río San Juan in the southeast. Scattered over the lowland from the Gulf of Fonseca to Lake Nicaragua are numerous volcanic cones, some still active, from which the outpourings of lava and outbursts of ash and cinders have given the northern part of the lowland an unusually productive soil. The fourth region is a narrow northward extension of the highlands of Costa Rica which separates Lake Nicaragua from the Pacific.

The highlands of Nicaragua are related geologically to the highlands of Honduras and, as part of the Central American-Antillean system

form prominent east-west ridges and valleys. They rise sharply from the Nicaraguan lowlands to a general elevation between 4000 and 5000 feet above sea level.

The entire eastern part of Nicaragua is rainy. The Miskito Coast and the highlands back of it receive the full effect of the easterly trades that sweep in from the warm Caribbean. This is the wettest part of Central America. Along the east coast the rainfall is well over 100 inches: at San Juan del Norte (Greytown) it averages more than 250 inches. Heavy rainfall occurs especially at the heads of the valleys where the rising moist air is concentrated as in a funnel. As a result, the highlands and part of the plain to the east are covered with a luxuriant evergreen rain forest. Only in the higher places is the evergreen broadleaf forest replaced by oak and pine, as in Honduras. The southern limit of pine on the highlands is just north of Jinotega. South of this the forests of the *tierra fría* are comprised of oak, with no pine.

Along the lowland back of the Miskito Coast, pine also occurs. This is an area of pine savanna that extends about 300 miles northward from Bluefields as far as the Río Plátano in Honduras. The southernmost stand of North American species of pine is a few miles north of Bluefields. The pines are associated with a deeply leached soil having almost pure quartz gravel or powdery quartz sand at the surface, with a hardpan beneath through which drainage is slow.[1] Burning by prehistoric Indians may have eliminated a previous broadleaf woodland, but at present the annual burning of savanna grasses has the effect of killing the young pine seedlings. Whatever its origin, this grassland is the largest pine savanna in the humid tropics.

The Lowland

The Nicaraguan lowland is of major importance in the geography of Central America. Not only

---

[1] J. J. Parsons, "The Miskito Pine Savanna of Nicaragua and Honduras," *Annals of the Association of American Geographers* 45 (1955): 36–63.

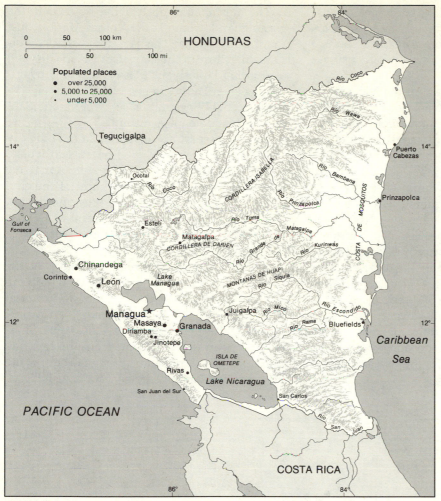

FIGURE 7-1

does it provide a low passage across the isthmus from sea to sea, but it is also a major geologic and biotic boundary, for it marks the southern end of the Central American-Antillean structures, and no North American pines are found south of it. Within the lowland are the two largest lakes in Central America. Lake Managua, the smaller of the two, is about 38 miles long by 15 miles wide. Its surface is 136 feet above sea level. It drains southeastward into Lake Nicaragua by way of the Río Tipitapa. With the growth of population and industry along the south shore, pollution of the lake has become a major problem.

Lake Nicaragua is about 100 miles long and 50 miles wide, and occupies an area of 3200 square miles. Its surface is only 106 feet above sea level, and drainage is via the Río San Juan to the Caribbean. This lake and river would serve as links in any trans-isthmian canal that might eventually be built through Nicaragua. It is believed that the lake itself was once part of the Pacific Ocean, from which it became separated by uplift of the land and volcanic eruptions to the west. Fish trapped in the newly formed lake had to adjust to their new habitat in order to survive. Today, the world's only fresh-water sharks, plus swordfish and tarpon, are found in Lake Nicaragua. Transportation on the lake is primarily by small boats which carry basic grains from the eastern shore to Granada and return with a variety of consumer goods.

The northwestern portion of the Nicaraguan lowland is the scene of intense volcanic activity. Mount Cosigüina, which stands about 2800 feet above the sea on the peninsula just south of the entrance to the Gulf of Fonseca, was formed by an explosive eruption in 1835. During that event the sun was blotted out within a radius of 35 miles, and dust from Cosigüina fell in noticeable quantities in Jamaica, 700 miles away. There are more than 20 volcanic cones between Cosigüina and Mount Momotombo north of Lake Managua. In Lake Nicaragua there are three volcanic cones of which the Volcán Concepción is more than 5000 feet high. Many of these cone-shaped, explosive volcanoes are still active, and deep ash falls are common. Fortunately, the volcanic ash weathers into a soil of high productivity, and the region has good potential for the development of geothermal energy.

The most continuously active volcano in the Nicaraguan lowland is the Volcán Masaya, just south of Managua. This is a dome-shaped volcano, like those of the Hawaiian Islands. Instead of erupting explosively, Masaya was built by upwelling lava. The crater in which the molten lava is always bubbling emits gases which stunt plant growth for some miles distant.

Earthquakes are also common. In 1931 the city of Managua was devastated by an earthquake and fire and had to be entirely rebuilt. A massive earthquake in 1972 destroyed most of the city, killing 10,000 people and leaving more than 300,000 homeless. About 10 percent of the nation's industrial establishments also were eliminated.

Rainfall in the Nicaraguan lowland is heavy near the Caribbean, but moderate to light in the northwest. Abundant precipitation and tropical rain forest extend along the San Juan Valley about as far as the shore of Lake Nicaragua. Since the rain-bearing winds come from the east, the driest part of the lowland is along the base of the northern highland, where a local "rain shadow"

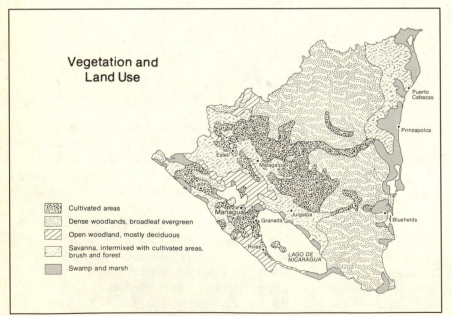

Vegetation and Land Use

Cultivated areas

Dense woodlands, broadleaf evergreen

Open woodland, mostly deciduous

Savanna, intermixed with cultivated areas, brush and forest

Swamp and marsh

FIGURE 7-2

is formed on the southwest-facing leeward slope. The northeast margins of Lake Nicaragua and Lake Managua are relatively dry and, because of high temperatures and rapid evaporation, are not suitable for crops without irrigation. This part of the lowland was originally covered by dry scrub woodland. The land southwest of the lakes, on the other hand, receives a moderate rainfall, most of it during the summer.

The Pattern of Settlement

The Nicaraguan lowland was first plundered and then settled by expeditions from Panama. As early as 1519 exploring parties moved westward from Panama City, seeking sources of precious metals or gems. Along the southwestern shores of Lake Nicaragua, a moderately dense Indian population was discovered, practicing a shifting cultivation of maize and supplementing its diet with wild game and fish. These people were peaceful and were quickly converted to Christianity, for which in gratitude they loaded the strange newcomers with gold ornaments. This display of what the Spaniards thought was only a part of a great accumulation of such gold objects led the conquerors to believe that here, indeed, was a land worth further attention. In 1524 another expedition from Panama founded two colonies in what is now Nicaragua. One was Granada, on the shore of Lake Nicaragua on some of the most productive land of the whole region; the other was León, on the Pacific slope near one of the great volcanoes. Granada became the more prosperous, and in it dwelt some of the Spaniards whose extensive properties gave them a very adequate living. The crops of this period were indigo, cacao, and sugarcane. The drier soils around León did not yield well enough to do more than provide subsistence crops of maize.

During the period of the California gold rush after 1849, one of the important routes of travel passed through the Nicaraguan lowland. It was the North American railroad builder Cornelius Vanderbilt who developed the Nicaraguan route. The interest of foreigners in this route, together with the continued political rivalry of conservative Granada and liberal León, kept the country in turmoil. In an effort to bring peace between the warring factions, a new capital city was established midway between the rival centers. This was Managua, founded in 1858. Warfare continued, however, fanned in part by the activities of Great Britain and the United States, both of which were eyeing the Nicaraguan lowland as a possible route for an interoceanic canal. From 1912 to 1925 and from 1926 to 1933 the U.S. Marines enforced order within the central area but were never able to stop the guerrilla warfare in the highlands. Rural areas, even those close to the big cities, were frequently raided by bandits.

Settlements and Products of the Lowland

Most Nicaraguans still live in the part of the lowland that lies between the Pacific Coast and the shores of the lakes. Managua was a city of more than 500,000 in 1978 and was growing rapidly. Like other cities of this part of the world, the capital was surrounded by a crowded slum area into which rural migrants came with the hope of finding jobs. Following the 1972 earthquake, the entire central part of the city stood empty and silent like some relic of the ancient past. Yet, Managua has continued to function as the nation's capital, its primary industrial center, and, increasingly, as its major cluster of population.

Cotton farming on the Pacific coastal lowland began in the 1950s. The chief cotton area in Nicaragua is between Chinandega and León, where the coastal lowland is covered with productive soils derived from volcanic ash. Insect pests, which for many years made cotton-growing in Nicaragua hazardous, are now controlled by frequent dusting with chemical spray delivered by airplane. The spray, however, contaminates milk and meat supplies when cattle graze in adjacent pastures or on cotton plants after the picking season. Cotton production has declined, but raw cotton is still one of the country's leading exports.

Since 1963, an area of intensive agriculture with irrigation has been cultivated between the western shore of Lake Nicaragua and the Inter-American Highway. Here farmers on small properties grow bananas, sesame, and sugarcane. Some of the plain east of the lake also has been placed under irrigation, and the land use has changed from cattle-raising to large-scale, mechanized rice cultivation. Because two crops can be harvested in a single year, this has become one of the major rice-producing areas of Nicaragua.

The crop that occupies more area than all others combined is maize. Maize, rice, and beans are the basic subsistence crops of the rural people and are raised wherever there are settlements. Much land is also used for the grazing of cattle, and Nicaragua shares in the export of processed beef from Central America.

A number of other commercial crops are grown in this central area. Since 1937 sesame has been cultivated, in some years with considerable success. The uncertainty of the rainfall makes cultivation of this moisture-loving crop a hazardous venture unless it can be irrigated. Ipecac root grows wild along the eastern margin of the lowland to the east of Lake Nicaragua, and some tobacco is grown throughout the area of concentrated settlement. Commercial banana cultivation, formerly concentrated in the eastern area near Rama, is now most highly developed between León and Chinandega. Until the *Sandinista* government expropriated their interests dur-

AN AIRPLANE FUMIGATES RICE ON A PLANTATION NEAR TIPITAPA, NICARAGUA.

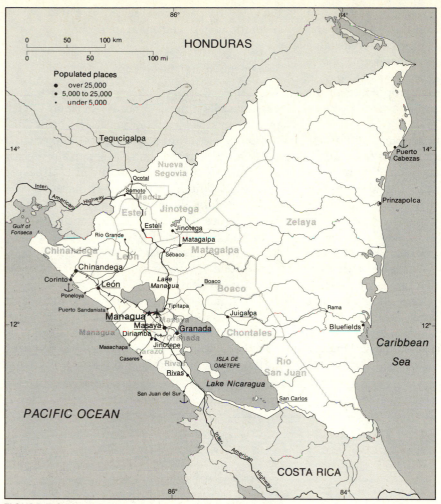

FIGURE 7-3

ing the early 1980s, both United Brands and the Standard Fruit Company were involved in the banana trade.

Settlement in the Highlands

The highlands of Nicaragua, like those of the other Central American countries, are used for the production of coffee. Coffee was first introduced into Nicaragua in 1850, at which time large plantations were developed in the hilly belt west of the northern end of Lake Nicaragua. The coffee plantations around Diriamba, about 20 miles south of Managua, are still productive. From this area comes about half of Nicaragua's coffee.

Coffee plantations were developed more recently in the highlands northeast of Managua. The Inter-American Highway, which has been completed through Mexico and Central America and on into Panama, climbs from the shores of the Gulf of Fonseca into the high country on the border of Honduras and Nicaragua. It passes through the Nicaraguan town of Estelí before descending to the Nicaraguan lowland east of Lake Managua. Not far from this highway, and connected to it by a paved road, are Matagalpa and Jinotega. This is the other major area of coffee production in Nicaragua, which also produces about half of the total crop. Here all the plantations are small and are worked by the owners.

Gold and silver mining supports an entirely different kind of settlement in parts of the highlands. Nicaragua is unusual among the countries of Latin America in that none of its gold ores was exploited during the colonial period. They were discovered in 1889–90 by men who pushed inland from the east coast, searching in the dense rain forest for sources of rubber. Gold was discovered in the valley of the Río Pis Pis, a tributary of the Río Coco, and La Luz mines were developed a little to the south. This mining district, about 90 miles west of Puerto Cabezas and 180 miles northeast of Managua, is now connected by regular air service with the capital and the east coast. The mining operations, previously owned by U.S. companies, were expropriated by the revolutionary government in 1979.

Settlement on the Miskito Coast

The entire Caribbean Coast from Yucatán to Panama was almost unoccupied when English pirates, seeking bases for their attacks on Spanish shipping, established the first settlements. In 1678 the governor of Jamaica set up a protectorate over the Miskito Indians, and in 1740 several colonies of Jamaican blacks were placed in what is now eastern Nicaragua. The two most important colonies were at Bluefields and Greytown (renamed San Juan del Norte). Eventually, the British recognized Spanish sovereignty along all the Caribbean Coast except for British Honduras, now known as Belize.

The small groups of settlers who did occupy the region developed a succession of export products, each of which has been largely exhausted. The list includes green and hawksbill turtles, sarsaparilla, mahogany, rubber, bananas,

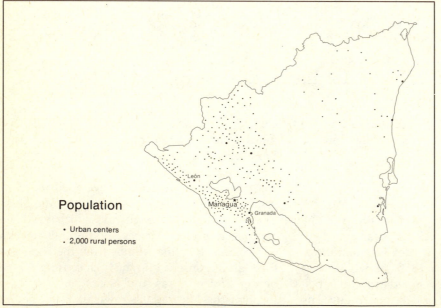

Population

· Urban centers
· 2,000 rural persons

León
Managua
Granada

FIGURE 7-4

pine and cedar lumber, animal skins, silver, and gold. Offshore and in coastal lagoons the exploitation of shrimp and lobster continues, but there is a danger that these resources are also being depleted.

Nicaragua as a Political Unit

The fact that Nicaragua has one single nucleus of concentrated settlement gives it greater unity than it might otherwise enjoy. There are no large clusters of people in conflict with each other, as in Honduras. The core area of settlement is not divided into an Indian segment and a ladino segment, as in Guatemala, and there is no lack of unoccupied potential cropland, as in El Salvador.

Nicaragua offers an interesting example of the dissemination of innovation. Part of the innovation is the transfer of people from subsistence farming outside of the money economy, into a new economic system of surplus production which involves buying and selling. Another part is the replacement of a military dictatorship devoted to the preservation of special privileges for the elite, by a government devoted to the idea of relieving poverty and eliminating inequities of status and opportunity. However, expropriation and management by government do not alone assure economic progress and social justice.

The Economic Situation

The problem of economic development in Nicaragua involves the shift of more and more people from subsistence living into the money economy. New jobs must be created at a rate faster than the growth of population, and this requires a heavy investment of new capital. A large proportion of the working force is employed in agriculture, and many people in the rural areas are either engaged in raising their own food or are only seasonally included in the money economy. The economic transformation of Nicaragua has started but has not progressed far.

Subsistence farmers of Nicaragua are not much different from people in similar economic conditions in other parts of Latin America. The

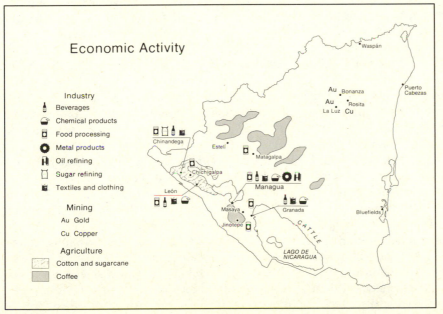

FIGURE 7-5

food crops they raise are maize, rice, and beans. They cultivate the poorest soils and steepest slopes, for much of the better land has been used for commercial crops or the grazing of cattle. Yields of food crops are so low on badly eroded land that these subsistence farmers are not only poor but are also getting poorer. Thousands of them have sought temporary seasonal employment on commercial plantations, picking cotton or coffee or cutting sugarcane. Few people move away from areas of concentrated settlement to pioneer zones in unoccupied parts of the country. The fact is that not many people choose to live in isolation. Most subsistence farmers, if they are forced to move, prefer city slums over remote areas of forest. Yet, when a road is built into potential agricultural land, pioneer settlement moves out along the road onto land not previously occupied. This has been the experience in Nicaragua along the road to Rama, which will eventually connect Managua with the Caribbean port of Bluefields.

Economic development includes the use of certain parts of the country for commercial crops. In the older areas of commercial coffee planting, estates are large and workers are paid wages, as around Diriamba. In coffee zones of the highlands, the plantations, or *fincas,* are small and are worked by the owners. The largest development of commercial farming, however, has been in the planting of cotton around León, with export via Corinto. The largest source of employment in agriculture is in this section of the country.

Nicaragua began to industrialize in the late 1950s. In 1958, in Managua, foreign capital helped build the largest meat-processing plant along the Pacific lowland of Central America. Processed meat is normally sent in refrigerator trucks over the Inter-American Highway to the Caribbean port of Santo Tomás de Castilla in Guatemala, where it is loaded on ships bound for Miami. Concern over dependence on cotton and coffee exports hastened an increase in the production of meat, tobacco, and bananas. When bauxite and copper deposits were discovered, new highways were planned to make these min-

erals accessible and to encourage also pioneer settlement in empty parts of the country. In an attempt to decrease dependence on the United States, trade with Japan and West Germany was expanded.

President Anastasio Somoza Debayle seemed to have tight control over the republic until 1972, but the earthquake of that year weakened the Somoza family's empire. Managua was destroyed, and it was claimed that Somoza and his friends diverted international aid intended for the earthquake victims. All classes of society expressed themselves by a wave of strikes, land seizures, and violent demonstrations.

The Political Situation

The unity of settlement pattern has not always meant unity among the Nicaraguan people. It has been said that there are no Nicaraguans, only liberals and conservatives, *costeños* and *españoles.* When the people who lived around León and Granada asserted their independence from Guatemala in 1838, they were united in only one thing: the desire for freedom from outside control, the right to make their own decisions. Immediately, the struggle for supremacy flared up between the liberals of León and the conservatives of Granada. When political parties occupy geographically distinct areas, internal conflict is difficult to avoid.

The political situation in Nicaragua has long been complicated by foreign intervention. In 1841 Great Britain established a protectorate over the Miskito Indians on the east coast and in that same year seized the port of San Juan del Norte. The United States began to turn its attention to potential passageways across the isthmus after the Mexican War resulted in expansion of its national territory all the way across North America to the Pacific Ocean. In 1850 the United States and Great Britain signed the Clayton-Bulwer Treaty in which both countries agreed not to occupy, fortify, colonize, or exercise dominion over any portion of Central America. Great Britain insisted that this did not include British Honduras (Belize), but it did involve the British settlements along the Miskito Coast. Still it was not until 1893 that Nicaragua regained control of this territory.

In 1893 the liberal General José Santos Zelaya became dictator, and in 1909 he signed a treaty with Great Britain giving that country exclusive rights to build a canal. The United States then withdrew recognition, and Zelaya was soon deposed. In 1912, when conditions seemed thoroughly out of hand, the United States landed 2600 Marines. After order had been imposed, the force was reduced to a legation guard of about 100 men. In 1916 Nicaragua granted to the United States exclusive rights to build a canal, for which the United States paid the sum of $3 million.

By 1925 the military occupation of other countries had become unpopular in the United States, and the Marines were withdrawn from Nicaragua. Immediately, civil war between liberals and conservatives erupted again, and in the spring of 1926 the Marines returned. Elections were conducted under U.S. control, and a liberal was elected. It was at this time that General César Augusto Sandino began his revolt against military intervention and against any government elected under foreign auspices. Sandino became the hero of many people throughout Latin America. When the Marines were finally withdrawn in 1933, Sandino promptly surrendered to the Nicaraguan government, after which, because of his large political following, he was promptly shot.

The year 1932 marked the beginning of the Somoza era in Nicaraguan history. The strong man General Anastasio Somoza García had been trained by the Marines and placed in command of the highly efficient National Guard. With this military backing he became the only real political power. In 1937 he seized the government and thereafter remained either as President or as the power behind the President until his assassination in 1956. Throughout this period he gave Nicaragua political stability by prohibiting all opposition parties. He treated the country as if it

were his own private estate, and he acquired extensive personal property. After his death, his son, Luis Somoza Debayle, assumed the presidency and in 1957 was duly elected to that office.

In that same year a dispute involving the easternmost part of the boundary between Nicaragua and Honduras again flared up. Both countries had been moving settlers into the disputed area since the 1950s and were finally forced to request the Organization of American States to seek a solution. In 1960 the International Court of Justice awarded the entire area to Honduras.

Anastasio Somoza Debayle, a younger brother of Luis, was elected to the presidency in 1967 and reelected in 1974. *Sandinistas*, who had been trying to undermine the Somoza family for some time, launched a major offensive in 1977 in an effort to overthrow the government. The brutal civil war that followed turned Managua into a city of terror, and thousands escaped to the countryside to avoid the conflict. Somoza finally resigned in 1979 and was assassinated the following year in Asunción, Paraguay. The *Sandinistas* assumed power and ruled through a cabinet, called the Government Junta for National Reconstruction.

The new government proved initially to be quite moderate. The Somoza enterprises were nationalized, however, as were domestically owned banks and insurance companies, mining, and foreign trade. The *Sandinistas* recognized the need for extensive investment and expanded production, if the revolution were to provide material gains for the people. Agricultural land reform received immediate attention, with the vast holdings of the Somoza family being the first to be divided among small-farmer cooperatives. Great emphasis also was placed on education and health. Student enrollment almost doubled from 1979 to 1982, and illiteracy was reduced from 50 to 13 percent through an intensive national campaign. A similar effort brought health facilities to rural areas throughout the country and sharply reduced the infant mortality rate.

Meanwhile, the war-ravaged country was turned far to the political left, favoring Cuba and the Soviet Union. Nicaraguan officials claimed that the United States had not given them enough aid, and thus they entered an agreement whereby Soviet technicians would assist in the modernization of areas such as agriculture and mining. Cuban technicians were conspicuous in the fields of health and education. The Soviet and Cuban governments contributed to a large-scale expansion of the Nicaraguan military forces which, in turn, provided moral and material support to leftist guerrillas fighting in El Salvador. Relations with the United States and with neighboring Central American nations were strained accordingly. It is certain that the Nicaraguan government needs help in rebuilding the country and the economy. What is less certain is whether alignment with Cuba and the USSR will benefit Nicaragua in the long term or will simply involve it more deeply in a worldwide struggle between the major world powers.

COSTA RICA

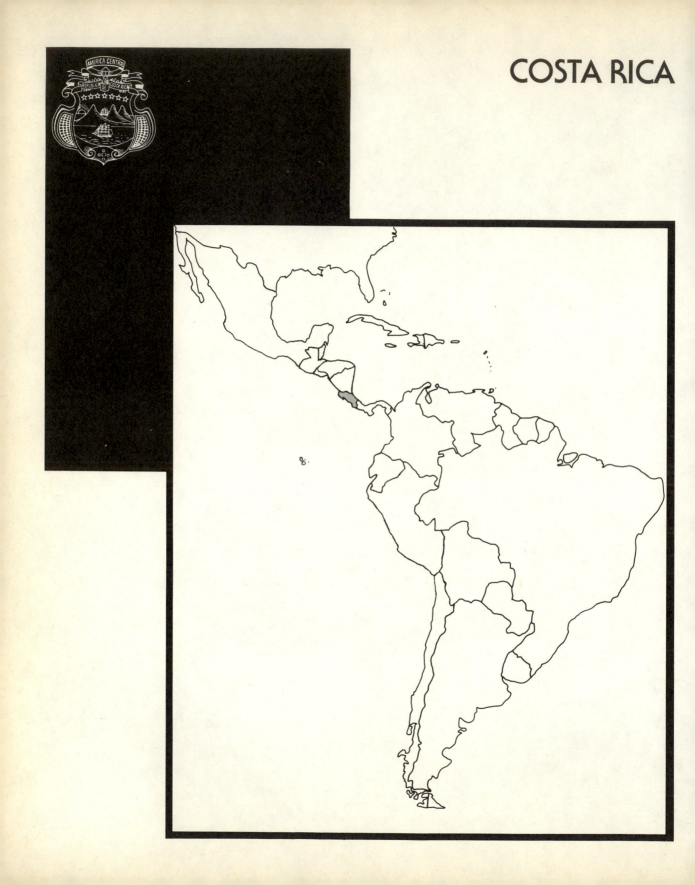

8

REPÚBLICA DE COSTA RICA

Land Area   19,652 square miles

Population   Estimate (1985) 2,600,000
             Latest census (1973) 1,871,780

Capital city   San José 867,800

Percent urban   48

Birth rate per 1000   31

Death rate per 1000   4

Infant mortality rate   19.3

Percent of population under 15 years of age   37

Annual percent of increase   2.7

Percent literate   93

Percent labor force in agriculture   33

Gross national product per capita $1020

Unit of currency   Colón

Physical Quality of Life Index (PQLI)   85

COMMERCE (expressed in percentage of values)

Exports

coffee   24
bananas   23
chemicals   8
beef   7
sugar   4

Exports to

United States   35
Nicaragua   12
West Germany   11
Guatemala   6
El Salvador   5

Imports from

United States   35
Japan   11
Guatemala   6
Mexico   6

Data mainly from the 1985 World Population Data Sheet of the Population Reference Bureau, the 1985 Britannica Book of the Year, and the 1985 World Almanac.

Costa Rica, among all the nations of Latin America, is noted for the devotion of its people to the ideals and practices of democracy and for its highly developed system of education. There is no place in Latin America like it.

The area of concentrated settlement that forms the nucleus of Costa Rica is in the highlands, where a small intermont basin with deep volcanic soils offers land of relatively gentle slope in the midst of the *tierra templada*. The densely populated core area, including this basin and the lower slopes of the bordering mountains, measures only about 15 by 40 miles. The people of the core area are of almost unmixed Spanish ancestry. With a birth rate of 31 per thousand (1984) and an exceptionally low death rate of 4 per thousand, the population of Costa Rica as a whole is increasing at a net rate of 2.7 percent per year.

There are many other ways in which the conditions of life in this area are distinctive. Costa Rica and only a few other countries of Latin America, such as Argentina, Uruguay, Chile, Puerto Rico, and Barbados, enjoy the prestige of a population that is more than 80 percent literate. Two-thirds of the population is concentrated in the core area, as is about 90 percent of the nation's industry. The great majority of farmers here own and operate their own farms. The traditional large estate with tenant workers is not common. There is no landed aristocracy to dominate the social life, manipulate politics with the support of an army, and collect a disproportionate share of benefits from the economy. In Costa Rica there is a notable feeling of equality.

Another distinctive characteristic of Costa Rica is the long existence of zones of pioneer expansion. Farm settlers have been moving out from this overpopulated core area for many decades, occupying previously empty land, and creating new communities carved from the forest. This kind of outward movement of pioneers has been generally lacking in Latin America until recently. Costa Rica has had its pioneer zones for hundreds of years. Even in the modern period many Costa Ricans are more attracted by the op-portunities of a pioneer zone than by the presumed amenities of city life.

Yet, Costa Rica also has internal variety. On the Pacific side, especially in the Province of Guanacaste, nearly half of the people are mestizo in origin, racially indistinguishable from the people of Nicaragua. In this part of Costa Rica, moreover, there are many large properties, a small landed aristocracy, and tenant workers who live in relative poverty. On the Caribbean side more than half of the people are blacks, mostly of Jamaican origin, who were brought in when the banana plantations near Puerto Limón were first developed.

The census of 1950 counted 800,875 people in Costa Rica, and in 1984 the population reached 2.5 million. For the country as a whole, about 80 percent are of unmixed Spanish ancestry, 17 percent are mestizo, 2 percent are blacks, and less than 1 percent are pure Indian.

The Land

The physical features of Costa Rica are not complex. The backbone of the isthmus is composed of a ridge of mountains that extends from northwest to southeast. This is a continuation of the belt of hills which, in Nicaragua, separates Lake Nicaragua from the Pacific Coast. The ridge slopes steeply toward the Pacific and more gently toward the Caribbean. The mountain ridge in this part of Costa Rica is called the Cordillera de Talamanca.

In central Costa Rica, just northeast of San José and Cartago, four great volcanic cones stand in a row, their bases merged into one massive pedestal. From northwest to southeast they are Póas (8930 feet), Barba (9567 feet), Irazú (11,260 feet), and Turrialba (10,974 feet). This commanding line of volcanoes stands parallel to the main crest of the mountain backbone, but about 20 miles to the northeast of it. Nestled between the steep slopes of the main cordillera and the gentle lower slopes of the four volcanoes is an intermont basin, the Meseta Central, between

FIGURE 8-1

3000 and 4000 feet above sea level. This basin is a structural depression, deeply filled with porous volcanic ash that remains generally level and undissected except close to the headwaters of the rivers that drain the basin. Near the rivers the surface is hilly and in many places too steep for agriculture. The southeastern part of the Meseta Central is drained by the headwaters of the Río Reventazón, which flows through a deep gorge at the base of Mount Turrialba and emerges on the Caribbean coastal lowland north of Limón. The northwestern part of the Meseta Central is drained by the headwaters of the Río Grande, which enters the Pacific near Puntarenas.

Another feature is the long narrow structural depression parallel to the Cordillera de Talamanca in southeastern Costa Rica. This is the Valle del General, a deep lowland drained by the Río General, which runs along the southwest side of the cordillera for some 50 miles and is separated from the Pacific Coast by another steep mountain ridge.

Bordering the Pacific Coast are patches of lowland lying between the base of the highland backbone and outlying mountain blocks. The lowland of Guanacaste at the head of the Gulf of Nicoya is the largest such area of plain. It extends northward almost to the border of Nicaragua, from which it is separated only by the hilly northwestern end of the central highland. The lowland is bordered on the southwest by steep block mountains on the Nicoya Peninsula which reach a maximum elevation slightly over 3000 feet. South of the Gulf of Nicoya the lowland is cut off as the central cordillera rises directly from the Pacific. In the southern part of Costa Rica, however, there is another lowland area where the Río Diquis emerges from the mountains. Here also a mountain block borders the ocean, and a small embayment provides a protected harbor for the banana port of Golfito.

The northeastern part of Costa Rica includes the end of the Nicaraguan lowland, which extends along the base of the highlands southeastward as far as Limón. Much of this plain is poorly drained, but along the base of the volcanic mountains there is a piedmont zone of gently sloping land in which drainage is not a serious problem.

Climatic conditions in Costa Rica differ notably between the Caribbean and Pacific sides, and in the highland area they are to a large extent governed by altitude. Rainfall is abundant on the Caribbean lowland and on mountain slopes that face the easterly winds. A belt of more than 100 inches of average annual rainfall extends from eastern Nicaragua, across Costa Rica and Panama into Colombia. On the Pacific side not only is the average annual rainfall somewhat less, but it is also less evenly distributed during the year. In the Valle del General and all along the Pacific Coast there is a dry season from December to April, when the prevailing winds are from the east. When the equatorial westerlies are onshore from the west and southwest, rainfall is heavy.

In the highlands the pattern of climate and vegetation is complicated. None of the mountains is high enough to reach the zone of permanent snow, but on some of the higher slopes there are tall grasses. The Meseta Central lies in the rain shadow of the four volcanoes and was covered with grass when the Spaniards arrived. Three grasslands near Cartago offered the first Spanish colonists the kind of open land they preferred as places on which to settle. Despite widespread deforestation, about 15 percent of the total national territory remains covered with trees. Included are commercial stands of rosewood, cedar, mahogany, and other tropical hardwoods.

The Pattern of Settlement

The Spaniards found no tribes of peaceful natives in Costa Rica such as they found in Nicaragua. A settlement was attempted on the Nicoya Peninsula as early as 1522, but it was soon abandoned because of the hostility of the natives. In the highland basins there were perhaps 7000 or 8000 sedentary Indians when the Spaniards arrived, but they were soon decimated by disease. Nor did the explorers find any gold. The name *Costa Rica* was probably given to the Caribbean Coast by Columbus because of the gold ornaments he found the Indians wearing. It proved, rather, to be a *Costa Pobre*. As a result, the area remained unoccupied for a long time.

Settlement of the Highlands

In about 1560 what is now Costa Rica was placed under the jurisdiction of Guatemala, and in 1561 the first group of settlers established themselves in the Meseta Central. In all of colonial Latin America there were not many settlers like these, except for the Chileans. These settlers came to establish their own farms and build homes, not just to find El Dorado and return to Spain. Furthermore, in the next year Juan Vásquez de Coronado arrived from Guatemala bringing with him the wives and fiancées of the pioneers. He also brought cattle, horses, and hogs, and in this area established what was probably one of the earliest large cattle ranches in the Americas. In this isolated spot, the people became small farmers, raising their own food and fibers, and bearing children of unmixed Spanish ancestry. There was no aristocracy. No individuals could claim special prestige because of position or status. The families that chose to come to Costa Rica were in a sense selected because of their unusual attitudes and objectives.

Expansion of Settlement

The first signs of settlement expansion appeared early in the eighteenth century, while Costa Rica was still a poor country. In 1736, pioneers from Cartago moved into that part of the Meseta Central that drains to the Pacific and established the town of San José. In 1790 Alajuela was founded not far from San José, and a little later Heredia, also in the vicinity of San José.

Costa Rica was still a poor country when the people of Mexico declared their independence, including in the new sovereign nation all of the area formerly administered as New Spain. The people of Costa Rica were not greatly concerned, however, for in their remote location they had always been independent if only because no one was interested in them. When Guatemala later broke away from Mexico, this also involved Costa Rica. Finally, in 1838, when El Salvador and Nicaragua separated from Guatemala, Costa Rica was left to go its own way. In 1821, there were some 60,000 people in all of Costa Rica, most of them in the central area. The population density around Cartago was about 260 people per square mile, one of the greatest densities then to be found in all of Latin America.

The Spread of Coffee Planting

The first country in Central America to begin the cultivation of coffee was Costa Rica. The plant was introduced as early as 1797, but several decades passed before it became the major commercial crop. When Costa Rica found itself independent, its government faced an urgent need for some source of money, some product to export and tax. At an early date the government sought to stimulate this form of commercial production by offering free land to anyone who would agree to set out coffee trees. The result was a "coffee rush." In 1825 the first few bags were shipped, and in 1829 coffee became the chief export from the country.

Until the cart road to Puntarenas was completed in 1846, transportation was by muleback, and no great volume of coffee could be moved. At first, the chief market was in the west coast countries of South America, but when the first bags were sent to Great Britain in 1845, the European market was also opened to the Costa Rican product. Large-scale coffee exports began about mid century. After the opening of the railroad to Puerto Limón in 1891, that port became the chief outlet.

After the middle of the nineteenth century, settlement expanded more rapidly. One result of the new prosperity was an increase in the birth rate. With the population increasing and the government adopting a policy of giving free land for new coffee plantations, continued pioneer expansion occurred around the margins of the original settlements, without an accompanying decrease of population density in the older settled areas. Pioneers first spread down the valley of the Río Reventazón, their settlements extending as far as the old mission town of Turrialba. In that region there were only 1068 people in 1883, but settlement continued rapidly after 1890 when the railroad from Puerto Limón reached Turrialba. New settlement also moved up the slopes of the volcanoes, a movement that has now gone beyond the upper limits of coffee. On Irazú, at an altitude of about 9800 feet, there are settlers who pasture dairy cattle and cultivate potatoes. Over a pass between Barba and Poas, settlement has spread to the northeast slopes of the volcanoes. Another current of settlement has gone to the west of Poas, almost completely encircling its base. Pioneers have advanced down the railroad line that connects San José and Puntarenas, and since 1910 there has been a movement northwestward from this line along the Pacific slope of the highlands.

One of the largest pioneer movements in

FIGURE 8-2

Costa Rica, however, has taken place since 1936 in the Valle del General. In that year a new gravel highway was completed between Cartago and San Isidro at the northwestern end of the Valle. This little frontier town became the jumping-off place for new settlers. By 1960 there were more than 50,000 farmers, and thousands more have come since. In 1962 an all-weather highway was completed across the Valle del General and through the gorge of the Río Diquis to the low-lands along the Pacific. The road was extended eastward into Panama, providing for the first time a regularly passable route of travel between the rest of Central America and Panama. This pioneer zone is now connected to markets by all-weather roads. As a result, the food products raised in the Valle are not just for subsistence. The people in this zone now account for about half of the total Costa Rican production of maize, beans, and rice.

In 1976 the Aluminum Company of America contracted with the government of Costa Rica to mine bauxite in the Valle del General. Mining and smelting on a major scale, however, will depend on favorable market conditions and completion of the 760-megawatt Boruca hydroelectric plant in southern Costa Rica.

Present Settlement of the Highlands

Costa Rica is no longer isolated from the rest of Central America. Not only is the central area well supplied with paved roads, but highways also have been extended to Puerto Limón and Puntarenas. The Inter-American Highway provides an all-weather connection all the way from the United States through Mexico, Guatemala, El Salvador, Honduras, and Nicaragua. In Costa Rica it passes through the Guanacaste lowland to Puntarenas, then climbs to San José and Cartago. From Cartago it descends through the Valle del General and passes through the gorge of the Río Diquis to the Pacific lowland, and on into Panama.

San José is Costa Rica's largest city and its capital. Culturally, it is the heart of the nation. It is also the center of manufacturing, which includes textiles, chemicals, paper, leather goods, and furniture. In 1984 the metropolitan area, which extends beyond the limits of the political city, had a population of almost 1 million.

Since the early 1960s the production of coffee in the central area has been reorganized. Coffee-growers were previously unable to obtain high yields per acre because their small holdings made it difficult to use modern agricultural techniques. After 1960 many small coffee-growers sold their *fincas* and moved to pioneer zones, and many of the small *fincas* were then consolidated. As a result, the annual per acre coffee harvest doubled, with increased efficiency and the use of pesticides and fertilizers. At the same time, Costa Rica as a whole began to experience some of the inequality of landholding that characterizes most other Latin American countries.[1] A census in

1973 revealed that the largest 1 percent of all properties occupied more than 25 percent of all the land and that half of all farmers with the smallest properties controlled less than 4 percent of the land. Perhaps more serious, only 22 percent of the economically active peasant population were landowners. Problems appear likely to become increasingly severe when unoccupied land with productive potential is no longer available.

Central America produces about 12 percent of the world's exportable production of coffee, of which Costa Rica produces about 88,000 metric tons. All Central American countries, except Nicaragua, levy a tax on coffee exports. In the case of Costa Rica, the tax is a fixed percentage of the coffee exports value which, of course, varies from year to year. Most of Central America's coffee is for future delivery; establishing projections for coffee prices is thus a complex procedure.

In the central area coffee is grown up to about 3200 feet above sea level. Between this and the cloud zone where there is a wet forest, most of the land is used for pasture, but within a system of crop rotation. The crops, namely potatoes, maize, and vegetables, are followed by a year of pasture, a year of fallow, and then crops again. The pasture is primarily for dairy cattle. Fine herds of purebred stock are used to produce milk and butter, which are transported by truck to the city markets.

To the west of Alajuela, within the central area, is a sugarcane district, where there are seven large sugar refineries. The greater part of the sugar is consumed within Costa Rica. With expanding acreage in sugarcane, Costa Rica also produces alcohol as a motor fuel.

One reason for the continued productivity of farms in the central area, even after 400 years of heavy use, is the deep, well-drained soil formed on volcanic ash. Some of the finest agricultural lands in the tropics are those with such volcanic soils, but volcanic regions are occasionally subject to disasters. In 1963, with no warning, Mount Irazú erupted, sending forth clouds of

---

[1] M. A. Seligson, "The Impact of Agrarian Reform: A Study of Costa Rica," *Journal of the Developing Areas* 13:2, (1979): 161–174.

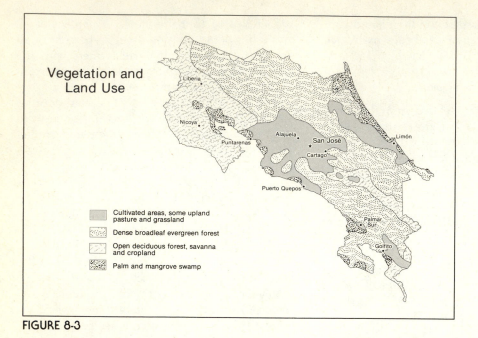

Vegetation and Land Use

Cultivated areas, some upland pasture and grassland

Dense broadleaf evergreen forest

Open deciduous forest, savanna and cropland

Palm and mangrove swamp

FIGURE 8-3

fine volcanic ash that fell over the land to the west of it. San José itself was covered with dust. In the rural areas coffee plantations were ruined as the leaves became coated with ash. When the rainy season began, the thick accumulation of ash on the slopes began to slide, at one time cutting the highway between San José and Cartago. Costa Rica became a disaster area. The only compensation was that this was how the fine soils of the central area had accumulated.

Settlement on the Caribbean Coast

A very different kind of settlement appeared on the Caribbean lowland of Costa Rica. In 1871, the government, seeking a more direct route for the increasing shipments of coffee, contracted with a North American company for the construction of a railroad to connect San José with Puerto Limón. Since the forested coastal lands were uninhabited, the company brought in workers from Jamaica. It took nine years to build the first 70 miles, and at least 4000 lives were lost to disease. By 1890 the line extended up the val-

ley of the Río Reventazón to Turrialba, however, and in the next year it was completed to San José.

One of the North Americans in charge of the railroad project was Minor C. Keith, who helped develop traffic on the lower part of the line in order to pay the high cost of construction. He imported rootstalks of the banana plant from the vicinity of modern Colón in Panama and planted them in Costa Rica along the new railroad. The first bananas were shipped from Costa Rica to New Orleans in 1874, and thereafter he established a regular export of bananas to both New Orleans and New York. In 1899 the United Fruit Company was formed by a consolidation of the Boston Fruit Company, which was operating in the Antilles, and several companies headed by Keith, who was growing bananas in Costa Rica, Panama, and the Santa Marta area of Colombia.

As skill in the solution of technological problems developed, the planting of bananas began to increase rapidly. By 1909, Costa Rica was the world's leading banana producer. In 1913, the peak year of production in Costa Rica, 11 million bunches were exported. By that time the United

Fruit Company had developed a second banana planting district along the Río Sixaola on the border of Panama. The bananas were moved across the border by train and exported from a small port in the Bocas del Toro area.

At first, the agricultural experts of the United Fruit Company rated the Caribbean lowlands of Costa Rica as almost ideal for tropical plantation crops. It soon became clear, however, that whereas some plantings of bananas remained productive for as long as 25 years, in most places yields declined rapidly after about 5 years. Despite new plantings, the peak year in the Caribbean lowland was reached in 1907. Exports from Costa Rica as a whole continued to mount until 1913, because of the new plantations along the Río Sixaola. Plantations, instead of lasting five years, began to die out in only two years. The difficulty lay in a fungus that caused the plants to rot and a second disease, Sigatoka, that stopped production entirely.

In 1942 the last banana ship was sunk at its dock in Puerto Limón by a German submarine, and all banana exports ceased. The eastern part of Costa Rica became a serious economic problem area. A proposal to repatriate some of the Jamaicans met an unenthusiastic response in that overcrowded island.

Since World War II, although bananas have been planted again on the Pacific side of Costa Rica, the Caribbean lowlands have not been entirely neglected. As early as 1914 some of the abandoned banana lands were planted experimentally with cacao. Now the United Brands Company operates a cacao plantation in this area, and small planters also cultivate cacao. An industrial enterprise to process cacao beans into a variety of cocoa products was established at Siquiries in 1975. United Brands has some land planted to abacá (Manila hemp), and the Goodyear Tire and Rubber Company has developed a small rubber plantation. Bananas also are once again a major crop. In Costa Rica as a whole, the area under bananas increased tenfold between 1964 and 1970, and banana exports in 1982 slightly exceeded those of coffee. However, by 1985, United Fruit had closed down its Costa Rican banana empire. United Brands will no longer own plantations, but it will buy the fruit from the Costa Rican growers. Thus ends an era.

The principal banana area is now focused on the modern banana port of Golfito, extending from Puerto Cortés to, and across, the border of Panama. In these plantations, which exceed 15,000 acres in each country, overhead irrigation systems have been built which also deliver copper sulfate through nozzles that spray in circular patterns like the watering systems of golf courses. The soil is treated with fertilizers and insecticides, and the bananas are dipped and washed before they are boxed and loaded on ships. In this area banana planting has ceased to be a temporary and shifting use of the land; it has become an intensive kind of agriculture, with relatively large capital and labor investment.

Settlement on the Pacific Coast

Economic development in Costa Rica has been concentrated on the Pacific Coast. The northwestern part of this region was the first to be reached by the Spaniards. North of Puntarenas the extensive plains of Guanacaste Province were divided into vast private estates and used for the grazing of cattle. Even now this province remains Costa Rica's chief area of cattle production. These are beef cattle, steers that are grass-fattened for a period of six to nine months. The dairy type are concentrated in the Pacific coastal provinces of San José and Alajuela.

The properties are large, and one estate may include areas of open range, forested areas in the mountains, and planted fattening pastures. The cattle are driven into the neighboring highlands on either side during the wet season and are brought back to the lowlands during the dry season. Most people of the area work for wages on the large ranches. Small bits of land near the villages or ranch headquarters are used for the production of basic foods: maize, rice, and beans. Since World War II there has been a considerable movement of new settlers into the southern part

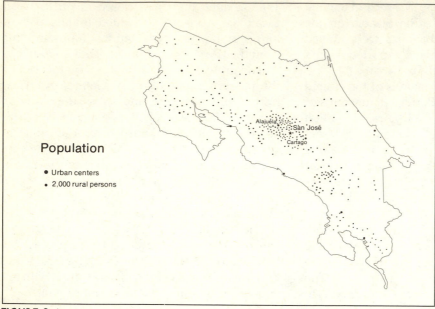

Population

• Urban centers
• 2,000 rural persons

FIGURE 8-4

of the Nicoya Peninsula where these crops are raised for local subsistence. Now, for example, Guanacaste normally accounts for 70 percent of Costa Rica's total rice production.

Around Liberia, near the northwestern end of the Guanacaste lowland, much of the land has been so severely gullied that its usefulness for pasture has been lost. In addition, logging operations in the forest have resulted in the clearing of watershed areas, thereby increasing runoff in the rainy season and intensifying droughts in the dry season.

Costa Rica as a Political Unit

Costa Rica has a well-developed sense of national identity. Costa Ricans are conscious of being different, and, because they share a clearly formulated state-idea, most Costa Ricans are less sensitive than their neighbors regarding matters of national prestige. There are at least five different ways in which Costa Rica has been distinctive: (1) the absence of a landed aristocracy; (2) the widespread support for democratic values

and procedures; (3) the high rate of literacy (93 percent), which is exceeded in Latin America by only a few countries; (4) a high rate of population increase (now reduced to a modest 2.7 percent per year); and (5) a movement of people to the frontiers of pioneer settlement exceeding movement to the cities. In addition, the population of the core area of Costa Rica is homogeneous in terms of racial and cultural background and therefore relatively free from the conflicting value systems that characterize many other Latin American societies.

The Economic Situation

In terms of gross national product per capita, Costa Rica stands midway between the countries of Latin America. In 1955 it ranked sixth in Latin America, but the rapidly accelerated economic development of such countries as Mexico and Peru pushed them ahead in the standing. Nevertheless, Costa Rica has been modernizing rapidly.

The increase in value of goods and services has come chiefly in the agricultural sector. Major

GEOGRAPHY STUDENTS FROM NORTH AMERICA? NO, THESE ARE STUDENTS FROM THE UNIVERSITY OF COSTA RICA IN SAN JOSÉ.

investments by the United Brands Company around Golfito have increased the value of exports, but more important has been the doubling of coffee yields. This has resulted from an increased use of fertilizer and insecticide, which is also related to the consolidation of the fincas into larger plantations. Consolidation could not have been achieved without a large migration of coffee-growers to the pioneer zones, especially to the Valle del General. Costa Rica merits special attention because of the importance of the frontier as a safety valve to absorb not only the farmers displaced from the central area, but also the rapidly increasing total population.

The country is still heavily dependent on two export products, coffee and bananas. However, other agricultural and manufactured products account for about 40 percent of the country's exports. Textiles, plus food and beverages, are the principal manufactures in Costa Rica, but automotive assembly, glass, and paper products are recent additions. Bauxite mining and the manufacture of aluminum are conducted on a small scale, and the expanding tourist and fishing industries provide additional income. Moreover, many Americans retire to Costa Rica when they learn that retirees are not subject to an income tax, and their pension checks subsequently contribute to the nation's economy.

As in many other countries, the single greatest economic problem in Costa Rica is the increased cost of energy. The cost of petroleum imports

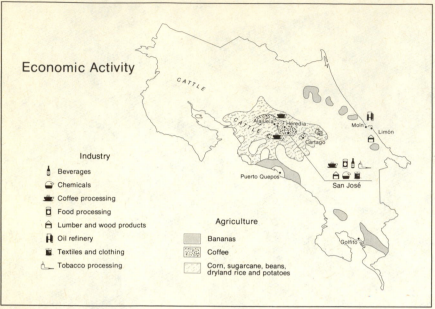

FIGURE 8-5

rose from $11 million in 1970 to $240 million in 1980. Meanwhile, a 60-pound bag of coffee worth 30 barrels of oil in 1970 bought only 3 barrels in 1980. Similarly, Costa Rica had to export 62 pounds of bananas to pay for 1 barrel of oil in 1973, but 924 pounds in 1981. As a consequence, increased attention is being given to the development of hydroelectric resources. The 156-megawatt Arenal plant in Guanacaste Province was completed in 1980, and the 174-megawatt Corobicí plant downstream from Arenal soon thereafter. Three major problems prevail, however, with hydroelectric generation: high initial costs, limited use in transportation, and the loss of fertile valley-bottom land in the reservoir areas. Petroleum exploration has been sporadically pursued in Limón Province since as early as 1915, and coal was discovered in the Talamanca Valley in 1978. At a huge new mill near Liberia, alcohol is being distilled from sugarcane for use in automobiles; exploitation of the Miravalles geothermal field, also in the Province of Guanacaste, will be initiated with a 55-megawatt power plant.

The Political Situation

In 1838, when the Central American countries separated, Costa Rica, without ever demanding it, found itself independent. Because of the lack of political interest, the new government fell into the hands of a dictator. The dictator had to earn his living like anyone else, however, for the national treasury contained no great wealth, and the use of printing presses to create money made little sense to the hard-working Costa Ricans.

A succession of governments, some more totalitarian than others, were all more clearly guided by public opinion than in any other Central American country. Costa Rica developed a reputation for freedom of speech, liberal concepts of government, and avoidance of force. In 1889 the first entirely free election in all of Latin America was held, and the ballots were counted honestly. Despite the existence of opposing political parties, the election results were respected and the majority party took control. Since 1889, except for only two brief periods, in 1917–1919 and in 1948, Costa Rica has operated as a democracy. The tradition of democracy has become an important part of the Costa Rican state-idea. Attempts to seize power by force in 1948 were met by a popular uprising. Since 1948, Costa Rica has had no army, only a national police force.

Since the late 1960s, the political trend in Latin America has been toward the emergence of authoritarian governments. An outstanding exception has been Costa Rica. In 1973 concern over the country's economic and political future prompted the *Plan Nacional de Desarrollo,* a series of four lengthy documents that analyzed the political economy, the agricultural and industrial structure, the present situation of dependency, future plans for development, and growth of the public sector. A reassessment took place in 1976 at an open symposium in which multiple aspects of Costa Rica's future were discussed. The Minister of National Planning and Political Economy took the position that the role of the state should be reduced rather than expanded, that there should be greater rather than less political participation, and that there should be more rather than less overall socioeconomic equality in society.

By 1984, Costa Rica was surrounded by dictatorships and revolution. It began the decade as an island of peace and tranquility in the midst of a stormy sea. It is to be hoped that this small nation can maintain its political stability, economic progress, and international image as a constructive example to the rest of the world.

PANAMA

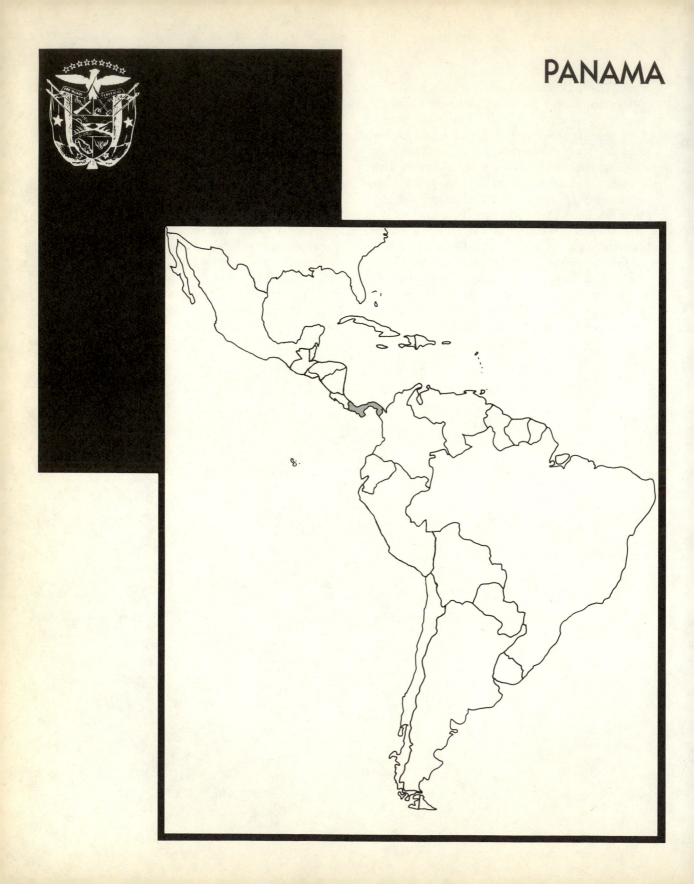

REPÚBLICA DE PANAMÁ

Land area   29,208 square miles

Population   Estimate (1985) 2,000,000
               Latest census (1980) 1,824,796

Capital city   Panamá 655,000

Percent urban   49

Birth rate per 1000   25

Death rate per 1000   5

Infant mortality rate   26

Percent of population under 15 years of age   38

Annual percent of increase   2

Percent literate   85

Percent labor force in agriculture   29

Gross national product per capita $2070

Unit of currency   Balboa

Physical Quality of Life Index (PQLI)   81

COMMERCE (expressed in percentage of values)

Exports

bananas	22
petroleum products	18
sugar	16
shrimp	14
coffee and cacao	5

Exports to

United States	20
Costa Rica	13
Ecuador	10
Colombia	5
Nicaragua	5

Imports from

United States	33
Venezuela	9
Mexico	8
Saudi Arabia	8
Japan	6
Trinidad and Tobago	4

Data mainly from the 1985 World Population Data Sheet of the Population Reference Bureau, the 1985 Britannica Book of the Year, and the 1985 World Almanac.

People elsewhere in the world commonly consider Panama as part of Central America, but residents of Panama and neighboring countries do not. This is because the historical background of Panama is different from that of any other Central American state. Until 1903 Panama was an outlying province of Colombia. Its present status as an independent country was achieved principally as a result of its strategic location. Since Vasco Nuñez de Balboa first revealed the geographical nature of the isthmus in 1513, human history in Panama has been concerned more with passage than with settlement. At one time or another most great maritime nations of the world have coveted this little strip of territory, and the forces that shaped the larger urban communities on either side of the isthmus have been international rather than local.

Present-day Panama differs from all other Latin American states in that is possesses no central area of concentrated rural settlement focusing on an urban core. The nucleus of the state is the city of Panama. The city is not, however, a product of the country. It came into existence,

and its importance has been maintained, because it controls a pass route. Meanwhile, the various small clusters of people along the Pacific Coast west of Panama City are distinctly minor ones. The greater part of Panamanian territory remains almost unoccupied.

The population of Panama was estimated at 2.1 million in 1984. About 49 percent of the people live in urban places, of which the largest is Panama City (655,000). Colón is the second city (117,000), and David (80,000) is third in size. The primacy of Panama City is expected to increase throughout the 1980s, as is the urban population generally, because of continued migration from the rural areas.

The racial composition of the country is highly diverse, for people from all over the world have been drawn together at the canal. It is estimated that 70 percent are mestizo or mulatto. About 13 percent are of African ancestry, mostly English-speaking descendants of people who came from Jamaica and Barbados to work on the canal construction. About 10 percent of the Panamanians are of European ancestry. Only 6 percent are In-

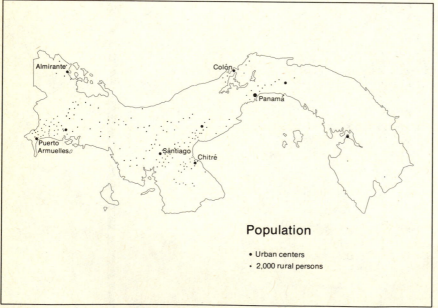

Population

• Urban centers
• 2,000 rural persons

FIGURE 9-1

dian, most of whom live along the Caribbean Coast east of the canal. Another 1 percent trace their origins to China, India, or the Near East.

The Land

Panama occupies a narrow, S-shaped strip of land varying in width from about 35 to 120 miles. Its mountain structures trend generally northwest to southeast. In the northwest the Cordillera de Talamanca, which forms the backbone of Costa Rica, extends into Panama. Near the border this range is surmounted by several inactive volcanoes, the highest of which is the Volcán Barú (11,400 feet). The range continues southeastward to the Gulf of Panama. Four peaks exceed 10,000 feet, but the general summit elevation is only about 3000 feet. The separate mountain block on the Azuero Peninsula also reaches an elevation of 3000 feet.

Two ranges of low mountains extend into eastern Panama from Colombia. One is the San Blas Range, which curves along the Caribbean Coast; the other is an extension of the Serranía de Baudó of Colombia which borders the Pacific. Both ranges reach elevations of about 3000 feet. Between them is a structural depression that is drained by rivers flowing into the Pacific and that is invaded in its middle portion by the Golfo de San Miguel.

There are only a few small areas of lowland plain. The largest is in the far southwest around Puerto Armuelles and David. Another is around the western side of the Gulf of Panama, south of the main range and north of the Azuero Peninsula. Here Penonomé, Chitre, and Las Tablas are the main towns. There are also small patches of lowland around the Laguna de Chiriquí in the northwest and the Golfo de San Miguel in the southeast.

The Isthmus of Panama forms a narrow land connection between the mountains of western Panama and the two systems of low mountains

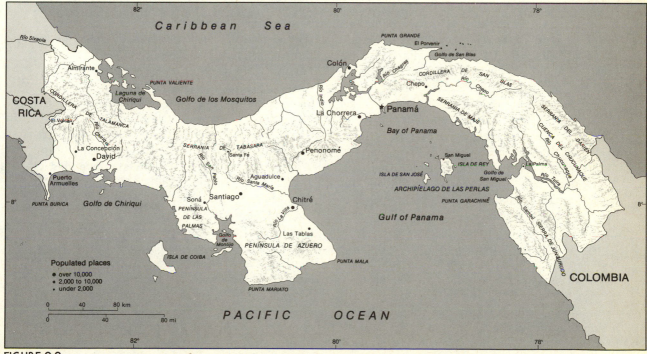

FIGURE 9-2

in eastern Panama. Here the strip of land is only 42 miles wide, and a passage from the Caribbean to the Pacific requires a climb of only 285 feet. The surface, however, is composed of steep, knobby hills. Because this narrow bit of land curves toward the northeast, in order to cross from the Caribbean to the Pacific one travels toward the southeast. At Panama City the sun rises over the Pacific Ocean.

Along the entire length of the Isthmus of Panama there is great contrast between the two sides. The Caribbean Coast is rainy, many places receiving more than 150 inches a year, concentrated in a season of especially heavy rains between May and December. At Colón the average annual rainfall is 128 inches. Such rainfall in a land where the temperatures average nearly 80°F results in deep decomposition of the rock and the growth of a luxuriant tropical rain forest. At the continental divide the moisture diminishes, and on the Pacific side it is distinctly less, although nowhere is it deficient. The annual rainfall averages between 60 and 120 inches. In western Panama there is a distinct dry season from January through March; in eastern Panama it is shorter, only February and March. As a result of the dry season the Pacific side of Panama has a seasonal forest, interrupted in places by wooded savanna. Spanish settlement went chiefly to the savannas, for not only did the Spaniards prefer open country, but also few parts of the world were as unhealthful, because of disease-carrying insects, as the Caribbean Coast south of Yucatán.

Along both sides of the isthmus there are many deep bays providing protected anchorages. Especially important is the shallow Gulf of Panama which separates the range of western Panama and the beginning of the Serranía de Baudó. On the Caribbean side near the opening between the ranges there are several small harbors, especially northeast of Colón. The problem of effecting a landing differs considerably between the Caribbean and the Pacific, for the maximum tide on the Caribbean shore amounts to no more than 36 inches, while that on the Gulf of Panama is 23 feet.

Passage Across the Isthmus

Although the Spaniards cruised along the Caribbean Coast of the isthmus as early as 1501, the strategic importance of this area did not become apparent until Balboa crossed it to the shores of the Pacific in 1513. In 1519 an expedition founded the first town of Panamá (now Panamá Vieja, five miles east of modern Panama City), and in 1524 Charles V of Spain ordered the first survey for a canal route. On the Caribbean Coast several small ports in succession were used as landing places. At first, Nombre de Dios was the main Caribbean port, and then Portobelo. Much later, when the railroad and then the canal were built, these places were all but abandoned in favor of Colón.

Panama City was of great importance to the Spaniards. From there expeditions set out to conquer the Pacific side of Central America as far north as Nicaragua and the entire west coast of South America. Although Lima became the primary settlement center of western South America, all lines of communication between Lima and Spain passed through Panama. Goods from the mother country were sent here in exchange for the treasures of the Americas. Then, as now, Panama City derived its importance from the convergence of overseas interests.

With the collapse of the Spanish Empire in the New World, Panama lost some importance for a time. Soon, however, the interest of a new maritime power became apparent. As a result of the War with Mexico, the United States extended its borders to the Pacific in 1848. Almost at once the world heard of the discovery of gold in California, and there ensued a rush to this new source of wealth. By many routes people not only from the eastern United States but also from Europe made the long trip to California. All the pass routes across Middle America were tried: Veracruz to Acapulco, the Isthmus of Tehuantepec, the rift valley of Honduras, the lowland of Nicaragua, and Panama. After 1850 increasing numbers of people arrived at Caribbean ports to make the difficult trip by stagecoach or horseback to Pan-

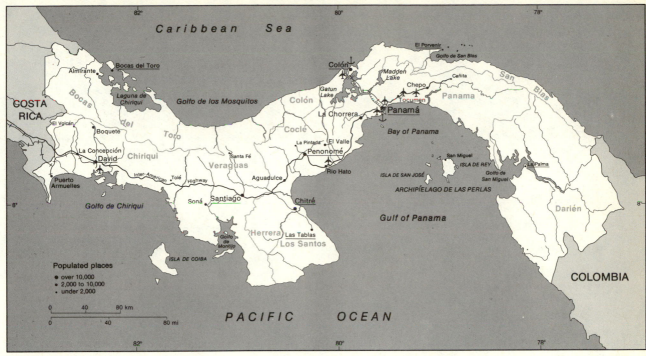

FIGURE 9-3

ama City. In 1855 America's first transcontinental railroad was completed from Cristóbal near Colón, to Ancón near Panama City.

The United States was not the only maritime power interested in Panama. The idea of providing a passage across the isthmus had such obvious justification that several nations gave serious consideration to canal projects. Many British activities in the Caribbean during the nineteenth century were influenced by a desire to control the strategic approaches to a canal. The French, successful in completing the world's first great canal, the Suez in 1869, were also first to undertake the work in Panama. In 1880 Ferdinand de Lesseps began the construction of a sea-level canal and continued until the collapse of his company in 1889. Diseases carried by tropical insects were most directly responsible for the failure, but poor management was also a factor.

The Spanish-American War made it clear to the United States that as a matter of defense alone the construction of a canal was of vital importance. Negotiations for the right to a canal zone were in process with the Colombian government when the people of Panama, believing the negotiations had failed and fearing that a Nicaraguan route would be selected, revolted and declared their independence. The United States, with conspicuous haste, recognized the government of Panama and was granted "in perpetuity" the use, occupation, and control of a zone five miles wide on either side of the canal, excluding the two cities on either end: Panama City and Colón. Within the Canal Zone the United States was granted "as complete authority as if it were under the sovereignty of the United States."

Work on the canal began in 1904. A widespread attack on the problem of sanitation preceded and accompanied the actual excavation. Workers on the canal were recruited not only in the United States and Europe, but also among the crowded populations of Jamaica and Barbados. A dam was constructed at Gatún, near Colón, im-

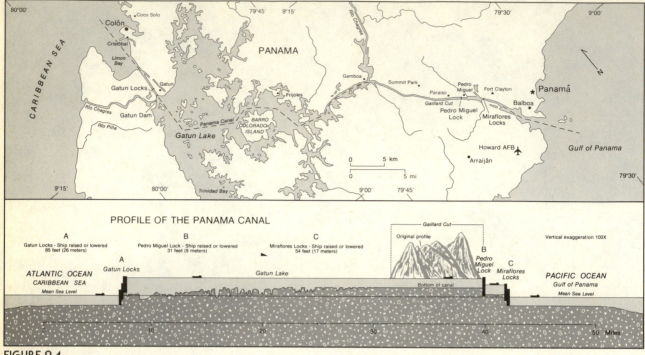

FIGURE 9-4

pounding the waters of the Río Chagres into what was then the world's largest artificial lake. Access to the lake, which is 85 feet above the Caribbean, is gained through locks. The Gaillard Cut carries the impounded water of Gatún Lake to the Pacific side, where two sets of locks permit descent to the Pacific Ocean. In August 1914, the first ship (the *SS Ancón*) passed through the completed canal.

From 1914 to 1977 more than 523,000 ships traversed the Panama Canal, making it one of the major "crossroads of the world." Individual ships set impressive records in passage. The longest ship was the oil-bulk ore carrier *Marcona Prospector* (973 feet), and the widest were battleships of the *Missouri* class (108 feet). The record cargo was 61,078 long tons (of coal) by the bulk carrier *Melodic*, and the fastest transit of the 50-mile canal was by the destroyer *USS Manley* in 4 hours and 38 minutes, only a fraction of the normal time required. The maximum toll paid (based on tonnage) was $68,499.46 by the passenger ship *Queen Elizabeth II* in 1977, and the smallest was 36 cents paid by the adventurer, Richard Halliburton, for swimming the canal in 1928.

By 1966 the canal was considered partially obsolete, and the existence of the Canal Zone was challenged by the people of Panama. The locks, 110 feet wide, could no longer accommodate the largest ships afloat. Each year some 15,000 ships pass through the canal, but the great bulk carriers and largest warships must be excluded. In 1966 the United States indicated its intention to build a new sea-level canal, and a commission was appointed to study alternative routes among at least 25 historically identified routes. At the same time, negotiations were undertaken to formulate a new treaty with Panama in which the perpetuity provision of the previous pact would be rescinded. The Canal Zone would be transferred to Panama in October 1979, and the canal and related facilities, in stages, by the

THE FAMOUS MIRAFLORES LOCKS NEAR THE PACIFIC END OF THE PANAMA CANAL.

year 2000. Ratification of treaties to this effect were completed in the United States and Panama by 1979.

The Central Corridor

A narrow strip of land that includes the Canal Zone, metropolitan Panama City, and Colón forms the core of the Panamanian nation. As a transportation corridor between the oceans, it is the focus of intensive commercial activity, making Panama City the "Hong Kong of the Western World." Business people from all over Latin America come to Panama to buy a wide variety of wholesale merchandise to be sold at retail prices in their home countries. Additional revenue ac-crues to Panama through payments made by the United States for rental of the canal and military bases within the area.

Panama City is the national capital, and government is the greatest source of employment. The city also has become a world-class financial center, being the site of more than 120 international banks. Since tax and labor laws, like banking regulations, are lenient, shipping companies of many nations have registered their vessels under the Panamanian flag. Thus, Panama ranks seventh among the world's nations in size of merchant fleet, although many of the ships registered in Panama City never dock at Panamanian ports. Panama City has likewise become an airline hub of the Americas. Its airport is served by 20 international airlines and is used by more than

1 million passengers annually. This volume of traffic has stimulated tourism, focusing on the city, the canal, and resort facilities along the Pacific Coast and on adjacent islands. The nation's fishing fleet is also based mostly in the Panama City area. Shrimp, lobsters, and fish abound in the Gulf of Panama and form a leading segment of the export trade. The world's largest shrimp farm is located in Aguadulce; with the addition of 25,000 acres of ponds, it is expected that Panama will produce nearly 37 million pounds of shrimp annually.

Colón is the Caribbean terminus of the Panama Canal, the Panama Railroad, and the Trans-Isthmian Highway. Its initial growth occurred during the canal construction period. Its survival as a major urban center is due in part to the establishment in 1948 of the Colón Free Trade Zone, now the largest in Latin America. There, raw materials are imported, processed, and exported as finished products tax-free, providing employment for the local population. In addition, a petroleum refinery was established nearby in 1962, importing crude oil from Venezuela and exporting refined products to other na-

tions. Refined petroleum is now the nation's leading export.

With the political and economic life of the nation revolving around the capital city and central corridor, the remaining 98 percent of the country has been little known by the average citizen and often neglected by the government. Although not centrally located, all of this outlying territory is commonly referred to as "the interior."

Settlement in the Outlying Parts of Panama

Eastern Panama is the "empty frontier" of the nation. A thin sliver of land along the Caribbean Coast is occupied by the San Blas Cuna Indians, who have maintained a substantial degree of political and cultural autonomy despite several centuries of contact with the outside world. The remainder of eastern Panama is largely unoccupied: a tropical, wet, forested wilderness including the Province of Darién which borders on Colombia. This area is so cloud-covered and rainy that only with the development of aerial photography employing "side-looking radar" has its

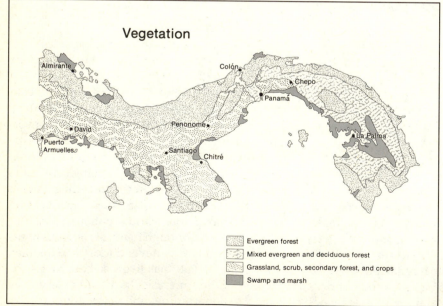

FIGURE 9-5

true physical configuration been revealed in recent years. The famous Pan American Highway has terminated here, with the unfinished segment known as the "Darién Gap." Along the completed portion of the highway there are commercial poultry farms near Panama City, as well as fields of maize and sugarcane. Farther eastward the land is devoted primarily to pastures and small-scale lumbering operations. There is some commerical fishing and subsistence agriculture around the Golfo de San Miguel, but most of Darién Province is "empty country."

Western Panama is generally considered to be the most typical, most traditional part of the country. It is also the primary agricultural and ranching zone. The principal commercial crops are bananas and coffee. Bananas are grown in two areas, both of which include extensive plantations that were developed by the Chiriquí Land Company, a subsidiary of the United Fruit Company. The first plantations were laid out in 1880 when banana planting spread along the Caribbean lowlands from Costa Rica. The chief port in Panama was Almirante on the Laguna de Chiriquí. Panama disease wiped out these plantations,

and the company shifted its operations to the Pacific Coast, inland from Puerto Armuelles. In the 1950s the plantations of this area produced about 60 percent of Panama's exports. Since 1960, however, Panama disease has been controlled largely through flood fallow, by which the land is kept under water for several months between plantings. The plantations inland from Almirante are producing again, and bananas are also grown commercially by small-scale independent farmers all along the Caribbean Coast.

Three small areas of commerical agriculture are used for coffee production. These are all on steep slopes of the mountains in western Panama, on slopes of the Volcán Barú, on the Pacific slopes north of Penonomé, and on the higher mountain slopes of the Azuero Peninsula.

Around David, and in a triangular zone between Penonomé, Santiago, and Las Tablas, food is produced to supply the urban markets of the central corridor. There are two intensively cultivated, irrigated rice-producing areas, one near David and the other near Penonomé. From these areas, where modern technology is used, yields are such that Panama has become self-

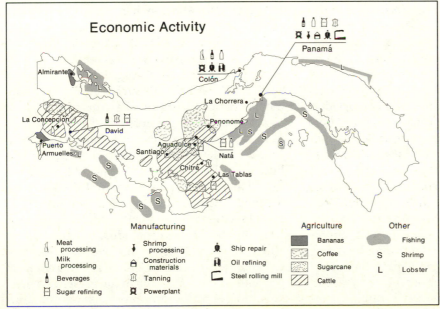

FIGURE 9-6

sufficient in this basic food. In the same areas there are also sugarcane plantations. Throughout the wooded savannas of the Pacific lowlands, there are large areas devoted to cattle-raising.

Also near David, the 300-megawatt Fortuna hydroelectric project has been installed. This facility was designed to double the nation's output of electricity and thus contribute significantly to its economic development. Farther westward, a major oil terminal has been built at Puerto Armuelles. Here Alaskan oil is transferred from supertankers to smaller oil carriers for passage through the Panama Canal or pumped via an 80-mile pipeline to a new port at Chiriquí Grande on the Caribbean for reloading.

The Challenge of the 1980s

During the 1980s, the Republic of Panama faces a number of major problems and opportunities. How these are dealt with will influence the lives of Panamanians for many years to come.

To an extraordinary degree the history of Panama has revolved around the central corridor of transportation between the two oceans. With acquisition of the Canal Zone this focus can only be intensified. During U.S. control some portions of the Zone were preserved essentially in their natural state, and none was open for occupation by the Panamanians. Now, the question for Panama is how to incorporate the Zone into its national territory, with due regard for ecological considerations. Expansion of the present canal offers some prospect, as does the construction of a totally new sea-level canal. Yet, these alternatives are enormously expensive, and external sources of funds are likely to be limited, since external control of the completed facility is doubtful. Meanwhile, efforts have been made to expand transportation across the isthmus by a modernized system of conveyance, including highway, railroad, and pipeline. As in the past, the central corridor across Panama appears likely to receive primary attention in the nation's future. Moreover, domestic events in Panama will continue to be of international concern, because of the country's strategic location.

The extraordinary success in linking the oceans via the Panama Canal has been matched by the failure to achieve a comparable line of terrestrial communication between the American continents. As early as 1884, a proposal was made to the U.S. Congress for the construction of a Pan American Railroad, and some work was subsequently initiated to achieve that goal. However, by 1925 plans were formulated for a Pan American Highway, and the railroad project was abandoned. One segment of the Pan American system, known as the Inter-American Highway, extends from the Rio Grande at Laredo, Texas, to the Bridge of the Americas over the Panama Canal. In eastern Panama sporadic construction has been followed by near abandonment of the Pan American Highway project. By 1980 the route could be traveled by passenger car 100 miles eastward from Panama City, still 30 miles short of the border of Darién Province.

The Political Situation

In Panama there seldom has been any important development, economic or political, that was purely indigenous. In 1739, when the Vice Royalty of New Grenada was established with its administrative center at Bogotá, Panama was a part of it. After Bolívar's successful effort to gain independence from Spain, Panama was included as a part of Gran Colombia. In 1841, when the new state was breaking apart, Panama declared its independence, but troops from Colombia had little difficulty in recapturing this outlying territory. Panama again seceded in 1853 but soon rejoined Colombia. During the period from 1898 to 1903, when British, French, and U.S. interests were promoting the canal idea, Panama was in almost constant disorder and revolt. The declaration of independence in 1903 was different from previous movements only in the presence of U.S. naval forces and in the prompt recognition afforded the new state.

To the extent that national coherence has developed in Panama since 1903, it has focused on reducing the country's dependence on the United States. The formulation of a more sophisticated state-idea has yet to be accomplished. In 1968 the elected President of Panama was overthrown, the National Assembly was dissolved, and political parties were abolished by National Guard forces under the direction of Brigadier General Omar Torrijos. The Torrijos era was characterized by increased nationalism, intensive negotiation over control of the canal and Canal Zone, and increased attention to conditions of the rural Panamanian. Torrijos remained the Chief of State throughout the treaty negotiations, until 1978.

The Presidential elections in 1984 were the first to be held in 16 years and resulted in a victory for 82-year-old Arnulfo Arias. He likewise had been elected in 1940, 1950, and 1968, only to be overthrown by the military long before the completion of his terms of office. The Panamanian electorate obviously is not easily discouraged in its quest for democracy.

THE ANTILLES
AND THE GUIANAS

THE ANTILLES: INTRODUCTION

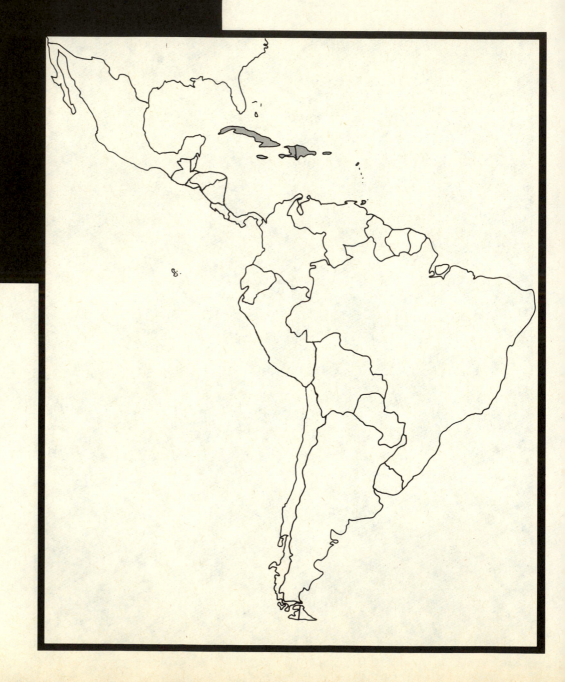

The Caribbean area is a distinctive subregion of Latin America. The languages of four European countries (Spain, France, Britain, and the Netherlands), as well as many local dialects, are spoken here. The economies range from petroleum-rich in Trinidad and Tobago to starkly poor in Haiti. The Antilles include all the islands of the Caribbean that lie between the southeastern part of North America and the northern part of South America.[1]

Physical Nature of the Islands

The islands of the Antilles are diverse in geologic structure and in their present surface features. The island chain might seem to be formed by the crests of one partly submerged mountain arc. Actually, several mountain systems are involved, and the islands are at various stages in the process of growth and denudation.

One of the principal mountain systems that produce the Antilles is the so-called Central American-Antillean system which extends from the ranges of Central America in the west to Anegada Passage east of the Virgin Islands. The main axes of mountain growth have the shape of a two-pronged fork. The handle of the fork is in the east, extending in a single line of uplift from the Virgin Islands through Puerto Rico and the eastern part of Hispaniola[2] to the Cordillera Central of the Dominican Republic. Here the whole system reaches its greatest elevation, exceeding 10,000 feet. In western Hispaniola the two prongs of the fork emerge, one forming the southern peninsula of the Republic of Haiti, the other the northern peninsula. The southern range continues under the Caribbean, emerging as the Blue Mountains of Jamaica and several banks and miniature islands between Jamaica and the northeast corner of Honduras.

---

[1] The name *Antilles* is derived from *Antilia*, the mythical islands that appeared on maps of the Atlantic Ocean before Columbus. Some writers prefer to call them the West Indies, an official name used by the British Colonial Office, but the French and Dutch use the name Antilles. Sometimes the Bahamas are considered to be separate from the Antilles. In this book the Antilles include two groups of islands: the Greater Antilles and the Lesser Antilles. For convenience of presentation, here Trinidad is considered along with the former group; the Bahamas with the latter.

[2] The U.S. Board on Geographic Names has adopted the term *Hispaniola* to apply to the island occupied by Haiti and the Dominican Republic. There is some historical precedent for this, although European writers refer to the whole island as Santo Domingo.

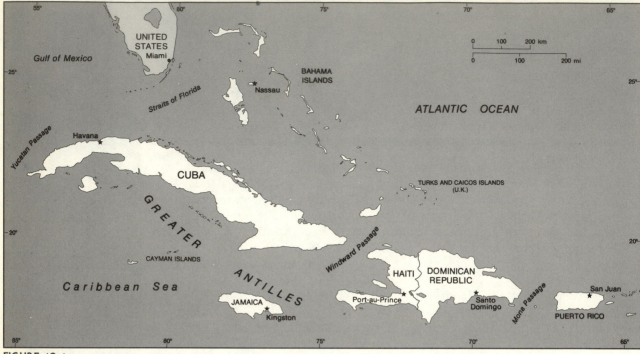

FIGURE 10-1

The east-west ranges of Honduras and the central part of Guatemala form the western extremity of this system. The nothern range, leaving the northern peninsula of Haiti, appears again in the Sierra Maestra in southeastern Cuba. It also is submerged under the waters of the Caribbean, appearing above the surface only in a few small islands such as Little Cayman and Grand Cayman. It reappears in Belize, north of the Gulf of Honduras, and extends westward through Chiapas and Oaxaca in Mexico to the shore of the Pacific.

North of the Central American-Antillean system are several areas which, although separated by stretches of water, are related structurally. These are the limestone platforms including the Peninsula of Yucatán, the main part of Cuba, most of the Bahamas, and the Peninsula of Florida.

From either end of the Central American-Antillean system, mountain chains of volcanic origin extend southeastward to connect with the continent of South America. In the west this mountain connection forms the backbone of Costa Rica and Panama, and the chain stands high enough to make the land continuous. In the east the mountain connection consists of a string of volcanic islands, with deep passages between them, known as the Windward and Leeward Islands.

The Antilles lie within that portion of the tropical seas in which coral reefs can form. Coral survives only where the ocean water is free from silt and where its temperature does not drop below 68°F. The corals attach themselves to the shores of islands where these conditions exist and form fringing reefs. According to the coral-reef theory set forth by Charles Darwin and William Morris Davis, the reefs are built higher and higher as an island sinks, with the result that they remain within the zone reached by the salt spray of ordinary waves; but submergence of an island reduces the land area and leaves the coral growth as a barrier reef, some distance offshore. In some cases the original island may be entirely submerged, leaving the coral formation as a circular atoll, with a shallow lagoon in its center. Where the islands have been raised rather than lowered, the reefs form a sort of collar of limestone, elevated above the sea, as has occurred on Jamaica. All stages of island growth and destruction, and of the development of coral reefs, can be observed in the Antilles.

Despite the existence of all these varied island forms, it is possible to group the Lesser Antilles into two categories. There are islands that are relatively low and include considerable areas of fairly level limestone reefs. These are found mainly on the outer or eastern side of the island arc. In contrast to these low-lying islands are the mountainous ones, some still in the process of growth. The principal islands of the Lesser Antilles, classified in this way, are presented in the following table.

LOW-LYING AND MOUNTAINOUS ISLANDS OF THE LESSER ANTILLES
(Maximum elevation in feet)

Low-lying Islands	Mountainous Islands
Anguilla (213)	Saba (2820)
Saint Martin (1360)	Saint Eustatius (1950)
Saint Bartelemy (992)	Saint Kitts (4314)
Barbuda (115)	Nevis (3596)
Antigua (1330)	Redonda (1000)
Grande-Terre (eastern part of Guadeloupe) (450)	Montserrat (3002)
	Guadeloupe (4869)
Désirade (912)	Îles des Saintes (1036)
Marie Galante (672)	Dominica (4747)
Barbados (1100)	Martinique (4428)
	Saint Lucia (3145)
	Saint Vincent (4048)
	The Grenadines (series of rocky islands, highest one about 1000)
	Grenada (2749)

Source: C. Schuchert, *Historical Geology of the Antillean-Caribbean Region* (New York, 1935).

Several of the mountainous islands are volcanic, with volcanoes that are still periodically active. Two of these volcanoes have erupted violently during the present century. Mount Sou-

frière on Saint Vincent killed more than 2000 persons when it erupted on May 7, 1902. The eruption of Pelée on Martinique the next day covered surrounding parts of the island with a deep layer of ash and destroyed the city of Saint Pierre with the loss of 30,000 lives. Pelée's ash has a mineral composition that provides little plant food; hence, the area covered by ash remains even today barren wasteland. The mass of material ejected was estimated to equal in bulk that of the entire island of Martinique.

Of different geologic origin are the continental islands related to South America. The most easterly of these is Barbados. This island, 21 miles long by 14 miles wide, is composed of a gently rolling limestone tableland, with a maximum elevation of 1100 feet. It is formed by an upraised portion of the continental shelf, and it stands on a wide platform only slightly submerged and bounded in places by barrier reefs.

Closer to the South American coast, the islands are even more directly related to mainland structures. The little island of Tobago is formed along a mountainous backbone, the highest elevation of which is 1900 feet. Trinidad has a range of mountains along its northern side which reaches 3000 feet in elevation, whereas the southern part of the island is composed of two hilly belts with mangrove-filled bays along the coasts. There are no corals on Trinidad, for this entire coast is bathed with silt from the Orinoco. The islands of Margarita and Tortuga off the Venezuelan north coast are part of the Caribbean coastal range. The three Dutch islands of Bonaire, Curaçao, and Aruba are formed of ancient crystalline rocks, similar in geologic structure to the Guajira Peninsula of Colombia.

Climate

In places like the Antilles one finds the truly "temperate" climates of the world. These islands are bathed by currents of warm ocean water and are swept by the easterly trade winds of the open sea. Temperatures are moderately high and vary little from season to season. Extremely high temperatures, such as are experienced in the Middle West of the United States, never occur in the Antilles; but neither are the low temperatures characteristic of mid-latitude winters experienced.

The easterly trade winds which blow day and night throughout the year produce great differences in rainfall on the eastern and western sides of the mountainous islands. From the warm ocean the air picks up large quantities of moisture, so that a very slight rise of the air with consequent cooling results in the formation of towering cumulus clouds and heavy downpours of rain, mostly of short duration. The rains are heaviest on the windward sides of the islands. In Jamaica, for example, the average annual rainfall at a station on the northeast side of the Blue Mountains is 222 inches; at Kingston on the south side of the island, some 30 miles distant, the average rainfall is only about 30 inches. The eastern sides of the islands, too, are exposed to the highest waves, so that especially for colonial sailing ships the protected western sides offered the safest anchorages. Almost all of the larger towns on the islands are on the leeward sides.[3]

Throughout most of the Antilles there are two rainy seasons and two dry seasons. The first usually comes in May, though sometimes in June or July; the second arrives in October or November. Trinidad, however, has only one rainy season, from June to December. Rain occurs in the form of violent showers, followed by rapid clearing. The showers come at shorter intervals as the day progresses, followed by clear skies at night and in the early morning.

The Antilles are occasionally disturbed by climatic violence. In this area of the world tropical cyclones, or hurricanes, are common occurrences. The hurricane season begins in August and lasts through October. Storms originate off the coast of Africa and sweep westward toward the Lesser Antilles, veering toward the north as they proceed. The island of Trinidad never experiences these violent storms, and the southern members of the Lesser Antilles only rarely, but the northern Antilles are frequently traversed by them. The hurricanes follow two chief tracks. One crosses the Caribbean to the Yucatán Channel and thence proceeds across the Gulf of Mexico, where the storms either ravage the Gulf Coast of the United States, rapidly losing violence as they proceed into the interior, or curve eastward across Florida. The other track follows the Lesser Antilles, passing east of Puerto Rico, across the Bahamas to the East Coast of the United States, after which it curves again toward the east, following the Gulf Stream.

A hurricane is a great whirl of air, like a dust whirl that forms over a country road or field on a hot summer day, but on a vastly larger scale. In the Northern Hemisphere it always rotates in a counterclockwise direction. Hurricanes cover hundreds of miles of territory, and even those places that do not experience the destructive violence of the storm's center are visited by high winds. Hurricanes also bring extremely heavy downpours of rain which may cause additional damage.

Course of Settlement

It was to the Antilles that the Spaniards came on that momentous first voyage of Columbus in 1492. Most scholars identify the land that Columbus first reached as the island of San Salvador, also known as Watling Island. To the everlasting confusion of succeeding generations, Columbus, believing that he had reached the eastern coast of Asia, used the term *West Indies* and designated the native inhabitants as *Indians*.

Columbus and the other Spaniards who came to the Antilles in the early years of the conquest were primarily interested in finding gold. The Indians had never placed much value on the element, although they used gold ornaments, and

---

[3] In this connection, it should be noted that the designation of the northern group of Lesser Antilles as "Leeward Islands" and the southern group as "Windward Islands" has no basis in terms of wind direction. The prevailing trades of this part of the world come from the northeast and east.

when the Spaniards pressed them to reveal the sources of gold they could only point to places where nuggets had been picked up. There was some gold in the stream gravels in the central part of Hispaniola, and nuggets were actually found —enough to lend support to stories of great wealth that spread in Spain.

The Spaniards established several settlements in the Antilles during the first two decades of the conquest. In 1496 Bartolomeo Columbus, brother of the Admiral, founded Santo Domingo on the southeastern shore of Hispaniola. This is now the oldest permanently occupied city in the Americas. Santo Domingo was the earliest Spanish primary settlement center, from which colonists were sent out to various places around the Caribbean. In 1509 a town was founded on Puerto Rico, and a colony was established on Jamaica in the same year. Havana was founded in 1515, and this port became the usual point of departure for the voyage back to Spain. From Havana, Cortés began his expedition which resulted in the conquest of Mexico.

Indians of the northern Antilles were the first to feel the destructive effects of the Spanish conquest. Once friendly, they soon turned hostile, for despite the efforts of the king and queen of Spain, the natives were forced into hard and unaccustomed labor in the mines. New diseases introduced by the Europeans had a devastating effect. When the Spaniards arrived in Hispaniola, they found more than a million native inhabitants on the island, and at that time only Cuba was thinly populated. Fifty years later there were just small remnants of the original Carib and Arawak peoples, and these were in isolated places in the mountains.

Sugarcane

When the ready sources of gold were depleted, some other basis for the economy had to be found. Early in the sixteenth century, Spanish leaders attempted to import slaves from Africa and begin the cultivation of sugarcane, which they had been producing for many years in the Canary Islands. However, the development of the Antilles as a major source of sugar did not come until a century later. By then, the main stream of Spanish settlement and the focus of Spanish interest had shifted to Mexico and Peru, and the older Antillean settlements were neglected.

The Portuguese in Brazil were the first to build a large-scale sugarcane plantation economy with African slaves. Before the sixteenth century, sugar had long been known in Europe as an expensive luxury, sometimes even prescribed in small quantities as a medicine. When the merchants in Amsterdam began to supply sugar to Europe from Brazil at much lower prices than had been possible before, the sale of sugar expanded rapidly. The Portuguese enjoyed their first period of speculative production and for nearly a century had a virtual monopoly on the rapidly growing market. Then in 1624 the Dutch invaded and occupied parts of northeastern Brazil, and for 30 years they held the entire coast from the Rio São Francisco to the mouth of the Amazon. During this time they learned the techniques of planting cane, extracting the juice, and preparing raw sugar for shipment. When they were driven out, the Dutch promptly occupied parts of the Guiana Coast and islands in the Lesser Antilles that had been neglected by the Portuguese and Spaniards. From the Dutch, sugar technology was passed on to the French and British, and along with it the importation of African slaves. Little by little during the latter part of the seventeenth century, the center of sugar production shifted from the northeast of Brazil to the Antilles.

Most of the islands changed hands several times. In 1655 the British launched an attack on the Spanish colony of Santo Domingo, which was repulsed. To save the expedition from complete disgrace, the British forces landed on Jamaica and successfully expelled the small Spanish garrison from most parts of the island. The Spaniards took refuge in the Blue Mountains and continued to raid the British settlements until 1660. Spain acknowledged British ownership of

Jamaica in 1670. During the century that followed, French Saint Domingue, in the western part of Hispaniola, and British Jamaica became the leading sugar producers. Trinidad remained the least developed. When Britain seized Trinidad in 1797, the island was still largely covered with virgin rain forest.

The End of Slavery

The sugar economy of the seventeenth and eighteenth centuries was based on the institution of slavery. This meant that sugarcane planting was a rich man's business, for the establishment of a plantation, the construction of sugar mills, and the purchase of slaves all required large capital investment. Profits for the plantation owners were in most places enormous, but life on a sugarcane island was anything but pleasant. Slave labor was inefficient, restless, and resentful. Not only were slaves expensive to buy, but they also had to be fed, clothed, and housed. Many of the smaller islands became totally dependent on imported foods, such as salt cod from New England. Only in Saint Domingue and Jamaica were slaves permitted to use land to grow their own food. On an island such as Barbados or Antigua, where the entire surface was suitable for cane, there was no land to waste on food crops.

During the eighteenth century, explosive ideas of equality before the law were being formulated. Many mulattoes of Saint Domingue, who were children of French planters and African women, were sent to France for an education. In Hispaniola and Jamaica educated mulattoes, rejected by both pure African and pure European groups, developed an undercurrent of revolt. It was a question whether the French colony or the British colony would most likely suffer a slave uprising. Meanwhile, in the home countries there were strong antislavery movements, and in one way or another the institution of slavery was doomed. In 1808 the British Parliament passed the Abolition Act, which declared the slave trade illegal. Other countries passed similar legislation: the United States in 1808, the Neth-

erlands in 1814, France in 1818, and Spain in 1820. Yet, illegal trade in slaves continued, and more slaves were probably taken out of West Africa for shipment to the Americas after 1808 than in all the years before that date.[4]

Slaves were given their freedom in Antigua in 1834 and in the remaining British possessions in 1836; other parts of America followed this lead over a period of 50 years.

The story in each of the islands of the Antilles was somewhat different as the plantation owners attempted to readjust the basis of the economy. The one thing the newly freed slaves most wanted to avoid was working on the sugarcane plantations, for this kind of employment was so clearly associated with slavery. On various islands, however, the opportunities to avoid such work differed. In the former Saint Domingue, where the slaves were successful in staging a revolt against their French masters and establishing the independence of Haiti in 1804, the former slaves fled to the mountainous interior, and sugar production collapsed. In Jamaica, which had plenty of unoccupied land suitable for peasant subsistence farming, the sugarcane plantations were also ruined. Africans left the cane fields and established themselves as squatters in the interior of the island. In Barbados, things were different. There was no place to go, no land not being used for sugarcane, no empty mountains in which to hide. During the decades after emancipation, sugar production in Barbados actually increased, as wage workers proved to be more efficient than slaves. In Trinidad, on the other hand, although there was plenty of unoccupied land, the governor levied a tax on any person settling on Crown lands. This kept the freed slaves at work on the cane plantations, but since there was a chronic shortage of workers on this island, the government supported the immigration of contract laborers from India.

Around 1900 the technology of producing

---

[4] J. H. Parry and P. M. Sherlock, *A Short History of the West Indies* (Macmillan Publishers, Ltd., London, 1956).

cane sugar was changed by the development of more efficient machinery for extracting the juice. This initiated the rise of Cuba to world leadership, with the aid of capital investment from the United States. For large-scale operations the island of Cuba, long a neglected Spanish colony, offered unusual advantages. After the Spanish-American War (1898), which gave Cuba its political independence (1902), this part of the Antilles forged ahead rapidly in sugar production, leaving the Lesser Antilles far behind.

The United States established quotas as early as 1934 to give a share to both cane and beet sugar in the American market and to favor sugar production in places such as Cuba and the Philippines. After 1959 when Cuba entered the Soviet sphere, its quota for the American market was reassigned to more friendly nations. Most Cuban sugar is therefore bartered to the Soviet Union, with the remainder sold on the international market.

Prospects

Diversity in terms of historical background, physical and cultural conditions, and economic potential among the Antilles islands is obvious. Some generalizations can be made, however, concerning the entire area. On most of the islands the population density is high relative to the very limited natural resources. Unemployment and dissatisfaction are widespread. Emigration is therefore common from island to island, to Europe, and to the United States. Those who remain tend to live in poverty more severe than that experienced in most other parts of Latin America.

Politically, the United States for a time sought to reduce its influence in the area by maintaining a low profile. Later, it reversed that position, recognizing that stability in the Antilles is an essential component of its own national security. Meanwhile, the influence of Cuba expanded, its government serving as an agent of the Soviet Union in seeking to promote revolution and further reduce U.S. influence. Two other nations, Venezuela and Mexico, have entered the political scene as major suppliers of petroleum to this area which so conspicuously lacks energy resources.

Within an environment of social stress and political instability, an array of newly independent states has been created, including single islands so small that their viability as self-governing nations must be seriously questioned. Political independence does not in itself secure economic prosperity, and the future of most islands of the Antilles therefore remains in doubt. This part of Latin America, however, will certainly play a role in world affairs quite out of proportion to its relatively meager land area.

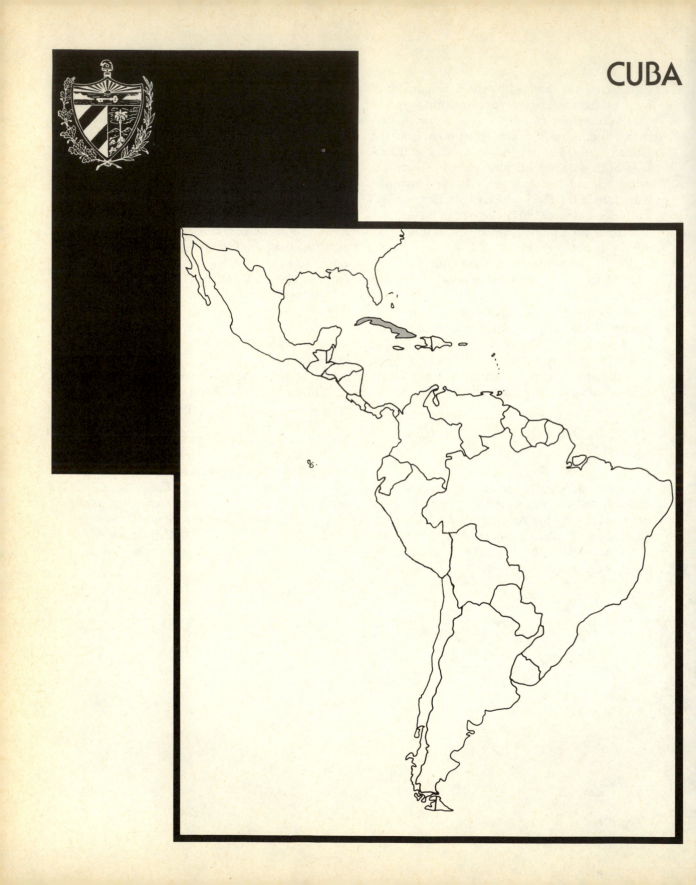

CUBA

REPÚBLICA DE CUBA

Land Area   44,827 square miles

Population   Estimate (1985) 10,100,000
Latest census (1981) 9,706,364

Capital city   Havana 1,952,300

Percent urban   70

Birth rate per 1000   17

Death rate per 1000   6

Infant mortality rate   16.8

Percent of population under 15 years of age   29

Annual percent of increase   1.1

Percent literate   96

Percent labor force in agriculture   34

Gross national product per capita $1410

Unit of currency   Cuban Peso

Leading crops in acreage   sugarcane, citrus, tobacco, rice

Physical Quality of Life Index (PQLI)   92

COMMERCE (expressed in percentage of values)

Exports

sugar   77
minerals   6
tobacco   2

Exports to		Imports from	
USSR	76	USSR	68
East Germany	3	East Germany	4
Bulgaria	32	Czechoslovakia	2
Czechoslovakia	2		

Data mainly from the 1985 World Population Data Sheet of the Population Reference Bureau, the 1985 Britannica Book of the Year, and the 1985 World Almanac.

Cuba, a Socialist country in the Caribbean, lies only 90 miles from the United States. Largest of all the islands of the West Indies, it was once a haven for drinking, gambling, and vacationing Americans. For nearly 40 years tourist dollars poured into Havana and offered Americans temptations they could not find at home. That was before Fidel Castro. Until 1959, Anglo-America exerted a strong cultural influence on Cuba. Many Americans think that the Cubans should have been grateful for the benefits of close attachments to the United States and for the prosperity that resulted from free access to the world's greatest market for sugar. However, many Cubans followed their great nineteenth-century patriot, José Martí, in demanding freedom from outside interference from whatever source. It was Martí who exclaimed: *"Nuestro vino es agrio, pero es nuestro vino."*[1] Because of Cuba's location, the principal source of foreign interference and domination in Cuban affairs has been the United States.

For many years Cuba was quite neglected as the main course of Spanish conquest was directed elsewhere. Except for a few North Ameri-

cans who coveted the tropical island to the south for its strategic importance, or as a potential addition to the list of slave states, Cuba remained of little interest to a people who were facing westward rather than southward. A Cuban insurrection against Spanish rule in the 1870s aroused little desire in the United States for the rescue and liberation of an oppressed people. Just 20 years later, however, another insurrection was met with a wave of sentiment in the United States demanding military intervention for the purpose of bringing freedom to the Cubans. The result was the Spanish-American War of 1898, by which the United States gained control of Cuba, Puerto Rico, Guam, and the Philippines.

As a result of the large subsequent investment of U.S. capital in Cuban sugar production and the services of technicians, Cuba could produce sugar at lower cost than most competing parts of the world. The extensive pre-industrial estate, with inefficient machinery and unskilled supervision, was replaced by a large-scale operation carefully managed by experts. By 1946 more than a third of all the agricultural land in Cuba was owned by about 900 large corporations, some with more than 600,000 acres. U.S. corporations owned 40 percent of the sugarcane lands, plus 90 percent of the public utilities and mines, and 50 percent of the railroads.

---

[1] "Our wine is bitter, but it is our wine." Martí has many followers, including Castro, who insist on the right to chart their own destiny, even if the results are sometimes bitter.

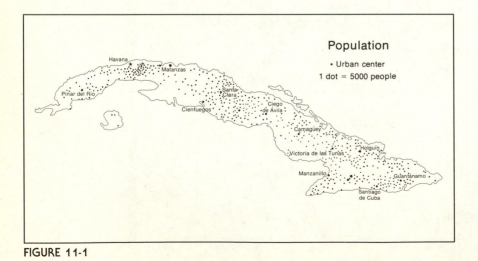

FIGURE 11-1

All was not well in Cuba. Some Cubans were wealthy, as a direct or indirect result of sugar production, but many more were poor. Workers were paid good wages during the harvest season from December to June, but during the rest of the year most were unemployed. In Havana there were luxury hotels and beautiful homes, but there were also crude shelters made of any available material. There were families living in poverty and ill-health, with no income even to purchase food. In the rural areas less than 10 percent of the homes were supplied with electricity or running water. Order was maintained by a dictator, but rumblings of revolt were heard from Cubans who had been forced into exile and from students in the universities. The revolution led by Fidel Castro gained control of Cuba in 1959 and resulted in radical changes.

Cuba has a population exceeding 10 million. The population density is not nearly as great as that of some of its Antillean neighbors, but the capital city, Havana, by 1985 was nearly 2 million. Some 66 percent of the Cubans are of unmixed European ancestry, while 12 percent are black and 21 percent are mestizo. The remainder includes Chinese, Filipinos, and other Asians. Literacy reached 96 percent in 1982, one of the highest rates in all of Latin America.

The Land

More than half of the land area of the Antilles is in Cuba. The 44,000 square miles of territory extend for 785 miles in an east-west direction, with a width that varies from 25 to 120 miles. At least half of the area is level enough for mechanized agriculture. The soils of Cuba are well adapted to a variety of crops, of which sugarcane is only one. The "temperate" tropical climate, with no frosts and adequate, well-distributed rainfall, is ideal for plantation agriculture. Hence, over the years Cuba has remained basically an agricultural country. The island is well drained and is swept by easterly winds which reduce the problems of sanitation and insect-borne diseases. For Spaniards of the colonial period, these advantages were outweighed by the disadvantages of a sparse native population; for North Americans in the twentieth century with money to risk on speculation in tropical plantations, the Cuban land had great appeal.

Cuba is surrounded by 1600 cays and islands, the largest of which is the Isle of Pines, now called the Island of Youth. Before Castro's revolution, this island was the location of *El Presidio*, a brutal prison where Castro himself was imprisoned for 20 months in 1954–55. About the size

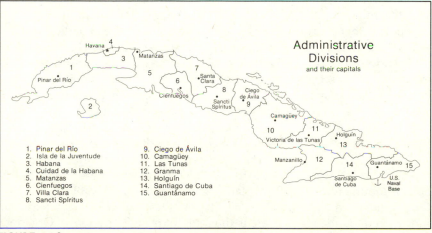

FIGURE 11-2

A MESSAGE AT A CONSTRUCTION SITE IN ALAMAR, CUBA.

of Rhode Island, the island has approximately 55,000 permanent residents, plus another 30,000 teenage students. One-third of the teenagers are from foreign countries, such as Nicaragua, the Cape Verde Islands, Ethiopia, Mozambique, and Namibia. They, along with young Cubans, are learning the fundamentals of mathematics, physics, chemistry, and Marxist ideology. The school routine is quite military in nature except for time spent working in the fields and grapefruit groves.

Surface Features

Only about one-fourth of Cuba is mountainous. The most rugged area is at the southeastern end. West of Guantánamo Bay and north of Santiago de Cuba, the steep slopes of the Sierra Maestra overlook the sea, rising to elevations of nearly 8000 feet. The northern side and eastern end of the Sierra Maestra are bordered by the Guantánamo Valley (Valle Central), a gently rolling hill country that leads out to the head of the bay. East of Guantánamo Bay stands a rough, stony highland, deeply dissected by streams and including few patches of flat land. In these highlands are found the Cuban manganese, nickel, chromium, and iron ores formerly exploited by North American companies; but from an agricultural or pastoral perspective, the district is of little value.

There are two other small mountainous areas. Near the middle of the island are the Trinidad Mountains, with maximum elevations just over 3700 feet. In western Cuba, west of Havana, there is the long, narrow Sierra de los Organos,

FIGURE 11-3

reaching a maximum elevation of about 2500 feet. At the western extremity of Cuba is also the rugged hill country known as the Guaniguánicos, which is of special interest because of its extraordinary scenery. This is a region formed late in a cycle of karst erosion, in which steep-sided limestone blocks, honeycombed with caverns, stand like great castles above irregular-shaped, flat-bottomed valleys.

The remaining three-quarters of Cuba is composed of gentle slopes. Partly on limestone and partly on other types of rock, a series of terraces has been formed, now somewhat dissected by short streams so that in certain localities the terrain is moderately hilly. Along many sections of the coast the land is swampy, but most of Cuba is well drained. Where terraces border the sea, the coast is cliffed, and there are many deep, pouch-shaped bays which form ideal natural harbors. Outstanding are the harbors of Havana, Santiago de Cuba, and Guantánamo.

Climate and Vegetation

Because of its generally moderate relief, Cuba shows less contrast between windward and leeward sides than do any of the other Antilles. The heaviest average rainfall occurs in the western part of the island, for this section lies closest to the hurricane track. No part of Cuba is deficient in moisture, and a notable feature of the rainfall is its dependability during the critical agricultural season from May to November. Temperatures normally are quite uniform, with no great extremes. During the summer, however, temperatures as high as 100°F are occasionally recorded. Although they never bring freezing weather to Cuba, cold air masses of the winter season do bring decidedly lower temperatures.

When the Spaniards first came to Cuba, they found the island covered in places with semideciduous forest and elsewhere with a patchwork of scrub woodland and grass. Leo Waibel recon-

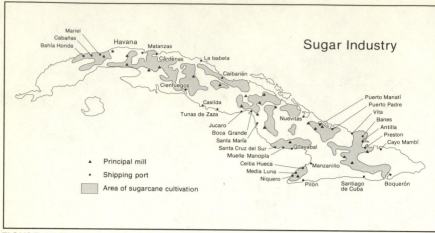

FIGURE 11-4

structed the vegetation of Cuba by a study of place names, many of which use the word *sabana,* suggesting the existence of open patches in the midst of the woods.[2] These grassy openings are not characteristic tropical savannas, for the grass is not tall. Waibel described them as a thorn-scrub steppe, in which pine and palmetto are intermingled with short grass. Rapid runoff on hard-packed soil, rather than lack of rainfall, accounts for the parched appearance of certain parts of the island.

The Pattern of Settlement

The settlements founded by the Spaniards in the sixteenth century were primarily for defense. The sparse Indian population was soon almost completely eliminated by epidemics, and there was little to attract Spaniards to the island. However, the fine harbor of Havana, with the commanding heights at its entrance, offered an opportunity for the construction of a strong naval base. Havana was established to guard the approaches to the Gulf of Mexico.

[2] Leo Waibel, "Place Names in the Reconstruction of Original Vegetation of Cuba," *Geographical Review* 23 (1943):376–396.

The long period of Spanish control produced only one main cluster of rural population. This was along the northern coast, inland from Havana, and lying mostly within a radius of 30 to 40 miles from the capital. There also were several little coastal towns, each with a small zone of rural settlement around it, but the interior of the island remained almost uninhabited, except for the widely spaced estate headquarters.

In 1850 the population of western Cuba, which included Piñar del Río, Havana, and Matanzas, totaled about 734,000, or nearly 65 percent of the population of the entire island. Of this population centered in Havana, 325,000 were people of pure Spanish descent, and 409,000 were Africans, most of them slaves. The main product of the area was sugar, raised on large estates, but there was also a considerable amount of tobacco and coffee.

As long as Cuba remained a colony of Spain, it was badly neglected. Roads were poor or nonexistent; methods of production were primitive and costly. The census of 1899 indicated that about 47 percent of the land in crops in all of Cuba was devoted to sugarcane; smaller percentages to yams, tobacco, bananas, maize, and other food crops; and 1.6 percent to coffee. At that time only 3 percent of the island was cultivated.

Spread of Sugarcane Cultivation

The treaty of 1901 between the United States and Cuba gave a tariff reduction of 20 percent on sugar produced in Cuba. Because Cuban sugar paid a lower duty than sugar imported from other countries, and because security was gained by the right of the U.S. government to intervene in Cuban domestic affairs, North American capital exceeding $1 billion poured into the newly created republic. Roads and railroads were constructed, Havana was modernized, and in the rural districts new sugar mills were built with the latest devices for grinding cane and extracting raw sugar. In terms of technical equipment and cost of production, no part of the world could compete with Cuba.

The first plantations to be developed around mills beyond the Havana area were located in an area of limestone soil known as Matanzas clay. This is one of the world's best sugarcane soils, in one of the world's best sugarcane climates, in a position where the sugar could move easily to a huge nearby market. The underlying limestone weathers into a deep red soil that shows no appreciable chemical or physical change for as much as 20 feet below the surface. This is a typical low-latitude soil: deficient in silica and high in iron and aluminum compounds. The chief element in its productivity for shallow-rooted crops is its porosity. This soil has a physical structure in which the clay particles are grouped together in floccules, leaving wide pore spaces for the downward percolation of water. As a result, there is little runoff and almost no erosion.

Production Problems

Sugar production in pre-Castro Cuba was an efficiently organized large-scale operation. Before World War II about 60 percent of the capital came from the United States; wealthy Cubans provided 22 percent; Spaniards contributed 15 percent; and the remainder came from Canada and other foreign countries. By the 1950s the U.S. share of the total investment in sugar production was down to 40 percent, but this included 7 of the 10 largest corporations. Most of the cane was grown by tenants who had contracts with the large mills, or *centrales*. Superintendents and inspectors from the *centrales* controlled the kinds of cane planted, the methods and standards of cultivation, the time of cutting, and the rate of delivery to the mills.

The labor requirement of sugar production, in Cuba as in most other cane-growing countries, has always been a problem. At harvest time, from December to June, there were never enough workers. The cane was cut by hand at a rate of more than 100 acres per day. Oxcarts carried freshly cut cane to railroads, which moved it to the *centrales* in a steady stream. The giant mills worked day and night. Immediate transportation of cane to the mills was essential because after cane is cut fermentation of the sugar sets in quickly, and at least 2 percent per day of sugar content is lost. Therefore, any efficiently run operation had to move the cane to the mill and extract the juice within 48 hours. So many workers were needed that they were recruited on a seasonal basis from nearby Haiti and Jamaica.

The picture was quite different between June and the end of November. This was what the Cubans called *el tiempo muerto*, the dead time. There were not nearly enough jobs to provide the workers with employment. Perhaps they were paid good wages during the harvest season, and there were fringe benefits such as health insurance and old-age retirement plans, but few people could spread their incomes adequately over the entire year. The great majority of sugarcane workers were hopelessly in debt, and their families were on the verge of starvation. Traditionally in Cuba, the period from June to November was one of social unrest.

Today, Cuba has a centrally planned economy. Yet, despite strenuous attempts at industrialization, the economy is still dominated by the annual sugar harvest. Cuba accounts for about 7 percent of the total world production of sugar and more than a fourth of world exports. There

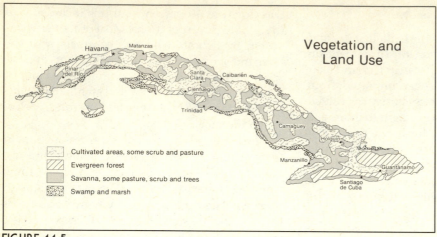

Vegetation and Land Use

Cultivated areas, some scrub and pasture

Evergreen forest

Savanna, some pasture, scrub and trees

Swamp and marsh

FIGURE 11-5

has been an attempt toward mechanization, and with the increased use of teenagers and women in the agricultural sector, the seasonal labor shortage appears reduced. Sugar processing itself has undergone widespread modernization. Cuba now has 150 sugar mills, and 13 more are to be built with Soviet aid.

Tobacco and Other Crops

Since the beginning of large-scale sugar production in Cuba, this one crop has dominated the nation's economy. Nevertheless, there is another commercial crop for which Cuba has long been famous: tobacco. The manufacture and smoking of cigars were taught to the Occidental world by the Indians, who had made Cuba a land of tobacco production long before the arrival of Columbus. The crop was formerly grown on small properties that averaged only about 40 acres, but the work of cultivating, harvesting, and curing was so great that on each 40-acre farm the services of at least 20 laborers were required. Tobacco for use as cigar filler is grown without shade, but tobacco destined for use in cigar wrappers must be grown under cheesecloth, protected from insect pests, and handled carefully so that the leaf is not torn or damaged. In 1980, 90 percent of the tobacco plants were destroyed by blue mold. This led to the closing of

nearly all tobacco factories, and cigar smokers of the world mourned the loss of their favorite Cuban brands.

Other major crops grown in Cuba are citrus fruit, coffee, and rice. Citrus has been expanded rapidly and appears likely to displace tobacco as the second agricultural export crop. Since 1975 irrigated areas have doubled, and the use of artificial fertilizers is growing.

In 1980 reform measures introduced the *Mercado Libre Campesino*, a free market. It allows farmers to sell that part of their harvest which exceeds a production quota set by the government, a profit idea borrowed from the capitalist world. Thus, the government has reestablished a legal, nonrationed system of farmers' markets that better serves the food needs of the Cuban population. It provides an incentive for farmers to improve their production techniques, and it may help to eliminate the long-existent black market for agricultural products.[3]

In 1979 and 1980 the livestock sector was affected by African swine fever, which may have arrived in Cuba with Haitian refugees landing on the Guantánamo Coast. In an effort to limit the spread of this disease it was necessary to slaughter 100,000 pigs, including the total swine popu-

---

[3] *Agriculture Abroad* 25, No. 4 (1980): Canada.

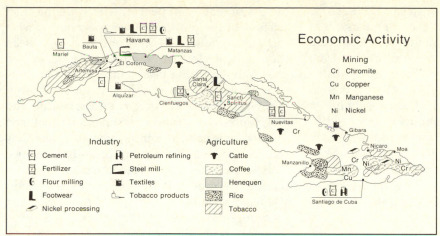

FIGURE 11-6

lation of Guantánamo as well as animals in five other provinces.

Mining

Cuba's new Five-Year Plan gives special attention to the development of mining, particularly nickel production. It is estimated that Cuba has about 10 percent of the world's known nickel reserves, and it is currently the world's fourth largest producer. Nickel has been mined in the highlands of eastern Cuba for many years. There are significant copper and iron ore deposits near Santiago de Cuba, west of Havana, and south of Camagüey; modest amounts of gold, silver, chromium, and cobalt are also mined. Cuba has produced some oil since 1916 from fields along the north coast, and recent discoveries have been made offshore from Havana.[4] More than 95 percent of Cuba's oil, however, is supplied by the Soviet Union.

Havana

The once glamorous capital of Cuba is now an aging city under Communist management. With a present population of almost 2 million, modern Havana has spread far beyond its outlines as a Spanish colonial town. The fortress that long

---

[4] *World Oil* 191, No. 3 (1980):96, Houston, Texas.

guarded the northern approach to the Gulf of Mexico still stands at the entrance to the bay. The downtown section of the city was laid out with the usual narrow streets on a rectangular plan, built around a central plaza. Along the ocean front, fine residential suburbs included the expensive homes of wealthy Cubans. Old tourist hotels still exist, as do impressive museums and beautiful old colonial buildings. On the north shore a vast apartment complex has been built to house workers, and major hotel construction is being planned in the hope that there will be a revival of tourism.

Cuba as a Political Unit

In 1959 the revolution headed by Fidel Castro was successful in overthrowing the Batista dictatorship. Since then the Cuban economy has been rebuilt along Marxist lines, with a centrally planned control of production and distribution of goods. Large numbers of older people who disliked the rigid discipline and loss of personal freedom emigrated from Cuba, chiefly to the United States, while the younger people who stayed were carefully indoctrinated.

Early in 1960 Castro expropriated large American landholdings, banks, and industrial concerns. When protests were ignored, President

Dwight Eisenhower canceled the U.S. sugar quota and imposed a total export embargo that severely damaged the economy. Relations were further strained in 1961 when about 1400 Cubans trained in the United States made an unsuccessful invasion of Cuba, at the Bay of Pigs, in an attempt to overthrow the Castro regime.

By 1980 Cuba was thoroughly involved in military maneuvers in Africa and subsequently lent active support to leftist forces in various Caribbean and Central American countries. At the same time, Cuba came to be regarded in some circles as a puppet of the Soviet Union.

The Economic Situation

Rebuilding an economic system is not easy. In accordance with Marxist principles the economy is now centrally managed, and the responsible agency is the *Junta Central de Planificación* (JUCEPLAN). This agency, together with a hierarchy of subordinate entities, allocates resources, determines production quotas, and directs the distribution of income among the workers.

The results of the JUCEPLAN program can now be assessed. At first, sugarcane production dropped to a record low in 1963. There were not enough workers to harvest the crop, and the seasonal migration of workers from Haiti and Jamaica had been stopped. Imported machinery was freqently idled by a lack of spare parts, and in 1963 "Hurricane Flora" did serious damage to central Cuba. In 1966, however, sugar production reached a record high level, and in 1967 the harvest was about a million tons greater than the average of pre-Castro Cuba. Agriculture also was being diversified by increased attention to basic grains, fruits, cotton, oilseeds, and other crops that help to provide year-round employment. Rice production almost doubled between 1957 and 1960 and continued to rise during the 1960s. However, the overall use of goods and services dropped 25 percent between 1957 and 1969.

An important factor in improvement of the agricultural sector was the construction of new, low-cost housing for farmworkers. There can be no doubt that these people, who were the chief victims of the pre-Castro sugar economy, were now much better off. They obtained new, clean homes with electricity and running water and were employed the year round. For them there was a marked increase in purchasing power. They could now eat rice, potatoes, wheat, and vegetables; they could buy shoes, and they had more simple cotton clothing available than they had before 1959. New capital investment, which was accomplished by decree, not persuasion, increased the production of some consumer goods. Demand still exceeds supply, however, hence washing machines, refrigerators, and even apartments are allocated by vote of the workers.

In 1968 the Soviet Union subsidized the Cuban economy at a cost of more than $400 million per year; by 1980 the cost was well over $2.5 million, with some estimates approaching $8 million, per day. An agreement reached in 1964 specified that Cuba would deliver 3 million tons of sugar to the Soviet Union in 1966, 4 million tons in 1967, and 5 million tons each year in 1968, 1969, and 1970. The quota was not met until 1970 when the harvest reached 8.5 million tons, Cuba's largest ever. It required twice the labor and twice the time to surpass a pre-Castro record of 7.2 million tons.

The Political Situation

Independence from Spain in 1899 did not bring real freedom to Cuba. U.S. and Cuban leaders worked out a new constitution which included the so-called Platt Amendment, by which Cuba agreed not to incur excessive debts but to continue the sanitary measures started by the U.S. Army, to lease a naval base to the United States at Guantánamo Bay, and to acknowledge the right of the United States to intervene if necessary in the domestic affairs of the island. Subservience to the United States, however, proved even more irritating than rule by Spain.

The Platt Amendment was renounced by the United States in 1934, but it did not bring an end either to political conflicts or to dishonesty and mismanagement by people in positions of power. From 1933 to 1959 Fulgencio Batista was the primary focus of political direction in Cuba, based on his ability to command the loyalty of the army. For several periods Batista permitted other military officers to become President, each of whom gained millions of dollars. There was no such thing as political liberty, honest elections, or equality before the law.

The rumblings of revolt, as usual, were centered on the university, where students, in accordance with Latin American tradition, remained free from police interference. In July 1953 a 26-year-old graduate of the University of Havana, Fidel Castro, led a small band of men in an attack on an army barracks near Santiago. Most of the attackers were killed, but Castro and his brother were placed in jail. When Castro was tried before a court later that year he acted as his own lawyer, taking the opportunity to speak for five hours on his plans for the social and political reform of Cuba.

Batista was so firmly in control of Cuba that in 1955 the Castro brothers were released from jail and permitted to depart for Mexico. There, they organized a band of 80 men, and in 1956 landed on the mountainous southeastern shore of Cuba. Most were killed by Batista's army, but Castro, his brother Raúl, and the Argentine revolutionary Ché Guevara, together with nine other survivors, escaped into the rugged Sierra Maestra.

From the mountains Castro conducted a guerrilla warfare campaign against Batista. Within a year he had recruited more than 2000 followers and was strong enough to make raids on the nearby lowlands, where sugar mills could be damaged or the cane set on fire. Batista's forces became increasingly oppressive and brutal. The people of Cuba, especially the large middle class of merchants and professional people, began to support Castro, as did many people in other parts of Latin America. Finally, even Batista's troops refused to oppose the guerillas. In January 1959 Fidel Castro assumed control of Cuba.

At first Castro's revolt was distinctively Cuban. The person most quoted in support of revolution was José Martí. Castro himself is reported to have said that "communism is only another form of slavery," but to keep control of Cuba especially after he expropriated land and seized foreign-owned property, Castro needed the support of the best organized political force in the country, the Communist Party. In 1961 he declared that his revolution was Marxist, and he proceeded to set up another tightly controlled dictatorship, accepting military and economic aid from the Soviet Union. He even permitted the establishment of Soviet missile bases in Cuba.

Castro was not content to confine his sphere of influence to the Caribbean, or even to Latin America. Between 1975 and 1985 thousands of troops were sent to Angola, Ethiopia, Somalia, Guinea, Mozambique, and Nicaragua. His continued revolutionary involvement in African and Central American affairs placed a severe strain on already tense relations with the United States.

Meanwhile, domestic affairs had reached a low point. Castro admitted that his country was "sailing in a sea of difficulties." Rationing was severe and strictly enforced. By 1980, even the staple diet of black beans was in short supply. Finally, in a dramatic protest against repression and the austere way of life in Cuba an estimated 10,000 people fled to Havana's Peruvian Embassy demanding asylum. Peru could not accommodate the thousands anxious to leave the country, but five other Andean Pact nations offered to take a certain number. The United States had already admitted more than 800,000 since Castro's coming to power in 1959 and offered to accept "a fair share of the refugees." As a *Freedom Flotilla* was organized between Florida and Cuba, Castro began emptying the jails and hospitals of criminals, the ill, political dissidents, and the mentally deficient. Within six weeks about 115,000 made the difficult journey to the United States in tiny boats, and Miami, Florida, became home to 600,000 Cubans. One-tenth of the population of Cuba fled.

Within Cuba, the greatest advances have been made in the fields of education, health, and sports. Education is available to most Cubans who wish to learn. The main objectives, after the revolution, were universal primary education and adult literacy. University education was redirected, with reduced enrollments in social science, law, and humanities, and expanded programs in agriculture and engineering. Besides career training, education in Cuba promotes revolutionary virtues. Nearly every factory and state farm has an adult education program, and some offer primary through pre-university courses. There are well-organized health education and immunization campaigns, and mass organizations that involve most of the population. Two organizations, the Committee for Defense of the Revolution and the Federation of Cuban Women, are the main distributors of health information.

Throughout Cuba there is great interest in physical fitness and sports. Moreover, all athletic events in Cuba are free. Cubans have a particular passion for the Yankee game of *beisbol* (baseball), at which they excel and which has become their national sport. In recent years, Cuban athletes also have competed successfully in the Olympic Games. Such achievements help to build national pride, an essential ingredient of the modern nation-state.

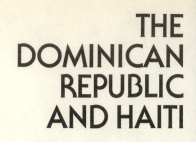

THE
DOMINICAN
REPUBLIC
AND HAITI

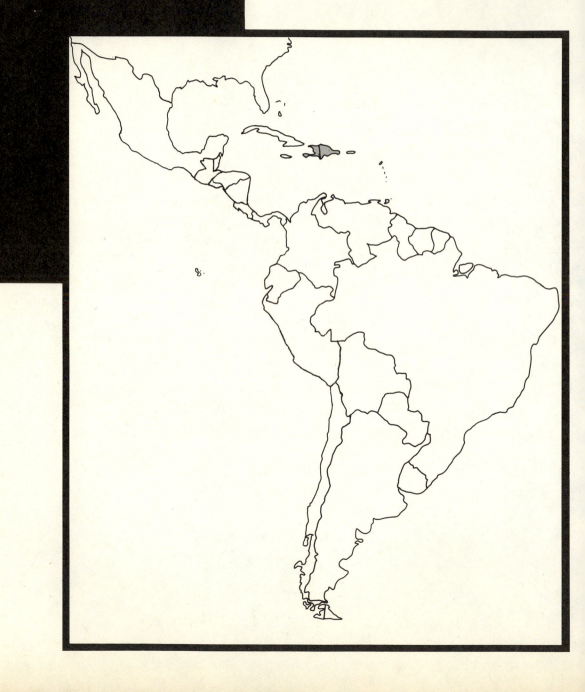

12

REPÚBLICA DOMINICANA

Land Area   18,816 square miles

Population   Estimate (1985) 6,200,000
Latest census (1981) 5,647,977

Capital City   Santo Domingo 1,313,200

Percent urban   52

Birth rate per 1000   33

Death rate per 1000   8

Infant mortality rate   64

Percent of population under 15 years of age   41

Annual percent of increase   2.5

Percent literate   62

Percent labor force in agriculture   47

Gross national product per capita $1380

Unit of currency   Dominican Peso

Leading crops in acreage   sugarcane, coffee, cacao, maize

Physical Quality of Life Index (PQLI)   64

COMMERCE (expressed in percentage of values)

Exports

sugar and honey	54	leaf tobacco	7
iron and steel	11	chemicals	4
		fruits and vegetables	3

Exports to

United States	54
Switzerland	12
USSR	8

Imports from

United States	39
Venezuela	18
Mexico	14
Japan	5

Data mainly from the 1985 World Population Data Sheet of the Population Reference Bureau, the 1985 Britannica Book of the Year, and the 1985 World Almanac.

RÉPUBLIQUE D'HAITI

Land Area   10,714 square miles

Population   Estimate (1985) 5,800,000
             Latest census (1982) 5,053,792

Capital city   Port-au-Prince 763,200

Percent urban   28

Birth rate per 1000   36

Death rate per 1000   13

Infant mortality rate   108

Percent of population under 15 years of age   44

Annual percent of increase   2.3

Percent literate   21.1

Percent labor force in agriculture   79

Gross national product per capita $320

Unit of currency   Gourde

Physical Quality of Life Index (PQLI)   57

COMMERCE (expressed in percentage of values)

Exports

coffee	39	chemicals	6
bauxite	11	essential oils	6
toys and sporting goods	10	electrical equipment	5
		cacao	5

Exports to		Imports from	
United States	59	United States	45
France	13	Netherlands Antilles	10
Italy	7	Japan	9
Belgium-Luxembourg	6	Canada	8
		West Germany	5
		France	3

Data mainly from the 1985 World Population Data Sheet of the Population Reference Bureau, the 1985 Britannica Book of the Year, and the 1985 World Almanac.

Locked together within the confines of the island of Hispaniola are two very different independent states. In the west is Haiti, Black African in many aspects of its culture but with a French tradition and language. In the east is the Dominican Republic, mostly mulatto but with a Spanish heritage. The western third of Hispaniola was occupied in 1985 by an estimated 5.8 million Haitians, with a population density of 513 per square mile. The eastern two-thirds is occupied by an estimated 6.2 million Dominicans, with a density of about 335 per square mile. Historically antagonistic toward one another, each country tries to maintain and foster its own type of nationalism. Much of this conflict has revolved around the international boundary, which has crept eastward into territory claimed by the Dominicans.[1] On the Haitian side of the border the rugged land is a patchwork of small farms, cultivated with hoe and machete, used for the production of food crops; on the Dominican side, behind a screen of agricultural colonies, there are large properties used for the grazing of cattle, areas thinly inhabited or entirely empty, and small areas of concentrated settlement. The presence of two dissimilar peoples, politically separate, within so small an island, creates the ever-present danger of overt conflict.

[1] J. P. Augelli, "Nationalization of Dominican Borderlands," *Geographical Review,* 70, No. 1 (1980):22.

The Land

The island of Hispaniola has the most rugged and complicated terrain of all the Greater Antilles. It is ribbed by steep-sided, narrow-crested ranges, oriented in various directions, and creased by deep valleys and pocket-like lowlands. In the Cordillera Central the mountains of Hispaniola reach their greatest elevations. This range extends northwest-southeast, with one end just within Haiti to the south of Cap Haïtien

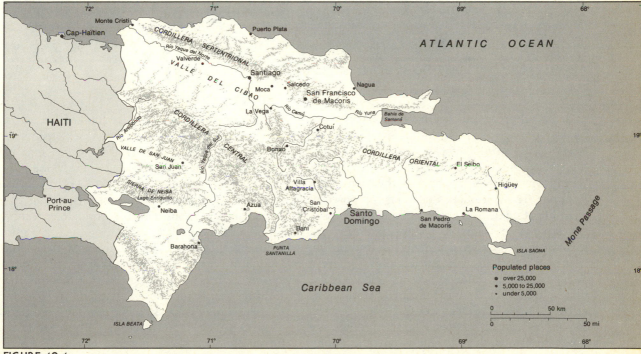

FIGURE 12-1

AVENIDA MERINO IN SANTO DOMINGO, DOMINICAN REPUBLIC.

and the other on the south coast just west of Santo Domingo. The summits are generally between 8000 and 9000 feet above sea level, but the highest mountain, Pico Duarte, rises to 10,206 feet in the west-central part of the Dominican Republic. The Cordillera Central is composed of a jumble of ridges and peaks threaded with rushing streams in narrow, steep-sided canyons. Most of the slopes are too steep for farming, yet there are small flat-bottomed valleys where a dense farming population is established in isolation from the outside world. From the northwestern end of the Cordillera Central, lower but still rugged mountain ranges extend through the northern Peninsula of Haiti where they point

across the Windward Passage toward the Sierra Maestra of Cuba. An eastward extension of steep-sided ridges reaches the eastern end of Hispaniola, where it points across the Mona Passage to the mountains of Puerto Rico.

In the southwestern part of Hispaniola another range of mountains forms the long southern Peninsula of Haiti, extending eastward into Dominican territory. This is the eastern end of the southern prong which makes up the forked system of ranges described in this book as the Central American-Antillean system. In Hispaniola this range is separated from the Cordillera Central by the deep structural Cul de Sac-Lake Enriquillo depression.

In the northeastern part of Hispaniola there is still another distinct range of mountains, the Cordillera Septentrional, extending from the Haitian border eastward across the Dominican Republic to form the small Peninsula of Samaná. The highest summits of this range are only 3000 to 4000 feet above sea level, but the slopes are steep and the valleys deeply cut. This range is separated from the Cordillera Central by the narrow east-west lowland known in the Dominican Republic as the Cibao and in Haiti as the Plaine du Nord.

Small lowlands and intermont basins are scattered throughout this array of ridges and summits and along the coast. The largest plain is in the southeast of the Dominican Republic, eastward from Santo Domingo. This is a limestone platform formed by corals and raised above the sea, similar in many ways to the limestone terraces of Cuba. The other lowland in the southern part of the island is the Cul de Sac-Lake Enriquillo depression partly in Haiti and partly in the Dominican Republic. In the middle of this lowland are two brackish lakes, the Étang Saumâtre

in Haiti and Lake Enriquillo in the Dominican Republic, the latter about 140 feet below sea level. This depression, a downfaulted block between the mountain ranges, was at one time filled with sea water. Streams at either end of the lowland built alluvial fans that eventually sealed it off from the ocean, while evaporation lowered the lake levels.

In Haiti there is a wedge-like area of lowlands, the Artibonite Plain, and in the midst of the mountain ranges on the border between Haiti and the Dominican Republic is the Plaine Centrale. This is an intermont basin about 1000 feet above the sea and is generally level to rolling except where cut into narrow ravines by headwaters of the Artibonite River.

In addition to these intricately arranged mountain axes and structural depressions, the surface of Hispaniola is complicated by many isolated mountain blocks, miniature valley lowlands, and coastal plains. Geologically associated with the surface features of the main island, too, are the bordering islands of Gonâve off the west coast of Haiti and Tortue to the north.

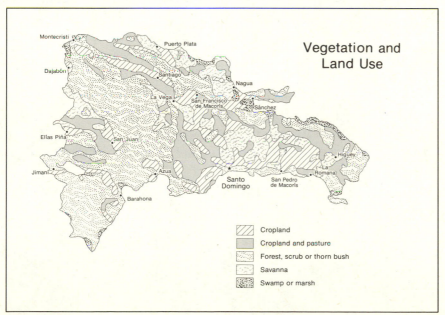

Vegetation and Land Use

Cropland
Cropland and pasture
Forest, scrub or thorn bush
Savanna
Swamp or marsh

FIGURE 12-2

Climate and Vegetation

A surface so complex in its pattern of slopes and basins could not fail to develop a complex pattern of climatic conditions and vegetation. Temperatures generally decrease with increasing elevation, but many protected pockets are so disposed that the heating effect of the sun is especially great, and exceptionally high temperatures occur. Port-au-Prince, on the protected western side of Hispaniola and in the embrace of minor spurs from the southern mountains, has one of the highest average temperatures of any major city in the Antilles.

The pattern of rainfall is also complicated. The north and east-facing slopes of the mountains are generally wetter than the south and west-facing slopes, although this generalization does not apply to the southern mountains of Haiti where both north and south slopes are wet. Among the lowlands, the Cibao receives abundant moisture, especially in its eastern part where there is an agricultural district of great productivity known as the *Vega Real*. The Plaine du Nord of Haiti receives somewhat less moisture than its eastern continuation in the Dominican Republic. The coastal plain in the southeast, on which Santo Domingo is situated, receives barely enough rainfall for crop production without irrigation, and in this area most of the sugarcane plantations are now irrigated. The Plaine Centrale of Haiti is also near the margins between humid and subhumid, whereas both the Artibonite lowland and the Cul de Sac are semiarid despite high atmospheric humidity.

The natural vegetation of Hispaniola closely reflects the underlying conditions of climate and surface. The wetter places were originally clothed with a dense rain forest, whereas the drier slopes and basins supported only a thorny scrub woodland which varied in density according to variations in drainage. The drier northwestern part of the Plaine Centrale was covered by open savanna, with trees only in the narrow, wet ravines; the wetter southeastern part supported a dense scrub woodland. Only in the Cordillera Central are the elevations sufficient to reach the zone of pines.

Such is the nature of the island of Hispaniola. This is a land occupied by two strongly con-

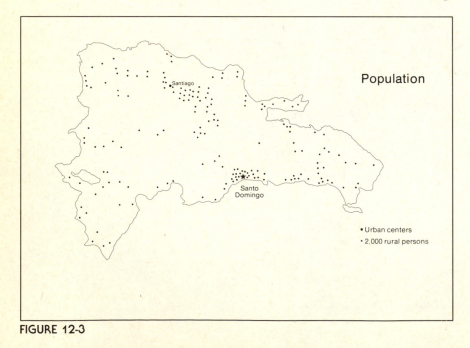

FIGURE 12-3

trasted peoples, whose traditions, technical abilities, and basic attitudes are so different that the political boundary dividing Hispaniola into two parts has become a sharp cultural boundary as well.

THE DOMINICAN REPUBLIC

The Dominican Republic, which occupies the eastern two-thirds of Hispaniola, is notably less populated than Haiti. Even the figure of population density (335 per square mile) is quite misleading, for the Dominicans are concentrated in certain areas only, while large parts of the republic remain thinly settled. About half of the population is concentrated in the eastern part of the Cibao and along the northern coast. Nearly one-third is in the sugarcane lands of the south and in the capital city of Santo Domingo. The most densely populated territorial division is the National District, with more than 1400 people per square mile.

When Columbus discovered the island in 1492, it was inhabited by Arawak and Carib Indians. Four years later, his brother, Bartolomeo, founded the city of Santo Domingo. It was from this city that the Spaniards explored and conquered other West Indian islands and the mainland. The colony prospered for some years, but its decline began after the conquest of Mexico and Peru, and after many Indians died from contagious diseases.

Of the total Dominican population, people of European ancestry comprise about 16 percent; mulattoes about 73 percent; and people of unmixed African ancestry about 10 percent. The remaining 1 percent includes a variety of peoples, mostly Japanese and Chinese.

The Course of Settlement

The eastern half of Hispaniola was the first part of America to be occupied by the Spaniards. After two failures to establish a port on the northern coast (Navidad, now Cap Haïtien, in

1492; and Isabela, 25 miles west of Puerto Plata, in 1493), Santo Domingo was made the primary Spanish settlement center in 1496. Here too, the newcomers led by Columbus were concerned mainly with finding gold, and they did find some gold in stream gravels of the Cibao. The large native population, which probably numbered more than a million, was pressed into service in the mines. Within 50 years most of the Indians had died from overwork and lack of food, as well as from epidemics of European diseases such as measles and smallpox.

The Indians of Hispaniola had lived in a kind of paradise, enjoying an abundant food supply and usually freedom from warfare. Their main food was manioc, supplemented by sweet potatoes, peanuts, and a variety of other crops. They grew maize, but it was less important on the islands than root crops. In the Cibao the Indians grew tobacco, which they smoked in cigars or took as snuff. Their diet was well balanced because they also ate fish, crustaceans, turtles, marine mammals, and waterfowl, all of which were found in abundance around Hispaniola.

When the Spaniards arrived, Indian agriculture was neglected. The introduction of wheat and grapes was a failure, but cattle and pigs did well. When Columbus first reached Hispaniola, he found the level lands mostly cleared and cultivated. Within 50 years, large areas once densely populated were abandoned, and a second growth of forest or woodland savanna had appeared. Cattle and hogs ran wild and found plenty to eat where once there had been Indian farms.[2]

With the decline of gold mining, the Spaniards who remained on Hispaniola tried other ways of making a living. As early as 1515, sugarcane plantations had been developed around Santo Domingo, and African slaves were imported to do the work. Tobacco was raised in the Cibao, where the Indians had been growing this crop for

---

[2] For a discussion of the changes introduced by the Spaniards all around the Caribbean, see Carl O. Sauer, *The Early Spanish Main* (Berkeley: University of California Press, 1966).

many generations. Cacao, grown in the wet eastern part of the Cibao, was an important export prior to 1800. Attempts also were made to grow indigo. By the end of the eighteenth century, the contrast between the western and eastern parts of Hispaniola was apparent. In the French colony there were 524,000 inhabitants, of whom 88 percent were African slaves; in the Spanish colony in eastern Hispaniola there were 103,000 inhabitants, of whom 30 percent were African slaves.

The last decade of the eighteenth century was one of violent disorder, leading to utter ruin of the colonial economy. Ownership of Hispaniola passed from one country to another. In 1804 Haiti declared its independence. When the Spanish colony declared its independence from Spain in 1821, the Haitians promptly invaded and took control of the entire island. Families of wealth and social position left the country as economic and intellectual activity came to a halt. It was not until 1844 that the Dominicans ousted the Haitians and finally established an independent country. For years thereafter a smoldering discord centered on the international boundary as thousands of Haitians crossed into Dominican territory. The Dominican government was helpless in its efforts to control the border regions. Since then the relatively small population of the eastern part of the island has lived in constant fear of expansion by the relatively dense population of the western part, whether by infiltration across the border or by actual conquest. For a brief period, while the United States was engaged in its Civil War (1861–65), Spain resumed control over its lost colony. In 1869 the people of the Dominican Republic voted to request adoption by the United States, but the United States refused to accept this added responsibility.

In the midst of such confusion, certain economic advances were being made. In 1865, a colony of people from the southern United States was established on the southern side of the Samaná Peninsula. On Hispaniola the colonists again began to plant cacao. As a result of their example, other farmers in the Vega Real planted that crop. During the next 20 years the Cibao developed a considerable production of cacao, sugarcane, coffee, cotton, tobacco, beeswax, and honey. In 1914 sugar passed cacao as the leading export.

U.S. Marines occupied the Dominican Republic during World War I to protect approaches to the Panama Canal. They landed at Santo Domingo in 1916 and soon brought the entire territory under their control. Order and security were established, roads and railroads were constructed, and a considerable gain in commercial production was effected. When the Marines withdrew in 1924, they left the country in the hands of an officer they had trained, Rafael Leónidas Trujillo Molina. General Trujillo became one of the strongest dictators in Latin America, invincible because he controlled the army. He imprisoned, killed, or exiled anyone who opposed his rule. Trujillo accomplished much in developing the economy of the country, which he managed as if it were his personal domain. He named the highest mountain Pico Trujillo (now Pico Duarte), and when Santo Domingo was rebuilt following the hurricane of 1930, he named it Cuidad Trujillo. When he was assassinated in 1961, the Dominicans promptly restored the previous names.

Under Trujillo, the government gave up most of its historical claims, accepted the present boundary with Haiti, and tried various schemes to nationalize the borderlands. This included the expulsion of thousands of Haitians, the prohibition of all immigration from Haiti except seasonal laborers, and the creation of an intensive educational and religious campaign. A strip of land, about six miles wide, parallel to the international boundary was reserved for Dominican settlers who were given land, houses, cash, and other inducements to become pioneers in this frontier zone.

Patterns of Settlement and Land Use

After 1930, under the dictatorship of General Trujillo, the Dominican Republic made important advances in economic productivity. By 1959 about 23 percent of the total area was under cultivation. The amount of irrigated land increased from 7500 acres in 1930 to 380,000 acres in 1959. Not only was agriculture expanded in those areas already settled, but also Trujillo undertook a program of colonization in parts of the country previously unoccupied.

The period from 1962 to 1964 were years of discontent and political unrest, until finally in 1965 an outbreak of violence occurred between the military who favored a return to a "constitutionalist" government and those who favored a military *junta*. Once again U.S. forces landed to protect the lives of American citizens and to aid in the evacuation of these and other foreign nationals. Communist leaders, many trained in Cuba, took an active part in the revolutionary movement. It was not until September 1966 that peace was restored.

The Cibao and the North Coast

The Cibao and the north coast constitute the most densely populated part of the country. The lowland is well settled for all 150 miles between the Haitian border and Sánchez on the Bahía de Samaná, but the densest population is in the Vega Real, east of Santiago de los Caballeros. This is a highly productive agricultural land. On fertile alluvial soil the farmers grow dry-land rice, maize, peanuts, beans, yams, and vegetables. The eastern end of the Vega is used chiefly for cacao, especially around San Francisco de Macorís. In many places bananas are grown to give shade to the cacao or to coffee on the lower slopes of the mountains to the north. Around Santiago there is a concentration of tobacco, while westward as far as Monte Cristi the floodplain of the Río Yaque del Norte is used for paddy rice. Along the northern slopes of the Cordillera Septentrional there are coffee plantations higher up and cacao plantations at lower elevations. Along the coast the crops include tobacco, bananas, manioc, and coconuts.

At the westernmost end of the Cibao, there is a zone of plantations from which bananas have been exported regularly since 1939. These plantations, developed by the Standard Fruit Company, are irrigated and equipped with modern spray systems for the control of banana disease. As a result, they are the most productive banana plantations in the Antilles.

Santiago, 85 miles northwest of the capital, is the Republic's second city. It has survived many disasters and now has a population of 300,000. It was destroyed by an earthquake in 1564 and was then rebuilt; it was burned and plundered by the French, and then burned again by the Haitians and Spaniards. Its industries include tobacco products, pharmaceuticals, furniture, pottery and baskets, rum and alcohol.

The Southeast

The southeast is separated from the Cibao by a densely forested but not very high series of limestone ridges. The plains of the southeast are developed on limestone, much like the Camagüey area in central Cuba. For a long time the land has been used for cattle-raising, and animals have been exported to other parts of the Antilles both for meat and for draught purposes. The animals are small but hardy and are well adapted to the conditions of tropical pastures. Under Trujillo the herds were improved, and a modern dairy farm was built near the capital. Milk was sent by airplane to San Juan, Puerto Rico.

Since World War II there has been a notable expansion of privately owned sugarcane plantations all along the southeastern coast. The focus of this activity is La Romana, which also features the world's largest sugar mill. Soils here are similar to the best in Cuba. Because it is a little drier and less cloudy than in Cuba, the area receives more sunshine. As a result, its cane has the highest sugar content of any grown in the Antilles. Because of lower rainfall, a large part of the cane lands must be irrigated.

Large-scale enterprises have promoted increased mechanization. Tractors are replacing oxen in transporting sugarcane to the mills. The southeastern region has depended on migrant workers from Jamaica, Puerto Rico, and Haiti to do the harvesting, but the need for extra labor has decreased with increased use of machinery.

The metropolis of the southeast, and of the nation, is Santo Domingo. It is an historical city featuring many "firsts": the first permanent European settlement, the first courtroom and city hall, the first monastery, the first mint, the first cathedral, and the first university in the Western Hemisphere. It also has the oldest hotel and the oldest street within its inner core. Here, in the Cathedral of Santa María la Menor is a small casket that contains what are said to be the remains of Christopher Columbus. Around the Cathedral are many restored structures from the Columbus era in history.

The sprawling "new city" extends westward from the Río Ozama. In addition to residential neighborhoods, there are tall, modern buildings which serve as highrise apartments, government offices, and tourist hotels. Still farther westward is a zone of light manufacturing. Heavy industry is concentrated at the port of Haina, about 10 miles west of Santo Domingo, where there is a petroleum refinery, cement factory, sugar mill, and electric power plant.

Santo Domingo, like many other cities of Latin America, has grown too fast. Migration to the city from rural areas throughout the nation has created serious shortages of housing, employment, and public services. Yet, this urban center has been the heart of the Dominican nation since 1496, a condition that defeats all attempts at decentralization.

Agricultural Colonies

With so much unused land, the Dominican government under Trujillo undertook to occupy the national territory more firmly by establishing agricultural colonies. Most of the colonists were Dominicans. Trujillo took care of the urban slum problem by moving people away from the cities, which also reduced the pressure on his government to provide property for landless tenant farmers. Land was given free, and the government built houses and provided seed, tools, machinery, and technical supervision. There were three major areas of colonization. One was in the lowlands west of Barahona and in the mountains to the south along the Haitian border. In 1942 a colony of Dominicans settled along the northern side of Lake Enriquillo on land supplied with irrigation water. The main crops were rice, coconuts, and bananas.

The mountains south of Barahona have also been occupied. On the eastern and northeastern slopes, coffee plantations have been developed in the shade of the taller trees of the forest which remain uncut. Coffee from this area is reputed to be the best in the country.

From Trujillo's perspective, the most important new colonial area was along the Haitian border in the mountainous interior. This is where dense populations of Haitian farmers, making a miserable living on badly eroded lands, looked across the border at almost empty country where cattle grazed on vast private estates. In 1942 an all-weather road was built near this border, and a string of agricultural colonies was established.

Roads and Railroads

The Dominican Republic has few railroads. Santo Domingo shares with Tegucigalpa the distinction of being a large capital city with no rail connections. Most of the Dominican railroads were built around 1875 by British investors to transport cacao from the Vega Real to the port of Sánchez. Some railroads were built about the same time in the older sugarcane plantations. Only about 350 miles of rail are currently in use.

The Dominican Republic is placing its dependence on all-weather roads and trucks. Now, there is no settled part of the Republic that is not within eight hours' driving time of Santo Domingo. All-weather roads have also reduced the number of ports. Haina and Santo Domingo predominate, but most products of the Cibao are shipped from Puerto Plata or Sánchez. Sugar, as in Cuba, is shipped from ports near the places of

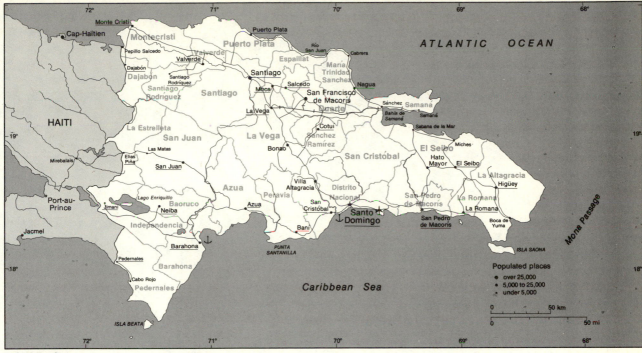

FIGURE 12-4

harvest, especially La Romana, San Pedro de Macorís, and Barahona.

In 1979, after "Hurricane David" and "Hurricane Frederick" devastated the Dominican Republic, a major effort was required to rehabilitate the nation's basic infrastructure. This included roads, irrigation, rural electricity, and water systems, as well as bridges, electric power, canals and dikes, and the repair of highways. The country is also evaluating its urban and interurban transportation systems. An estimated 13,000 automobiles and 1900 buses are used throughout the country. The annual cost of fuel for these vehicles amounts to $550 million.

The Dominican Republic as a Political Unit

After the assassination of Trujillo, who ruled from 1930 to 1961, turmoil prevailed as various leaders sought to gain control of the country. The conservative Joaquín Balaguer was then demo-

cratically elected to three terms from 1966 to 1978. Thus, during 43 years of a 50-year period, only two men held executive power in the Dominican Republic. The inauguration of Antonio Guzmán Fernández in 1978 represented the first peaceful transition of power from one political party to another in this century. Another peaceful transition occurred in 1982, when Salvador Jorge Blanco became President. Yet, the permanence of democratic government was far from assured, given the country's critical economic conditions.

The Economic Situation

Trujillo did much to build the Dominican economy. On a small island well endowed with resources, he was able to increase economic production so that the country had a gross national product of $232 per capita in 1955, which placed it seventh among 20 Latin American republics. To be sure, income was not widely distributed

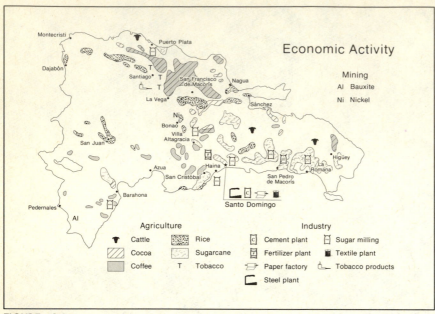

Economic Activity

Mining
Al Bauxite
Ni Nickel

Agriculture
- Cattle
- Cocoa
- Coffee
- Rice
- Sugarcane
- T Tobacco

Industry
- Cement plant
- Fertilizer plant
- Paper factory
- Steel plant
- Sugar milling
- Textile plant
- Tobacco products

FIGURE 12-5

among the people, and the largest share of a profitable economy went to the dictator and a small group of wealthy landowners, military, and government officials. Furthermore, domestic tranquility was gained by strict military rule, including forced resettlement of slum dwellers in agricultural colonies. Despite the relatively high productivity of the economy, most people remained poor.

One-third of the population in the Dominican Republic is supported by the sugar industry, and about 15 percent of arable land is occupied by sugarcane.[3] Despite hurricane damage to sugar mills and plantations, the country expects to increase sugar production in the coming years. Emphasis is also being given to food crops, such as bananas, rice, beans, sweet potatoes, and peas. Consumption of rice, the mainstay of the Dominican diet, is increasing as the population and per capita income increase. Another important item in the local diet is the tomato. Tomatoes for processing are grown chiefly in the northwest

and on the south coast near Azua, where tomato-processing plants make paste, catsup, and juice, with paste accounting for 90 to 95 percent of total production.

The country's livestock sector places particular emphasis on beef and poultry. At one time pork was important commercially, but an outbreak of African swine fever curtailed production. In 1980 all 1.5 million pigs in the Dominican Republic were rounded up and slaughtered, since no vaccines had yet been developed to protect the animals against this virulent disease.[4]

At one time the Dominican Republic ranked eleventh in world nickel production, but as the demand for nickel fell, it temporarily closed its Falconbridge Mine. This facility is capable of producing annually more than 54,000 metric tons of ferronickel, which is used in steel making. To compensate for such losses in revenue, the government granted a 15-year exploration concession for copper, gold, and silver along the Río Yujo near Jarabacoa. These minerals are now

---

[3] R. Chardon, "Sugar Plantations in the Dominican Republic," *Geographical Review* 74, No. 4 (1984),: 441–454.

[4] "African Swine Fever Strikes," *Agenda* 4, No. 6 (July/August 1981).

being mined, as is iron ore, bauxite, and amber. After the Baltic Sea area, the Dominican Republic has the world's largest supply of amber.[5] Near Barahona, a 10-mile block of almost solid salt is thought to be the largest salt deposit in the world.

The Political Situation

Despite many problems, the Dominican Republic has managed to retain a certain degree of political stability since the death of Trujillo. An election in 1962 gave overwhelming support to a liberal government, but, the liberal regime was promptly labeled Communist by those whose privileged positions were threatened. When the country was in confusion and no one army officer could command enough support to ensure control, the United States again dispatched an occupation force in 1965. This was made part of an international force, responsible to the Organization of American States, which remained in the Dominican Republic until 1966. The occupation was justified as necessary to protect the lives of foreigners in Santo Domingo; it also was recommended by representatives of the United States in the Dominican Republic who feared a Communist takeover such as that of Castro in Cuba. There are volumes of published reports proving beyond doubt that the Communists were, in fact, responsible for the conflict and were about to seize control; there are other volumes that prove, also beyond doubt, that the whole Dominican affair had little to do with communism.

When the conservative Joaquín Balaguer became President in 1966, his appointment of women to the governorships of all 26 provinces in the country caused astonishment in political circles. Cynics labeled this a political ploy, since "governors do not have much autonomy."[6] Traditionally, public affairs and politics have been almost exclusively male property in Latin America as in the rest of the world.

President Antonio Guzmán was inaugurated in 1978 and gave special attention to tourism, industrial free zones, and commercial farming. The tourist sector was to be free from income tax for 10 years, manufacturing would have no tax for 20 years, and new incentives were implemented for commercial farming. In 1982 Guzmán was followed by Salvador Jorge Blanco whose administration faced a severe financial crisis as world prices for the export commodities on which the Dominican Republic depends plummeted. On the bright side, tourism increased dramatically, and oil drilling in the Lake Enriquillo depression showed promising results.

Continued political stability will depend largely on a high rate of economic growth. The importance of agriculture will remain, but increased attention must be given to mining, manufacturing, tourism, energy, and transportation. Given the strategic location of the Dominican Republic within the Caribbean area, political conditions there will continue to attract close external attention.

HAITI

More than a decade ago, a Belgian nutritionist described Haiti as one of the most disease-ridden, poverty-stricken, malnourished, and illiterate nations in the Western Hemisphere.[7] This is still true, for Haiti is the poorest of the poor. More than 5.8 million people are crowded into territory about the size of New Jersey, in a land that is two-thirds mountainous. Eighty percent of the population under five years of age suffers from malnutrition. A lack of water purification and proper sewage disposal is responsible for such

---

[5] *Américas* 32, No. 10 (1980):33–41.
[6] S. B. Tancer, "La Quisqueyana: The Dominican Woman, 1940–1970, in A. Pescatello, ed., *Female and Male in Latin America: Essays* (Pittsburgh: University of Pittsburgh Press, 1973).

[7] I. Deghin, "Les problèmes de santé et de nutrition en Haiti: Un essai d' interprétation," Académie Royale de Sciences d'Outre-Mer, Classe des Sciences Naturelles et Medicale, N.S., 17:6 (1969).

AN ELEMENTARY SCHOOL IN A POOR RURAL SECTION OF HAITI.

communicable diseases as influenza, diarrhea, intestinal parasites, and tuberculosis.[8]

Ninety-five percent of the people are blacks; the remaining 5 percent are mulattoes of French ancestry. It is the latter group that comprises the Haitian aristocracy. Most Haitians live outside of the economic system of international trade and even to a certain extent outside the Haitian political system, which touches their lives only remotely. There is a cosmopolitan and well-educated minority of French-speaking landowners and intellectuals who are set apart in striking contrast to the peasant farmers whose lives are dominated by religious ceremonies and other traditions inherited from Africa.

About 80 percent of Haiti's 5.8 million people live in the countryside, where population pressure is intense and arable land is scarce. There is but a meager existence for many people. If there is unrest, order in this heterogeneous society is maintained by the military. Not everyone in Haiti is poor, for approximately 5 percent of the population accumulates more than 50 percent of the national income. About 7000 families have an average annual income of nearly $100,000.[9] This powerful elite exhibits a lifestyle that is envied by a large majority of the people.

---

[8] *News Release,* Inter-American Development Bank, 3699F, NR-8/81, Washington, D.C., 1981.

[9] R. Maguire, *Bottom-up Development in Haiti* (Rosslyn, Va.: Inter-American Foundation, 1979).

FIGURE 12-6

Sequence of Settlement

In the early seventeenth century, western Hispaniola offered a fine unoccupied land base for the English and French pirates who preyed on Spanish ships bringing silver and gold back to Spain. In about 1625 the Île de la Tortue (Tortuga) became one of the chief pirate strongholds of the Antilles. Bands from this base would come over to Hispaniola to hunt wild cattle and hogs that had escaped from the Spanish settlements in the east. Large fires were built, and over these the carcasses, laid on grills *(boucans)*, were processed for tallow. Here in the hills of Haiti, the pirates came to be known as *boucaniers*, or buccaneers. In time the French drove out the English and, supported by the French colonies on other nearby islands, established settlements on Haiti, especially along the northern coast. In 1697 Spain recognized France's claim to the western third of Hispaniola. The new French colony, now officially established, was known as Saint Domingue.

The French Period

A century of French ownership witnessed the rise of Saint Domingue to the status of one of the world's richest colonies. In those days, when the Antilles produced sugar for Europe's growing markets, colonies were far from being financial liabilities. Destructive exploitation of land and slave labor could be made, for a time at least, to yield enormous profits to the owners.

The settlement of Haiti by French sugarcane planters was concentrated on the lowlands. The first district to be developed was on the Plaine du Nord, in the territory served by Cap Haïtien, then known as Cap Français. This settlement was the social and shipping center for the central Caribbean and became France's wealthiest colonial

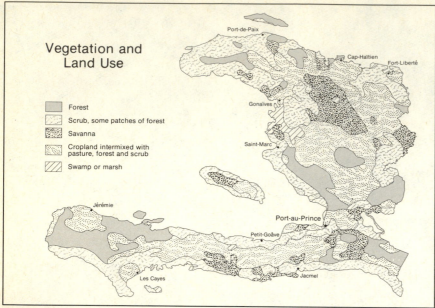

Vegetation and
Land Use

- Forest
- Scrub, some patches of forest
- Savanna
- Cropland intermixed with pasture, forest and scrub
- Swamp or marsh

Port-de-Paix
Cap-Haïtien
Fort-Liberté
Gonaïves
Saint-Marc
Jérémie
Port-au-Prince
Petit-Goâve
Les Cayes
Jacmel

FIGURE 12-7

capital. The fertile Plaine du Nord and the Arti-bonite Valley supplied half of Europe with sugar and cocoa. On immense indigo, cotton, cacao, sugar, and coffee plantations were innumerable slaves who cultivated the land, whereas the French Creole planters lived like kings until the French Revolution. The productive lowlands were divided into a rectangular pattern of well-kept roads and large properties, all neatly bordered by hedges. The mansions of the planters were luxurious, and the prosperity of the colony became famous. Seeking more space, the French extended their plantations southward to the Cul de Sac. In 1749 the town of Port-au-Prince was laid out, and in 1770 it was made the seat of government.

The shift of the center of French authority to the south resulted in part from the enormous productivity of the Cul de Sac. With the aid of slaves the French built elaborate systems of irrigation, including long stone aqueducts, some of which are still in use. Because of the larger amount of sunshine where the rainfall is not so heavy, these dry plains, when supplied with irrigation water, proved to be better producers of

sugar than the Plaine du Nord. Soon the Cul de Sac became the main center of sugar cultivation.

As the prosperity of the colony grew, other commercial crops were added. In the lowlands, indigo was grown along with sugarcane. Smaller areas were devoted to bananas, yams, manioc, cacao, coconuts, and cotton. Late in the French period, coffee was introduced, and several important plantations were developed, especially on the slopes of the Cordillera Central south of the Plaine du Nord.

The social situation in Saint Domingue was becoming explosive, however. Africans greatly outnumbered Europeans, but this condition might not have led to disastrous revolts had it not been for the mulatto class. The mulattoes, made free by a decree of the French government but not accepted on terms of equality by either the pure Africans or pure Europeans, became increasingly discontented.

Period of Independence

The liberal political doctrines of the French Revolution had a special meaning for the mulattoes of

Saint Domingue, many of whom had been sent to Paris for an education. The expectation of freedom and equality, along with the existing political disorder, led to a revolt of the former slaves, destruction of the Saint Domingue estates, and the hurried escape of those French landowners who were able to avoid death. In 1804 the Africans of Hispaniola declared independence and adopted the Indian name of Haiti, thus creating the world's first black republic.

Changes in the distribution of people and in their form of economy produced changes in the Haitian landscape which can still be observed. Today the French aqueducts, the mansions of the sugar planters, old stone sugar mills, and many old churches remain as ruins. The rectangular field patterns, dear to the hearts of the French-

men, have disappeared under the haphazard and irregular trails and fields of the carefree Haitians, but the old rectangular French patterns have not been entirely lost. From the ground they are no longer visible, but from the air one can still observe the faint trace of straight lines crossing at right angles. One result of the overlap of patterns is an utter confusion of land titles.

It has been demonstrated that the peasant farmer in Haiti has become mired on a road of economic regression. Lack of education, a diminution of arable land, and a general decline in living standards have all contributed to this situation.[10]

---

[10] M. Lundahl, *Peasants and Poverty: A Study of Haiti* (New York: St. Martin's Press, 1979).

CULTIVATION OF MOUNTAIN SLOPES OVERLOOKING PORT-AU-PRINCE.

Haiti, however, is alive with the activity of organizations and agencies: international, multinational, governmental, and nongovernmental. Many of these are backed by considerable amounts of money, and funds are being channeled into various sectors of the economy and regions of the country. There are projects to improve roads, to plant trees that will stop erosion, and to build drainage and water supply systems; there are educational programs in industrial management and craft industries; and there is a whole new field in telecommunications. Nevertheless, Haiti has had difficulties in attracting private foreign investments and commercial loans.

Vital to Haiti's economy is the continuation of tourism. The focal point of tourism is the port of Cap Haïtien with its international traffic of cruise ships and cargo vessels. Nearby are the remains of King Christophe's Palace of Sans Souci and awe-inspiring fortress known as the *Citadelle*. As a result of the deterioration of existing port facilities at Cap Haïtien, a much needed port development project was undertaken in 1981. This project also included the construction of five small shipping ports along the coast: Baraderes, Corail, Anse d'Hainault, Port-à-Piment, and Anse à Galets.[11]

Haiti as a Political Unit

When Haiti is considered as a political unit, the most striking aspect of the country is its poverty. Its gross national product per capita of $320 in 1983 was one of the lowest in the world. Haiti

---

[11] World Bank *IDA News Release* 81, No. 63 (1981).

A SHANTYTOWN IN PORT-AU-PRINCE.

has no tradition of democratic processes, despite the influence of the French Revolution on the Haitian mulattoes. Haitian independence was gained by a revolt against slavery; when independence was gained, the people were happy to be rid of their French masters and did not welcome new ones.

The Economic Situation

When Haiti became an independent nation, it had fewer than 500,000 inhabitants. From 1789 until 1950 no census was conducted, nor was it thought that a census would be possible. For many years in the present century, the Haitian population was regularly estimated to be about 3 million, but no one could be certain of the figure. In 1950 Haiti actually joined with most other Latin American countries in taking a census. In that year it counted 3,097,000 people, of whom only 8 percent lived in cities and towns. Port au-Prince, the capital and largest city, had a population of only 143,000 in 1950 but had grown to 763,200 by 1984. The population is so widely spread over the national territory that the overall

figure has some meaning—almost 500 people per square mile. The rate of increase is not at all certain, owing to the lack of past censuses, but seems to be less rapid than that of the Dominican Republic. A new census in 1982 gave an official population count of 5,053,792.

Haiti's recent economic history has been bleak. Throughout the many revolutions, the country's coffee, hemp, bauxite, and copper production suffered badly while farm laborers struggled to survive. Crop yields have declined with the continued use of steep mountain slopes that results in soil erosion, and the despair of those who can barely avoid starvation is apparent. Yet, agriculture continues to be the life blood of the economy, and Haiti will continue to grow coffee, sugar, sisal, cacao, cotton, and bananas in the years to come. As in the Dominican Republic, African swine fever resulted in the need to exterminate the hog population. The slaughter of 1.2 million animals in 1982–83 further reduced the food supply, but new and improved breeds are now being introduced.

An important item in Haitian peasant life is charcoal used for cooking, but even charcoal is

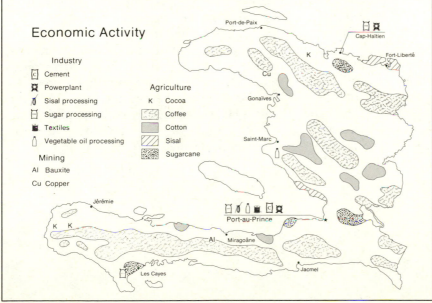

FIGURE 12-8

becoming scarce as the trees used to make this valuable product are cut and often not replaced. Cutting of the forests to make charcoal fosters continued erosion, but an attempt is now being made to plant new forests with bamboo, teak, mahogany, and tung.

The Political Situation

Since independence, the mulattoes of Haiti have had the advantage of education and have taken a keen interest in politics. In the course of time, the government was effectively concentrated in the hands of not more than 300 mulatto families, and the majority of former slaves found that they had new masters who also expected to be obeyed.

The political factions that developed brought chaos to Haiti. Each group that came to power raided the public treasury, and political corruption brought the country to financial ruin. Insecurity in the rural districts was a result not only of banditry, but also of the recruiting system for the army. Able-bodied men were "conscripted" wherever they could be found. As a result, men feared to venture forth on the trails, and attendance at the markets was largely restricted to women.

From 1843 to 1914 Haiti had 24 presidents, only one of whom completed his term in office. The 1915 assassination of President S. D. Sam brought in the U.S. Marines, and Haiti was occupied until 1934. The chaotic internal conditions in Haiti during the early years of World War I were considered a threat to the defense of the Panama Canal. After the war, the Marines remained to keep peace among rival political fac-

tions. Troops from the United States were finally withdrawn in 1934 in the era of the Good Neighbor Policy announced by Franklin D. Roosevelt.

During 1957–71, Haiti was ruled by Francois Duvalier who, like Trujillo in the Dominican Republic, ruled the nation as if it were personal property. In 1964, he declared himself "President for Life" and continued to maintain control by severe oppression. After his death in 1971, his 19-year-old son, Jean-Claude, also became President for Life and continued the oppression of black Haitians that characterized his father's regime. As a result, what was once a trickle of emigrants from Haiti became a mass exodus to the Bahamas and later to Florida.

While the number of Haitians arriving in the United States is smaller than the number of Cubans, it is substantial. At times, as many as 150 per day have arrived in rickety boats along the Florida coast. Most arrive without skills, very few are old, many are pregnant women, and some are children of school age. Like the Cubans, they prefer to settle in Miami, although some have been located in adjacent areas. Much as they lived in Haiti, they cluster in certain neighborhoods, and their social life may center on the local Winn-Dixie Store. In 1981, 16,000 Haitian children entered the public school system of Dade County, Florida.

Within Haiti, life goes on. Poverty is more severe than anywhere else in the Americas, but the hope of survival and a better future prevails. It will be a long while into the future, however, before Haiti compares favorably on an economic or a political basis with most other countries of the Western Hemisphere.

PUERTO RICO

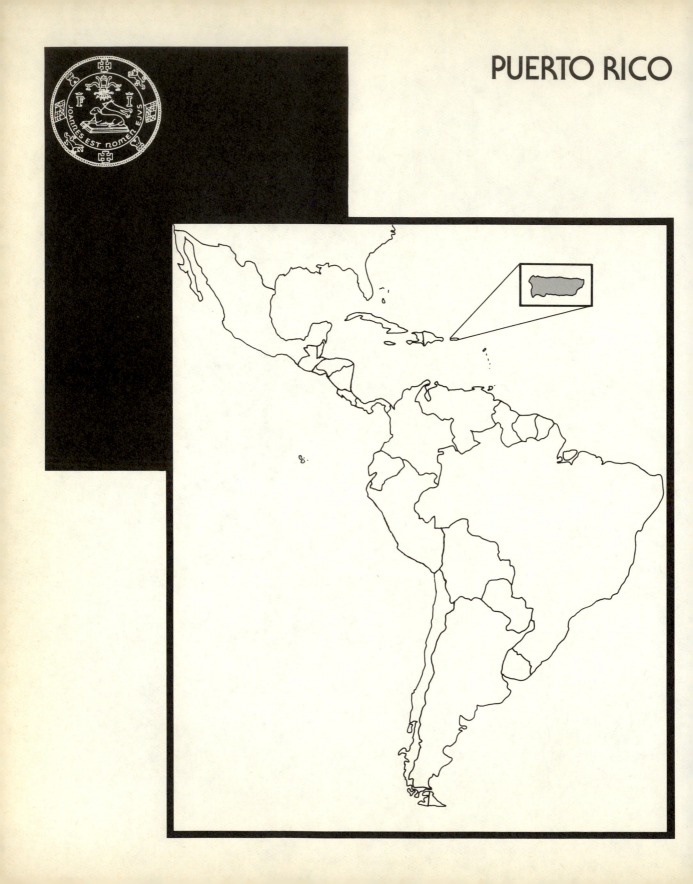

COMMONWEALTH OF PUERTO RICO

Land Area   3435 square miles

Population   Estimate (1985) 3,300,000
Latest census (1980) 3,196,520

Capital city   San Juan 434,849

Percent urban   67

Birth rate per 1000   20

Death rate per 1000   6

Infant mortality rate   16

Percent of population under 15 years of age   32

Annual percent of increase   1.3

Percent literate   91

Percent labor force in agriculture   6

Gross national product per capita $2890

Unit of currency   U.S. $1

Physical Quality of Life Index (PQLI)   90

COMMERCE (expressed in percentage of values)

Exports

chemicals	25	clothing	11
machinery	15	fish	8
petroleum products	14	tobacco	4

Exports to

United States	84
Virgin Islands	3
Dominican Republic	3

Imports from

United States	64
Venezuela	7
Netherlands Antilles	3

Data mainly from the 1985 World Population Data Sheet of the Population Reference Bureau, the 1985 Britannica Book of the Year, and the 1985 World Almanac.

Slowly, but surely, the gross national product of Puerto Rico has risen, as its industry has been modernized and diversified. The trend of emigration to the mainland has been reversed as many Puerto Ricans have returned home. This island has provided the world with a demonstration of how a once very poor country can rise by its "bootstraps." In the 1930s Puerto Rico was described as the "greatest concentration of destitute people under the flag of the United States." After "Operation Bootstrap" proved successful, thousands of visitors came to see for themselves the miracle that had taken place.

The transformation of Puerto Rico was no miracle, however. It was the result of inspired leadership and hard work. Here a Latin American political leader, Luis Muñoz Marín, adopted a program of economic and social reform instead of seeking solutions in the traditional way through political change. Instead of demanding independence from the "Yankee imperialists," Muñoz laid before the Puerto Ricans an alternative. In 1952, with approval of the Congress of the United States and with the support of more than 80 percent of the voters of Puerto Rico, this small island became the Commonwealth of Puerto Rico, voluntarily associated with the United States.[1] Approximately 3.5 million Puerto Ricans maintain citizenship in the United States, travel freely to and from the mainland, enjoy a position within the huge area of free commercial exchange, and exercise all the rights of self-government. A program of economic development led to the rapid increase of industry, until, beginning in 1956, the value of manufactured goods became greater than the value of agricultural products. The problems of poverty have not been completely solved, for Puerto Rico's population has increased to 990 per square mile, 33 percent being rural. However, life is no longer hopeless.

The state-idea in Puerto Rico is not a simple one, but is rather a complex of ideas that make the people feel a love of country, or *patria*. As Muñoz Marín once said:

To the Puerto Rican, patria *is the colors of the landscape, the change of the seasons, the smell of the earth wet with fresh rain, the voice of the streams, the crash of the ocean against the shore, the fruits, the songs, the habits of work and leisure, the typical dishes for special occasions and the meager ones for everyday, the flowers, the valleys, and the pathways. Even more than these things* patria *is the people: their way of life, spirit, folkways, customs, their ways of getting along with each other.*[2]

Puerto Ricans can now look at the changes taking place in their land and say with feeling, *"Es bueno y es nuestro":* it is good, and it is ours.

The Land

Puerto Rico is a rectangular island about 35 miles wide by 105 miles long. It is formed by a tightly folded and faulted arch, the eastern end of the Central American-Antillean system of structures. The highest peak on the backbone of Puerto Rico is only slightly under 4400 feet in elevation, while just north of Puerto Rico is the deepest place in the Atlantic Ocean, a fault trough that lies 30,246 feet below the surface.

Very little of the island is level, and almost all the flat places that do exist are along the coast. On the northern side, behind a coastal lowland, is a zone of limestone terraces, standing not very much above the sea. The mountains begin abruptly south of these terraces, which are especially wide in the western part of the island. Both the north-facing mountain slopes and terraces are deeply dissected by many streams. The main crest of the Cordillera Central, which is also the divide between the streams that flow north and those that flow south, is only 10 miles from the

---

[1] The relationship with the United States is expressed in Spanish: *Estado Libre Asociado.*

[2] Luis Muñoz Marín, "Development Through Democracy," *Annals of the American Academy of Political and Social Science* 283 (January 1953): 1–8.

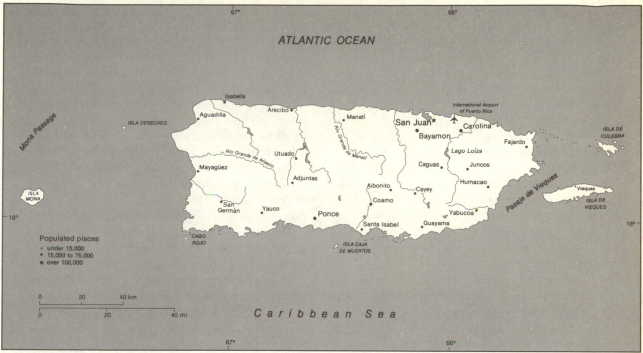

FIGURE 13-1

southern coast. On the rainy north side the streams are cutting vigorously, but on the drier south side many of the stream channels are filled with water only after a shower. Below the rugged belt of foothills, including a series of tilted limestone cuestas (rock layers), the south coast, like the north coast, is fringed by plains.

Because of the simple arrangement of these structural features, the contrast between rainy northern and eastern slopes and dry southern and western slopes is pronounced. On the northeast coast, San Juan, the capital and principal city, receives a rainfall of 60 inches, which in these latitudes may be considered moderate. Ponce, on the southwest coast, receives only about 36 inches, which is inadequate to support more than a sparse scrub woodland and does not permit agriculture without irrigation. Ponce has somewhat higher temperatures in summer and about the same temperatures in winter as San Juan, although both enjoy the "temperate" tropical climate characteristic of the trade wind is-

lands. The green, lush interior of the island receives well over 100 inches of rainfall at some of the higher elevations on the northern slope.

The Course of Settlement

Puerto Rico (known as Borinquén) was a paradise for the Arawak Indians, who reaped abundant harvests from its productive lands; when the Spaniards saw a prosperous and contented Indian population, they thought they, too, had found a paradise. When Europeans first settled on the island in 1508, there were perhaps 30,000 native people living on it. By 1515, dreadful epidemics of imported diseases had reduced the Indians to no more than 4000. As a result, plantations were abandoned, as was the search for precious metals. A few groups of colonists remained in San Juan and Ponce, while most of the interior was used for cattle-raising on large estates.

The Indian population had practically disappeared by the end of the sixteenth century. Black slaves, imported from Africa, replaced them as agricultural workers.[3] By the eighteenth century, Puerto Rico, along with other Spanish possessions in the Antilles, shared moderately in the sugar prosperity. Most of the sugarcane plantations were located on lands of low relief near the coast, and in these areas the black population soon outnumbered the Europeans. The poorer Europeans were forced out of the sugar districts, for free workers could not compete in the same area with slaves. In the mountainous interior the "poor whites" settled as squatters on the vast, unfenced cattle range of large landowners and supported themselves with a shifting cultivation of maize and beans.

Between 1800 and 1825 the island received a considerable number of immigrants of European ancestry. Some came from Spain: from Galicia, Asturias, and the Balearic Islands. They no doubt selected Puerto Rico because it was one of the few Spanish possessions in the New World in which the spirit of revolt from the home country had not developed. A number of immigrants to Puerto Rico came from other Spanish possessions in America from which they were forced to flee because of their loyalty to the Spanish Crown. Consequently, the proportion of people of European ancestry increased, and the proportion with African ancestry decreased. It is now estimated that 99 percent of the Puerto Ricans are of Hispanic origin.

During the nineteenth century the colonies left to Spain in the American Hemisphere suffered from neglect and poverty. During nearly four centuries of Spanish rule in Puerto Rico, only 166 miles of roads were built, and these mostly in the sugarcane districts along the coast. The interior was all but inaccessible and therefore limited in the potential cattle market. The large landowners introduced coffee and made

use of the poor white settlers of the interior as tenants and sharecroppers. So fine was the aroma of the Puerto Rican coffee that it commanded a special place on the Spanish market, much as Haitian coffee commanded a special place on the French market. Transportation costs were high, however, so that even with the high prices the Puerto Rican coffee could bring, only a small net profit was left to the landowners, very little of which was passed on to the tenants.

At the end of the nineteenth century, Puerto Rico showed all the worst aspects of the Spanish colonial system. There was the usual concentration of landownership and wealth in the hands of a very small group, which maintained its position of prestige and economic security through the exploitation of a much larger laboring class. In Puerto Rico the rural workers were not Indians, and only in the sugar plantations along the coast were they blacks. In the interior the tenants were almost pure European. Political, social, and commercial life was centered in San Juan and involved the participation of only a small fraction of the total population. Such was the condition of Puerto Rico in 1898 when this last remaining Spanish colony became a possession of the United States.

A Territorial Possession of the United States

The United States set to work immediately to provide the new colony with all sorts of public works. By 1919 the mileage of all-weather roads had been increased from 166 to 739, and, as a result, the landowners even in remote districts could bring products to market cheaply enough to make a profit. Schools were established, and in some the newer techniques of agriculture and animal husbandry were taught. The literacy rate increased greatly. Sanitary measures were undertaken, and diseases, such as yellow fever, were virtually eliminated. As a result, the death rate was reduced rapidly, and the birth rate increased. Whereas in 1920 Puerto Rico had a population of 1.3 million, it is expected to reach 4.2

[3] R. Picó, *The Geography of Puerto Rico* (Chicago: Aldine Publishing Co., 1974), p. 226.

million by the year 2000. The population density will be more than 1300 per square mile, which is a little less than that of present-day Bangladesh.[4]

A strictly agricultural economy could not provide support for such population densities. Accordingly, emigration to the mainland of the United States was encouraged, and by 1977 the number of Puerto Ricans living there totaled 1.4 million. Beginning in 1970, the rate of population growth on the island rose sharply to an average of 3 percent per year, owing largely to an unpredicted net return migration of 31,000 per year. The reason for the high return may have been unemployment in the United States at that time or the equality of the welfare program in the United States and Puerto Rico. By 1985, the rate of increase had declined to 1.3 percent.

Poverty in Puerto Rico was complicated by the nature of the sugarcane industry. For most people who were crowded around the sugar district, there was no employment at all for several months each year, followed by a period of employment and steady income during the harvest from January to June. When wages were paid, they were higher in the cane plantations than elsewhere. As a result, in the period from 1900 to 1949, the coffee-growing district of the interior lost about 13 percent of its rural population; the population of the cane areas in the same period increased by 31 percent. In 1940, the average annual income of a cane worker was approximately $250—all concentrated in a few months of the year.

Most employees who worked part of the year for the sugar companies lived with their families in poor districts of the cities, especially San Juan. They occupied shacks made from any available material and crowded together without water, lights, or streets. One of the densest slums in San Juan was *El Fanguito* (the mudhole) which covered two miles along the mud flats at the water's edge.

---

[4] *Intercom* 7, No. 4 (April 1979), Population Reference Bureau, Washington, D.C.

The Transformation of Puerto Rico

The transformation of Puerto Rico began in 1940. Beginning in 1929, at a time of the Great Depression, widespread destitution caused considerable unrest. However, by 1940 the economy was gradually transformed from one based on agriculture to that of a modern industrial society. In 1948 Muñoz Marín became the first elected governor of the Commonwealth.

Improvement of Agriculture

When the Popular Democratic Party came to power in the Puerto Rican Congress in 1940, reforms in the land-tenure system were begun immediately. Land redistribution and the improvement of agriculture were not done blindly in Puerto Rico. A first step was an inventory and evaluation of the resource base and existing land use. This was the Puerto Rico Rural Land Classification Program directed by Clarence F. Jones and Rafael Picó.[5] Between 1949 and 1951, detailed maps (1:10,000) of land quality and land use were prepared for the entire island. With the information thus made available, planning for economic development could be conducted effectively. Maps of recommended land use were prepared, roads were constructed, rural electrification was undertaken, and specific sites for new residential and industrial developments were selected.

Agriculture today employs less than 7 percent of the labor force. Sugar is no longer "King." The typical farm is a medium-sized unit employing few workers, but it calls for high capital investment and mechanized equipment. The most important products are milk, meat, chickens, and eggs, with milk ranking first in value. Although sugarcane, coffee, and tobacco are still grown,

---

[5] Reported in Clarence F. Jones, et al., *The Rural Land Classification Program of Puerto Rico*, Northwestern University Studies in Geography, No. 1, Evanston, Ill., 1952; and Clarence F. Jones and Rafael Picó, *Symposium on the Geography of Puerto Rico* (Río Piedras, P.R., University of Puerto Rico Press, 1955).

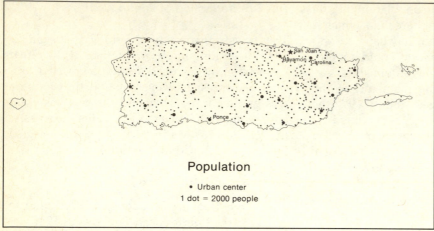

Population

• Urban center
1 dot = 2000 people

FIGURE 13-2

they do not contribute as much to the economy as in the past. There also has been a change in the Puerto Rican diet. Former favorites, such as sweet potatoes, bananas, and taro, have been drastically reduced, while food consumption itself has risen sharply because of food stamps. In 1982 more than half of the total Puerto Rican population was supported by the U.S. Food Stamp Program.

Much of the demand for food is met by imports, yet there has been a growth in dairy and livestock production which is recognized as the principal agricultural activity.[6] The most important commercially produced fruit is pineapple. In Barceloneta, about 30 miles west of San Juan, a modern canning plant ships almost its entire output to continental markets. Other fruits exported from Puerto Rico are coconuts, oranges, grapefruit, mangoes, and avocados. Other crops include maize, beans, peppers, tobacco, and coffee, while the byproducts of sugar (rum and molasses) are also important.

The rainiest and most rugged part of the mountains of Puerto Rico is in the west-central area. This is the main coffee zone, for coffee production is a form of land use that offers maximum protection for steep slopes in wet climates.

Although coffee has been grown on the island since 1736, the concentration in this one district is notable. In the period just before 1940, coffee exports comprised scarcely 1 percent of total exports, yet some 20 percent of all farms and plantations listed coffee as the chief source of income. Today, the most widespread coffee in Puerto Rico is *café arabica*, which requires shade but is also easily destroyed by hurricanes. Yields from the coffee plantations are low, and Puerto Rican growers are unable to compete in the world market. As a result, the U.S. Department of Agriculture has been promoting a type of coffee that will grow without shade and that will produce more and take less time to grow.[7]

Tobacco is grown in the east-central mountains, where temperatures are mild and the humidity is relatively low. This tobacco is used as cigar-filler. On the mainland, Puerto Rican tobacco is more expensive than that of other countries. It is of good quality but cannot compete, as is the case of coffee. However, the industry is relatively stable.

Some people believe that Puerto Rico should seek self-sufficiency in foodstuffs, since food imports amount to almost $1 billion per year. It has been estimated that 60 percent of the arable land lies fallow, while the amount of land under culti-

[6] "Puerto Rico," *ABECOR Country Report* (London: Barclay's Economics Department, June 1980).

[7] Picó, *The Geography of Puerto Rico.*

vation is half of what it was 20 years ago. Conservationists complain that the use of agricultural land for speculative residential expansion is unwarranted. To others, this is modernization.

Puerto Rico does not support a major fishing industry, but its people are large consumers of fish. A great quantity must be imported because Puerto Rico does not have good fishing banks. Fishermen work primarily along the north coast, where most of their customers live. The government has encouraged them also to work away from home, with the result that some have fished for tuna in places such as West Africa and Peru. The tuna is then processed and canned in Puerto Rico, and shipped to large eastern markets of the United States. Puerto Rico is also becoming well known as a fishing area for sportsmen.

Manufacturing

Manufacturing requires cheap electric power. One of the first efforts of the government was the construction of new hydroelectric and steam-electric plants. By 1958 a National Planning Association study of Puerto Rico reported that here was a country in which the hydroelectric potential was fully developed.[8] A grid of transmission lines now provides electricity throughout the island.

The industrialization of Puerto Rico began when the government established a modern cement plant in 1939. This plant and others subsequently erected provide building materials for many new construction programs: for factories, highways, and low-cost housing. The government has built factories that are offered to manufacturing concerns as an inducement to move to Puerto Rico. Industries also were offered a 10- to 30-year tax exemption, which proved an attractive incentive for U.S. companies. Another inducement was that there were no federal income taxes on personal income earned from local

sources. These schemes have been replaced by a more complex system of taxation, and cheap labor no longer exists; the U.S. minimum wage now also applies to Puerto Rico.

Puerto Rico has various kinds of industry, including textiles, screen wire, electric coils, neckties, brassieres, and ballpoint pens. The Commonwealth also continues to be the world's largest producer of rum. New growth industries, such as metallurgical products, chemicals, and electronics, are also expanding. In total there are about 2400 manufacturing establishments, nearly half of which are located in the San Juan metropolitan area. More than half of the employees are women. The government has offered incentives to companies that locate in other parts of the island, with the result that nearly every town in Puerto Rico has at least one factory.[9]

In the late 1960s a whole new group of industries was attracted by capital investment in petrochemicals. Petroleum refining, based on imported oil, permitted the establishment of dozens of satellite industries, giving employment to thousands of workers, while producing a great variety of things ranging from fertilizers to plastics. Emphasis on the manufacture of pharmaceuticals and electronics, plus recent discoveries of copper, nickel, and offshore oil, could lead to further industrial production and bolster the island's economy.

Puerto Rico has maintained a vast program of slum clearance. Low-cost housing provides homes for those whose shacks have been destroyed by the bulldozers. Expressways, tall buildings, and shopping centers have been constructed, as have roads that extend throughout the island. No place is far from a paved highway. There is a modern airport in San Juan and smaller airports at most other cities. Puerto Rico has no railroads.

The tourist business attracts more than 2 million visitors a year. With the closing of tourist attractions in Havana, San Juan became a popular visiting place for people from the mainland,

[8] William H. Stead, *Fomento — The Economic Development of Puerto Rico,* Planning Pamphlet 103 (Washington, D.C.: National Planning Association, 1958).

[9] Picó, *The Geography of Puerto Rico.*

A RECENT SUBURBAN HOUSING DEVELOPMENT IN SAN JUAN, PUERTO RICO.

mostly because of its accessibility. Puerto Rico is making an all-out effort to maintain and protect this industry which brings in about $600 million a year. Three-fourths of the tourists come from the United States; hence, tourism will flourish as long as the economy of the United States remains healthy.

Puerto Rico as a Political Unit

The people of Puerto Rico have a state-idea: loyalty to their *patria* brings coherence and unity such as has been achieved in few other parts of Latin America. Yet, whereas Puerto Rico has certainly benefited from its association with the United States, the leadership, planning, and hard work of transformation have been strictly Puerto Rican.

The Economic Situation

Many statistical data document the favorable economic and demographic conditions of Puerto Rico. In 1966 the value of the gross national product per capita exceeded $1000. Although this placed Puerto Rico in the lead in Latin America, it was only about half of the value of economic production per capita in the poorest state of the United States. By 1984 the gross national product of $3010 was among the highest in Latin America, but it was still lower than that of the poorest state at that time.[10] It appeared that the Commonwealth was not keeping up with projected industrial growth. One reason has been a

---

[10] In 1966 the poorest state in the United States was Alabama; in 1981 it was Mississippi.

HOTELS LINE THE WATERFRONT OF SAN JUAN.

continuing competition from countries such as Japan, Taiwan, South Korea, Haiti, and Mexico. This is especially true of certain manufactures, including shoes, textiles, and women's clothing. Another reason has been a failure to reinvest in plants and equipment.

The Political Situation

The transformation of Puerto Rico has been accomplished by thoroughly democratic processes, with ample opportunity for issues to be discussed publicly, and with decisions left to a plebiscite secretly recorded and honestly counted. Puerto Rico is a demonstration to the world that the problems of poverty and inequality, and of the denial of human dignity, which loom so large throughout Latin America, can be solved within the framework of democracy.

Despite any ill effects of the period of administration by the United States between 1899 and 1948, Puerto Ricans during that time learned to place a higher value on economic development than did most other Latin Americans. Incorporation as a state was not popular in Puerto Rico, and, although there are some who think statehood must eventually be achieved, such a move would not yet receive the endorsement of a majority of the people. The more they develop the feeling of loyalty to *patria*, the less they are likely to accept any plan of incorporation into the United States.

A suggestion by Muñoz Marín seems to permit Puerto Ricans both to have their cake and to eat it, too. The Commonwealth has elected its own governor since 1948; it remains essentially self-governing with regard to domestic matters; it has its own constitution developed within the framework of the constitution of the United States; the people enjoy all the rights of U.S. citizenship, except that they may not vote for the President, and consequently, they pay no income tax on money earned in Puerto Rico; and taxes collected in Puerto Rico remain in the Commonwealth treasury. Puerto Rico is a separate, self-governing country; yet it is an integral part of the United States. Puerto Rico's present status was approved by the U.S. Congress and accepted by the people of Puerto Rico in 1952 and reaffirmed in 1967.

In the elections of 1976 the pro-statehood party secured 48 percent of the vote, while parties opposing statehood won 52 percent. This 52 percent was composed of those who favored independence as well as those in favor of the present status. By 1980 the margin had narrowed; the incumbent Governor Carlos Romero Barcelo received 47.2 percent of the votes to Hernández Colón's 47.1 percent. The winner had campaigned on a pledge that he would call a 1981 plebiscite on the question of statehood if the electorate returned him to office with a conclusive majority. Actually, the voters made it clear that they did not want to be hurried into any decision.

The issue of political status is one that will remain for some years to come. Many people, especially the young, feel that they are losing their Spanish heritage by not being independent. Those who favor statehood tend to ignore the fact that Puerto Rico would lose its tax-free position, and that many American and foreign investments would be lost. On the other hand, it would receive a larger share of welfare funds if unemployment were to rise. If independence were to succeed, there would be a complete change in Puerto Rican living habits, as the absence of free access to U.S. markets would alter exports. Social security also would be lost, and a lower standard of living would prevail as $600 million in food stamps would disappear. Since most of the Puerto Rican population receives direct benefits from social welfare programs of the United States, this aspect of the statehood versus independence question cannot be ignored.

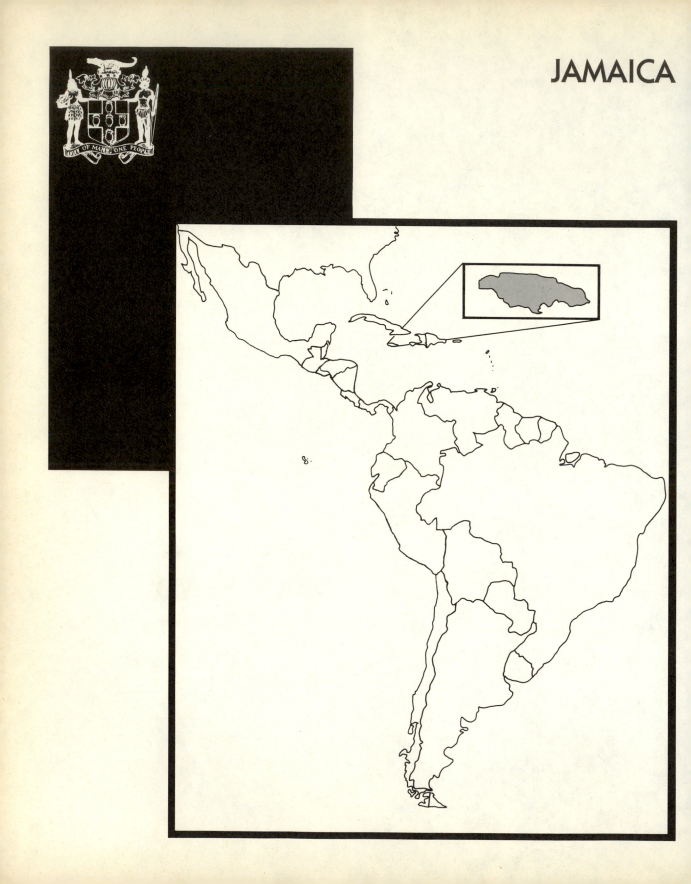

JAMAICA

14

JAMAICA

Land Area   4244 square miles

Population   Estimate (1985) 2,300,000
Latest census (1982) 2,095,878

Capital city   Kingston 684,000

Percent urban   54

Birth rate per 1000   28

Death rate per 1000   6

Infant mortality rate   28

Percent of population under 15 years of age   37

Annual percent of increase   2.2

Percent literate   82

Percent labor force in agriculture   36.4

Gross national product per capita $1300

Unit of currency   Jamaican dollar

Physical Quality of Life Index (PQLI)   87

COMMERCE (expressed in percentage of values)

Exports

alumina   56
bauxite   21
sugar   6

Exports to		Imports from	
United States	33	United States	35
United Kingdom	18	Venezuela	14
Canada	12	Netherlands Antilles	13
Norway	8	United Kingdom	8
Trinidad-Tobago	7		

Data mainly from the 1985 World Population Data Sheet of the Population Reference Bureau, the 1985 Britannica Book of the Year, and the 1985 World Almanac.

Jamaica is third in size and fourth in population among the four islands of the West Indies that comprise the Greater Antilles. Unlike Cuba, Hispaniola, and Puerto Rico, this island has primarily a British historical background. When independence was imminent, Great Britain attempted to bring most of the British West Indies together under one government, but the attempt proved unsuccessful. Now, the former British-oriented confederation has broken up into independent states, associated states, and remaining colonial possessions.

The Federation of the West Indies was established in 1958. This was an unusual kind of political unit. The national territory included scattered small islands extending in an arc from Sombrero in the Anegada Passage for some 700 miles southward to Trinidad off the eastern coast of Venezuela. Also in the Federation were Barbados, 100 miles to the east of the arc, and Jamaica, 1000 miles to the west. Not included were the British Virgin Islands and the Bahamas. With a total population in 1958 of about 3 million people, the density varied among the islands from 1378 per square mile on Barbados to only 207 per square mile on Dominica. Jamaica, with an overall density of 330 per square mile, and Trinidad, with 360, were near the average for the Federation as a whole.

The Federation faced several problems that eventually proved impossible to solve. Jamaica and Trinidad were relatively prosperous compared with the other islands, and the people of these two islands were less than enthusiastic about accepting responsibility for the economic development of their very poor neighbors. Trinidad, where less than half of the population was of African descent, feared a flood of immigrants from overcrowded Barbados. Political leaders on the several islands could not reach an agreement about the location of a capital, but it was eventually awarded to Trinidad. In 1958 the Federation of the West Indies became an independent state and a member of the Commonwealth.

The basic difficulty in forming a coherent state out of such widely scattered pieces was the lack of any commonly held state-idea. In 1961 Jamaica withdrew from the Federation, and in 1962 both Jamaica and Trinidad were granted independence. Barbados became independent in 1966. In 1967 some of the smaller islands of the Windward and Leeward groups were granted self-government, leaving the conduct of foreign affairs and defense to Great Britain.

During most of its brief history as an independent nation, Jamaica has experienced economic growth within an atmosphere of political stability. In addition to their democratic institutions, the Jamaicans have a long record of self-reliance and have demonstrated an exceptional capacity for hard work. Emigrants from Jamaica helped dig the Panama Canal and develop the banana plantations in Costa Rica. As migrant workers they helped harvest the sugarcane in pre-Castro Cuba. Some 85 percent of the Jamaicans are of unmixed African ancestry, and the remainder are mostly mulatto, with a small proportion of Europeans. Yet, the people of Jamaica have preserved much less of the African heritage than have the people of Haiti.

The Land

The island of Jamaica, exceeded in size among the Antilles only by Cuba and Hispaniola, is part of the Central American-Antillean system. Its mountains, however, are deeply covered with upraised coral limestone, and in only a few places do the underlying geologic structures appear at the surface. The highest elevation, 7388 feet, is in the Blue Mountains, near the eastern end of the island—an area of narrow ravines and sharp, knife-life ridges that descend from the central peak like the spokes of a wheel. However, the Blue Mountains, and two much smaller mountainous areas where the underlying formations appear at the surface, are almost completely surrounded by limestone formations which have accumulated to depths of many thousands of feet. The limestone plateau, which at its highest

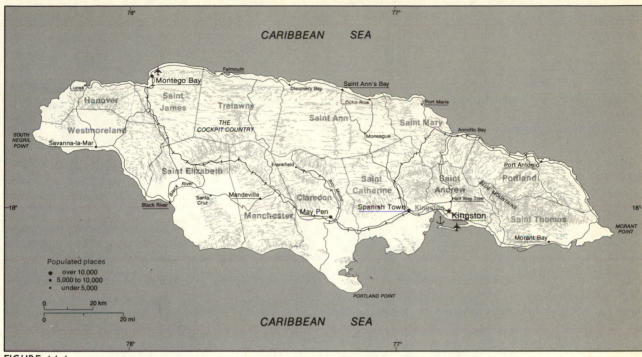

FIGURE 14-1

point is about 3000 feet above sea level, extends from the easternmost point to the westernmost point of the island.

The surface of the limestone plateau is by no means flat, and not more than 14 percent of Jamaica can be classed as level. Streams have cut deep valleys back from the coast, and the limestone itself is honeycombed with caverns and pitted with sinks. In the section known as the ''cockpits,'' some of the many sinks are as much as 500 feet deep. In the midst of the plateau are several large solution basins, their bottoms deeply filled with a residual red soil similar to the Matanzas clay of Cuba. The largest basin is the Vale of Clarendon, 50 miles long by 20 miles wide. These solution basins contain some of Jamaica's most productive agricultural land and are densely populated. Similarly, the coastal lowlands are generally areas of concentrated settlement, especially the large lowland that lies north and west of Kingston.

Rainfall is heaviest on the northeastern side,

particularly in the John Crow and Blue Mountains, and is lightest on the southern and western sides. The sea-border plains of southern Jamaica, including the one near the capital, are too dry to support agriculture without irrigation. Kingston, for example, receives an average of only about 29 inches a year. More than 100 inches are received in the plains on the northeast side. The top of the Blue Mountains, almost always buried in great billowing cumulus clouds, receives between 150 and 200 inches.

Although Jamaica in many ways qualifies as a tropical paradise, the tranquility is periodically shattered by disaster. The southern hurricane track passes over or near Jamaica. Hurricanes are likely to come at least once a year between August and October, and if they do not pass directly over the island, they come close enough to cause damage. Occasionally, too, there are earthquakes. In 1692 an earthquake destroyed most of Port Royal, with the result that Kingston became the most important south coast settlement center.

In 1907 Kingston also was destroyed by a quake but subsequently was rebuilt.

The Course of Settlement

Jamaica was already well populated by Taino Indians, of Arawak culture, when discovered by Christopher Columbus on his second voyage to America, in 1494. On his fourth voyage, Columbus and his crew came to know the island better, as they were marooned there for slightly more than a year. No gold or silver was found, however, and few Spaniards elected to settle there permanently during the next 200 years. Hence, the island was easily conquered by an English army in 1655.

Cattle-raising was the principal activity of the early Spanish settlers, but historically agriculture has been Jamaica's chief concern. Unprofitable attempts were made at growing indigo, tobacco, and cotton before sugar became the main crop. During the seventeenth and eighteenth centuries, the plantations were located on plains near the coast. The source of labor was slaves, and by the end of the eighteenth century Jamaica's population consisted of about 20,000 whites and 300,000 African slaves. The remainder was composed of "free coloureds" born of slave women and white men.

The chief concentration of people was around Kingston and on the lowland that extends westward. The interior of the island was used for cattle-raising or for the slaves to raise their own food. When slavery came to an end, there was an exodus of former slaves from the sugarcane lands to the interior. The idea was to escape as completely as possible from work that was so closely associated with slavery.

Agriculture and Land Tenure

The resulting pattern of rural land tenure persisted until the 1960s. The property granted to new farm settlers remained the property of all the descendants of the first occupants and could not be sold without their consent. By 1961 about 70 percent of the farms on Jamaica were less than five acres in size. To combine farms or sell farms would require the agreement of thousands of rel-

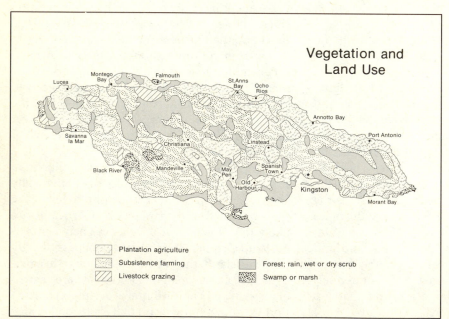

FIGURE 14-2

atives, scattered all over the world. Furthermore, many city people looked on the system as a kind of social security, for if a city job were lost the individual could always move back to the farm and claim a portion of its produce. Most of the people actually working the land were old folks, and crop yields were low in part because of poor farming techniques and in part because of disastrous soil erosion on steep slopes. Only 15 percent of the total area of Jamaica can be considered suitable for crops, and this includes some land that should never have been cleared of its original forest cover.

By 1965 tree crops, such as coffee or citrus, had replaced the traditional crops of maize, yams, and bananas. It was estimated that the value of crops produced on the small farms exceeded the value of crops grown on plantations for export. Much of interior Jamaica is unsuited for agriculture. In places this is due to steepness of the slopes, but in large areas where solution caverns in the limestone drain water underground the surface is too dry for crops. This is excellent pastureland, and traditionally Jamaica has been an important producer of beef for local consumption.

As in many other countries of Latin America, the concentration of wealth in the hands of a few people has been notorious. In 1980 a report by the Interior Ministry revealed that 21 upper-class families of English, Chinese, Syrian, and Jewish backgrounds dominated the cultural, economic, and intellectual life of Jamaica. This group controlled 104 companies with total assets worth several hundred million dollars, involving such sectors as electricity, banking, foreign trade, and natural resources. In the agricultural sector, the group controlled the production of sugar, rum, and bananas.[1]

Commercial Agriculture

Some 40 percent of the agricultural area of Jamaica is held in a small number of large estates,

operated by corporations. In the slave period, Jamaica was able to produce a record sugar crop of 100,000 tons; this figure was not equaled until the 1930s. The all-time record crop year was 1964–65, when 500,000 tons were produced. Output was estimated at about 280,000 tons in 1980, at which time the industry was plagued by smut and rust diseases as well as mill breakdowns, labor disputes, and unfavorable weather. Some sugar is always set aside for the manufacture of a distinctive dark rum that is unique to Jamaica. Production reached 6.5 million gallons in 1981.

Another traditional commercial crop in Jamaica is the banana. In 1870 an enterprising businessman from New Jersey shipped a load of bananas to the United States from Jamaica and realized a large profit from the sale of this then unfamiliar fruit in New York. The company he set up to grow and ship bananas was one of those that combined in 1899 to form the United Fruit Company. Before World War II about 32 percent of the cropland of the island was used for this crop, and the annual export of bananas exceeded 25 million stems, constituting between 50 and 60 percent of the colony's exports. The unusual fact about banana-growing in Jamaica was that most of the crop was grown by small independent farmers; only 30 banana plantations were larger than 30 acres.

During the 1960s banana production in Jamaica increased greatly. A new variety, immune to the Panama disease, was introduced, and Sigatoka disease was kept under control by the use of chemical sprays. Bananas were produced mostly on small plantations, where they fit in easily with the production of other commercial crops, but there were also a few areas of large plantations, as along the rainy northeast coast. Recently, torrential floods, rain, and hurricane damage have caused 95 percent destruction in the main banana-producing areas. Other commercial crops produced in Jamaica include high-quality coffee from the Blue Mountains, a premium type sold almost entirely to Japan; cacao from the wetter places along the coast; and to-

[1] *Comercio Exterior de México* 26, No. 11 (1980):413.

bacco, cotton, citrus, rice, maize, and root crops. It is the small farmer, however, who holds the key to self-sufficiency for Jamaica.

Today, a new breed of agricultural agent is being trained in rural Jamaica, educated in composting, crop rotation, and nutrition. Many of those involved in the program are young women who travel from farm to farm, listening to small farmers and teaching them how to get more from the land. The government, staggering under the forces of population growth, migration to the city or another country, deforestation, and economic chaos, has tried to solve some of the problems by encouraging small farmers to plant cash crops to be sold to hungry people in the city.

While most farmers struggle to make an honest living, others grow the illegal *ganja* (marijuana). Those who grow it earn about $250 million annually, more than the government collects from its principal legal export, bauxite. To the mobsters who transport it to the United States, it represents about $750 million yearly, which is deposited in banks in Miami, Grand Cayman, and the Bahamas. It is believed that the *ganja* trade is controlled by a mystic organization that calls itself the "Coptic Church" and considers *ganja* smoking a part of its religion. In any event, the Coptics are becoming wealthy as they buy up supermarkets, real estate, and trucking firms.[2] Unfortunately, the poor suffer the brunt of the tragic marijuana trade.

Mining

In certain parts of the world, mostly in the rainy tropics, the continued downward percolation of rainwater removes the soluble minerals from the soil, leaving a residue in which there is a concentration of insoluble iron and aluminum compounds. Where aluminum oxide occurs in sufficient concentration to make mining worthwhile, the ore is called bauxite. In Jamaica deep accumulations of bauxite occur in the bottoms of solution basins. Jamaican bauxite has a relatively low alumina content, but the size of the bauxite accumulation makes large-scale mining possible, which greatly reduces the cost of production. It is estimated that Jamaica has the world's fourth largest reserves of bauxite, some 2000 million tons. From 1957 to 1971 Jamaica was the world's largest supplier. It has subsequently been surpassed by Australia and several other countries, including Guinea, West Africa.

Six mining corporations operate in Jamaica, including Kaiser and Reynolds Aluminum, Alcan (Aluminum Corporation of Canada), and Alcoa (Aluminum Company of America). Jamaica's capacity of 15 million tons of bauxite per year is expected to increase substantially within the near future. The Jamaican government has acquired 51 percent of the capital of companies exploiting

STUDENTS BEING TRAINED IN A FARM PROGRAM SPONSORED BY THE WORLD BANK.

---

[2] *The Wall Street Journal*, October 6, 1980.

A BAUXITE OPERATION ON THE NORTHERN COAST OF JAMAICA.

the bauxite mines, and this nationalization program has been extended to cover certain other sectors, including agricultural exports.[3]

Jamaican law requires that after bauxite has been extracted by huge mechanical shovels, the land be smoothed over and made ready for crops or pasture. The aluminum companies have returned their mined-out areas to pasture grasses and have imported high-grade beef and dairy cattle as a contribution to the island economy. Jamaica also has a small but important source of gypsum, located near the coast about 10 miles east of Kingston, which is used for cement production. The port of Kingston is also the site of a free-trade zone.

_____

[3] *Comercio Exterior de México* 26, No. 11 (1980).

Jamaica as a Political Unit

The world has become full of independent states, like Jamaica, struggling to survive in the midst of conflict and revolutionary innovation. The question remains, however, whether any group of people with feelings of nationality can form a viable state, or whether viability requires a minimum number of people. It may be that the survival of small states will depend on the organization of federations, or at least of free-market areas. Jamaica, still seeks to establish its viability as a nation.

The Economic Situation

Jamaica has major economic problems yet to be resolved. With an overall population density of

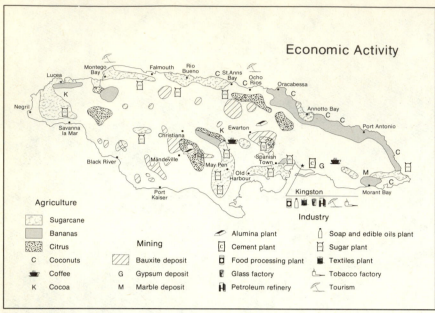

Economic Activity

Agriculture

	Sugarcane
	Bananas
	Citrus
C	Coconuts
	Coffee
K	Cocoa

Mining

	Bauxite deposit
G	Gypsum deposit
M	Marble deposit

Industry

	Alumina plant		Soap and edible oils plant
C	Cement plant		Sugar plant
	Food processing plant		Textiles plant
	Glass factory		Tobacco factory
	Petroleum refinery		Tourism

FIGURE 14-3

565 people per square mile and a much higher density of population per square mile of cropland, many rural Jamaicans live in poverty. There is immediate need to improve the productivity of agriculture and to reduce the import of basic foods, but any program of agricultural improvement must include a reduction in the number of farmers. In 1985 about 36 percent of the working force was employed in agriculture, and in the cities displaced farmers remained unemployed. Unfortunately, many men and women over 40 years of age have never been employed or trained in any way.

Jamaica, like Puerto Rico and Cuba, has been the source of a steady flow of emigration. For years Jamaicans emigrated to cities such as Birmingham, England, and to East Coast cities of the United States in search of jobs. As a result, for the first time in its history the British came face to face with problems of race relations. Because the Jamaicans are dark-skinned, they are easily identified and easily become victims of discrimination. In 1965 Great Britain began to restrict the immigration of West Indians, but problems of race relations have continued to the present day.

In the 1950s Jamaica began a program of industrial development modeled after that of Puerto Rico. Tax incentives were offered to attract manufacturing firms from the United States and the Commonwealth countries. There were new jobs in the bauxite and sugar refineries. Factories were built to produce textiles and clothing, batteries, paint, soap, porcelain, and cement. There was even an oil refinery near Kingston, a cement plant, and a small steel mill using imported scrap metal. Jamaica's great disadvantage is its distance from potential customers. Consequently, the principal source of new jobs has been the tourist industry.

The Political Situation

While still filled with intoxicating dreams of a prosperous future following independence, Jamaica was put under great stress by the "OPEC war." The price of oil was raised almost overnight from about $2.50 per barrel to more than $11. Simultaneously, there was a worldwide depression. Michael Manley, who was swept into office as Prime Minister in 1972, imposed such

You don't have to get pregnant

Family Planning Clinic

EFFORTS TOWARD POPULATION CONTROL IN JA-MAICA.

huge levies on bauxite that it became noncompetitive in the world market, and hostile employees in the government-run hotels almost destroyed the tourism industry. The results were

disastrous. Manley's aggressive brand of "democratic socialism," which included expropriation of private farms, abolition of private education, and heavy taxation on business, was a complete failure. The Jamaican dollar off the island became worthless, and everyone's standard of living declined. This resulted in an escalation of smuggling and black market operations. It became obvious that a tremendous amount of capital investment would be required in the island if a bleak future was to be avoided.

In 1980, almost a million people went to the polls to vote for a new Prime Minister. Opposing Manley, of the Peoples National Party (PNP), was Edward Seaga, of the Jamaican Labor Party (JLP), a conservative organization. Seaga's party won by more than 75 percent of the votes. In his inaugural address, Seaga declared that Jamaica was neither a capitalist nor a Communist country. According to Mexico City press reports, his economic views are founded on a free-market concept, the "Puerto Ricanization" of Jamaica.[4] Emphasis was placed on agriculture and education, both of which were greatly in need of attention. The National Family Planning Board established two major objectives: to educate Jamaicans against having more than two children per family, and to keep the total population under 3 million in the year 2000. Also important was the expressed need for increased political stability after some years of severe internal strife.

[4] *Comercio Exterior de México* 26, No. 11 (1980).

TRINIDAD
AND TOBAGO

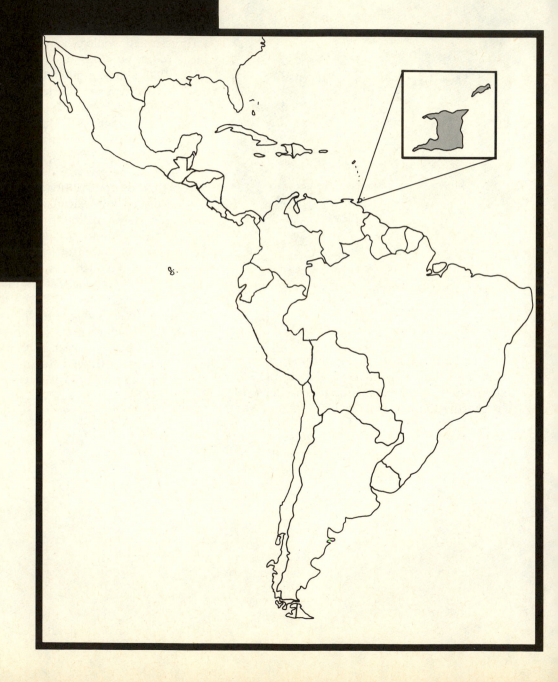

15

TRINIDAD AND TOBAGO

Land Area   1980 square miles

Population   Estimate (1985) 1,200,000
             Latest census (1980) 1,059,825

Capital city   Port-of-Spain 250,000

Percent urban   23

Birth rate per 1000   25

Death rate per 1000   7

Infant mortality rate   28

Percent of population under 15 years of age   34

Annual percent of increase   1.9

Percent literate   96.7

Percent labor force in agriculture   10

Gross national product per capita $6900

Unit of currency   Trinidad and Tobago dollar

Physical Quality of Life Index (PQLI)   87

COMMERCE (expressed in percentage of values)

Exports

petroleum products   88
chemicals   5
transport equipment   3

Exports to		Imports from	
United States	50	Saudi Arabia	26
Netherlands	7	United States	26
Italy	4	Indonesia	10
Guyana	4	United Kingdom	9
Suriname	4	Japan	6
Honduras	3		

Data mainly from the 1985 World Population Data Sheet of the Population Reference Bureau, the 1985 Britannica Book of the Year, and the 1985 World Almanac.

Trinidad was discovered by Christopher Columbus in 1498, and its name derived from three
low mountain peaks he sighted on the south
coast. Captured from Spain in 1797, Trinidad
became a British colony and has been linked with
the small island of Tobago since the latter's sugar
industry failed in 1888. The two islands were
granted independence in 1962 after the breakup
of the Federation of the West Indies and remained a member of the British Commonwealth
of Nations until 1976 when they became an independent republic. Trinidad itself has a cosmopolitan, officially English-speaking society composed of blacks (43 percent), East Indians (40
percent), mixed (14 percent), and Chinese and
others (3 percent). With only 1980 square miles
of territory, Trinidad and Tobago's 1.2 million
people provide an overall density exceeding 600
per square mile.

The population is concentrated in and around
the capital, Port-of-Spain. Smaller concentrations occur in San Fernando to the south, Sangre
Grande to the east, and Rio Claro toward the
southeast.

TRINIDAD

Geologically, Trinidad belongs to the continent
of South America rather than to the Antilles. Its
seven-mile separation from the easternmost
point of northern Venezuela is the result of a
geological fault. Beyond the narrow passage, the
same mountain structures that form the coastal
range of Venezuela continue across the entire
northern side of Trinidad, reaching a maximum
elevation of 3085 feet. Tobago, 22 miles north-
northeast of Trinidad, is also related geologically
to these continental mountains. The southern
part of Trinidad is composed of two ranges of low
hills, not more than 1000 feet in elevation, crossing diagonally from northeast to southwest, and
two lowland areas, now slightly uplifted and dissected by streams. Mangrove-filled swamps,
separated from the sea by wide sandy bars, characterize parts of the eastern and western coasts.

Trinidad lies in the path of the trade winds. At
the eastern end of the Northern Range the rainfall is as much as 150 inches. Over this area large
cumulus clouds begin to form early every day,

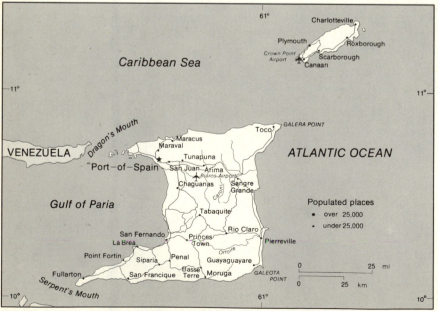

FIGURE 15-1

even during the drier part of the year, and drift westward across the island bringing brief but heavy downpours. The rainy season is between June and December, when on an average, two days out of three on the western side of the island feature at least one shower. On the eastern side of the island, there are few days that do not have frequent showers interspersed with periods of brilliant sunshine.

The temperatures on Trinidad, which lies only about 10° north of the equator, are even more "temperate" than those experienced on the northern islands of the Antilles. The range between the average of the warmest and the coldest months at Port-of-Spain, on the western side of the island, is only about 3°, and the average temperature for the year is about 77°F. Cold air masses from North America do not reach Trinidad, nor do hurricanes touch this fortunate island, for they reach the Antilles on their way across the ocean from western Africa at least two degrees of latitude farther to the north.

Trinidad was originally covered almost completely by a heavy tropical rain forest. Only the dry westernmost end of the Northern Range had a sparse vegetation of xerophytic character. The highest peak of the Northern Range, Cerro Aripo (3085 feet), is clothed to its summit with forest.

LUSH TROPICAL GROWTH IN THE MARACAS BAY AREA OF TRINIDAD.

Settlement and the Economy

Trinidad was for many years a possession of Spain. Because it contained none of the resources of land or people that were attractive to the Spaniards, it was never effectively settled by them, however. Port-of-Spain was established at the southern base of the Northern Range, where protection from the easterly winds together with a deep anchorage and a firm landing place offered ideal conditions for a port. Although Trinidad remained a Spanish possession, many Frenchmen came to the island, introducing African slaves in considerable numbers. In 1797 the British seized control of Trinidad and of numerous other French and Spanish possessions in the Antilles. In 1802 most of these conquered territories were returned to their former owners, but Trinidad remained British until 1962. The population of the island still reflects this background of early settlement, for there are many families of French and Spanish descent as well as those that came later from Great Britain.[1]

Sugarcane

The importation of East Indian contract laborers, beginning in 1845, made possible the rapid expansion of sugarcane. Sugarcane spread not only around the original plantations along the Gulf of Paria, but also to other parts of the island. All over the island the forest was cleared, cane was planted, and small sugar mills were established.

The technological revolution that changed sugar production from a primitive small-scale, high-cost operation to a large-scale industry came to Trinidad during the last quarter of the nineteenth century. At that time planters in the Antilles were faced with serious competition from the beet-sugar growers of Europe. Failing to secure government aid in the form of subsidies and price controls, Trinidadian planters had to go out of business or invest more capital to make

their production more efficient. The first modern, large-scale mill was built in 1871, just inland from San Fernando, which is still the main center of sugar production. Whereas each small mill had been supplied with cane from fields immediately around it and was operated directly by the planter, the new large mills needed at least 38 square miles of cane to operate with maximum efficiency. As a result, most small mills ceased production, and sugarcane planting was concentrated again in the part of the island best suited to it.

Today the sugar industry is in trouble and will remain so unless drastic action is taken to improve its performance. It consists only of two government-owned companies that employ about 15,000 people and occupy 100,000 acres of land. A survey of cane farmers and sugar workers conducted in 1978 indicates that most of these people have only an elementary education or none at all. Their children are being educated to find employment elsewhere. More than 90 percent of the workers are of East Indian descent and are either Hindu or Moslem. Religion, race, the unions, and the "sugar vote" are so tied into the sugar industry that decisions affecting the industry have tended to be political rather than business decisions.[2] In former years sugar accounted for about half of total agricultural output. Production in 1976 was 200,000 tons; by 1980 it had declined to only about 110,000 tons. The industry is affected by international problems as well. The price on the international market is determined by external forces, whereas on the local market it is determined by the Prices Commission. Ironically, Trinidad must now import sugar because of increasing domestic consumption.

Crops

Trinidad has long been famous for its cacao, but by 1950 much of the cacao area had been aban-

---

[1] Preston E. James, "Changes in the Geography of Trinidad," *Scottish Geographical Magazine* 73(1957):158–166.

[2] V. Ramlogan, "The Sugar Industry in Trinidad and Tobago: Management: Challenges and Responses," *Inter-American Economic Affairs* 33, No. 4(1980):29–59.

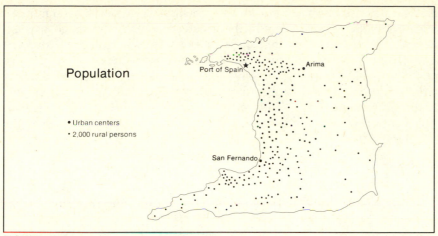

Population

• Urban centers

· 2,000 rural persons

Port of Spain

Arima

San Fernando

FIGURE 15-2

doned. Cacao is still grown but in small quantities. Other crops include coffee, citrus, bananas, and rice. Efforts are being made to expand the production of dairy products and to broaden the agricultural base by growing more food crops such as sweet corn, squash, peanuts, peppers, onions, garlic, carrots, and tomatoes, as well as fruit crops like papayas, avocados, and mangoes.

Oil, Asphalt, and Manufacturing

The economy of Trinidad is dominated by petroleum production, but there also has been a considerable development of manufacturing. The island is well known for its Pitch Lake, a natural occurrence of asphalt resulting from the seepage of oil from underground into a shallow depression at the surface. Distillation of the volatile parts of the oil leaves a residue of asphalt, which continues to accumulate as rapidly as it is dug out. In 1595 Sir Walter Raleigh made use of the pitch to caulk his ships. Now, the asphalt is extracted and carried by cable line to a port on the southwestern shore of Trinidad on the Gulf of Paria. For many years Pitch Lake was the world's only source of asphalt, but during the twentieth century its product has met heavy competition from asphalt produced in oil refineries. Since World War II Trinidadian asphalt has again become competitive, and large amounts have been shipped to Great Britain where it is used for street pavement. Pitch Lake is the largest lake of its kind in the world, some 100 acres in size, with reserves of 10 million tons.

Near Pitch Lake and extending southward are numerous oil wells. Wells are also scattered throughout the southeastern part of the island and offshore to the west. During World War II and thereafter, new wells were drilled, even into the Gulf of Paria. Still, production has declined, because no major oilfields have been discovered in recent years. It is estimated that at present rates of recovery the proven crude reserves will last only 10 more years.

Production of natural gas, unlike petroleum, is increasing rapidly. Proven reserves amount to about 12 trillion cubic feet. This production will satisfy local needs and leave a surplus for export for at least another century. Thus, in the long term natural gas may replace oil as the principal source of revenue.[3] A $6 billion heavy-industry complex at Point Lisas, about 25 miles south of Port-of-Spain, is geared to supplies of natural gas. Activities include the manufacture of petrochemicals, iron and steel, fertilizers, pulp and

[3] "Trinidad and Tobago," *ABECOR Country Report* (London: Barclay's Economics Department, October 1980).

paper, methanol, and liquified gas. There is also a new bauxite smelter, producing 180,000 metric tons of aluminum per year.

One of the oldest industries in Trinidad is the manufacture of Angostura Bitters. This bittersweet liquid gives a zip to certain alcoholic drinks and is made only on this island. Its formula has been jealously guarded for generations, and some call it "Trinidad's secret weapon."

TOBAGO

Untouched as yet by developers, life on the little island of Tobago is slow-moving and relaxed, but it has not always been that way. Before 1803, the government changed hands 31 times, and the island experienced murderous raids, retaliatory

attacks, and rebellion. Because of a sugar monopoly in the nineteenth century, Tobago was extremely prosperous, but when disaster struck the financial operations of the industry, the island's economy collapsed. Then, Tobago came under the protectorship of Trinidad. At the present time Tobago has considerable authority over its own affairs. It is but one-sixteenth the size of its neighbor, and its five "cities" are really just miniature settlements along the wayside.

Trinidad and Tobago as a Political Unit

Although Trinidad has the highest gross national product per capita in all of Latin America, exceeding even that of Venezuela, it has not solved all problems of creating a viable state. There are

THE PEACEFUL GREAT COURLAND BAY OF TOBAGO.

serious difficulties to be faced, and these are both economic and political in nature.

The Economic Situation

Since Trinidad became independent in 1962, the economy has grown rapidly, but it has not yet reached a level that provides general prosperity to the people. Some 31 percent of the population is under 15 years of age, and unemployment exceeds 13 percent. The population of Trinidad is too large to be supported by an underdeveloped economy and yet is far too small to provide the mass market that supports modern industry. The nation is able to assist some of the less developed Caribbean countries, however, by supplying them with low-cost oil and financial aid.

Trinidad was the first Latin American country to scrap its railroads and to try to meet its land transport needs with trucks and buses. The change was made easier in Trinidad because of its small size, because of short land hauls to ports on the Gulf of Paria, and because it had low-cost asphalt with which to surface highways. With the extremely high world oil and gas prices, Trinidad has prospered in recent years, and with a massive investment in capital projects, the economy is expected to show steady progress.

The Political Situation

Problems that lie beneath the surface of Trinidad's political stability are related to the nature of its inhabitants. Trinidad is much less homogeneous in its population than is Jamaica. Its racial diversity is conspicuous, and while English is the official language, Spanish, French, German, Chinese, and Arabic are spoken, as well as Hindi, Creole, French, and Portuguese. Furthermore, there are important religious differences: 66 percent of the people are Christian, 23 percent are Hindu, 6 percent are Moslem, and the remainder are mostly Buddhist. To get Christians, Hindus, and Moslems to work together to form a coherent national group is not easy, but, fortunately, in Trinidad these diverse religious and racial groups are intermingled. The problems would be much more difficult to solve if each of the religious groups occupied separate parts of the national territory.

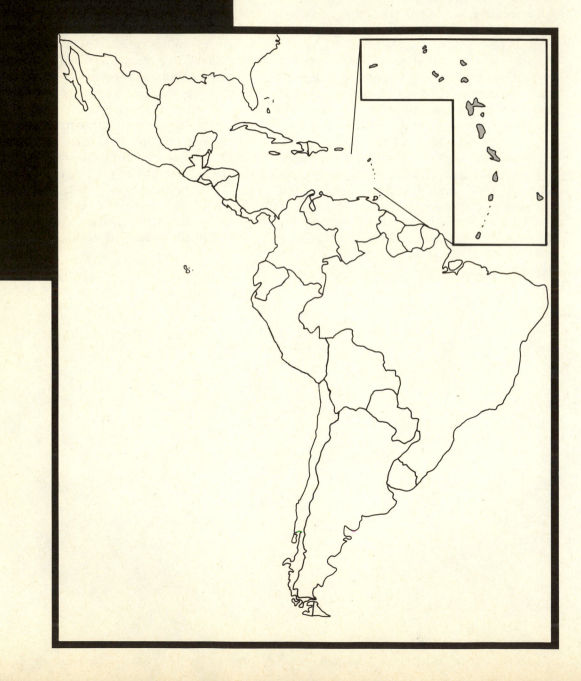

16

BARBADOS

Land area	166 square miles
Population	Estimate (1985) 300,000
	Latest census (1980) 248,983
Capital city	Bridgetown 7600

GRENADA

Land area	133 square miles
Population	Estimate (1985) 111,000
	Latest census (1970) 92,775
Capital city	Saint George's 7,500

SAINT VINCENT AND THE GRENADINES

Land area	150 square miles
Population	Estimate (1984) 134,000
	Latest census (1970) 86,314
Capital city	Kingston 32,600

SAINT LUCIA

Land area	240 square miles
Population	Estimate (1984) 122,000
	Latest census (1974) 92,000
Capital city	Castries 48,782

DOMINICA

Land area	290 square miles
Population	Estimate (1985) 100,000
	Latest census (1970) 69,549
Capital city	Roseau 8,300

ANTIGUA-BARBUDA

Land area	171 square miles
Population	Estimate (1985) 100,000
	Latest census (1970) 64,794
Capital city	Saint John's 30,000

SAINT KITTS-NEVIS

Land area	101 square miles
Population	Estimate (1984) 47,200
	Latest census (1980) 44,500
Capital city	Basseterre 14,700

BAHAMAS

Land area	5353 square miles
Population	Estimate (1984) 260,000
	Latest census (1980) 209,505
Capital city	Nassau 143,148

MARTINIQUE

Land area	425 square miles
Population	Estimate (1984) 333,800
	Latest census (1982) 326,536
Capital city	Fort-de-France 97,800

GUADELOUPE

Land area	687 square miles
Population	Estimate (1984) 328,400
	Latest census (1982) 328,400
Capital city	Basse-Terre 13,700

NETHERLANDS ANTILLES

Land area	383 square miles
Population	Estimate (1984) 258,400
	Latest census (1972) 223,196
Capital city	Willemstad, on Curacao 156,000

The term *Lesser Antilles* in a general sense can be applied to all islands of the Caribbean area other than Cuba, Hispaniola, Jamaica, and Puerto Rico. It also commonly refers in a more limited way to just the arc of small islands that extends like a string of pearls from Puerto Rico to Trinidad, separating the Caribbean Sea from the main body of the Atlantic Ocean. Here, the Lesser Antilles are considered as five island groups: the independent nations, the British dependencies, the French Antilles, the Netherlands Antilles, and the American Virgin Islands.

The Independent Nations

All of the islands within the Lesser Antilles that are now independent nations, eight in number, are former British possessions. These are Barbados, Grenada, Saint Vincent, Saint Lucia, Dominica, Antigua-Barbuda, Saint Kitts-Nevis, and the Bahamas, stretching across a distance of more than 1300 miles. First among these to become independent, and in many ways the most important, is the nation of Barbados.

Barbados

Easternmost of the Antilles, Barbados is part of the outer arc of islands composed largely of limestone formed by coral. The terrain is generally level or gently rolling, which has facilitated the cultivation of sugarcane. Mount Hillaby, in northcentral Barbados, rises to 1104 feet above sea level, the highest point on the island, and all around the coast are white sandy beaches that favor the development of tourism. Although this country is less than one-seventh as large as the state of Rhode Island, its population density is exceeded by that of few places on the surface of the earth. In contrast with most other parts of Latin America, Barbados has not become highly urbanized. Most of the population is concentrated in the southwestern part of the island around Bridgetown, but the national capital itself remains only a small town. Elsewhere the inhabi-

tants live along rural roads in tiny but well-kept wooden houses or mainly in a multitude of villages that dot the landscape.

In the early seventeenth century when British explorers first landed on this remote island, they found it unoccupied, and in 1627 a British colony was established. Actually, the island was discovered by the Portuguese who named it *Os Barbados,* meaning "the bearded ones," because the scraggy-looking banyan trees reminded them of men with beards. The island was never claimed for Portugal, and no other country except Britain has ever ruled it. By 1640 its population consisted of 37,000 whites and 6000 African slaves. The introduction of slavery led to a mass exodus of white workers so that in 1667 there were but 20,000 whites and by 1786 only 16,000 while blacks numbered 62,000. The steady increase of Africans in the nineteenth century brought their numbers to 180,000 in 1921, whereas the white population remained more or less steady at 15,000. At present, the population is about 92 percent black, 4 percent of European origin, and 4 percent mixed.

Barbados is unusual among the Lesser Antilles in that three-fourths of its total area is in farms. Sugar production has been the mainstay of the economy for many generations but is declining steadily, partly because of a shortage of labor at harvest time and partly because of the larger problem of depressed sugar prices worldwide. Still, sugar is the leading item among the island's agricultural exports.

Agricultural diversification is a primary objective of the government. To reduce the cost of importing food, the cultivation of green vegetables, potatoes, yams, and citrus has been encouraged and, along with dairying, has shown some success. Barbados is still largely dependent on imported fruit, except for locally grown mangoes, avocados, and papayas, while items such as lettuce and tomatoes are usually air-freighted from Florida and are quite expensive when available. Despite the surrounding ocean, even fish are commonly in short supply.

Barbados has been quite successful in its ef-

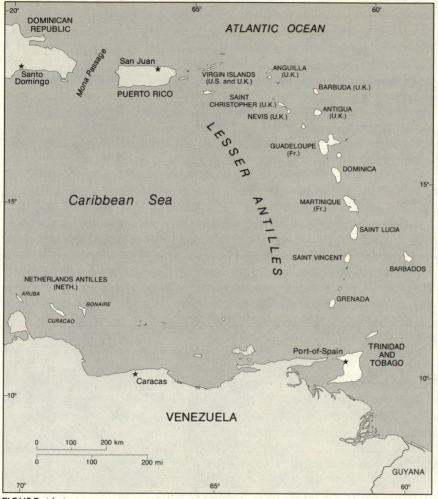

FIGURE 16-1

forts to promote industrial development. Companies based in the United States have established branch plants in Barbados to manufacture products such as electronic components, toys, inexpensive jewelry, and sporting goods. Exports to the United States predominate, but European markets are also important. More than a fourth of Barbados' exports are to the European Economic Community (EEC). Companies in Barbados benefit from preferential trading agreements that allow certain locally made products to enter the United States and EEC countries free from tariff duties. Especially attractive to international busi-

ness are the legal system based on the British model and an efficient communication system, including shipping through a new deep-water port at Bridgetown and worldwide airline connections.

Barbados possesses a small oil and gas field that supplies about a third of its petroleum needs. Offshore geophysical exploration has been undertaken with the hope of adding to the existing onshore reserves of about 1 million barrels.

Barbados, along with Jamaica and Trinidad, is one of the three major political units that were formerly parts of the Federation of the West

THE WATERFRONT AND DOWNTOWN AREA OF BRIDGETOWN, BARBADOS.

Indies, which lasted only from 1958 to 1962. With the breakup of that union of widely dispersed and diverse entities, Barbados went its own way in 1966 as the first independent nation among the Lesser Antilles.

The people of independent Barbados have enjoyed a relatively good living. The gross national product is not as high as that of Trinidad, but it does exceed that of any of the other nations in the Lesser Antilles, and Barbados has the highest literacy rate in all of Latin America. This neat, pleasant, attractive island has enjoyed political stability, which helps to make it the ideal tourist spot in the Caribbean. There are more than 3000 hours of sunshine annually, and the tropical climate is favored by the prevailing trade winds. It is therefore not surprising that tourists return again and again. In 1979 they left a record $166 million for the Barbadian treasury, which amounted to more than a third of the country's total foreign exchange earnings and exceeded the value of sugar exports.

The major challenge for Barbados is the pressure of population. It has been assumed that emigration from the Caribbean area has been the primary curb on population growth rates. In the case of Barbados this is not true. In 1955 Barbados became the second country in the world, after India, to support a National Family Planning Program. It was thus one of the first countries to recognize the need to control population growth before it reached crisis proportions. Billie Miller, the first woman to become a Minister of government in Barbados, has declared that "Family planning is not a political issue; it is a question of survival, or at least survival with a decent standard of living for everyone."

Along with a strongly supported state-idea, the people have a deep sense of loyalty to Barbados which helps to make it a politically stable and strongly coherent country.

Grenada

Grenada is the "Spice Island" of the Caribbean, having long been a major exporter of nutmeg, cloves, cinnamon, and mace. Cacao, bananas, plantains, papaya, yams, and coconuts are also grown on the island's fertile volcanic soils. In recent years, however, tourism has been increasingly the key to economic development. Politically, the news from Grenada focused on its "Revolutionary" government and close ties with the Castro regime of Cuba until 1983, when the island was invaded and occupied by U.S. armed forces.

Southernmost of the Windward group, Grenada lies only 90 miles north of Trinidad and maintains close trade relations with that island. Imports from Trinidad include a wide variety of products, such as condensed milk, crackers, soap, and, particularly, gas and petroleum.

Grenada is part of the rugged "inner arc" of islands and, although small in size, reaches an elevation of 2757 feet at Mount Saint Catherine. Its main mountain mass extends north-south and consists of ridges with volcanic craters. The largest of these contains Grand Étang Lake. The island is thickly wooded, with many streams and natural springs. Saint George's, the capital, is one of the most beautiful little harbor cities in the Caribbean. The Carenage, a land-locked inner harbor, has formed in the depths of an extinct volcano. Many of the buildings that line its shore are hundreds of years old and are still in use. South of Saint George's is a two-mile stretch of incomparable white sandy beach along Grande Anse Bay.

About 30 percent of the population dwells in and around the capital, while the remainder is located on small farms throughout the island. Some 6000 people live in the southern Grenadines dependency on the islands of Carriacou, Petit Martinique, and Ronde. In 1980 the population of Grenada included blacks (84 percent), mulattoes (11 percent), East Indians (3 percent), whites (1 percent), and a few Caribs.

Columbus sighted the island in 1498 on his third voyage to the New World and named it Concepción, but since 1523 the name Grenada (pronounced Gren-*ay*-da) has prevailed. British interest originated in 1609, when a group of 200 English adventurers attempted to establish a settlement. Subsequently, both the British and French sought control over the island but met determined resistance from the Caribs. After several battles, the desperate Carib men threw their women and children over the Morne des Sauteurs precipice, and then hurled themselves to the craggy rocks below. Not until the Treaty of Versailles of 1783 did the British gain firm control, and the island was proclaimed a Crown Colony in 1877. Its early prosperity depended primarily on the production of sugar and coffee.

The economy of Grenada remains basically agricultural. Bananas constitute the principal export by volume, but cacao and nutmeg exports are greater in value. About 44 percent of the land is under cultivation, 12 percent is in rain forest, and 4 percent is in pasture. Sugar is no longer a major crop, following a plague of voracious ants and a hurricane in 1980. Manufacturing includes fruit canning, cocoa and spice processing, and nutmeg oil distillation. Wooden furniture, garments, rum, soap, and lime juice are also produced. Tourism ranks second in the nation's economy but has suffered the adverse effects of local political events.

THE LANDLOCKED HARBOR OF ST. GEORGE, CAPITAL OF THE SPICE ISLAND. GRENADA.

Grenada achieved full independence from Britain in 1974 but, unlike Barbados, has not been blessed with political tranquility. Fierce conflicts in the early 1970s between anti- and pro-independence factions resulted in a strike that shut down the island. Electricity and water were cut off, and trade and tourism ceased. Following independence, the island struggled toward normalcy under the leadership of Prime Minister Eric Gairy. Gairy, a one-time school teacher who had also worked for a Dutch oil refinery in Aruba, became interested in labor organization, an avocation he continued in Grenada where he became Premier in 1962. It was then that he began to recruit police aides, known as the Mongoose Gang, some of whom were alleged to be criminals. Gairy's opponents were intellectual left wingers, wealthy landowners, businessmen, and religious leaders.[1] With his election, Gairy became increasingly dictatorial until 1979, when a 36-year-old London-trained lawyer, Maurice Bishop, with the aid of 13 armed men toppled the Gairy regime and set up the

[1] *Britannica Book of the Year,* 1975, Encyclopaedia Britannica, Inc.

People's Revolutionary Government of Grenada. Most of the islanders welcomed Gairy's overthrow. After the first year, however, enthusiasm for the revolution waned.

In foreign policy Grenada became an adherent to its closest ally, Cuba. Shortly after the 1979 coup, Cuba was quick to send arms, military advisors, doctors, and technicians. The Bishop government responded by voting against a U.N. resolution condemning the Soviet invasion of Afghanistan and backed Castro-supported guerrillas and liberation movements in Central America and the eastern Caribbean. The government sought close relationships with its neighbors but was rebuffed by Trinidad and the United States.

Grenada desperately needed a new airport to support its faltering tourist industry. Construction was well underway on a $71 million airport at Port Salines by 1981. However, before Grenada applied for funding, Cuba established itself as the power behind the airport, supplying 250 workers and technicians as well as construction equipment and materials. A gnawing suspicion regarding possible military use of the airport brought criticism not only in Grenada, but also in foreign quarters.[2]

The U.S. invasion in 1983 verified the presence of almost 900 Cuban, Soviet, and East European personnel, including military advisors, plus many tons of arms and ammunition presumably intended to support revolutionary movements in Latin America. For a brief moment, too, attention was focused on about 600 American students at one of the numerous offshore medical schools in the Caribbean where training is provided to those unable to gain admission to medical schools on the mainland.

The occupation of Grenada was brief, and the military problems were quickly resolved. What remained were high expectations by the local populace for an equally prompt resolution of the island's economic problems. However, political stability was far from assured.

Saint Vincent

The island of Saint Vincent is identified with arrowroot, of which it is the world's largest producer, and with disastrous volcanic eruptions, of which it has experienced many. Arrowroot is an edible starch derived from an herbaceous perennial plant, similar to manioc or cassava, which grows to a height of five or six feet and concentrates the starch in its tubers. The starch is easily digested and therefore noted as a food for babies in the form of arrowroot cookies. More important is its use in the manufacture of computer paper, pharmaceuticals, jellies, and custards. Yet, arrowroot is far surpassed by bananas as an export crop and source of foreign exchange.

The rugged little volcanic island is still largely forested, as a result of the relatively sparse population and abundant rainfall from the prevailing northeast trade winds. On the Atlantic side of the island, along the Windward Highway, are plantations of coconuts and bananas and fields of arrowroot. These are the chief exports. Hurricanes are not frequent, most passing to the north of Saint Vincent, but "Hurricane Allen" destroyed 95 percent of the banana crop in 1980. The banana plant recovers quickly from this type of damage, however, reproducing within one year. Only 56 percent of the island is suitable for crops, but there are intensively cultivated plots of land clinging to the mountainsides and extending down into the valleys. Breadfruit remains a staple food of the islanders, and vegetables, fruit, and cotton are grown for local consumption.

Kingston, the capital and largest city of Saint Vincent, has a population of about 33,000. It is a busy little port where fishermen and small farmers meet to exchange their produce. Located in Kingston is the Botanic Garden, founded in 1765. This is the site of a breadfruit tree said to be grown from the original seedling plant brought to Saint Vincent from Tahiti by Captain William Bligh in 1793.

---

[2] R. Freedman, and R. B. McColm, "Grenada Stirs Discontent Among Socialists," *The Wall Street Journal,* August 24, 1981.

The original inhabitants of the islands were peaceful Arawaks whose males were exterminated by the cannibalistic Caribs. During a hurricane in 1675, a slave ship capsized between Saint Vincent and the small neighboring island of Bequia. Instead of killing them, the Caribs helped the blacks ashore. The subsequent union of blacks and Caribs resulted in a race called the Black Caribs. In the eighteenth century, after Spanish influence in the area had waned, the British and French fought for possession of Saint Vincent, and it changed hands three times before it finally became a British Crown Colony. It remained so until 1969, when it became a West Indies Associated State.

After two centuries of British rule, Saint Vincent became an independent nation in 1979. The population totaled about 134,000 in 1984 and has varied little in its spatial distribution since the early nineteenth century. Included in its national territory are the northern Grenadines, a group of islands and islets extending southward toward Grenada and whose beaches are more attractive to the development of tourism than those formed by black volcanic ash on Saint Vincent. These islands are particularly attractive to wealthy foreigners with private yachts; many of these foreigners have built luxurious homes and in some cases have purchased entire islands.

On Saint Vincent itself, unemployment and poverty are widespread. About 40 percent of the workforce is unemployed, and the trade deficit in 1979 was three times greater than the total value of exports. Related problems include political unrest and a high rate of emigration.

The government of Saint Vincent has sought to maintain a policy of cooperative relations with France, Canada, Great Britain, and the United States, while remaining firmly nonaligned in the broader East-West confrontation. The young state has no army and plans to remain part of the British Commonwealth of Nations.[3]

---

[3] *News from Latin America,* 25, No. 12 (1979).

Saint Lucia

While Saint Lucia is one of the largest of the Windward Islands, it encompasses only 240 square miles. It is part of the inner volcanic arc, with a main ridge of mountains extending north-south across most of the island. The highest peak, Morne Gimie (3117 feet), is in the southwest, as are the island's best known and most photographed features, Gros Piton (2619 feet) and Petit Piton (2461 feet), which rise abruptly like steep pyramids from the Caribbean shore.

As in the other islands of the Windward group, the climate of Saint Lucia is generally mild, sunny, and pleasant. Rainfall varies from 40 to 150 inches per year, the greater amounts occurring in the mountains, especially on east-facing slopes. Temperatures average from 70 to 90°F year round. The island is still largely forested, although the most valuable timber was removed long ago.

Christopher Columbus visited Saint Lucia in 1502, on his fourth voyage to the New World. The fierce Carib Indians remained in complete control until after the first French settlement in 1650, after which the French and British competed for the island almost continuously until 1814. In that year, the Treaty of Paris was signed, by which France ceded all its claims over Saint Lucia to Britain. The island became self-governing in 1967 and independent in 1979, while remaining within the British Commonwealth of Nations.

Settlement on Saint Lucia is concentrated in the northwest, near Castries, and along the south coast. Castries, the capital, is the chief trading center and has one of the finest natural harbors in the Caribbean. The harbor is located within the submerged crater of an old volcano, and around it are many of the island's principal hotels and tourist attractions. For many years, Castries was a major coaling station for ships plying the West Indies. Its economy was adversely affected in the early 1930s when the fuel for ocean-going vessels changed from coal to oil. Fires swept the city

LOOKING DOWN ON CASTRIES IN ST. LUCIA.

in 1948 and 1951, destroying most of the buildings. Reconstruction since that time has given Castries a conspicuously modern appearance.

Despite the claim that Saint Lucia is an unspoiled island, development is underway. A free-trade zone is being built, and an oil transshipment terminal is being installed. North of Castries, Pigeon Island has been made into a national park, while at Cap Estate land development goes on in the form of a golf course and luxury hotel. In recent years the manufacturing sector has expanded to include clothing, petrochemicals, and electrical components. These are primarily for export within the Caribbean Community.[4] The island has achieved a strong economic position through import duties, income taxes, fees, and licenses.

Agriculture remains the primary economic activity, but bananas have replaced sugar as the principal export. Rum, molasses, limes, mangoes, spices, and bay rum also enter the export trade, whereas a lively local trade in papaya, plantains, breadfruit, and fish is carried on by native merchants. Tourism is the second most important source of foreign exchange earnings, and additional income is gained from leasing military bases to the United States at Vieux Fort and Gros Islet. Vieux Fort is also to become the island's industrial center.

Saint Lucia retains its strong African heritage from the era of slavery and sugarcane production but is influenced as well by the many years of British colonial administration. French influence

[4] "Policies of St. Lucia," *IMF Survey* 9, No. 12 (1980):21–22.

is also conspicuous, as in the local dialect and the close proximity of the French island of Martinique.

All of the problems inherent to the newly independent ministates of the Caribbean and to the rest of the world are exhibited in Saint Lucia. Nations as small as Saint Lucia can never be truly independent within the modern world. Yet, this island nation has managed to maintain a substantial degree of political stability and freedom from foreign political influence.

Dominica

The island of Dominica, located between the French islands of Martinique and Guadeloupe, is rugged, rural, and remote. Approached from any direction, it is impressive in terms of the physical landscape. It rises abruptly from the ocean to a height of 4747 feet in Morne Diablotin peak, and there is almost no land sufficiently level for modern agriculture. The total population of Dominica is only about 100,000 and the capital city, Roseau, is still a quiet small town. It is, however, a national capital and the island's principal port.

Volcanic in origin, Dominica forms part of the inner arc of islands that comprise the Lesser Antilles. Volcanic vents and hot springs are common. Because the rugged terrain interrupts the west-flowing currents of air in this part of the world, rainfall is heavy. It averages as much as 250 inches annually on windward slopes and 70 inches along the coast. It is said that Dominica has 365 rivers, one for each day of the year.

The island was named by Christopher Columbus, who discovered it on a Sunday in 1493.

THE TINY FISHING VILLAGE OF SOUFRIERE, DOMINICA.

The French subsequently settled there, but in 1748 a French-British treaty ceded the island to the native Carib Indians. A reservation for the Caribs remains to this day in the northeastern part of the island. Dominica became a British colony in 1783 and remained so until 1958, when it joined the short-lived Federation of the West Indies. Full independence was achieved in 1978.

The population is primarily of African origin; the economy is based on agriculture and tourism. Bananas, citrus fruit, and coconuts dominate the export trade, but most farms are small and produce fruit and vegetables only for the domestic market. There is no mineral wealth and essentially no industry. Tourism is underdeveloped but offers some potential, as does forestry.

Development remains a distant objective, and independence has not been synonymous with progress. The main road, which cannot be called a highway, is 36 miles in length and connects Roseau with the island's only airport. It is narrow and rutted, requiring travel time by automobile of approximately two hours. Poverty abounds, and "Hurricane David" in 1980 added to the island's misery. Many buildings in Roseau were destroyed, and crops and trees throughout the island were devastated.

Politically and economically, Dominica remains practically "an island unto itself." Yet, even here there are civic organizations concerned with human welfare in general and with specific issues such as unemployment and population control. Some external aid arrives from Europe, the United States, and Trinidad, but threats against that government are also organized abroad. Twice during 1981 there were attempted coups that appeared to involve Dominican soldiers, American mercenaries with Ku Klux Klan and Mafia connections, and Rastafarians who grow marijuana in the Dominican highlands. Prime Minister Eugenia Charles, the Caribbean area's first female Chief of State, sought increased regional integration through the Caribbean Community as the best means to "survive the rigors of this modern age."

Antigua-Barbuda

In 1981 Antigua became the thirty-fourth independent nation in the Western Hemisphere, along with the adjacent islands of Barbuda and Redonda. Both Antigua and Barbuda are parts of the outer arc of islands formed by coral growth and are generally low in relief. They are consequently hot and dry, although some moderation of temperature results from the prevailing trade winds.

Antigua was discovered by Columbus in 1493 and was later subject to dispute between the British and French naval forces. The British prevailed and built at English Harbor one of the finest naval stations and fortresses in the West Indies. Among the distinguished officers to serve here was Horatio Nelson, who would later become the greatest British naval hero of all time.

The population of Antigua totals about 100,000 and is mostly of African descent. The island is heavily dependent on tourism, with more than 100,000 visitors arriving annually. Coolidge International Airport, near the capital city of Saint John's, serves numerous international airlines, making Antigua an important transportation hub for the eastern Caribbean.

Light industry employs about 1000 people, who are engaged in the manufacture of engines and aircraft parts. Efforts are being made to revitalize a petroleum refinery to process crude oil from Venezuela and Mexico. Likewise, investments to promote tourism are planned. Agriculture accounts for less than 10 percent of the gross national product, and prospects for its expansion appear limited.

A major political problem in the years ahead will be the successful integration of Barbuda with Antigua. Barbuda, only 62 square miles in area and with a population of only 1200, has already indicated its opposition to the union. The integration of Redonda appears to be less of a problem. It is merely an uninhabited rock rising to a height of 1000 feet within the inner arc of volcanic islands, lying 25 miles south of Antigua.

Saint Kitts-Nevis

Saint Kitts (Saint Christopher) and Nevis are easily recognized among the Leeward group of the Lesser Antilles because their physical outline is like that of an exclamation mark. The "slash" is Saint Kitts, which rises to an elevation of 3792 feet at Mount Misery and tapers off to a lowland, with its Great Salt Pond, pointed toward Nevis. The latter, just two miles distant across a narrow strait, forms the "dot." It rises in one steep volcanic cone, Nevis Peak, to 3232 feet above the sea. In 1983 Saint Kitts and Nevis became the world's newest nation, with a total population of less than 50,000.

Both islands were discovered by Columbus in 1493 and were at that time occupied by Carib Indians. Both prospered during the eighteenth century, when sugarcane cultivation prevailed throughout the West Indies. The American patriot Alexander Hamilton was born on Nevis during this period, giving the island an added claim to fame. Sugar continues to be the dominant crop on Saint Kitts, whereas on Nevis it has decreased in relative importance. Small farmers on Nevis now concentrate largely on vegetables, cotton, and coconuts, although some cane is still grown and shipped to Saint Kitts for refining. An attempt has been made here, too, to foster the growth of tourism.

Saint Kitts and Nevis were both formed by volcanoes, the tops of which emerge from the sea. There is some area of forest at the upper levels on both islands, whereas the lower slopes have been cleared for agricultural pursuits. Industry on Saint Kitts includes a single modern sugar mill, a brewery, and shops to produce local crafts.

Saint Kitts also possesses one of the most impressive forts in the New World, "Brimstone Hill," which in the eighteenth century became known as "the Gibraltar of the West Indies." Basseterre, the island's capital and principal town, will soon become a deep-water port. With arrival of the cruise ship trade, plus expanded international flight service, it is expected that tourism will reach new dimensions.

The Bahamas

The Bahamas are a group of about 700 islands lying just east of Florida and extending southeastward for a distance of 750 miles. They rise only slightly above sea level from a broad submarine platform and reach a maximum elevation of about 200 feet. The Bahamas have a total area of 5353 square miles and a population estimated to be 260,000 in 1984. More than half of the population is concentrated in and around the capital city, Nassau, on New Providence Island. Locally, all of the islands other than New Providence are referred to as the Family Islands.

The Bahamas are too low to receive much rainfall; plant growth is therefore generally sparse. The original vegetation was a mixture of pine and palmetto, similar to much of the vegetation of Florida. Stands of Caribbean pine were once extensive and supported commercial lumbering, especially on Andros Island, the largest within the archipelago. In 1984 efforts were initiated to prepare a forestry inventory and management plan that might aid in expansion of the remaining forest resources.

San Salvador, or Watling Island, is generally considered to be the first land sighted by Columbus on his original voyage to the New World in 1492. Soon thereafter the Arawaks who inhabited the Bahamas in small numbers were seized to work the Spanish mines in Cuba and Hispaniola. In 1647 the first British landed on the island of Eleuthera, meaning "freedom," having fled from Bermuda because of religious persecution. With the beginning of British settlement, African blacks, whose descendants now comprise 85 percent of the population, were brought to the islands as slaves.

The invasion of the Bahamas by pirates, who remained more than 30 years, introduced a chaotic period in history. The main business of the islands became raiding, looting, and destruction of any ships that ventured into the area. In 1718 the British claimed the entire group of islands, mainly because of its strategic importance in relation to the Caribbean Sea. Order was achieved under an appointed Royal Governor, but it required eight hangings to enforce law and order and to inaugurate a parliamentary form of government. From 1629 to 1973, when the islands became independent, the Bahamas were under almost continuous British control.

For many years the Bahamas were a major source of sponges, but this product almost disappeared after the sponge beds were attacked by disease in 1930. The manufacture of artificial sponges was perfected shortly thereafter. Now, crude oil is the country's leading import, by value, and petroleum products dominate the exports. Yet, tourism provides about 75 percent of the gross national product and employs two-thirds of the total labor force.

Tourism is increasingly a year-round activity in the Bahamas. The nearest island is only slightly more than 50 miles from the coast of Florida and, therefore, most of the 1.5 million visitors who arrive in the Bahamas each year are from the United States. Favored destinations are Nassau, where huge cruise ships come to dock, and Freeport, which is a free port with an International Bazaar offering goods from all over the world.

Because of increased tourism and a rapid expansion of "offshore" banking activities, substantial economic growth has occurred. Such growth is adversely affected, however, by illegal immigration from Haiti, along with a high level of unemployment. Poverty does exist in the Bahamas. There is even a slum that dates from the 1700s; it is called Over-the-Hill and is located on New Providence Island adjacent to Old Nassau. It was reported in 1903 that more than half of the Bahamian population was concentrated on New Providence and that blacks in Over-the-Hill numbered more than 4600. During colonial times this slum was defined as a community of slaves and indentured servants. Over-the-Hill is still occupied by poor blacks,[5] many of them illegal Haitians who numbered an estimated 80,000 in 1981.

In 1968 the government promised to make the Bahamas self-sufficient in food within a decade. This ambitious schedule was not achieved, but an agricultural research, training, and development program was established, and the goal of self-sufficiency remains. The islands already produce enough poultry, eggs, and certain vegetables for local needs. A fisheries development program was started in 1977, when the fisheries zone was extended to 200 miles. Now, the Bahamians are fishing in local waters previously fished by Cubans and Americans. Within the next few years 3000 additional workers are to be involved in this lucrative industry.[6]

A limited amount of manufacturing has been developed, mainly in Freeport, largely because of investment from the United States. Included are oil refining and the manufacture of pharmaceuticals, cement, and steel pipe. There are also concerns making processed foods, clothing, furniture, aluminum products, and beverages. Oil has been discovered recently and may have important implications for the future, for there is geological evidence that gas and oil deposits may be fairly extensive beneath the waters of the South Florida-Bahamas Basin. A Bahamian Petroleum Act has already been passed to protect the natural environment, because oil spills on the beautiful beaches of the Bahamas would mean economic disaster.[7]

The most serious problem facing the Bahamas is the use of the islands as a major center for the marketing of marijuana and cocaine in the United States.

---

[5] M. F. Doran, and R. A. Landis, "Origin and Persistence of an Inner City Slum in Nassau," *Geographical Review,* 70, No. 2 (1980):182–190.

[6] Bahamas, *ABECOR Country Report* (London: Barclay's Economics Department, 1980).

[7] J. R. Morton, "News and Notes," *Journal of the Developing Areas* 13, No. 2 (1979):191–192.

The British Dependencies

Only a small number of scattered islands in the Caribbean remain under the administration of the British Colonial Office. Included, from east to west, are Montserrat, Anguilla, the British Virgin Islands, Turks and Caicos, and the Cayman Islands.

Montserrat

Part of the inner arc of volcanic islands, Montserrat is only 35 square miles in area but rises to a height of 3000 feet. Its total population is less than 15,000. Formerly one of the "sugar islands," it contains the remains of sugar estates and abandoned mills. There is little level land, but some small-farm agriculture continues. In recent years production has focused largely on vegetables, limes, and sea island cotton. However, serious problems of erosion and continued emigration have substantially reduced the export of even these crops.

Like many other islands of the Caribbean, Montserrat has become increasingly dependent on tourism. However, the black volcanic sand beaches on Montserrat are less attractive than those formed elsewhere from coral. Consequently, emphasis has been placed on local advantages for retirement, and it is certain that the concerns of the rest of the world seem remote when viewed from this small island in the Lesser Antilles.

Anguilla

The most northerly of the Leeward Islands is Anguilla, just 3 miles wide and 16 miles in length. Its name, appropriately, derives from the French word for "eel." The island rises only 200 feet above sea level and has a population of about 5000. The main town is known as The Valley. The island is low and arid; hence, agriculture is limited. Boatbuilding and fishing have consequently been prominent occupations. Some cotton and food crops are produced, as is salt from the evaporation of sea water.

Anguilla does not normally figure prominently in international news, but it did in 1967, when British parachute troops landed on the island to quell an insurrection. In that year, five years following the collapse of the ill-fated West Indies Federation, Anguilla, along with Saint Kitts and Nevis, became a state voluntarily associated with the United Kingdom. This union proved unpopular, however, and 250 armed Anguillans overthrew the local authorities. A British invasion drew criticism worldwide, and perhaps most severely in London. Not until 1971 was a settlement reached, and that provided for continued British control, with a large measure of local autonomy.

The British Virgin Islands

The British Virgin Islands include nearly 50 islands and islets, with a total population of about 12,000. The largest individual island is Tortola, with an area of 21 square miles and a population of 9000. Many years ago Tortola was a haven for pirates because of the cays and coves where they could conceal themselves as they waited to raid trading ships. With the exception of Anegeda, the islands are steep and mountainous. The highest elevation, on Tortola, stands 1780 feet above sea level. Anegeda, on the other hand, is a piece of coral reef reaching no more than 30 feet above the sea.

Since the end of the seventeenth century, British planters have tried to produce crops on the islands. With African slave labor they raised sugarcane, cotton, tobacco, indigo, and other crops. Their plantations were high-cost producers, however, and in the nineteenth century they ceased to operate. The people who remain on the islands, mostly blacks, now use the mountainous areas for a shifting cultivation of vegetables, followed by pasture and then second-growth scrub woodland again. The clearing of the woodland results in the production of small quantities of charcoal.

Today, 80 percent of the labor force of the British Virgin Islands is employed in government offices, hotels, and restaurants. Many workers also go to the American island of Saint Thomas in search of employment. The main exports, mostly re-exports, are motor vehicles, timber, beverages, and fish, as well as iron and steel. More than half of the total export trade is with the U.S. Virgin Islands. The economic ties are so close that the U.S. dollar is the principal unit of currency exchanged in both sets of islands.

Turks and Caicos

The miniature Turks and Caicos islands, physiographically an eastward extension of the Bahamas, are inhabited by about 7700 people. These people make their living from tourism, fishing, and the sale of postage stamps. While the British government would readily grant independence to the islands, the people themselves are more interested in their immediate social and economic problems.

Locked within a coral reef, the islands consist of only 268 square miles. The first inhabitants were the Arawak and Lucayan Indians, who founded a salt industry in the Caicos. Later, in the seventeenth century, the Caicos became hideouts for pirates. After the American Revolution, the islands became home to loyalists who fled from the Southern colonies.

An Old World atmosphere still prevails along the narrow streets, on the tranquil beaches, and in the quaint homes of Turks and Caicos. However, there are few jobs, and those that do exist are mostly government-sponsored. The largest industry is the harvest and export of spiny lobsters and the queen conch, which are sold to France and the United States. Unemployment causes many people to emigrate; those who are employed, and remain, experience advantages such as the absence of noise, pollution, and personal or corporate income taxes.

Cayman Islands

The Cayman Islands are emerged portions of the submarine Cayman Ridge, which continues southwestward from the Sierra Maestra Mountains of Cuba. Some 150 miles to the southeast lies Jamaica, to which the Cayman Islands were attached politically until 1959, when they became a self-governing British colony. There are three islands: Grand Cayman, Little Cayman, and Cayman Brac; all are low, arid, and sandy. The total population is about 12,000.

The Caymanians have always prospered. For generations the men between the ages of 18 and 50 went to sea, serving on Caymanian or American ships. Some stayed on in the United States, raised families, and sent for their relatives. Then, many elderly persons who had emigrated returned home to retire on social security benefits and savings. Most had become American citizens.

The sea still plays a major role in the economy. Some former fishermen are now involved in boatbuilding, or in the rental of boats for sailing and charter fishing for a flourishing tourist industry. The world's first successful green turtle farm, established here in 1968, has tried to preserve the species and at the same time supply turtle meat, shell, oil, and other products to consumers in a regulated manner. Raised in captivity, the turtles are fed high-protein pellets and can attain a weight of 600 pounds.

A crisis arose in 1980 when environmentalists in the United States objected to the farming of turtles and secured a ban on the import of turtle meat, shell, soup, and leather. The farm subsequently curtailed operations and released thousands of young green sea turtles into the Caribbean. The environmentalists opposed this, too, claiming that release of the farm's turtles, as descendants of turtles in Costa Rica, Suriname, Nicaragua, Ascension Island, and Guyana would cause destruction of the turtle-gene pools in the Caribbean.[8]

Agriculture is limited by the shortage of arable land. The land available is devoted largely to the production of poultry, beef, dairy products, veg-

---

[8] *The Wall Street Journal,* July 16, 1981.

etables, and tree crops. These are intended for the local market, although a preference for imported food presents marketing problems.

The Cayman Islands have no income tax, corporate tax, capital gains tax, or inheritance tax. Such advantages, plus political stability, have helped to attract international businesses, including major banking concerns. The economy has therefore boomed, as have real estate sales and the construction industry.[9] Government remains the chief source of employment, with health, education, and social services providing most of the jobs. Tourism is also well developed and contributes significantly to the Caymanian economy.

The French Antilles

French possessions in the Antilles were once far more numerous than they are today. Before the downfall of Napoleon, French authority at one time or another extended over most of the islands between Hispaniola and Tobago, except for Barbados. On Trinidad, which never came under French rule, there are still many people of French ancestry. Now, only two principal islands remain, each considered to be politically a *département*, like the divisions of mainland France. These are Martinique and Guadeloupe, and they are truly French. The inhabitants vote in French elections and elect senators and deputies to the French Parliament. They participate in free health programs and are eligible for French social security benefits.

The French landed on both islands in 1635. From the beginning Martinique prospered, developing a certain charm of its own as it blended French and West Indian traits. Guadeloupe, on the other hand, experienced difficulties as it struggled under inexperienced leaders and unskilled settlers.

[9] "Cayman Islands," *ABECOR Country Report* (London: Barclay's Economics Department, March 1981).

Martinique

The island of Martinique, between Saint Lucia and Dominica, is rugged and mountainous. It is about 40 miles long by 16 miles wide and includes along its coast several well-protected harbors, such as the one on which Fort-de-France, the capital, is located. Its older volcanoes are deeply dissected by torrential streams, but Pelée, near the northern end of the island (which erupted violently in 1902), has all the forms of a newly built volcanic cone.

Saint Pierre, north of Fort-de-France, was one of the major ports of the Antilles before the eruption of Pelée. It was the island's capital and had a population of 30,000. Within three minutes on May 8, 1902, all but one person perished in the burst of superheated steam and gas that spewed from the eruption of Pelée. The sole exception was Ludger Sylbaris, who had been held in solitary confinement at the local jail and consequently survived the ordeal. This is not the only natural calamity that has disturbed the peace of Martinique. The island has averaged almost one natural disaster every five years, which in an area so densely populated causes great distress and poverty among its populace.

Alcohol and rum are noteworthy products of Martinique. Sugarcane was introduced from Brazil in 1654, and by the beginning of the eighteenth century this colony had become one of the most wealthy of the sugar islands. At present about 80 percent of the better lands are devoted to cane. The Plain of Lamentin, bordering the Bay of Fort-de-France, is the only extensive area of level land in Martinique. It is covered with canefields and is devoid of habitations or of land used for subsistence crops. Almost all of the cane lands are held in large estates. So specialized is the interest in sugarcane planting on Martinique that no other commercial crops have ever received much attention. Since World War II, however, there has been some increase in the production and export of bananas and pineapples. The chief manufacturing plants are engaged in sugar refining, the distillation of rum, or the canning of pineapples. Exports, mainly to France, include petroleum products, bananas, and rum.

Martinique is densely populated by a largely black peasantry. There has been a substantial emigration to France since the colony was made a part of the French state in 1946. Martinique, along with Guadeloupe, has a remarkable literacy rate of 95 percent, because of France's compulsory educational regulations.

Guadeloupe

Guadeloupe, which lies between Dominica and Montserrat, is composed of two islands separated by a shallow arm of the sea, the Rivière Salée, filled with mangrove. To the east is the low-lying limestone platform of Grande-Terre, with an area of 219 square miles. To the west is Basse-Terre, with an area of 364 square miles. Basse-Terre, despite its name, is rugged and mountainous. Its highest volcanic peak, Mount Soufrière, 4869 feet in elevation, is the highest mountain in the Lesser Antilles.

The *département* of Guadeloupe also includes six small islands. Close to Grande-Terre, and like it, formed of coral limestone, are La Désirade, Marie Galante, and Petit-Terre. To the south of Basse-Terre are the Îles des Saintes, the tops of deeply submerged peaks. Some 125 miles to the northwest, between Saint Kitts and Anguilla, is Saint Barthélemy (9 square miles), and 12 miles beyond is Saint Martin, of which the northern part (20 square miles) belongs to France. Both Saint Barthélemy and Saint Martin are coral limestone on top of underlying volcanic rock.

The population of the entire *département* of Guadeloupe in 1982 was 328,400. Most of Guadeloupe is occupied by people of mixed African and French ancestry, but on Les Saintes the people are mostly French. The density of population on Les Saintes is more than 500 per square mile, which is high even for the Lesser Antilles.

Guadeloupe has long been used for the planting of sugarcane. When the French first settled on the island, the chief area of cane planting was around the margins of Basse-Terre, but the principal area of production is now on Grande-Terre. There has been the usual change from small-scale to large-scale mills. Sugarcane, once established as it has been on Guadeloupe, persists in part because of the large investment necessary to begin operations, and in part because of tradition and the reluctance to take a chance on other crops, even when sugar yields only a meager living. At Beauport, plantations of sisal and other fiber crops have been started and seem to promise good results. On Basse-Terre, where rainfall is sufficient, there has been an increase in the planting of bananas and pineapples. Because of frequent hurricanes, those crops that require several years to mature are risky. Only sisal can survive these storms: sugarcane, bananas, and pineapples are destroyed but can easily be replanted.

The economy of Guadeloupe is still based on sugarcane. The principal industries are sugar refining and the distillation of rum. Exports, over 80 percent of which go to France, are sugar, rum, molasses, bananas, wheat meal, and flour.

The capital of the *département* is the city of Basse-Terre (population 13,700), but Pointe-a-Pitre (population 82,500) is the main port and commercial center of Guadeloupe. Located on Grande-Terre, Pointe-a-Pitre has the appearance of a French city. Its busy harbor is lined with boats and modern apartments that contrast with the historic nineteenth-century cathedral. Huge jumbo jets land at the nearby international airport, with passengers stopping en route between European and South American destinations. Roughly midway between the larger cities of Port-of-Spain, Trinidad, and San Juan, Puerto Rico, Pointe-a-Pitre is a metropolis among the Lesser Antilles and serves to some extent as a regional trade center.

The outlying islands, except for Marie Galante, are too steep or too dry for sugarcane. Production from Marie Galante has declined. La Désirade, Saint Barthélemy, and Saint Martin produce some cotton, sisal, coconuts, and food for local consumption. Les Saintes are too steep for much farming and are occupied by people engaged chiefly in fishing and the production of charcoal.

The Netherlands Antilles

The Netherlands Antilles include six islands, with a total area of 383 square miles and a population estimated at 258,400 in 1984. The islands are in two groups. Curaçao, Aruba, and Bonaire are located 20 to 50 miles off the northwest coast of Venezuela; the other three, Saba, Saint Eustatius, and the southern part of Saint Martin, lie 500 miles away, between Saint Kitts and Anguilla. Politically, the six islands form an autonomous state within the Kingdom of the Netherlands. They are administered from the capital city of Willemstad, on Curaçao, which is also the major seaport.

The two groups of islands differ in their economies and in their problems. The one activity that gives support to the Netherlands Antilles as a whole is the refining of petroleum on Curaçao and Aruba. These two islands now have about 94 percent of the total population. None of the islands possesses resources of soil or climate to support a very large population, but when major oil corporations built refineries to handle the crude oil shipped from the Lake Maracaibo region of Venezuela, a new type of economic life began. The Shell Oil Company plant was built on Curaçao in 1915, and Standard Oil established a refinery on Aruba in 1924. For a time, all Venezuelan oil was refined here, and both refineries were substantially enlarged. Since World War II, however, Venezuela has built numerous refin-

AN OIL REFINERY IN WILLEMSTAD, CURAÇAO.

eries of its own, and further expansion of those on Curaçao and Aruba seems unlikely.

None of the islands is high enough to receive much rainfall, and that of Curaçao averages only about 20 inches annually. Thus, the original vegetation was mostly scrub woodland. On Curaçao a distinctive kind of agriculture emerged, however, since dwarf orange trees could survive and produce a bitter, miniature fruit. This fruit is still ground up and fermented to produce the world famous orange liqueur known as Curaçao.

With the development of oil refineries and the growth of Willemstad as one of the major trading centers of the Caribbean area, some source of fresh water had to be secured. The refineries

themselves use large quantities of water, and there was a great increase in demand for domestic purposes. In 1957 distillation equipment was installed on both Curaçao and Aruba for the desalinization of sea water. The facilities have been expanded and modernized regularly, so that a plentiful supply of fresh water is now available for both industrial and domestic use.

Bonaire, like Curaçao and Aruba, is formed of coral rock and has an arid climate. It has no petroleum refinery, but the Antilles International Salt Company produces large quantities of salt from sea water on the southern shore of the island. Birds abound, particularly flamingoes, herons, parrots, and pelicans. Tourism has been pro-

HEADING INTO SABA IN THE NETHERLANDS ANTILLES ON A WINDJAMMER CRUISE.

moted, but emigration, especially by the youth, still provides better economic opportunity.

The northern islands are of slight economic importance and have only small populations. Saba, a simple volcanic cone rising 2887 feet above the sea and occupying only 5 square miles, rises so abruptly from the water's edge in cliffed slopes that no protected anchorages exist around its shore. Within the crater are several little villages, each containing a few hundred people. For a long while, Saba was fairly prosperous, despite difficulties of access, its primary economic activity being the construction of small sailing boats. The liberation of slaves in 1863, and the change to steamboats, brought an end to this industry. Tourism now predominates, and Saban women contribute their part by producing handicrafts such as drawn threadwork and silk screen printing on cotton.

Saint Eustatius, or Statia, was once one of the busiest ports of the Caribbean. It was here in November 1776 that a warship of the new United States first received foreign recognition, by a salute from cannons at Fort Oranje standing on cliffs above the beach. Part of the prosperity of Saint Eustatius came from smuggling items such as sugar and cotton in exchange for firearms, and for a time it was a very elegant place to live. Now, the island is dependent largely on tourism, about 3000 visitors arriving each year. Yet, opportunities for employment are few, with the result that emigration has been continuous. In 1790 Saint Eustatius had a population of 7830; the figure is now about 1300.

Saint Martin, or Saint Maarten, is a unique place. It is said to be the world's smallest parcel of land that is owned by two independent nations, in this case by France and the Netherlands. Although the boundary between the northern (French) and southern (Dutch) portions of the island is well defined, there is no customs office or guard on either side of the border. The island has an international airport, excellent beaches, and good facilities for boating, fishing, and golf. It also offers the opportunity to buy quality goods from Britain, France, and the Netherlands at bar-gain prices, which helps to account for an influx of 200,000 tourists annually.

The American Virgin Islands

The Virgin Islands of Saint Thomas, Saint Croix, and Saint John are among the most attractive in the entire Caribbean area. They, along with about 50 smaller islets in the group, were purchased by the United States from Denmark in 1917 for $25 million. The 133 square miles of land, just east of Puerto Rico, are now administered by the U.S. Department of the Interior, but with substantial local autonomy.

Like so many other islands of the West Indies, the American Virgin Islands were discovered by Columbus, after which ownership rotated among the major European powers. Under Danish rule, this island had more than 100 sugar plantations, but the sugar industry disappeared following the abolition of slavery in 1848. Saint Croix, the largest of the three major islands, was purchased from France in 1733 by the Danish king. Today, Saint Croix boasts an oil refinery, an alumina plant, and a rum distillery among its industrial establishments.

In 1980 the population of the American Virgin Islands was about 100,000. About half of that number is concentrated around the capital city, Charlotte Amalie, on Saint Thomas, and almost half is on Saint Croix. Saint John has fewer than 2500 regular inhabitants. Charlotte Amalie, with its well-protected, deep-water harbor, is the leading commercial center for the entire Virgin group, both American and British. Charlotte Amalie is now a transfer station for bauxite shipped by barge from the Guianas and destined to continue northward to the United States or Canada. Other products are brought to Charlotte Amalie from all the islands: sugar and alcohol from Saint Croix, bay leaves from Saint John, and vegetables and charcoal from the British Virgin Islands.

The major economic support for the islands is tourism. In 1960 the number of visitors numbered about 100,000 annually; by 1980 the number exceeded 1 million, who spent more than $250 million. Charlotte Amalie is the primary port of entry, and many people remain in or near the city. Others spread out to tourist facilities all around the shores of Saint Thomas and Saint Croix, where they may pass several months enjoying the delightful climate. Saint John is less frequently visited and less commercialized. In 1956 Laurance S. Rockefeller bought up nearly half of the island and donated 5000 acres of it to the U.S. government to form the nucleus of what has become the largest national park in the entire Caribbean area.

A major problem on Saint Thomas and Saint Croix is to provide enough water for the rapidly growing population. Groundwater is inadequate, and the small amount of rainfall provides little surface runoff. A plant for the desalinization of sea water has operated for some years and, as a subsidiary activity, a steam-generating station provides electric power for Charlotte Amalie. Yet, fresh water is commonly barged from Puerto Rico, and electric power remains in short supply.

The per capita income of the American Virgin Islands exceeded $5000 in 1981, a condition that attracts a growing number of immigrant aliens. About 50 percent of the population is composed of Puerto Ricans and immigrants from other islands of the West Indies. This, plus the annual influx of tourists, leaves the native population noticeably outnumbered and has created a variety of social problems. New industries, such as watchmaking and textile manufacture, have been attracted to the islands, yet more than a third of the labor force is employed by government. As elsewhere in the Lesser Antilles, and throughout the Caribbean, overpopulation and underemployment pose chronic problems for which there are few immediate solutions.

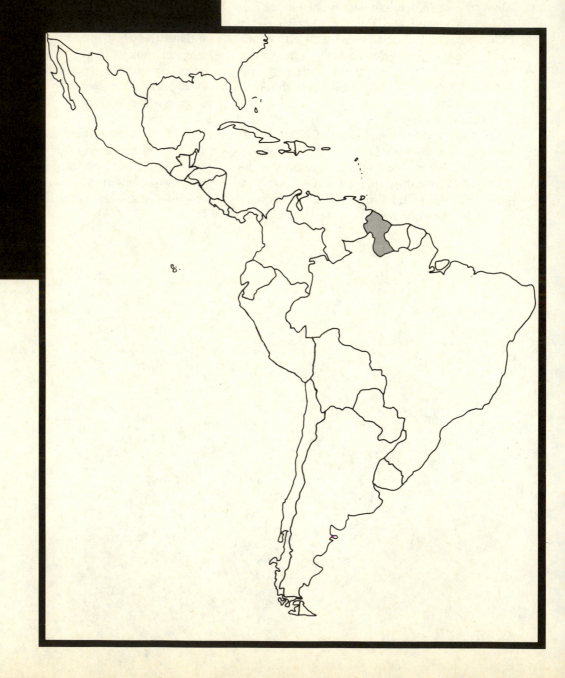

17

COOPERATIVE REPUBLIC OF GUYANA

Land area   83,000 square miles

Population   Estimate (1985) 944,000
Latest census (1970) 699,848

Capital city   Georgetown 182,000

Percent urban   32

Birth rate per 1000   28

Death rate per 1000   6

Annual percent of increase   2.2

Infant mortality rate   35

Percent of population under 15 years of age   37

Percent literate   96

Percent labor force in agriculture   33

Gross national product per capita $520

Unit of currency   Guyana dollar

Physical Quality of Life Index (PQLI)   84

COMMERCE (expressed in percentage of values)

Exports

sugar	36	rice	8
bauxite	33	alumina	5

Exports to		Imports from	
United Kingdom	26	Trinidad-Tobago	34
United States	22	United States	25
Trinidad-Tobago	9	United Kingdom	16
Jamaica	6	Canada	4
Canada	5		

Data mainly from the 1985 World Population Data Sheet of the Population Reference Bureau, the 1985 Britannica Book of the Year, and the 1985 World Almanac.

About a century after the Portuguese and Spaniards started the European colonial movement, the Dutch, French, and British also began to look for lands around the world in which to establish colonies. By this time Spain and Portugal had divided the world between them and had occupied those parts of the Americas in which there were large native populations or sources of gold and silver. The other European nations could only take what was left.

The Guiana Coast was first sighted by Amerigo Vespucci in 1496. Columbus also sailed along this coast, in 1498, but did not attempt a landing. Other explorers were discouraged from taking a closer look at the Guianas by the absence of any concentration of native peoples and by the dense cover of rain forest that made the search for gold very difficult. The Portuguese founded settlements along the Amazon, and the Spaniards established a few settlements along the Orinoco, but the stretch of low coast between these rivers remained empty and unclaimed.

Guiana is a regional name applied to that part of South America that is surrounded by water, namely, by the Orinoco River; the Río Casiquiare, which drains part of the upper Orinoco southward to join a headwater tributary of the Río Negro; The Río Negro itself; the Amazon; and the Atlantic Ocean. This chapter deals with Guyana, an independent nation since 1966, formerly known as British Guiana.

Surface Features

Guyana has three major kinds of surface features: a low, swampy coast, a hilly upland developed on crystalline rocks, and conspicuous flat-topped tabular uplands.

The low coastal lands are formed largely of mud, silt, and sand discharged by the Amazon River. The Amazon carries such a huge load of alluvium that the ocean is discolored for 200 miles offshore. This alluvium is swept northwestward along the Guiana Coast by a strong ocean current moving from southeast to northwest. As a result of wave action on the margins of the alluvial plain, the immediate shore is an almost continuous sandbar only a mile or so wide. The bar is broken where rivers draining from the interior keep channels open to the sea. In back of the sandbar is a belt of marshy lagoons across which the rivers meander on their way to the ocean. This wet coastal lowland with its bar, lagoons, and numerous rivers varies in width from 5 to 30 miles.

A hilly upland on crystalline rocks rises step by step toward the interior. In places the gently rolling surface is surmounted by steep-sided pinnacles of rock. In the farther interior are irregular areas of low mountains with massive rounded outlines. These mountains reach elevations of

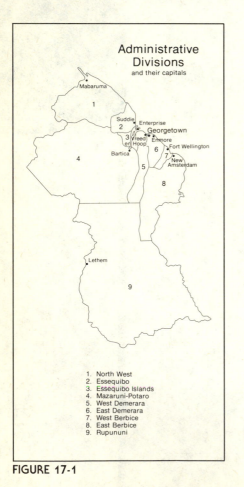

Administrative
Divisions
and their capitals

Mabaruma

1

Suddie
2
3 Vreed
en Hoop
Bartica
4
5
6
7
8

Enterprise
Georgetown
Enmore
Fort Wellington
New
Amsterdam

Lethem

9

1. North West
2. Essequibo
3. Essequibo Islands
4. Mazaruni-Potaro
5. West Demerara
6. East Demerara
7. West Berbice
8. East Berbice
9. Rupununi

FIGURE 17-1

FIGURE 17-2

4000 feet above sea level, but the general elevation of the hilly upland is only about 1000 feet.

In Guyana and Venezuela the hilly uplands are topped by table-like plateaus supported by horizontal layers of sandstone. In fact, the highest elevation in this part of South America is Mount Roraima, where the borders of Guyana, Venezuela, and Brazil meet. Roraima is a great flat-topped plateau that reaches 9212 feet above sea level. The rivers that rise on these sandstone plateaus drop over their cliffed sides in spectacular waterfalls. Angel Falls, in Venezuela, highest in the world, drop 3212 feet to the surface of the hilly upland below. In Guyana, the Potaro River, a tributary of the Essequibo, drops 741 feet at Kaieteur Falls.

Rivers descending from these mountains and plateaus and crossing the crystalline hilly upland toward the Atlantic are all interrupted at frequent intervals by falls and rapids. Only short stretches are navigable, and then only for shallow-draught boats and canoes. Four of these rivers in Guyana have been of major importance; from southeast to northwest they are the Courantyne, the Berbice, the Demerara, and the Essequibo.

Climate and Vegetation

The climate of the Guyana Coast is tropical. The heat equator, which is a map line drawn through the climatic stations with the highest average annual temperatures (but not the highest extreme temperatures), follows the Caribbean Coast of South America, passing through Maracaibo, La Guaira, and other points in northern Venezuela. It then follows the coast southeastward to about latitude 5° N, where it crosses the Atlantic toward Africa. The climate of Georgetown, capital of Guyana, is representative of the climatic conditions of the entire coastal region. The average annual temperature at Georgetown is 80°F, ranging from an average of 82° in the hottest month (September) to an average of 79° in the coolest months (January and February). In

FIGURE 17-3

Guyana people must adjust to monotonously high temperatures: temperatures that are never excessively high but that are never lowered by spells of cool weather.

Rainfall and humidity also are high. At Georgetown the average annual rainfall exceeds 87 inches, and the relative humidity averages 79 percent. The constant winds that sweep across the coastal region make life quite comfortable despite the humidity.

The vegetation is mostly a heavy rain forest. Along the coast, the forest is interrupted by wet savanna which borders the lagoons and tidal marshes. The crystalline hills and inner side of the lowland generally are covered with forest, in which there are few natural openings. Along the headwaters of the Essequibo and of the Río Branco, which drains southward toward the Amazon, is a wide belt of grassland known as the Rupununi Savanna. As is common in dry savannas, the rivers are fringed by galeria forest.

The Course of Settlement

Occupation of the Guiana region by European colonists provides a clear demonstration of an important geographic principle: that the meaning of any habitat differs for people with differing attitudes, objectives, and technical skills. The three European colonies, each of which developed in essentially the same natural surroundings, differ markedly in what the inhabitants have done with these surroundings.

The first Europeans to settle in Guiana were the Dutch. By 1613 they established three colonies in what later became British Guiana — on the Essequibo, the Demerara, and the Berbice rivers. The Dutch went upstream as far as the rivers were navigable for their sailing ships, which was just about to the edge of the crystalline hilly upland. Here their African slaves made clearings in the forest and planted tobacco and sugarcane. These early settlers, like many people even today, had the idea that any soil that would grow such a luxuriant cover of forest must be fertile. Only slowly did they discover that the soil outside of the river floodplains was actually poor for shallow-rooted crops. After a few years, during which the yields of crops steadily declined, the Dutch abandoned their up-river settlements and moved back to the alluvial plain along the coast.

The alluvial soils of the coastal lowland are not leached and are potentially much more productive than those of the uplands. Yet, much work was required to drain the marshes and keep out the ocean water at high tide. In the steamy humidity of the lagoons back from the coast, infested with mosquitoes, the work could scarcely

have been done without the labor of African slaves. Nevertheless, the plantations along the Essequibo, the Demerara, and the Berbice became quite prosperous.

Dutch settlements began to spread in either direction along the coast. When the Dutch advanced beyond the Essequibo toward the Orinoco, the Spaniards began to resist, and the Dutch authorities, wishing to avoid trouble, closed this part of the coast to further settlement. Settlement thereafter spread to the Courantyne and the Suriname rivers. Meanwhile, other parts of the Guiana Coast were being occupied by the French and British.

The Guianas changed hands many times. The British repeatedly raided the prosperous Dutch colonies and in 1796 took possession of the Essequibo, Demerara, and Berbice settlements. Finally, by a series of treaties negotiated in Europe between 1812 and 1817, the present boundaries of the three Guianas were fixed. British Guiana extended westward from the Courantyne far beyond the Essequibo, almost to the delta of the Orinoco, including a large area that is today in dispute between Guyana and Venezuela.

The abolition of slavery in British Guiana, as elsewhere, created severe shortages of labor. The British attempted to solve the labor problem by bringing in workers under contract. First, they tried the Portuguese from Madeira, where sugarcane had long been grown. Then they tried Chinese and East Indians (Hindus and Moslems). The result is that today Guyana is a country of highly diverse ethnic origin.

The Population and Settlement Pattern

The population of Guyana has increased slowly, reaching 900,000 by 1980, and is highly concentrated in a narrow coastal fringe. East Indians comprise about 52 percent of the total. Although they have given up traditional Indian costumes and speak English rather than Hindi, they tend to remain unmixed with other elements of the population. They are concentrated largely in villages in which there are no other types of people. Africans and part-Africans make up 42 percent of the total, and 1 percent are Portuguese, Chinese, British, and North Americans. These ethnic elements occupy villages where they are mixed together, as they are also in Georgetown. Another 5 percent of the population includes native Indians living in the forests of the interior.

Today there is an almost continuous zone of agricultural settlement along the coastal bar from the Suriname border on the Courantyne River to and beyond the Essequibo River. The pattern of settlement is distinctive. The land is divided into thousands of long, narrow strips running from the ocean into the drained lagoons. Each strip is perhaps a half mile wide and 10 miles long. People who work the land live along a highway on the sandbar in an almost continuous string of houses.

The various ethnic groups have tended to cluster in different places. The East Indians supply most of the plantation workers where sugarcane is grown. The mixed settlements of the other ethnic groups are found where other kinds of crops are grown. An exception is along the Berbice River inland from New Amsterdam, where an almost solid East Indian population is engaged in growing paddy rice.

Only 10 percent of the population lives inland from the coast, and is sustained primarily by forestry, grazing, or the mining of bauxite. In 1970 the town of Markenburg was created by the amalgamation of three bauxite mining communities: Mackenzie, Wismar, and Christianburg. Its population is a little more than 30,000. Another important town is Lethem in the Rapununi Savanna, from which beef is air-freighted to the coast and to Suriname, French Guiana, and Antigua.

The Economic Situation

Agriculture provides the main support for the economy of Guyana, although less than a third of the population is engaged in farming. The

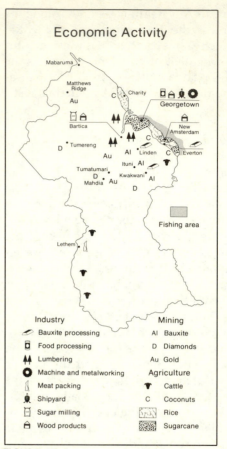

FIGURE 17-4

leading agricultural products are sugar and rice. Cane plantations are found all along the coast, many of them occupying strips of farmland mixed with other kinds of plantations. Near Georgetown, as well as around Guyana's second town, New Amsterdam, there are a few large plantations not divided into strips, and from these properties comes most of the sugar for export. Modern mills grind the cane and refine the sugar, and nearby are rum distilleries. Sugar and sugar products usually comprise nearly a third of the value of all exports, and they are sold mostly in the Commonwealth countries.

Only a small part of the total national territory is used to grow crops, and almost all of this land is along the coast. Here, agriculture has been diversified by the introduction of coconuts, cacao, coffee, citrus, oil palms, peanuts, bananas, and vegetables, all of which grow well in the rich alluvium of the coastal lowland. A part of this program of diversification has been the development of paddy rice along the Berbice, where enough can be grown for export.

The economy of Guyana is also supported by minerals, all of which come from the interior. The most important is bauxite, which was first produced at Linden in 1916. During World War II the bauxite mines around Mackenzie, some 55 miles up the Demerara River from Georgetown, produced around 20 percent of the world's supply and were the principal source of this ore for the aluminum industry of the United States.

The development of new bauxite mines in Jamaica after World War II reduced the proportion of the world total coming from Guyana. Although some of the ore is still imported by the United States, increasing amounts are shipped to Canada. A hydroelectric plant on the Demerara River now provides power for the refining of bauxite in Guyana. Oil, another source of power, has been discovered near the border of Brazil.

Guyana also produces gold and diamonds. These are found in crystalline rocks along the northeast-facing front of the plateau and in stream gravels of rivers that drain away from the plateau. The chief diamond sources are in stream gravels near the base of the escarpment, where manganese also has been discovered.

Another valuable resource of Guyana is timber and wood products. In the interior are vast forests that cover more than 80 percent of the national territory. There are the usual difficulties in cutting and transporting tropical timber, because any one kind of tree is widely scattered among other trees, and much of the most valuable wood is too dense to float in the streams.

Tourism is little developed, and the country is not well prepared to handle any large influx of tourists. There are few accommodations outside Georgetown, roads into the forested interior are poor, and there are few scenic attractions near the coast. Adverse publicity resulting from the Jonestown massacre of 1978, in which more than 900 members of a North American religious cult committed suicide, and continuing political unrest have further discouraged the development of a tourist industry.

The Political Situation

Since gaining independence in 1966, Guyana has faced serious problems. The social and geographical separation of the East Indians from the rest of the population is especially perplexing. Not only the East Indians have tended to remain separate, identifying themselves as Hindus or

THE TOWN HALL OF GEORGETOWN.

Moslems rather than Guyanians; so also have the other ethnic elements. This is in contrast to Trinidad, where most people, long before independence was gained, thought of themselves as Trinidadians. In the 1950s a remarkable change came to Guyana, and in the process a kind of state-idea was forged. This was brought about by a widely held desire for independence and for the development of a self-sufficient economy.

A political leader appeared who was able to command the support of the East Indians. This was Cheddi Jagan. He took advantage of a situation quite common in sugarcane areas: the great difference between the labor demand at harvest time and during the remainder of the year. Jagan promoted the organization of cane worker unions, and the unions pressed demands on the plantation owners for better pay and unemployment and health benefits. Jagan, who became the principal political leader of British Guiana, formulated a program of social change to be funded by taxes imposed on the estate owners. The estate owners responded by introducing machinery for the cultivation and harvesting of cane, which greatly aggravated the unemployment problem among the East Indians. Jagan succeeded in setting the workers against the landowners and the East Indians against people of other ethnic origin. Disorder threatened the program for granting independence, as Great Britain moved in troops to control rioting that broke out in 1964 between the East Indians and other groups.

In late 1964 Jagan was removed from office, and a coalition government was organized in an effort to achieve coherence among the diverse ethnic groups. Such coherence was made possible by the widely supported feeling that British forces should be removed and independence granted. In 1966 independence was proclaimed.

By 1976 the government had assumed control of all foreign trade, all of the nation's educational system, and 70 percent of the national economy. Guyana soon became a member of the Soviet bloc's Council for Economic Assistance (COMECON). Great political unrest continued, however, and in 1979 a new political coalition, the Working People's Alliance, was formed. The economy remained in a precarious state, with low prices prevailing for bauxite and sugar.

Guyana has a more varied resource base than either Trinidad or Jamaica, yet in 1984 the gross national product per capita was much less than that in either of those island nations. Its political stability is clearly correlated with economic development. The hope is that social change and a more equitable sharing of the GNP will bring security to the majority of Guyanians.

The most serious problem facing Guyana in the immediate future is resolution of Venezuela's claim to the Essequibo region, which constitutes more than two-thirds of Guyana's total national territory. This will require not only integration of the region through road construction, settlement, and resource development, but also alliance with hemispheric powers such as Brazil and the United States. Clearly, the current Marxist-Socialist orientation of the Guyana government does not lend itself to such accommodation with the region's leading capitalistic nations.

SURINAME

REPUBLIC OF SURINAME

Land area   70,060 square miles

Population   Estimate (1984) 420,000
             Latest census (1980) 352,041

Capital city   Paramaribo 150,000

Percent urban   66

Birth rate per 1000   28

Death rate per 1000   8

Annual percent of increase   2

Infant mortality rate   31

Percent of population under 15 years of age   43

Percent literate   80

Percent labor force in agriculture   29

Gross national product per capita   (GNP) $3520

Unit of currency   Suriname guilder

Physical Quality of Life Index (PQLI)   83

COMMERCE (expressed in percentage of values)

Exports

alumina	56	rice	8
bauxite	13	bananas	1
aluminum	10	plywood	1
shrimp	8		

Exports to		Imports from	
United States	35	United States	31
Netherlands	14	Netherlands	19
Norway	13	Trinidad-Tobago	15
United Kingdom	7	Japan	7
		West Germany	5

Data mainly from the 1985 World Population Data Sheet of the Population Reference Bureau, the 1985 Britannica Book of the Year, and the 1985 World Almanac.

The first settlements in present-day Suriname were established by the English and French between 1630 and 1640. These settlements were occupied by the Dutch in 1667, who in exchange for this territory ceded their colony on Manhattan Island (now New York City) to the British. Dutch Guiana, or Suriname, extended from the Maroni River to the Courantyne.

A plantation economy was developed in Suriname, based on slavery and the cultivation of sugarcane, coffee, cotton, and cacao. However, during the seventeenth and early eighteenth centuries the Dutch focused their attention on their more valuable possessions in the East Indies and South Africa. Ownership of the colony shifted periodically among Dutch, English, and French forces but finally reverted to the Dutch in 1815. Meanwhile, slave revolts were common and contributed to the general lack of stability and progress.

The Dutch in Suriname, in contrast with Dutch settlers in what is now Guyana, had great difficulty keeping their slaves from escaping into the forests of the interior. Escaped slaves formed independent African communities in the forests, practicing a shifting cultivation of subsistence crops. These were the Bush Negroes, who for many years raided Dutch settlements when they were extended inland from the coast. As a result, the Dutch did not build such extensive dikes and canals along the Courantyne, Saramacca, and Suriname rivers as they had constructed earlier along the Berbice, Demerara, and Essequibo. When slaves were freed from the plantations of Suriname in 1863, many fled to the forest to join the Bush Negroes. Others moved to Paramaribo, the colony's principal settlement.

The abolition of slavery resulted in severe labor shortages on the plantations. To resolve this problem, indentured workers were brought from India, China, and Java, of whom many remained as permanent residents. In the process, population density was increased in the coastal plain, and Suriname developed one of the most ethnically diverse societies to be found anywhere in the world.

Physical Conditions

The surface features of Suriname parallel those of Guyana. Along the Atlantic Coast a low, swampy plain varies in width from 50 miles in the west to 10 miles at the eastern border with French Guiana. This is the nation's agricultural district, a land of dikes and canals and high rural population densities. Inland from the coastal plain is the hill country, largely forested but including areas of savanna grassland. Here the soil is less fertile for agriculture, and the grasslands are devoted to the grazing of cattle. Also in this region are Suriname's principal deposits of bauxite. Farther inland, and largely unoccupied, is a vast area of low mountains culminating in the Wilhelmina Range, with peaks reaching 4000 feet in elevation.

Climatic conditions at Paramaribo are almost indistinguishable from those of Georgetown in Guyana. Temperatures are high throughout the year, averaging 81°F, with a diurnal range exceeding the seasonal differences in temperature. The average annual rainfall at Paramaribo exceeds 90 inches and is still greater on the slopes of interior mountains. With such abundant rainfall, about 90 percent of the country's total land area is forested.

Population and Pattern of Settlement

The population of Suriname is even more diverse than that of Guyana. Moreover, the various ethnic groups have retained their traditional forms of dress, their language, and their customs. About 31 percent of the total population is described as *Creole*, which refers to people of African ancestry born in America.[1] East Indians comprise 37 percent of the total. They are known as *Hindustani*, regardless of whether they are

---

[1] The term *Creole* is used in the southern United States to refer to people of European (French or Spanish) ancestry born in America, but in the Guianas the term refers to people of African or mixed African and European ancestry.

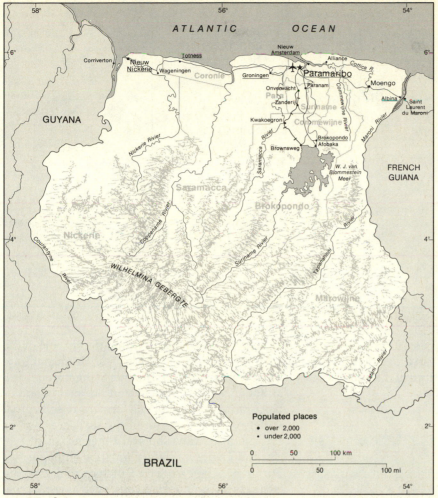

FIGURE 18-1

Hindus or Moslems. Some 14 percent of the population are Moslems from Java. In the interior are the Bush Negroes who make up about 10 percent of the total. Perhaps 3 percent of the population are native Indians, known as *Amerindians* to distinguish them from Hindustani. Another 2 percent are Chinese, and still another 2 percent include Spaniards, Portuguese, Dutch, Lebanese, Syrians, English, and even some North Americans.

In Suriname there is no common tongue. Each ethnic group continues to use its own language. If there is any one language that almost every one

can understand, it is a kind of patois or mixture of African and English words, used nowhere else but in Suriname. The official language is Dutch, but a large proportion of the people cannot use it. Lacking a common tongue, each ethnic group remains in separate communities, each usually performing its own kind of work.

The pattern of settlement is one of isolated clusters. There is no continuous string of settlement tied together by a road. Rather, there are scattered and isolated villages. Nor is the agricultural land continuous, for thriving plantations are often bordered by abandoned land growing

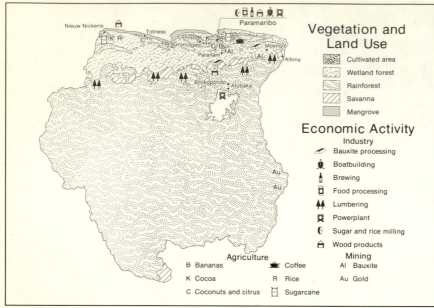

Nieuw Nickerie
Totness
Groningen
Paranam
Brokopondo   Atobaka
Paramaribo
Moengo
Albina
Au
Au

Vegetation and Land Use

Cultivated area
Wetland forest
Rainforest
Savanna
Mangrove

Economic Activity

Industry

- Bauxite processing
- Boatbuilding
- Brewing
- Food processing
- Lumbering
- Powerplant
- Sugar and rice milling
- Wood products

Agriculture

- B Bananas
- K Cocoa
- C Coconuts and citrus
- Coffee
- R Rice
- Sugarcane

Mining

- Al Bauxite
- Au Gold

FIGURE 18-2

up in brush. About 85 percent of the total population of 420,000 lives within a 25-mile radius of Paramaribo, and nearly a third of the total resides within the capital city itself.

AN ORE BOAT APPROACHING A BAUXITE PLANT.

Each ethnic group plays a different role in the economic and political life. The Hindustani are mostly small farmers with title to their properties, although some are tenants on rice paddyland or dairy farms. These small farmers raise a great variety of food crops for their own use, as well as for sale in Paramaribo and other urban places. The Indonesians are the plantation employees of Suriname. They work for Dutch owners growing a variety of commercial crops, such as cacao, coffee, citrus fruits, and sugarcane. Most of the Creoles are concentrated in Paramaribo, where they dominate the political life, operate small retail stores, and comprise the professional class. The Chinese are mostly shopkeepers, restaurant owners, or bank employees.

The Economic Situation

The economy of Suriname is based primarily on the mining and processing of bauxite. Bauxite deposits were discovered along the Cottica River near Moengo in 1917, and since 1929 this mineral has completely dominated Suriname's exports. The bauxite occurs in beds 15 to 20 feet thick, under an overburden of varying depth. It is usually found near the edge of the hilly upland, and access to the mines is gained by navigating the meandering rivers that cross the coastal lowland. These rivers, although narrow, are deep enough for ocean-going vessels.

Of greatest significance is a dam completed by the Aluminum Company of America (ALCOA), just south of Brokopondo on the Suriname River. It impounds one of the world's largest man-made lakes and provides electricity for the smelting of bauxite into alumina and aluminum at Paranam. Bauxite and its refined products now normally constitute 70 to 85 percent of Suriname's total export trade.

There has been some development of industry in Suriname, in addition to the processing of bauxite, especially in and around Paramaribo. A plant has been established to manufacture cardboard containers for use throughout the Caribbean in shipping bananas. Several other factories make plywood, flooring, paper products, and prefabricated houses. There is a plant for canning orange juice and pineapples, and other plants extract coconut oil. Most establishments are small, however, and serve only the domestic market with items such as food and clothing.

Less than 1 percent of Suriname's land is classified as arable, and less than 0.5 percent is under cultivation. Rice is the principal crop and is grown throughout the coastal plain, both as the basic food staple and for export. Also produced are bananas, citrus, and coconuts.

THE GOVERNMENT SQUARE, SURROUNDED BY OFFICIAL BUILDINGS IN PARIMARIBO, CAPITAL OF SURINAME.

Forestry offers great potential but currently contributes only 3 percent of the gross national product. There has been some expansion of fisheries in recent years, based primarily on the export market for shrimp, while tourism is virtually nonexistent. Much of the country is believed to contain valuable mineral and forest resources, but their development will depend on effective planning, political stability, and a rational policy toward foreign investment.

The Political Situation

Suriname became a self-governing Dutch colony in 1951, a member of the tripartite Kingdom of the Netherlands in 1954, and a fully independent republic in 1975. Meanwhile, it became the world's leading exporter of bauxite and still ranks among the five major suppliers of bauxite and aluminum products. Independence and a solid resource base have not, however, assured the new nation a future of peace and prosperity.

The political parties of Suriname are based largely on ethnic affiliations. Thus, coalitions must be formed to represent the populace, and real democracy is difficult to attain. In fact, the government of independent Suriname functioned as a democracy for only five years before it was overthrown by a military coup in 1980. Rule was subsequently by military decree. The constitution was suspended, political parties were banned, and efforts were made to develop closer ties with the Socialist world. Boundary disputes with both Guyana and French Guiana remain to be resolved. Especially critical is the need to develop political and economic conditions sufficiently favorable to stem the emigration of skilled personnel. Illustrative of the problem is the fact that, by 1985, about a third of the entire Surinamese population had migrated to the Netherlands. The majority were successful business people.

FRENCH GUIANA

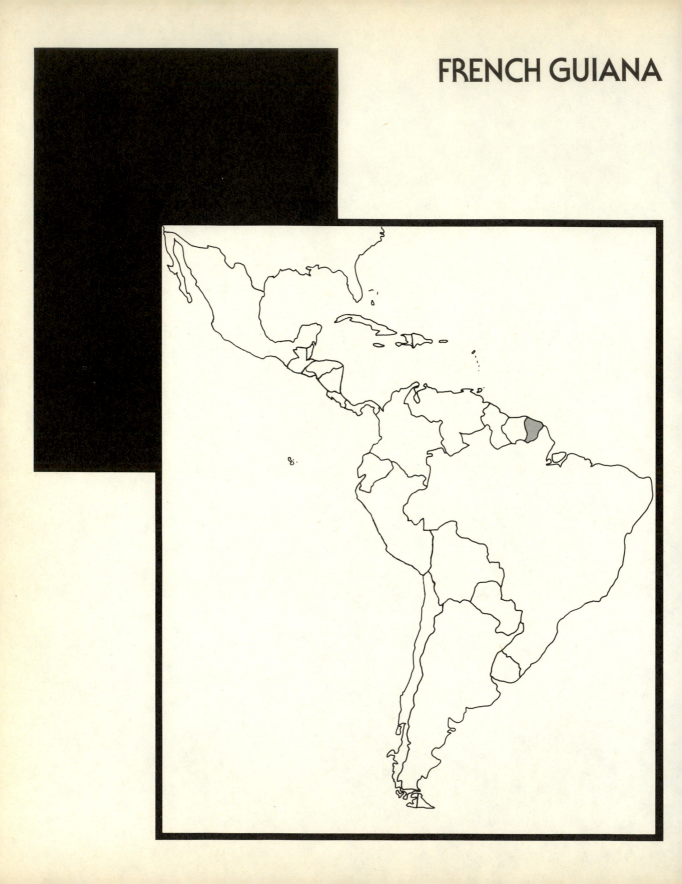

FRENCH GUIANA

Land area   34,750 square miles

Population   Estimate (1985) 73,022
             Latest census (1982) 73,022

Capital city   Cayenne 38,135

Percent urban   73.4

Birthrate per 1000   30.2

Death rate per 1000   6.2

Infant mortality rate   56

Percent of population under 15 years of age   34.9

Gross nation product per capita $3143

Percent literate   81.5

Unit of currency   French franc

COMMERCE (expressed in percentage of values)

Exports

shrimp   74
timber   5

Exports to		Imports from	
United States	48	France	52
France	21	Trinidad-Tobago	19
Japan	13	United States	9
Martinique	7	Japan	5
Guadeloupe	4		

Data mainly from the 1985 Britannica Book of the Year.

The third of the Guianas is an overseas *département* of France, similar to Martinique and Guadaloupe. This is French Guiana, the part of the Guianas with the smallest population and least economic development. It is also smallest in area among all the political divisions of South America, less than 35,000 square miles, yet larger than Austria and comparable in size to the countries of Central America.

Physical Conditions

French Guiana lies on the northeastern margin of the ancient crystalline rock mass known as the Guiana Highlands. Eons of time have left eroded surfaces as low mountains and hills over most of the country. The highest peaks, along the Brazilian border, scarcely reach 2600 feet in elevation. Along the Atlantic Coast is a narrow, swampy coastal plain, but a projection of the rocky uplands extends all the way to the ocean at Cayenne and reappears in the Îles du Salut a short distance offshore.

Most of the territory is exposed to the full effects of the northeast trade winds as they sweep across the equatorial portions of the north Atlantic Ocean. Rainfall averages 150 inches annually at the capital city, Cayenne, on the north coast and is even greater on windward slopes of the interior mountains. The result is a dense growth of tropical rain forest covering 90 percent of the land area, with savanna grass bordering the swamps and lagoons of the coastal plains. Maximum precipitation is from December to July, although there is no real dry season. Temperatures are also high throughout the year, averaging about 80°F at Cayenne.

Population and Settlement

Cayenne was founded in 1637 by French merchants, and, as elsewhere along the Guiana Coast, a plantation system of agriculture based on slavery soon spread across the lowland. However, the French lacked the talent or inclination of the Dutch to manage swampland agriculture, and the colony itself was subject to frequent change of control among the French, English, Portuguese, and Dutch. The abolition of slavery in 1848 brought ruin to the plantation system, and from 1852 to 1945 French Guiana was most noted as a French penal colony.

The worst of the prison camps was located up the Maroni River, about 125 miles from Cayenne, at a place called Saint Laurent. Health conditions were bad, and a lack of wind made the steamy humidity exhausting. More favored prisoners were sent to Devil's Island, one of three small Îles du Salut, about 8 miles offshore from Kourou. Here, there is always some wind, and health conditions are much better. The policy of exiling French convicts, begun in 1852, was terminated in 1944, and by 1946 the last of the prison camps had been abandoned. The Îles du Salut have subsequently become a tourist attraction of modest importance.

The entire population of French Guiana totals only about 73,000. More than half of that number reside in Cayenne, and most of the remainder live within the immediate vicinity. People of the coastal lowland are mostly of African ancestry, hence *Creoles*, whereas minority elements of the population include French and other Europeans, Chinese, Vietnamese, and Indonesians. There are some *Amerindians* in the interior, as well as some Bush Negroes, who escaped from slavery in Suriname during the colonial plantation era. The vast interior of French Guiana however, can be described as "empty country" in terms of human occupance.

The Economic Situation

After the prison camps were closed, the economy of French Guiana stagnated. The chief products came from the sawmills at Saint Laurent. In 1967, however, this part of France experienced

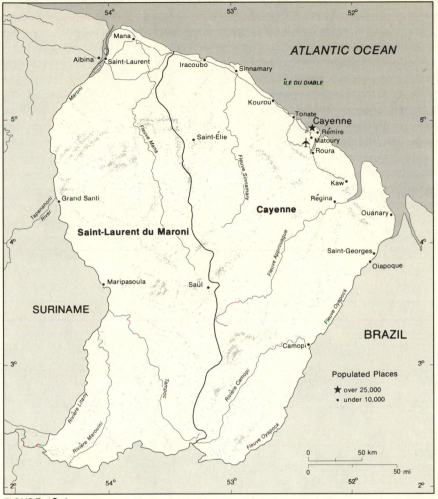

FIGURE 19-1

some revival of its economic life. French businessmen from Africa brought new projects to French Guiana, including the manufacture of plywood. Sugarcane plantations around Cayenne and Saint Laurent were expanded, and cacao, pineapples, and rice were introduced. To these have been added maize, manioc, and bananas. There is also some placer gold mining and export of bauxite.

Cayenne has a rum distillery, a pineapple cannery, and a factory for the manufacture of ball-point pens. Shrimp-processing is also one of the newer industries. About 17 percent of the value of all exports is from timber and wood products, and 11 percent is from gold mined at small scattered localities of the interior.

The most significant economic impact on French Guiana during recent decades has come from establishment of the Guiana Space Center, near Kourou, on the Atlantic Coast. This is a major base for the European Space Agency and is used for the launching of commercial satellites. Its location provides a special advantage in that the earth's rotation at the equator gives rockets

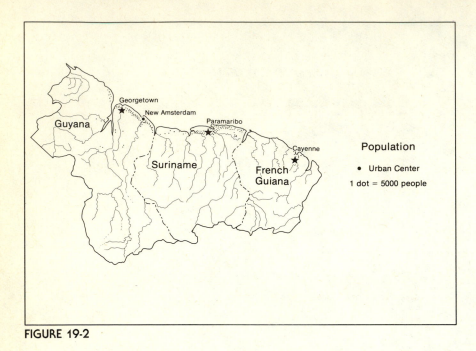

FIGURE 19-2

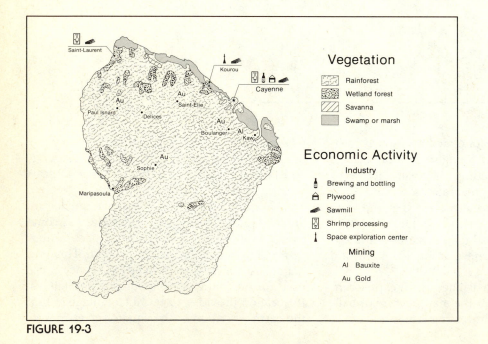

FIGURE 19-3

an additional boost at takeoff, thus achieving a higher orbit with less expenditure of energy. The little fishing village of Kourou has become a modern community of 9000 inhabitants, many of whom work at the Space Center, while other employees have found housing in nearby Cayenne.

Future economic opportunity appears abundant, for there are extensive known fishery, forestry, and mineral resources to support the sparse population of French Guiana. The incentives required to achieve greater economic development appear less certain.

The Political Situation

French Guiana, after several centuries as a minor possession of France, became politically an inte-

gral part of the French Republic in 1946. As an overseas *département*, along with Saint Pierre and Miquelon, Guadaloupe, Martinique, and Réunion, it conveys to its inhabitants all the rights and privileges of French citizenship. Three-fourths of all external trade is with France, and about two-thirds of the labor force is employed by government. Local industry is almost nonexistent, although there is some small-scale agriculture and enough ranching to serve the domestic market of Cayenne. Poverty and underdevelopment exist, but these conditions must be viewed within a context of general "comfort" at the expense of taxpayers in France. Hence, French Guiana is the one political unit in all of South America, and one of the few anywhere in the world, which has yet to experience a significant demand for political independence.

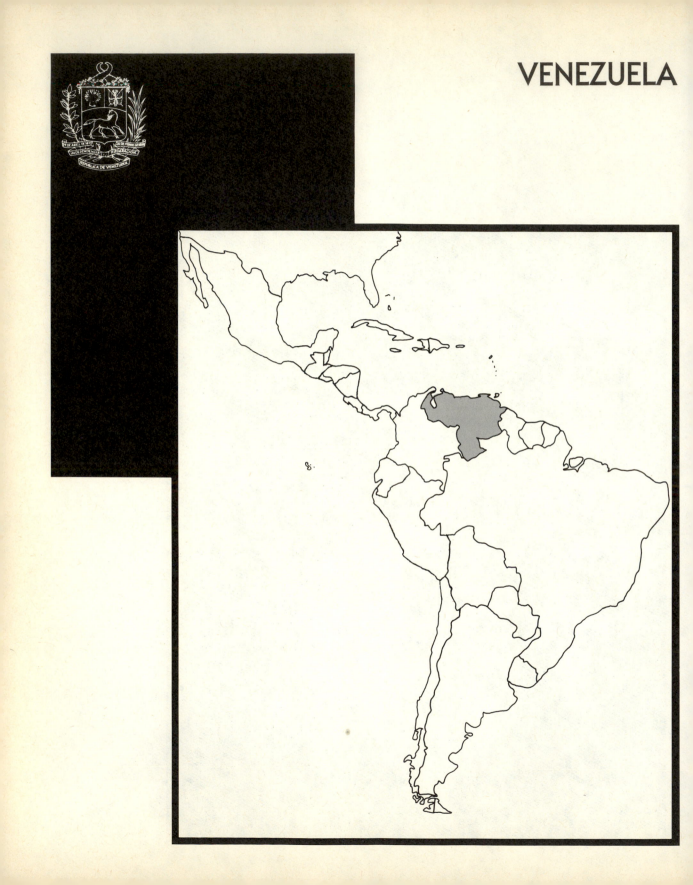

VENEZUELA

20

REPÚBLICA DE VENEZUELA

Land area   352,143 square miles

Population   Estimate (1985) 17,300,000
Latest census (1971) 10, 721, 522

Capital city   Caracas 3,300,000

Percent urban   76

Birthrate per **1000**   33

Death rate per **1000**   6

Infant mortality rate   39

Percent of population under **15** years of age   41

Annual percent of increase   2.7

Percent literate   86

Percent labor force in agriculture   15

Gross national product per capita $4100

Unit of currency   Bolívar

Physical Quality of Life Index (PQLI)   79

COMMERCE (expressed in percentage of values)

Exports

| crude oil | 60 |
| petroleum products | 32 |

Exports to		Imports from	
United States	37	United States	46
Netherlands Antilles	22	Japan	8
Canada	10	West Germany	7
United Kingdom	2	Canada	4

Data mainly from the 1985 World Population Data Sheet of the Population Reference Bureau, the 1985 Britannica Book of the Year, and the 1985 World Almanac.

To most people, the term *Venezuela* is automatically associated with wealth and, more specifically, with prosperity derived from rich deposits of petroleum. As one of the world's greatest oil-producing and exporting nations, Venezuela has also become one of the richest countries in Latin America as measured by gross national product per capita and related types of criteria. At the same time, this nation has achieved a notable degree of political stability and basic democracy. For more than 20 years, it has had freely elected popular governments, and its long struggle against autocratic rulers appears to have been won.

Historical Setting

The independence movement among the Spanish colonies in America appeared early in Venezuela. Simón Bolívar, who led the movement to liberate northern South America, came from an aristocratic family in Caracas. His ideal was the freedom of American-born leaders to govern without interference from Spain. Freedom for him meant freedom from foreign authorities. Bolívar hoped to set up a "United States of South America," and for a few years he was able to hold together in Gran Colombia the territory now included in Venezuela, Colombia, Ecuador, and Panama.

Bolívar's dream of a united South America was shattered by rebellion. The lesser political leaders in the separate clusters of population, having been freed from Spanish authority, turned against authority from Bogotá where Bolívar established his capital. In 1830 Bolívar's own chief lieutenant in the wars against Spain declared Venezuela to be free from Gran Colombia. Bolívar, in despair, said: "Our America will fall into the hands of vulgar tyrants; only an able despotism can rule America." This was more than 150 years ago, but the prophecy has been fulfilled quite literally during most of the succeeding years.

Between 1830 and 1935 Venezuela had more than a dozen rulers, but three "able despots" stand out above all others: José Antonio Páez, a mestizo peon, who declared the independence of Venezuela in 1830; Antonio Guzmán Blanco, who first came into office in 1870; and Juan Vicente Gómez, who ruled from 1908 to 1935. Under each Venezuela was run like a vast private estate, but order was enforced and there was an increase of material prosperity. In the intervening periods, the country was ravaged by the conflict of warring elements, no one of which was strong enough to force the others into submission.

Gómez was not descended from the aristocracy. He was a man of tremendous ambition, a ruthless fighter who ruled his country with an iron hand. His opponents were killed or exiled. When geologists found that Venezuela possessed important oil resources, Gómez, in the years immediately following World War I, made arrangements with the large oil corporations to undertake the expensive business of producing petroleum. The Venezuelan treasury was suddenly filled with money. In the Depression years after 1929, Gómez brought his country through without any foreign debt.

Since the death of Gómez in 1935, Venezuela has been the scene of a mighty struggle between those who favor democratic procedures, including rule by law, and those who seek to continue the traditional system of personal power. In 1958 another dictatorship was overthrown, and since that time the forces of democracy have dealt successfully with threats of violence from both the right and the left.

The People

Bolívar and Gómez dramatize the contrasts that exist among the people of Venezuela and some of the difficulties involved in establishing order among diverse elements. The population includes Europeans, Indians, and African blacks. Estimates indicate that 20 percent of the Venezuelans are of unmixed European ancestry, prob-

Administrative Divisions
and their capitals

1. Zulia   13. Aragua
2. Falcón   14. Miranda
3. Lara   15. Guárico
4. Trujillo   16. Anzoátegui
5. Mérida   17. Sucre
6. Táchira   18. Monagas
7. Barinas   19. Territorio Delta
8. Portuguesa   Amacuro
9. Yaracuy   20. Bolívar
10. Carabobo   21. Territorio Amazonas
11. Cojedes   22. Nueva Esparta
         23. Distrito Federal

FIGURE 20-1

ably not more than 2 percent are of unmixed Indian ancestry, and another 9 percent are black. About 69 percent are mestizos, persons of mixed European and Indian parentage in varying ratios. One-fourth of the population is composed of immigrants, many of whom have entered illegally from Colombia.

Racial percentages are not uniform throughout Venezuela. Europeans are concentrated in the larger towns and cities, such as Caracas, Maracaibo, and Valencia. Pure Indians survive only in the more remote places, such as in the Guiana highlands south of the Orinoco River or in the forests west of Lake Maracaibo. A black component predominates along the Caribbean Coast, including the ports of La Guaira and Puerto Cabello. Even among the mestizos there is a notable difference between people from the highlands and those from the lowlands.

The Indians and the European Conquest

Contact between the Spaniards and the Indians began along the coast during the first part of the sixteenth century. The first European settlement on the continent of South America which has survived to the present time was established at Cumaná in 1523. Four years later another colony was planted near the base of the Paraguaná Peninsula at Coro. From these two communities exploring parties pushed inland. West of Coro the Spaniards entered the Maracaibo lowland, and, coming upon Indian villages built on piles in the shallow waters of the lake, they named the country "Little Venice," or Venezuela.

The Spanish explorers pressed inland searching for gold, and after 1838 they discovered many places where stream gravels yielded the precious metal. The placer mines of the valley of Caracas were worked by Indian slaves, and for a time the Venezuelan mines seemed to promise great wealth. However, none of the sources of gold were better than low-grade deposits, and the miners were forced to turn to agriculture.

More than 20 years elapsed after the settlements at Cumaná and Coro before the first permanent town was established in the highlands. The entire highland region had been explored, however, and many valley bottoms were dug up in search of gold-bearing gravels. In 1555 Valen-

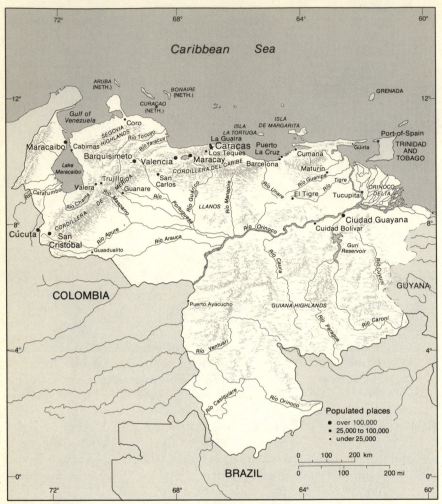

FIGURE 20-2

cia was founded in the intermont basin which subsequently became the country's leading agricultural district. The good agricultural land was quickly divided into large estates and awarded to the Spanish officers. On these estates the Indians, Christianized and enslaved, were set to work for their new masters. Barquisimeto was also established about this same time in the midst of a farming district west of Valencia, and in 1567 Caracas was laid out. Spaniards who had invaded what is now Colombia, and who founded Bogotá, Colombia, again turned northward and established San Cristóbal and Mérida in the mountains south of Lake Maracaibo.

Meanwhile, intermarriage proceeded rapidly. The Spaniards did not generally bring women with them to the Americas but took wives from among the Indians they conquered. While great numbers of the native peoples died from epidemics of measles and smallpox, the mestizo children showed a greater degree of immunity than did their Indian ancestors. Yet, the landowners soon found themselves faced with a shortage of agricultural labor. Where sugarcane was planted, African slaves were introduced to perform the hard work of harvesting and grinding the cane. Since intermarriage was unrestricted either by law or by custom, people of the sugarcane districts became considerably darker in complexion than the inhabitants of the rest of the country.

Present Population

The population of Venezuela increased slowly. At the beginning of the nineteenth century the famous German geographer, Alexander von Humboldt, estimated the number of Venezuelans to be about 1 million. In 1920 another estimate placed the figure at 2.4 million—not a great increase in more than a century. After World War II, however, modern public health programs, including insect control, resulted in a sharp decline in the death rate. As a result, Venezuela's net rate of increase is now almost 3 percent per year. In 1985 it was estimated that the Venezuelan population exceeded 17 million.

Like all population data for large areas, these figures obscure important differences in density and rates of growth or decline between one area and another. There is a notable current of internal migration toward the cities. In 1981 Caracas had a population of 3.3 million in its metropolitan area, and there were eight other cities of more than 100,000. To understand the significance of the population pattern, it is necessary to examine, region by region, the ways in which inhabitants have occupied the land and exploited its resources during the various periods of Venezuelan history.

The Regions of Venezuela

Venezuela is composed of four distinct regions, only one of which is densely populated. The backbone of the country is formed by the Venezuelan highlands, a branch of the Andes. This highland backbone is composed of four major divisions: the Central highlands, in which Caracas and Valencia are located; the Northeastern highlands; the West-Central highlands, north of Barquisimeto, including the Paraguaná Peninsula; and the Cordillera de Mérida, south of Lake Maracaibo.

The other three major divisions of Venezuela are much less densely populated and much less developed. These are the Maracaibo lowlands, the Orinoco Llanos, and the Guiana highlands.

The Venezuelan Highlands

The main branch of the Andes Mountains does not connect with the Rocky Mountains of North America. Rather, it swings eastward from Colombia into Venezuela, where it parallels the Caribbean Coast and continues eastward to the Paria Peninsula and beyond that into Trinidad and Tobago.

The Central Highlands That part of the highland backbone that borders the Caribbean between Puerto Cabello and Cape Codera has become the core area of Venezuela. Here, the political interests come to a focus, and here, too, are the densest rural populations and the largest city.

The Caribbean Coast which borders the Central highlands is hot and dry. Some 11 inches of rainfall at La Guaira are insufficient to support more than a sparse cover of xerophytic plants, among which the cactus is prominent. This Caribbean Coast holds the highest temperature records of the American tropics. La Guaira, for example, has an average annual temperature of 81°F.

The mountains of the Central highlands rise abruptly from the coast to elevations of 7000 to 9000 feet. Arid conditions are restricted to the first few hundred feet of the mountain slopes. Above that, rainfall is sufficient to support a cover of forest that continues to the tree line between 6000 and 7000 feet above sea level.

The Central highlands are composed of two distinct ranges, separated by the intermont basin in which Valencia is located and farther eastward by the deep valley of the Río Túy. In the center of the Basin of Valencia is a shallow body of fresh water known as the Lake of Valencia, which was visited in 1800 by Alexander von Humboldt. The lake has no surface outlet, but Humboldt noted that at some previous time it drained southwestward to the Orinoco. He used this as an example of the unfortunate results of excessive forest clearing. The level land around the lake was devoted to indigo, and the steep mountain slopes around the basin had been cleared for the cultivation of maize. Therefore, the rainwater ran off in torrents instead of sinking into the ground and seeping slowly toward the lake. After torrential floods there was less water underground, and the level of the water table declined. Since 1900, the level of the lake has lowered 16 feet. As a result, the margins of the lake are bordered by a wide belt of marshes, especially on the eastern and western sides. This basin, about 1500 feet above sea level, is partly drained to the east by the Río Túy, which has cut back westward and captured some of the drainage that once entered the lake.

The mountain ranges on either side of the Basin of Valencia and of the Río Túy are too narrow to include any extensive areas of gentle slopes. A few small basins do occur, however, nestled in the midst of the mountain country. Most important is the valley of Caracas. About six miles south of the Caribbean Coast the northern range is broken by a narrow structural depression, or rift, which extends about 15 miles east-west. In this rift valley is a narrow strip of gently sloping land, surrounded by steep mountains and now completely occupied by the city.

Climatic conditions and natural vegetation in the Central highlands form an extremely intricate pattern. Temperatures decrease, in general with increasing altitude, and averages of less than 65°F in the coldest month begin at 3000 feet above sea level. Throughout the area, rainfall and humidity are high on slopes that face eastward toward the prevailing winds. The east-facing valley of the Río Túy and that of the Río Yaracuy (west of Puerto Cabello) receive sufficient rainfall to support a forest cover.

Failing to discover any abundance of precious metals in the Basins of Valencia, the Spaniards turned to commercial crops, making use of cheap land and slave labor. The highland region and much of the more remote country elsewhere were divided among a relatively small number of Spaniards. The crops raised included a mixture of those native to America and others introduced from the Old World. The native food grain, maize, was cultivated widely as food for the workers, but only on land unsuited for the production of commercial crops. Sugarcane was the most valuable of the new crops, and for a time Venezuela participated in the rapidly expanding sugar market in Europe, which was then being supplied mainly from Brazil. In 1784 coffee was introduced. The landowners discovered, however, that they could tempt the foreign market with new American products, such as cacao, tobacco, and indigo. The latter crop brought such good returns during most of the colonial period that many areas of settlement were supported by indigo plantations. After chemical dyes captured the market, indigo was no longer profitable, and the plantations were put to other uses.

The Basin of Valencia continues to be one of the major agricultural regions of the country. Sugarcane is grown extensively on the irrigated lands and is processed in modern mills. Cotton is cultivated on the drier lands of the lake plain and supplies the textile mills of Valencia, Maracay, and Caracas. Food crops include maize, rice, beans, and vegetables. There are also extensive pasturelands for dairy and beef cattle. Farmholdings are smaller than in the past, since many large private estates have been expropriated under a land-reform program of the Venezuelan government and distributed to small-scale agriculturalists.

The core area in the Central highlands is also the major focus of urban and industrial growth. The city of Caracas, by far the largest urban agglomeration in Venezuela, serves as the national

A SECTION OF CARACAS, VENEZUELA.

capital and primary center of industry, commerce, education, and tourism. Until 1940, the valley of Caracas was still an important agricultural area with the densest rural population in Venezuela.

The population of Caracas in 1950 was 495,000; it is now more than 3 million. Not only is the entire valley floor occupied by the city, but the neighboring slopes as well. The parks and monuments of Caracas reflect the country's past, and its modern *autopistas* (highways), highrise buildings, and subway system reflect the nation's prosperity and dynamic character. At the same time severe problems have developed, in large part related to the city's geographic conditions.

Both government and private investments have for many years focused on the primate city, with the result that Caracas has experienced a great influx of migrants not only from other parts of the republic but also from abroad. More than 1 million immigrants from Spain, Portugal, Italy, Eastern Europe, and certain countries of Latin America have entered Venezuela since World War II, and many of these have made their home in the capital city. It has not been possible to build housing fast enough to accommodate the influx; hence, slum housing covers the hills surrounding the city. Traffic congestion is monumental, land prices are exorbitant, and there are periodic shortages of water and electricity.

Since the late 1950s the government has encouraged decentralization of urban-industrial activities from Caracas to other parts of the country, and especially to the Basin of Valencia and the valley of the Río Túy. Valencia in particular has become a major industrial center. Numerous food and textile plants are located within the city, and its extensive industrial park includes factories producing automobiles, tires, plastics, furniture, paint, soap, cement, feeds, foods, and structural metal products.

Maracay has always been an agricultural center, and its major industries still have an agricultural base. These include factories for the manufacture of textiles, foods and beverages, plus paper, stone and clay products, chemicals, and metalware. Even the smaller towns of the Basin of Valencia, such as La Victoria, Cagua, Guacara, and Los Guayos, have experienced substantial industrial expansion.

One negative factor in the industrialization of the Basin of Valencia is its effect on the lake. The Lake of Valencia, which has no external drainage, has become increasingly polluted with industrial and agricultural wastes. Many species of fish and plant life have been destroyed, and the decreasing lake level has resulted in rising air temperatures and decreasing humidity in the surrounding area.

The core area is also the focus of Venezuela's

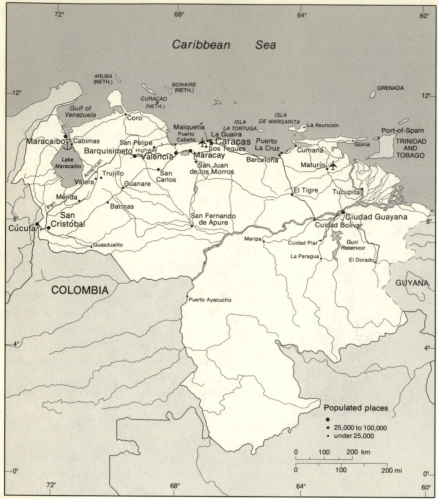

FIGURE 20-3

transportation network. Airlines connect Caracas with all parts of the national territory, and journeys that once took weeks now require only a few hours. The main national and international airport for Caracas is located at Maiquetía, just west of the principal seaport at La Guaira. Others serve all of the major urban centers throughout the republic.

It was not until 1950 that Venezuela's modern period of road building began. First was the problem of connecting Caracas with its seaport of La Guaira. The two are little more than six miles apart, but Caracas is about 3000 feet above

sea level, and to reach a pass over the coastal range required a climb of 3412 feet. Moreover, the slopes were so steep that the railroad had to cover 23 miles, and the old highway 21 miles, to make the climb. In 1953 a new superhighway was opened between La Guaira and Caracas. Bridges and tunnels carry this modern highway through the mountains, rather than over them, and the automobile trip from La Guaira to Caracas now takes about 15 minutes.

Paved highways connect the core area with all other centers of population and economic activity throughout the country. A highway crosses the Llanos to Ciudad Bolívar, where a bridge carries traffic to the south side of the Orinoco and to the industrial complex at Ciudad Guayana. Another highway reaches San Fernando de Apure. A branch of the Llanos Highway reaches Barcelona and Cumaná. Three highways extend westward from the core area. The Pan American Highway passes through Barquisimeto, and then along the northern base of the Andes to the southwestern edge of the Maracaibo lowland. There it climbs into the Andes to San Cristóbal and continues into Colombia. A second paved highway has been built along the coast through Coro and on to Maracaibo, crossing the northern end of the lake on a bridge. Another paved highway runs along the southeast-facing piedmont of the Andes to Barinas and beyond toward the Colombian border. An all-weather road has been built southward from the Río Orinoco to reach the border of Brazil near Mount Roraima. The total network now exceeds 29,000 miles of first-class paved highways.

Most of Venezuela's railroads were built between 1877 and 1893, during the rule of Antonio Guzmán Blanco. By attracting foreign capital, the dictator equipped the country with the most efficient type of transportation of that era. The expensive construction of these railroads, and the lack of bulk traffic, necessitated high freight charges. Commerce therefore continued to move over rough trails on mule back. Most railroad lines of the Central highlands have been abandoned and by 1977, there were only 261 miles of rail in all of Venezuela.

The Northeastern Highlands While the Central highlands region became increasingly a focus of political, social, and economic activity, and while its main city was growing rapidly, the northeastern part of Venezuela, beyond the Gulf of Barcelona, remained static. In this region the coastal range, which forms the two peninsulas of Paria and Araya, is only about 2600 feet high. The central mountain core of the northeast lies south of the structural depression in which Cumaná is located. Here, the summits reach 6700 feet above sea level. Abundant rains over the eastern part of this region supported an original cover of selva, or tropical rain forest, in which small clearings are used for the production of cacao. The western

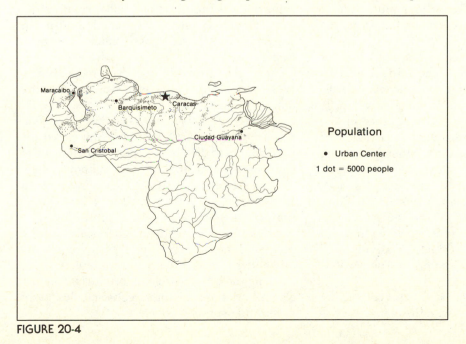

FIGURE 20-4

part of the region is relatively dry and was originally covered by scrub woodland and savanna.

Most inhabitants of the Northeastern highlands live on the drier western side, particularly in Cumaná, Puerto La Cruz, and Barcelona. Cumaná, the oldest European settlement in South America, is the capital of Sucre state, seat of a regional university, and a center for the fishing industry. Its fishing grounds are among the best in Latin America. Coffee, cacao, tobacco, and sugarcane are cultivated within its trade area. Near Barcelona there are mines of lignite and semibituminous coal, which is exported through the port of Guanta. Puerto La Cruz is the terminus of oil pipelines from the Llanos, which now account for more than a third of Venezuela's total petroleum output. Large-scale refineries are located nearby.

Offshore is the island of Margarita, which has become one of the leading tourism centers of Venezuela. A free-trade zone has been established, and fresh water is piped from the mainland. Margarita was discovered by Columbus in 1498 and quickly became known as a source of pearls. Some pearls are still recovered and supplement local income from the manufacture of ceramics, hats, and salt evaporated from sea water.

The West-Central Highlands The third section into which the highlands of Venezuela may be divided is the West-Central highlands, which lie north of Barquisimeto. This is another of the poorer sections of Venezuela. It is a region of recurring droughts that limit settlement to the permanently moist areas along the river valleys. Most of the region is thinly populated, although small clusters of population are located at Coro, around oil refineries on the Paraguaná Peninsula, in the valleys of the Río Tocuyo and Río Yaracuy, and near Barquisimeto. Dairying is of particular importance in the upper Tocuyo Valley, near Carora.

Coro, at the base of the sandy spit connecting the Paraguaná Peninsula with the mainland, has never prospered. The adjacent territory is sparsely inhabited by people who produce so little that they can purchase almost nothing from the outside. The rural population is grouped in small clusters wherever water is available even in dry years. The inhabitants devote small patches of wet land to maize and manioc, but most of the region is used only for the grazing of goats, sheep, and scrubby cattle.

The densest populations of the West-Central highlands are between Tucacas and Barquisimeto. Barquisimeto, known for its commerce, industry, and agriculture, is the third largest city of Venezuela. Irrigated lowlands surround the city, producing sugarcane, maize, and sisal. In the nearby highlands, wheat, oats, coffee, and cacao are grown. Between the ridges stretching westward toward Lake Maracaibo, cattle-raising and dairying predominate.

There is only one mining community in this district. The copper mines of Aroa, which were owned at one time by Bolívar, have an erratic history of production. Low prices in the world copper markets after World War I caused the abandonment of these mines, although in 1960 some ore was produced manually, chiefly for its sulphur content, and then stockpiled for potential future use in the petrochemical industry at Morón.

After 1950 major changes occurred within the West-Central highlands. First was the construction of two huge oil refineries on the southwestern coast of the Paraguaná Peninsula by transnational corporations. Oil from the Maracaibo area was brought to these refineries by pipeline. Later, a whole new industrial center was built about 15 miles west of Puerto Cabello, at the town of Morón. This site was selected for a huge industrial complex based on petrochemicals. New industries produce fertilizers, chemicals, plastics, pharmaceuticals, synthetic rubber, and many other items, whereas an oil refinery produces gasoline and other petroleum products. There are also factories to manufacture insecticides, herbicides, and synthetic fibers.

Cordillera de Mérida The southwestern segment of the Venezuelan highlands is formed by the high Cordillera de Mérida, in which several small intermont basins are occupied by clusters of people, primarily mestizo. Difficult access retarded occupation of the region by large landowners interested in commercial production. All of the more remote parts of Venezuela have suffered in comparison with the Central highlands in terms of access to ports on the Caribbean Sea. In the Central highlands contact with external markets has been continuous. There, the rulers of the country, no matter where they originated, finally established themselves. The almost complete isolation of the Cordillera de Mérida was finally ended in the nineteenth century with the rise of coffee as an important crop.

The Cordillera de Mérida is the only location in Venezuela where permanent snow is found. In the vicinity of Mérida itself are three snow-capped peaks, including the highest mountain in Venezuela, Pico Bolívar, which rises to 16,423 feet in elevation. The valleys and basins, however, are at relatively low elevations, and the transverse trench through which the boundary between Venezuela and Colombia passes permits a crossing from the Llanos to the Lake Maracaibo lowlands with a climb to only 4600 feet above sea level.

After the first towns were established by people from Bogotá, there was little further activity because of isolation and the lack of precious metals. Planters of sugarcane and indigo were unable to market their products as cheaply as could those of the more accessible Central highlands. Not until coffee became an important commercial crop did the people of Mérida find a product that could support the high cost of transportation. Today the Cordillera de Mérida is the principal coffee-producing region of Venezuela, and coffee remains a major agricultural export. This is partly due to an increase in planted area, but also to improvements in farm practices. Venezuela has joined the "Bogotá Group" of coffee-producing countries. This marketing organiza-

tion seeks to stabilize the coffee market and prevent speculation. Other member countries include Mexico, Guatemala, Honduras, El Salvador, Costa Rica, Colombia, and Brazil.

The three main clusters of population in the Cordillera de Mérida are all located in basins and valleys at altitudes between 1800 feet at Valera and 5380 feet at Mérida. San Cristóbal, the urban center of the cluster that crosses the border to the area around Cúcuta, Colombia, is 2700 feet above sea level. Because of the absence of intermont basins at higher altitudes, these zones are only thinly populated.

The agrarian reform program undertaken in recent decades has included a major project in western Mérida state. Initially, more than a million acres of land were made available for resettlement to more than 5000 farm families that settled in this area. Some 70 percent were formerly tenants on large private estates around Trujillo, Mérida, and Táchira, and another 30 percent came from the Maracaibo lowlands. This program has created an entirely new cluster of people.

The agrarian structure has changed radically here and elsewhere in Venezuela. Some forms of sharecropping and tenancy have been virtually eliminated, and some ambitious *campesinos* have become small or medium-scale commercial farmers. However, many have also become marginal producers who finally leave the rural setting and migrate to the cities.[1]

The Maracaibo Lowlands

The highest average annual temperatures in any part of Latin America, have been recorded in the Maracaibo lowlands. The city of Maracaibo has an annual average of 82.5°F, ranging from an average of 80.6° in January to 84.4° in August. On the shores of the lake, after sunset, the hu-

---

[1] P. Cox, "Venezuela's Agrarian Reform at Mid-1977," *Research Paper 71*, Madison, Wisconsin, Land Tenure Center, 1978.

OIL DERRICKS IN LAKE MARACAIBO.

midity is oppressive. The southern sky is illuminated with vivid flashes of lightning among the towering banks of cumulus clouds along the mountain slopes. Throughout the southern part of the lowlands, and especially on the lower slopes of the Cordillera de Mérida, heavy rainfall supports a dense growth of selva. Toward the north the decrease of rainfall produces a gradual transition from tropical rain forest, through semideciduous forest, to dry scrub.

Until 1918, settlement in the Maracaibo lowlands was of little importance. After highland plantations began to send coffee from the Cordillera de Mérida to the coast, Lake Maracaibo became a waterway of local significance. Transportation was by sloop, and connection with ocean vessels was made at Puerto Cabello or La Guaira.

The people who gained a meager living from the land knew nothing of the potential wealth beneath their feet, nor did the fishermen appreciate the black sticky substance that sometimes contaminated the water of the lake and fouled their nets. Geologists, however, long ago reported the existence of oil and oil-bearing formations throughout the northern part of the region and across the southern half of the West-Central highlands. Not until after World War I did the petroleum companies definitely turn their attention to Maracaibo. Soon thereafter, and especially after the Mexican oilfields were nationalized in 1938, oil derricks spread along the entire

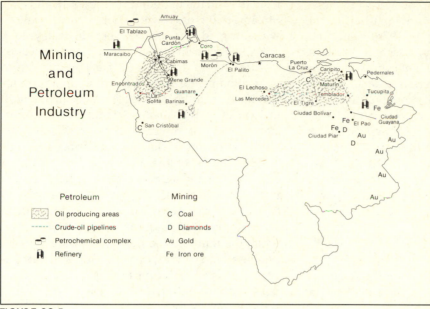

Mining
and
Petroleum
Industry

Petroleum

▨ Oil producing areas
---- Crude-oil pipelines
⬕ Petrochemical complex
🛢 Refinery

Mining

C Coal
D Diamonds
Au Gold
Fe Iron ore

FIGURE 20-5

northeastern shore and well out into the lake itself.

When the first productive oil well was drilled by the Royal Dutch Shell Company in 1917, and when the enormous potential of the Maracaibo Basin became clear, the construction of an oil refinery near the source of oil became necessary. For obvious reasons, the Shell Company decided to build its refinery on the nearby Dutch island of Curaçao. Later, when Creole Petroleum started operations, its refinery was built on Aruba. Because only small, shallow-draft tankers were able to reach the lake, pipelines to the refineries were also built. Large ocean-going tankers picked up refined oil from the Dutch islands for transport to Europe and the United States.

In the late 1950s, the government of Venezuela undertook to make Maracaibo and the entire lake accessible for ocean-going ships by dredging a channel through the sandbar at the lake's entrance. When completed, it permitted the passage of ships up to 28,000 tons. Subsequently, a five-mile bridge was completed across the lake narrows just south of Maracaibo to facilitate land transportation to and from the rest of

the republic. Industry, such as food processing, textiles, and the manufacture of fiber products, has also been promoted, and a major university established. Once a sparsely settled trading center, Maracaibo is now a thriving oil metropolis exceeding 650,000 in population, making it the nation's second largest city.

Venezuela possesses one of the most modern dairy industries in Latin America, and much of it is centered in the Maracaibo lowlands west of the lake, especially around Machiques. Dairy cattle abound on the grasslands of the transition zone between forests to the south and semi-desert lands to the north. Fresh whole milk is transported in modern glass-lined tank trucks for processing not only in Maracaibo but also in the Basin of Valencia and Caracas. As with most other farm products, however, Venezuela has yet to achieve self-sufficiency in its supply of milk, despite subsidies to dairymen; substantial quantities of powdered milk are therefore imported from abroad.

Oil remains the cornerstone of the economy of the Maracaibo lowlands and of the nation as a whole. Despite rapid development of the oilfields of eastern Venezuela, the Maracaibo lowlands continue to supply more than half of the nation's total output. Early in 1976, Venezuela nationalized its oil industry, which at that time was composed of 21 individual companies. The new national company, *Petroleos de Venezuela* (PETROVEN), is a state oil monopoly but continues to contract with foreign companies to obtain needed technical expertise. Deep drilling in Lake Maracaibo under such contracts has identified substantial new reserves of oil, and vast deposits of natural gas have been found in the Maracaibo lowlands and offshore in the Gulf of Venezuela.

The Orinoco Llanos

The Orinoco Llanos, the third major region of Venezuela, occupies nearly a third of the total national territory. The Llanos is the Spanish name given to the wide expanse of plains lying between the Andes and the Orinoco, covered by a mixture of savanna and scrub woodland. It slopes very gradually from the base of the Andes toward the great river that separates the Llanos

from the Guiana highlands to the south and east. Streams that cross it wind about in broad valleys with low gradients, and between the valleys low, mesa-like interfluves are the most conspicuous features of the landscape. If all of this region were good pastureland it might support 50 million head of cattle, but the total Venezuelan herd in 1981 scarcely exceeded 10 million with a substantial flow of contraband cattle crossing the border into Venezuela from Colombia.

The climate of the Llanos is divided into a rainy season and a dry season, and during these two extremes the landscape undergoes an extraordinary transformation. The wettest part of the rainy season occurs from June to October, when the rainfall is so heavy that the rivers are unable to carry off all of the water, and vast expanses of land are inundated. At this time, animals are concentrated on the low mesas that stand as islands above the floods. After they have eaten one of these islands bare, they are driven to another, sometimes being forced to wade many miles through shallow water.

In October the rains begin to abate, but occasional showers continue until November or De-

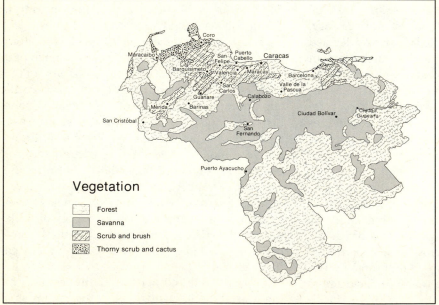

Vegetation

- Forest
- Savanna
- Scrub and brush
- Thorny scrub and cactus

FIGURE 20-6

cember. The real dry season begins in January and continues with no sign of rain until the end of March when floodwaters recede. Only the larger rivers continue to flow freely, and the smaller ones are gradually reduced to chains of swamps and pools along the valley bottoms. The tall savanna grasses turn brown and hard and become inedible. At the end of the dry season, the dry grass is regularly burned to make young growth accessible to hungry cattle at the onset of the next rains. Even then the native grasses are low in nutritive value.

Few regions are plagued with such a variety of insects as a tropical savanna. Mosquitoes and flies of many species breed in the stagnant waters and rank grasses. They make life uncomfortable for both man and beast, and also facilitate the spread of various diseases. Bird life is equally abundant and includes many migratory types.

Settlement of the Llanos   The native Indians of Venezuela had no domestic animals to utilize the savanna grasses and could therefore occupy the lowlands only where there was sufficient rainfall to support crops, or where permanent rivers contained enough fish. Indians lived on the forested delta of the Orinoco but avoided the grassy plains.

The Europeans introduced cattle to the Llanos in 1548. One important advantage which compensated for the handicaps to cattle-raising was a paucity of carnivorous animals. A century later herds of wild cattle numbering perhaps 140,000 were reported as grazing on these plains. By 1812 the number had increased, under the care of seminomadic cattlemen, or *llaneros*, to about 4.5 million. Immediately thereafter the Wars of Independence reduced not only the number of cattle but also the number of llaneros, for these hardy cattlemen were in great demand as fighters. By 1823 the cattle population had dropped to 256,000.

During the remainder of the nineteenth century, the number of animals rose or fell in accordance with the country's political stability. Guzmán Blanco took a personal interest in the cattle

business. He introduced better breeds of cattle and insisted on better methods of caring for them. General Gómez also took a direct interest in the pastoral activities of Venezuela. He introduced zebu cattle, crossing them with native stock to gain greater resistance to insect pests. This breed of cattle, which originally came from India, provides better meat and dairy products in the wet tropics than the common European breeds, for the rather oily hide of the zebu provides a defense against flies and ticks which infest the pastures.

Gómez gained a virtual monopoly over the cattle business. Large foreign-owned ranches on the Llanos, which during World War I sent cattle to a British packing plant at Puerto Cabello, were forced out of business. This was accomplished simply by taxation on all cattle driven across state boundaries. Since no clear distinction was made between his private funds and funds in the federal treasury, the taxes Gómez paid on his own herds were, in reality, paid to himself. He raised taxes so high that no one could compete. He also owned the fattening pastures around the Lake of Valencia.

Since World War II and especially since 1958, agricultural settlement of the Llanos region has developed rapidly. In 1956 a large resettlement project was completed in the Llanos near Calabozo. The Gúarico River was impounded near Calabozo by an earthen dam 9 miles long and 98 feet high. Behind this dam a lake was formed, covering 94 square miles. Downstream the land was marked off into 550 new farms, averaging about 500 acres. The farms are used primarily to raise cattle on planted and irrigated pastures enclosed with wire fences.

Agriculturally, the most dynamic portion of the Llanos is in the northwestern states of Cojedes and Portuguesa, where irrigation and land settlement projects have resulted in extensive fields of maize, rice, sesame, sugarcane, cotton, and sorghum. Farther westward along the Andes-Llanos frontier, in Barinas state, forestry and cattle-raising are more pronounced. Apure state, to the south, is almost exclusively cattle country, including some of the most modern ranching operations in Venezuela. San Fernando de Apure is the capital and trade center of the state, from which fresh meat is delivered regularly by plane to the urban markets of Caracas.

There are also important sources of oil and gas under the Llanos, especially in the eastern portion near Maturín and El Tigre. Pipelines have been constructed northward to Puerto La Cruz to refineries and export facilities, and a major natural gas pipeline extends to Caracas, the Basin of Valencia, and the petrochemical plants at Morón. Of greatest importance for the future is the so-called Orinoco heavy oil belt, which is expected to produce one million barrels per day by the end of the century.

The Guiana Highlands

About half of the national territory of Venezuela lies south of the Río Orinoco in a region known as the Guiana highlands. The surface is composed of rounded hills and narrow valleys formed on ancient crystalline rocks. Standing conspicuously above the general upland level, especially

in the far south along the border of Brazil and Guyana, are groups of plateaus and mesas, capped with resistant sandstone. These flat-topped tablelands reach the highest elevation in Mount Roraima, some 9219 feet above sea level. The crystalline hilly upland begins immediately south of the Orinoco. The river in several places flows across spurs of the upland, each of which produces a narrowing of the channel and limits navigation. The first of these narrows above the mouth of the river is just upstream from Ciudad Bolívar, formerly called Angostura, meaning narrows.

The Guiana highlands are covered with intermingled savanna and semideciduous forest. The grassy openings are extensive and irregular in outline. One of the finest popular descriptions of this mixture of forest and savanna was written by W. H. Hudson in *Green Mansions*.[2]

The Guiana highlands are less subject to extremes of flood and drought than are the Orinoco

---

[2] W. H. Hudson, *Green Mansions* (New York: Random House, 1944).

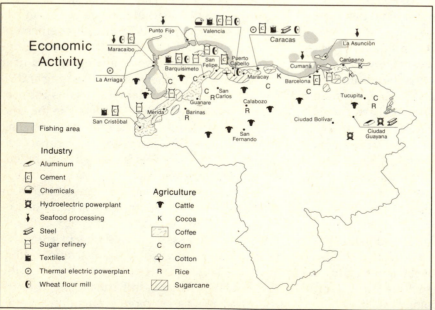

FIGURE 20-7

MINING IRON ORE AT CERRO BOLÍVAR.

Llanos. The year is sharply divided, as in the Llanos, into a rainy season and a dry season. Yet, the hilly nature of the terrain reduces the area that is subject to floods, and the cover of forest reduces surface runoff. Except for the factor of isolation, the Guiana highlands might be physically better suited for grazing than is the savanna-covered plain north of the Orinoco.

The remote Guiana highlands are still in large part outside of the effective national territory of Venezuela. Most of the small settlements are around mission stations along the rivers. An estimated 30,000 native Indians occupy the forests along the border between Venezuela and Brazil. The highlands, however, are not lacking in resources. A rich vein of gold was discovered in 1853 at El Callao, and by 1885 the mine at El

Callao was the world's largest producer of gold. Reactivated and modernized, it is now operated by a state-owned mining concern. For a long time the only way to reach El Callao was by muleback over a primitive trail from Ciudad Bolívar. During the 1930s the first regular airlines were established, and access to El Callao became much easier. A highway now reaches El Callao from Ciudad Bolívar and is projected to continue southward to the Brazilian border, and eventually to the Amazon at Manaus.

The Guiana highlands contain billions of tons of iron ore. After World War II, two large steel companies from the United States began operations near the lower Río Caroní. The U.S. Steel Corporation mined its ore from Cerro Bolívar, west of the river, and transported it by railroad to the Orinoco. There it was loaded on ocean-going ore carriers for export to the United States. Bethlehem Steel Corporation operated at El Pau, on the east side of the Caroní, and also shipped ore by rail to the Orinoco. Its exports were sent by barge to Puerto Hierro, on the southern side of the Paria Peninsula, where the ore was transferred to ocean-going vessels. The Venezuelan government nationalized iron mining operations and related facilities in 1975.

The Venezuelan government has invested large sums of oil income to develop an industrial center at the junction of the Orinoco and the Caroní. There, some 60 miles downstream from Ciudad Bolívar, existing communities were merged and Ciudad Guayana was formed. This booming industrial complex is now surrounded by heavy industries. One is a huge government-owned steel mill that has nine electric furnaces, making it one of the largest such plants in the world. It is supplied with iron ore from its mines in the Guiana highlands to the south. Coal is imported in the returning ore carriers by which iron is exported, and limestone is brought in from the port of Guanta, near Barcelona. The steel plant turns out a variety of products, such as structural steel, reinforcing rods, rails, steel sheets and tubes, and seamless steel pipes for the petroleum industry.

Power to run the electric furnaces comes from the Macagua hydroelectric plant on the Caroní River 30 miles to the south. In 1965 a huge new hydroelectric project was started at Guri, a few miles upstream from Macagua. The Río Caroní probably has the greatest power potential of any Latin American river, and the Guri Dam is designed to harness much of that natural resource. The dam, 500 feet high, will supply water to 24 electric generators. Taking advantage of abundant electric power are two major aluminum plants. Bauxite is currently imported, but deposits within the Guiana highlands will eventually be developed. In 1979 aluminum overtook iron ore as the second leading export of Venezuela, exceeded only by petroleum. Subse-

quently, a multibillion dollar project has been developed to connect Ciudad Guayana with the more populated parts of Venezuela by rail, to construct a shipyard, and to build a heavy vehicle plant. An oil refinery and petrochemical complex have already been established. Thus, a major "growth pole" has developed on the northern fringe of the Guiana highlands, while most of this vast territory remains as Venezuela's almost unoccupied frontier.

Venezuela as a Political Unit

Within the present century Venezuela has enjoyed a unique position among the countries of Latin America. Its population is small in relation to the total national territory, and its natural resources are almost beyond measure. Yet, it is recognized that all of the world's resources are finite, and wealth earned from their exploitation must be judiciously reinvested if what is now a relatively prosperous nation is to avoid poverty sometime in the future.

After many decades of dictatorship, this nation emerged in 1958 as a bulwark of democracy within a continent dominated by autocratic regimes. At the same time, this privileged status has placed the Venezuelan government under a special set of circumstances both internally and abroad. Probably nowhere in Latin America is the disparity between the rich and the poor more conspicuous. Wealthy families live in opulence much like those in all other OPEC (Organization of Petroleum Exporting Countries) nations, while the poor experience a degree of poverty matched in few other parts of the hemisphere. In a democratic nation such disparities cannot be ignored. Consequently, ambitious development plans are formulated and to a substantial degree actually implemented.

After 60 years of oil production, Venezuela still has proven reserves of 27 billion barrels. Only 10 percent of the country has been explored for oil, but now the entire nation will be explored for the last drop.[3] Between 1979 and 1988, $25 billion will be invested in the industry. The heart of the oil industry has been in the Maracaibo lowlands; however, Venezuela is beginning to develop Orinoco's heavy oil belt, one of the largest untapped oil reservoirs in the world. This is in a 16,000 square mile area just north of the Río Orinoco.

In 1982 petroleum accounted for 70 percent of all government revenue, 26 percent of the gross national product, and 95 percent of all foreign exchange earnings. Despite efforts to diversify its sources of income, Venezuela appears likely to depend heavily on its petroleum resources for many years to come.

The Economic Situation

Venezuela's 17 million inhabitants are very unevenly distributed over the national territory, and the national income is unevenly shared. So rapid is the movement of rural people to the city that chronic unemployment prevails. The inability to control this growth has led to a severe housing shortage, congestion, pollution, and inadequate supplies of water and electricity. These problems have spurred the government to encourage the relocation of business and industry from Caracas to the secondary cities. Ciudad Guayana has consequently prospered with the manufacture of steel, aluminum, and cement, reaching a population in excess of 300,000. Barcelona has become a center for automobile assembly, and giant steel and petrochemical industries are to be developed in the Maracaibo lowland.

Agriculture remains a problem. Despite more than 20 years of agrarian reform, Venezuela's rural population has declined to less than 20 percent of the total labor force. Not more than 20 percent of the country's land area is under cultivation or in pasture, and imported food constitutes 40 percent of the nation's total consumption.[4] Farm labor, both skilled and unskilled, is in critically short supply, which accounts in part for a substantial flow of illegal immigration from Colombia and other countries.

The greatest economic gains have been in the development of infrastructure — hydroelectric plants, power lines, pipelines, and highways — and the expansion of proven mineral reserves, such as petroleum, natural gas, iron ore, and bauxite. Venezuela has not been aggressive in promoting tourism, but the government does operate a chain of tourist hotels and has established a series of national parks. Major attractions include Angel Falls, the world's highest (3212 feet), in the Guiana highlands; Pico Bolívar, with cablecar service, in the Cordillera de Mérida; and the beaches, historical sights, and duty-free shopping on Margarita Island, in the northeast.

The Political Situation

After the death of Gómez in 1935, the army officers who succeeded him in office brought Venezuela a measure of political freedom and law. Political parties were permitted to organize, and a strong liberal democratic movement began to emerge. In 1946 a new liberal constitution was adopted, and in 1947 a national election was held. To overcome the problem of illiteracy, ballots were printed on paper of different colors. The vote was secret and the count honest. The result was that the democratic candidate received about 70 percent of the votes.

In 1948 the army again seized control of the government. For eight years, 1950–58, Venezuela was ruled by a brutal military dictatorship, under which civil liberty was eliminated. The prevailing philosophy of the Marcos Pérez Jiménez regime was like that of the dictator Gómez, who said, "My country is not ready for

---

[3] *World Oil* (January 1980): 91.

[4] "Venezuela" ABECOR Country Report (London: Barclay's Economics Department, December 1980).

the kind of democracy that brings abuses." In 1957 Jiménez had himself reelected by a ballot on which no other candidates were offered. A bitter revolt broke out in 1958, with support from units of the army, air force, and navy. For the first time in Venezuelan history, younger officers were fighting for civil liberty, not just for the removal of older colonels and generals to make promotion more rapid. The dictator was forced to flee, and a military-civilian government was established. A new election in 1958 returned the political leadership of Venezuela to the forces of democracy.

The leader who emerged as President in 1958 was Rómulo Betancourt. For several decades Betancourt had been active in the movement to end the traditional system of power and privilege for the few, with the great majority held in place by a powerful army. As head of the popularly elected government in Venezuela, he expressed support for similar movements elsewhere. In 1958 he pledged support for Castro in his fight against Batista, but when Castro was victorious in 1959 and visited Venezuela, the Cuban revolutionary expressed scorn for Betancourt's liberal ideas. He and Che Guevara began to organize guerrilla forces to overthrow Betancourt and to make the Andes of western Venezuela into a kind of Sierra Maestra from which to wage war throughout northern South America. Betancourt, however, was as much opposed to a dictatorship of the left

as he was to a dictatorship of the right. In 1961 Venezuela broke relations with Cuba.

In 1964 an important event took place in Caracas. Betancourt completed the term of office to which he had been elected, the first such completion of an elective office in Venezuelan history. He turned over the position to his duly elected successor. With the 1968 election the presidency was, for the first time, transferred to an opposition political party, marking another milestone in the country's progress toward representative democracy.

Since 1970 the Venezuelan government has not only been concerned with domestic issues such as economic and social development, but has also played an expanded role in international affairs. As a member of OPEC, Venezuela benefited from a fourfold increase in world crude oil prices and became a major financial aid donor and political force in the Caribbean area and Central America. Cordial relations prevailed with neighboring Guyana until 1982, when Venezuela announced the intent to pursue its long-standing claim to Essequibo, a 64,000 square mile area which Guyana considered an integral part of its own national territory.

By the time President Jaime Lusinchi took office in 1984, the country was deep in debt and the economy was stagnant. Venezuelans were again ready to accept a change.

COLOMBIA

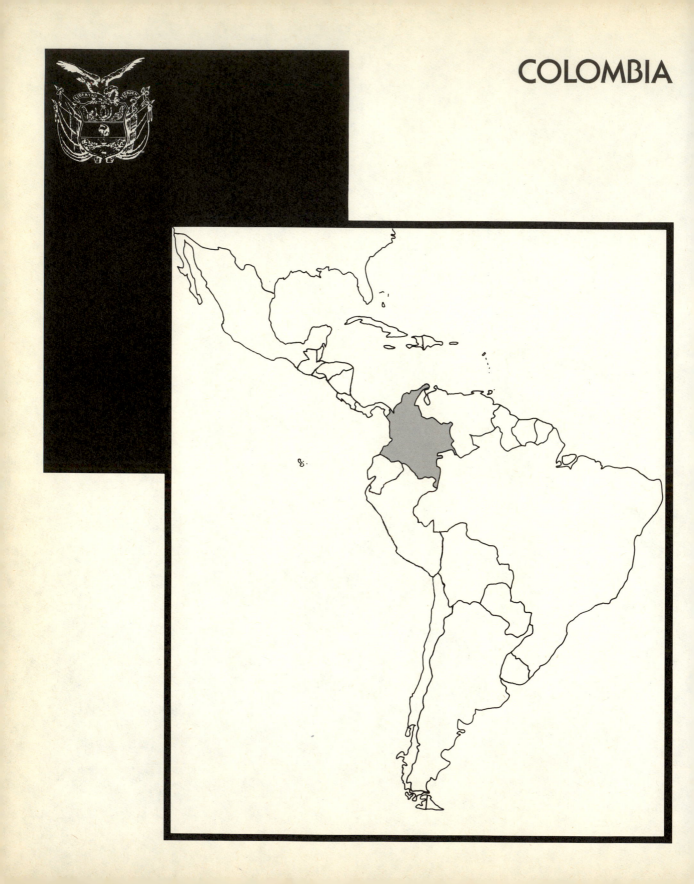

21

REPÚBLICA DE COLOMBIA

Land area   439,734 square miles

Population   Estimate (1985) 30,000,000
             Latest census (1973) 22,915,229

Capital city   Bogotá 4,584,000

Percent urban   67

Birthrate per 1000   28

Death rate per 1000   7

Infant mortality rate   53

Percent of population under 15 years of age   37

Annual percent of increase   2.1

Percent literate   82

Percent labor force in agriculture   27

Gross national product per capita $1410

Unit of currency   Colombian Peso

Leading crops   coffee, maize, wheat, rice, bananas

Physical Quality of Life Index (PQLI)   72

COMMERCE (expressed in percentage of values)

Exports

coffee	49
combustibles, minerals	14
fruit and vegetables	5

Exports to		Imports from	
E.E.C. Countries	38	United States	35
United States	29	E.E.C. Countries	14
Venezuela, Ecuador, Peru	6	Venezuela, Ecuador, Peru	14
		Japan	8

Data mainly from the 1985 World Population Data Sheet of the Population Reference Bureau, the 1985 Britannica Book of the Year, and the 1985 World Almanac.

If one were to identify the country most representative of the vast region known as Latin America, Colombia would be a worthy candidate. Historically, it experienced the development of an advanced Indian civilization and the Spanish conquest. Subsequent forms of settlement and land tenure were typical of the colonial era, as are the agrarian reform program and the urban migration typical of today. The people are primarily mestizo, but blacks, mulattoes, Caucasians, and native Indians are also represented. A wide gulf, both social and economic, separates the rich and the poor. The nation has experienced both dictatorship and democracy, as well as periods of intense political violence. Colombia possesses a substantial and varied mineral wealth, has developed a national iron and steel industry, and has relied heavily on a single commodity to provide foreign exchange earnings. The present population is concentrated in basins of the great Andean mountain chain, where the climate is characterized by vertical zonation, and there remain quite extensive areas of empty frontier. Yet, a prominent feature of Latin America is that each nation within the region is in many ways unique. In that respect, too, Colombia is representative.

Only the western third of Colombia is composed of mountains and valleys, but within this third are different kinds of land than are to be found in any comparable area in Latin America. There are peaks and ranges so high that their summits are permanently white with snow; high basins where the air is always chilly; forest-clad slopes where tropical showers feed torrential rivers; and lowlands, alternately baked in the tropical sun and drenched with violent rains. This western third of Colombia is the part in which almost all the Colombians live. The eastern two-thirds, which is mostly outside the effective national territory, is composed of a portion of the Guiana highlands, a portion of the Orinoco Llanos, and a portion of the Amazon Basin.

The diversity of western Colombia is not solely a matter of mountainous terrain and varied climates. It is also a matter of diverse people. There are six distinct regions of concentrated settlement that differ not only because of the land, but also because of differences between the people and their forms of economy. There are districts occupied by mestizo people, not unlike the inhabitants of highland Venezuela; there is one district where most inhabitants are nearly pure Indian, not unlike the people of highland Ecuador; and there are districts where a large proportion of the people are of African ancestry.

The major features of western Colombia are boldly marked. Four great mountain ranges separated by deep longitudinal valleys extend north-south. Along the Pacific Coast between Panama and Buenaventura lies the Serranía de Baudó. To the east is a broad lowland extending from the Caribbean to the Pacific, drained in the north by the Río Atrato and in the south by the Río San Juan. East of this lowland, and bordering the Pacific south of Buenaventura, is the Cordillera Occidental, or western range. Farther to the east is the highest of the Colombian ranges, the Cordillera Central. From the border of Ecuador to Cartago, the Cordillera Occidental and the Cordillera Central are separated by a wide structural depression, a rift valley, drained in the south by the Río Patía and in the north by the Río Cauca. From the northern end of this trench, downstream from Cartago, the Cauca makes its way toward the Caribbean through a series of profound gorges cut through the rugged but not very high country where the Cordillera Central and the Cordillera Occidental are joined. The Cordillera Central is the easternmost of the Colombian ranges between the border of Ecuador and approximately latitude 2°N. Here the eastern cordillera, or Cordillera Oriental, has its beginning. This wide cordillera continues northward and northeastward into Venezuela. About latitude 7°N it separates into two branches, one forming the western rim of the Maracaibo lowland, and the other the southern rim. Between the Cordillera Oriental and the Cordillera Central, and drained by the Río Magdalena, is a deep structural valley—a lowland that merges at its northern end with the lowlands along the coast of the Caribbean. An individual mountain group,

FIGURE 21-1

the Sierra Nevada de Santa Marta, stands prominently on the eastern edge of the Caribbean lowlands and towers above the Caribbean itself. This mountain group is separated from the end of the Cordillera Oriental by a structural depression drained by the Río César.

The People

The people who occupy this diverse terrain are of European, Indian, and African ancestry. The approximate proportions of the racial ingredients are as follows: European, 20 percent; black, 4 percent; Indian, 1 percent; mestizo, 58 percent; and mulatto, 14 percent. Such data can be misleading, however, for within the major racial groups there are many variations and racial proportions differ widely from one part of the country to another.

Native Indians

Before the European discovery of America what is now Colombia was occupied by Indians of many cultures. There were tribes whose way of living was primitive, but there was one group of tribes with a culture almost as advanced as that of the great Indian civilizations of Peru and Mexico. This was the Chibchas, a sedentary agricultural people who occupied the high basins of the Cordillera Oriental. In this remote mountain region the Chibchas were brought together politically by two chiefs: the *Zipa*, whose capital was near the present city of Bogotá, and the *Zaque*, whose capital was on the present site of Tunja. The political ability of the Chibchas far exceeded that of the other tribes in Colombia.

Like most highland Indians of America, the Chibcha tribes were dependent on the basic food staples: maize and potatoes. They also derived part of their food from the guinea pig which they had domesticated. Like other highland Indians from Mexico to Chile, the Chibchas had no concept of private property in land. These Indians established fixed settlements, and in places favorable to their form of agriculture the density of populaton was comparable to that of highland Mexico and Peru.

Indians with an agriculturally advanced technology lived in the Sinú Valley and along the shores of the gulf of Urabá. Ridged fields along the lower Magdalena bear witness to a relatively dense population of sedentary Indian farmers, but by the time the Spaniards arrived, the lowlands had been abandoned.

The European Conquest

The Spanish conquest of Colombia brought a great increase both in the area that could be used for human settlement and in the variety of land use. The Spaniards brought with them cattle, horses, and sheep. They also introduced wheat, barley, sugarcane, and a number of improved farm practices. These imports increased the productivity of the high basins because the European grains gave better yields at these altitudes than did the Indian maize, and domestic animals made possible the spread of settlement above the upper limit of potatoes into lands previously considered uninhabitable.

The first Spanish settlements in Colombia were along the Caribbean Coast. Balboa founded a colony on the west side of the gulf of Urabá which was perhaps the first European settlement on the continent, but it was soon abandoned. The oldest surviving Spanish colonies are Santa Marta (founded in 1525) and Cartagena (founded in 1533). The first expeditions to enter the highlands to the south were organized to search for mineral wealth, which the stories of El Dorado had magnified. When the Spaniards discovered the relatively dense populations of peaceful, sedentary Indians in the eastern highlands, they soon realized that here was the real wealth of the country — Indians to work the land and the mines, and to be converted to Christianity.

The Spanish conquest produced great changes in Colombia. The sendentary Chibchas soon learned to care for European domestic animals and to cultivate European grains. Indians in the high basins became serfs attached to large estates owned by officers of the conquering army, and new Indian communities were established in the *páramos*—the high country above the upper limit of agriculture but below the limit of permanent snow. In these higher regions the Indians remained predominant in numbers, but the new landowning aristocracy accumulated wealth in terms of the European economy.

Meanwhile, the more primitive Indians elsewhere in Colombia were inadequate to meet the labor demands of the conquerors. By the end of the first century after the conquest, the more primitive tribes of Colombia had been eliminated by epidemics or had withdrawn to the remote selvas of the Pacific slope. The Spaniards, therefore, resorted to African slaves, who eventually outnumbered the Europeans by a wide margin in certain parts of the lowlands.

Development Since the Conquest

By 1770 Colombia had a populaton of 800,000. A century later the population had grown, mainly by natural increase, to about 3 million. Well into the nineteenth century, gold remained the chief economic interest of the ruling group, and even today Colombia is the leading gold-producing country of Latin America. In the twentieth century oil and platinum were added to the list of mineral products. Yet, during all this time most Colombians were engaged in agriculture. Maize was by far the leading crop, but commercial crops

AN INSTRUCTOR TEACHES CHILDREN HOW TO RECOGNIZE RIPE COFFEE BEANS BEFORE HARVESTING IN LA CORONA, COLOMBIA.

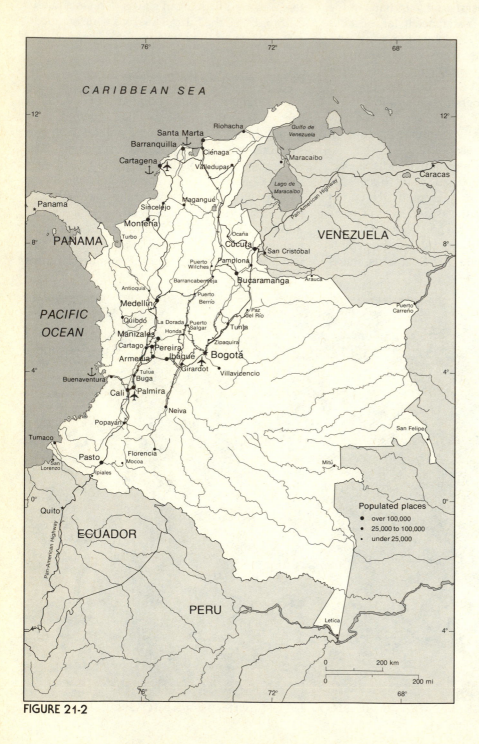

CARIBBEAN SEA

Riohacha

Santa Marta

Barranquilla

Ciénaga

Cartagena

Valledupar

Gulfo de Venezuela

Maracaibo

Caracas

Panama

Magangué

Sincelejo

Montería

PANAMA

Turbo

Ocaña

VENEZUELA

Cúcuta

San Cristóbal

Pamplona

Puerto Wilches

Barrancabermeja

Bucaramanga

Arauca

Antioquia

Puerto Berrío

PACIFIC OCEAN

Medellín

Quibdó

La Dorada

Puerto Salgar

Paz del Río

Tunja

Puerto Carreño

Manizales

Honda

Cartago

Pereira

Zipaquira

Armenia

Ibagué

Bogotá

Tulúa

Girardot

Buenaventura

Buga

Villavicencio

Cali

Palmira

Neiva

Popayán

San Felipe

Tumaco

Florencia

Pasto

Mocoa

San Lorenzo

Mitú

Ipiales

Quito

ECUADOR

Populated places

● over 100,000

• 25,000 to 100,000

· under 25,000

PERU

Letica

0   200 km

0   200 mi

FIGURE 21-2

of sugarcane, tobacco, indigo, and cacao were also cultivated. During the second half of the nineteenth century, cinchona bark, the source of quinine, was gathered in the forests.

The cultivation of coffee added an important factor to the economic life of Colombia. Only after 1880 did the mild, high-grade coffee produced in the highlands find a preferred place in the markets of Europe and North America. Gradually, coffee became even more important than it was in Venezuela, and Colombia today is the world's second largest producer. The spread of coffee planting in the *tierra templada*, on slopes too steep for most other forms of agriculture, brought increased productivity and a rapid growth of new settlement to parts of Colombia that had previously been of little economic importance. Today, coffee is produced on more than 300,000 farms and employs 3 million people.

These various economic activities served to delimit the Colombian territory into divergent regions, but even more striking diversification has been produced by the rapid development of manufacturing. Medellín, in the Department of Antioquia, became the leading textile manufacturing center of Colombia, and in the 1950s a rapid expansion of agriculture, electric power generation, and transportation facilities in the Cauca Valley made Cali the nation's third largest city in both population and industry. Bogotá remains the primate city and leading manufacturing center, with a population well over 4 million, but there are 14 cities of 200,000 or more inhabitants.

The Land

The four ranges of the Colombian mountains include a variety of geologic structures and surface forms. The westernmost range, the Serranía de Baudó, is by far the lowest and narrowest. The highest summit is less than 6000 feet above sea level, but intense erosion produced by heavy rainfall on sharply tilted layers of stratified rocks has resulted in some of the most rugged terrain in all of Colombia. The Serranía de Baudó is so narrow at one place that this part of Colombia has been studied as a possible interoceanic canal route.

The next two ranges to the east, the Cordillera Occidental and the Cordillera Central, are alike in their geologic structure, being composed of massive crystalline rocks. Together they form the western and eastern flanks of a great arch that extends southward from the Caribbean coastal lowlands almost to the southern border of Ecuador. In Colombia both cordilleras have crests that remain unbroken by stream valleys except where the Río Cauca has cut a way out to the north and the Río Patía, one to the west. Otherwise the crests of the two ranges are the divides between streams that rise on either side. Both cordilleras lack large intermont basins.

The Cordillera Central extends like a wall for more than 500 miles, forming a massive pedestal of crystalline rocks 30 to 40 miles wide, above which rise several volcanic cones, with their snow-clad summits more than 18,000 feet above the sea. The Cordillera Occidental, on the other hand, is relatively low. Its summits are about 10,000 feet in altitude, not high enough to reach the snow line, and between Cali and Buenaventura there is a pass only a little over 5000 feet. This pass has facilitated transportation between the Cauca Valley and the Pacific Coast, and the development of Buenaventura as Colombia's leading seaport.

The Cordillera Occidental and the Cordillera Central are separated, south of Cartago, by a deep rift valley. Along the crest of the arch of crystalline rocks, the keystone has broken into blocks and fallen, forming a depression that continues as a valley lowland or series of high basins from Cartago southward to Cuenca in Ecuador. Between Cartago and Cali the valley is only about 2300 feet above sea level. South of Cali, however, it is filled to great depths with ash from the bordering volcanoes. In the southern part of

THE HILLY, ISOLATED COUNTRYSIDE OF COLOMBIA.

the valley in Colombia, in the drainage area of the Río Patía, the floor is at an elevation exceeding 8000 feet.

Within the Cordillera Oriental are three groups of surface features. The highest crests, which form the first group, are not continuous as in the Cordillera Central. There are many short ridges, in echelon following the axes of the folds in a general northeast-southwest direction. Many of these crests are high enough to reach the zone of permanent snow, and some still have small glaciers, remnants of much larger ones that sculptured the high surfaces during the Pleistocene period.

The second group is composed of high basins in the central area around Bogotá. Between 8000 and 9000 feet above sea level is a surface of gentle gradient, forming three large intermont basins and various smaller basins and valleys. The margins of these basins are bordered by alluvial fans, but after crossing the fans, the streams meander with sluggish currents through broad valleys, forming swamps and even lakes in the centers of the basins.

Deeply dissected lower slopes below the high basins compose the third group of surface features. As streams reach the border between the basins and the dissected lower slopes, they plunge over spectacular falls. In the well-known Falls of Tequendama, the Río Bogotá drops more than 400 feet. The valleys in this part of the cordillera are narrow and steep-sided, and in only a few places are there patches of level land. East of the Cordillera Oriental, the northeastern part of

AN AERIAL VIEW OF BOGOTÁ, COLOMBIA WITH CORDILLERA ORIENTAL IN THE BACKGROUND.

Colombia includes a continuation of the Llanos of the Orinoco.

Vertical Zonation of Climate

In mountainous country such as highland Colombia, climatic conditions and the natural vegetation that reflects these conditions present an intricate pattern. Variations in exposure to the sun, in hours of sunlight, and in rainfall are characteristic. In general, however, all this intricacy of detail resolves itself into broad vertical zones.

Vertical zonation has more meaning in terms of human settlement in Colombia than in any other part of Latin America. Three principal facts account for this. First, the Colombian Andes are near the equator and high enough to reach the snow line. This permits the maximum amount of vertical differentiation, for as the snow line descends in higher latitudes so also do the other altitude limits. The second principal fact is that in Colombia areas of relatively gentle slope are to be found at various elevations from sea level to the snow line. Finally, the Colombian Andes are occupied by people whose many different ways of gaining a living make possible the use of lands at all altitudes.

The vertical zones are similar to those of the Cordillera de Mérida in Venezuela. The *tierra caliente* has a general upper limit of about 3000 feet. The *tierra templada*, or zone of coffee, lies between 3000 and 6500 feet. The *tierra fría* extends from 6500 to just over 10,000 feet. Above the *tierra fría* are the treeless *páramos*, which extend to the snow line at about 15,000 feet above the sea.

A common error is to believe that the *tierra fría* has climatic conditions comparable to those of places farther poleward from the equator. This is true in terms of average annual temperatures, but it is far from true in relation to seasonal variation of temperature or variety of weather. In tropical regions, even at sea level, the range of average temperature between the warmest month and coldest month is only about three or four degrees. At higher elevations, the range of temperature decreases. At Bogotá, 8660 feet above sea level, the average annual temperature is 58°F, exactly the same as the average annual temperature of Knoxville, Tennessee. In Bogotá, however, the difference between averages of the warmest and the coldest months is only 1.8°, while the difference at Knoxville is 38°. To describe Bogotá as having a "perpetual spring climate," as is frequently done, is to create a false impression, for it has none of the weather variety characteristic of a mid-latitude spring.

Regions of Settlement

The people who occupy these different kinds of land are grouped in six major regions of concentrated settlement. With few exceptions, each of

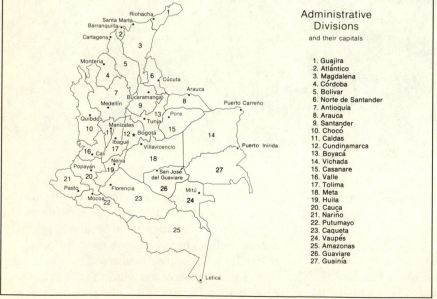

FIGURE 21-3

the departments into which the Colombian national territory is divided has a core of concentrated settlement, and the department boundaries pass through areas that are thinly populated. The six regions of settlement may be outlined as follows:

1. The High Basins of the Cordillera Oriental.
   Cundinamarca, Distrito Especial, Boyacá

2. The Valleys at Lower Altitudes in the Cordillera Oriental.
   Santander, Norte de Santander, Cundinamarca, Huila

3. The Antioquia Region.
   Antioquia, Caldas, Quindío, Tolima

4. The Cauca Valley.
   Risaralda, Valle del Cauca, Cauca

5. The Pasto Region.
   Nariño

6. The Caribbean Coastal Lowlands.
   Bolívar, Córdoba, Sucre, Antioquia, Magdalena, El César, Atlántico

The High Basins of the Cordillera Oriental

The high basins of the central Cordillera Oriental are among the most densely populated parts of Colombia. The largest is the basin of Cundinamarca. Bogotá, the capital of Colombia, is located near the southeastern margin of this basin. Bogotá is now more than the urban core of this one cluster of people; it is the political, social, and artistic focus of the entire country. It is also Colombia's leading commercial and industrial center. Farther north, in Boyacá, are smaller basins, each densely populated, which include the urban centers of Chiquinquirá, Tunja, and Sogamoso.

When the Spaniards invaded the Cordillera Oriental, they found the high intermont basins occupied by a relatively dense population of Chibchas. The tribes in the Basin of Cundinamarca had been united under the *Zipa.* However, the *páramos* between Cundinamarca and neighboring basins were uninhabitable as far as the Indians were concerned, and political authority had not been extended across to the territory of the *Zaque.* In 1538 the Spaniards founded Bogotá on the slopes of the alluvial fan overlooking the Basin of Cundinamarca. The basin itself, together with much of the surrounding high country, was divided into large estates on which the Indian communities remained the principal source of wealth for the owners.

The economic activity of most people in the high basins is still the production of food. Emerald mines, a little southwest of Chiquinquirá, and various localities, such as Zipaquirá, where salt was mined before the arrival of the Spaniards, remained active after the conquest. In the present century the emerald mines have been worked only sporadically, yet Colombia is normally the world's leading producer. The Europeans extended the area of habitable land and increased food-producing capacity by the introduction of wheat, barley, and cattle. Maize, the grain used by the Indians, can be cultivated up to an elevation of almost 9000 feet and, as in the Cordillera de Mérida in Venezuela, can be harvested twice a year. Wheat is grown up to 9800 feet, and barley and potatoes to about 10,500 feet. The *páramos* are utilized for pasture.

The high basins of Cundinamarca and Boyacá are now more densely populated than in the period before the conquest. Small villages are scattered over the basin floors and bordering alluvial fans. The mestizos and Indians who comprise the majority of rural people are still primarily subsistence farmers. Wheat, barley, and potatoes are important crops, especially in the highlands of Boyacá. Near Bogotá, in the Basin of Cundinamarca, flower production has become an important form of commercial agriculture. By 1980, almost 80 million dozen flowers, mostly carnations and mostly for export by air transport to the United States, occupied 2500 acres of land. Some 400,000 people were employed, two-thirds of them women.

In contrast to the traditional rural scene of the Cordillera Oriental is the steel mill at Belencito, near Sogamoso, in Boyacá. This is Acerías Paz del Río, a fully integrated steel producer. Its location has one great advantage over any other place in Colombia: basic raw materials are available from the immediate vicinity. Iron ore comes from low-grade deposits near the town of Paz del Río, 21 miles north of Belencito. Less than 20 miles away are some of the best coal measures in this part of South America, from which good- quality coking coal is obtained. Half a mile from the steel plant is an extensive outcrop of limestone, and the large quantity of water needed for steel production comes through an 18-mile pipeline from Lake Tota, south of Sagamoso.

Another striking contrast to the rural areas is Bogotá itself. Despite the difficulty of reaching it from other parts of the country, many elements of the national life are gathered in Bogotá. It is with justice that Bogotá is frequently called the "Athens of America." Its cosmopolitan atmosphere, its distinctive homes, churches, and museums reflect a rich cultural heritage.

Like Caracas, Bogotá has experienced massive immigration. In 1972 the city had a population of 2.8 million; this figure had increased to more than 4 million only a decade later. As elsewhere in Latin America the flow of urban migration, in addition to the high natural population increase, presents severe problems in the provision of housing, social services, and public utilities. Thousands of migrants crowd into rooming houses where primitive conditions are intensified by the peasant's lack of knowledge about urban sanitation. Another type of community, called the *zonas piratas* (pirate settlements), shelters 50 percent or more of Bogotá's total population.[1] In response to the problems of growth and congestion, large public housing projects have been implemented, urban expressways are being expanded, and major dams are being constructed on streams in the Cordillera Oriental to supply water and power. Unlike Caracas, which occupies a much smaller mountain basin, Bogotá has ample room to expand.

[1] H. Handelman, "High Rises and Shantytowns: Housing the Poor in Bogotá and Caracas," *American Universities Field Staff Reports* 9, No. 3 (1979).

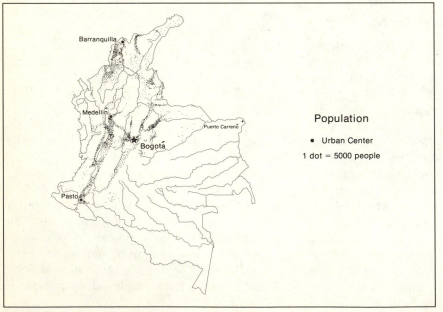

Population

● Urban Center

1 dot = 5000 people

FIGURE 21-4

The Valleys at Lower Altitudes in the Cordillera Oriental

The second of the six regions of settlement in Colombia is in the valleys of the Cordillera Oriental at elevations below 7000 feet. This part of Colombia, together with the Magdalena Valley to the west, was thinly populated by relatively primitive Indians before the arrival of the Spaniards, for there was little to attract settlement. The main course of conquest was directed to the higher Chibcha areas, and here a pattern of large estates with tenant workers has remained. The main concentrations of small-farmer settlements developed at four places at intermediate altitudes: around Bucaramanga in Santander; around Cúcuta and Ocaña in Norte de Santander; along the west-facing lower slopes of the Cordillera Oriental in Cundinamarca; and around Neiva in the Department of Huila.

The Settlements of Santander The big movement of European population into the northern Cordillera Oriental occurred in the nineteenth century. Because this region contained little gold and few commercial crops, Santander and Norte de Santander remained of little importance during most of the colonial period. Between 1860 and 1885, however, a resource was found in the forests of this region that would break the barriers of isolation. This was the bark of the cinchona tree, a source of quinine. Bucaramanga became a major collecting and shipping point for cinchona bark until Javanese cinchona plantations captured the market.

A new commercial product began to enter the export trade of Colombia soon after 1850. This was coffee, a crop that also occupied the steep mountain slopes in sparsely populated parts of Venezuela. Coffee more than compensated for the loss of the cinchona trade to Java. Today, coffee is grown on many mountain slopes around Bucaramanga, where there are numerous small owner-operated *fincas*. Production in the Bucaramanga district has been diversified by the establishment of plantations of cacao, tobacco, and cotton below the zone of coffee.

The metropolitan area of Bucaramanga numbers more that 400,000 and is the center of an important agricultural and livestock region. Industrial development is limited, but there are textile factories and small establishments that manufacture straw hats. Markets for these products are found throughout Colombia. Deep dissection of the Cordillera Oriental in Norte de Santander has produced a surface so rugged that transportation between Bucaramanga, Ocaña, and Cúcuta has been conducted with great difficulty. Cúcuta found its easiest outlet northeastward across the Maracaibo lowlands in Venezuelan territory. In both Ocaña and Cúcuta the products that paid for the first connections with the rest of the world were cinchona and coffee. Also significant at present is petroleum which is located in the southwest corner of the Maracaibo lowlands, where a pipeline has been built to the Caribbean Coast near Cartagena.

The Settlements Bordering the Magdalena Valley Since the introduction of coffee after 1850, other areas of concentrated settlement developed at intermediate altitudes bordering the deep valley of the Magdalena. In more recent times, with the introduction of insecticides, settlers have moved to the valley bottom. From Honda to Girardot, on either side of the river, are small farms where food crops and cotton are grown. For many years Colombia was an exporter of cotton, but production has slowly declined.

Closely allied with cotton production is Colombia's textile industry, which was fully developed in 1907. While the center of the textile industry is Medellín, there are some 460 firms scattered throughout the country in cities such as Manizales, Bogotá, Cali, and Barranquilla. Individual mills specialize in various aspects of the industry, producing a variety of cotton yarns, fabrics, rayons, polyesters, wools, acetates, and knitwear. This industry dates back to a time when Indians were expert weavers. They knew

A SYNTHETIC TEXTILES PLANT IN MEDELLÍN.

how to develop basic fabric materials and were skilled in weaving blankets and ponchos. Colors were extracted from cochineal and Brazilwood, and the juices of certain fruits and the roots of bushes were also used.

The Antioquia Region

The Antioquia region includes clusters of people in the departments of Antioquia and Caldas on either side of the Cauca Valley and on the eastern side of the Cordillera Central in Tolima. The central nucleus of this area of concentrated settlement is the city of Medellín, in a small valley about 5000 feet in altitude in the Cordillera Central. The people in the departments of Antioquia, Caldas, Quindío, and Tolima, which comprise the Antioquia region not only assumed leadership in the economic life of Colombia, but also expanded the frontiers of original settlement around Medellín and southward along the mountain slopes. The Antioquia region merits attention not only as one of the diverse elements in the Colombian scene, but also as one of the few places in mainland Latin America where ex-

pansion occurred before World War II without a decline of older settlements behind the frontier.[2]

The Settlement of Antioquia   The first penetration of this rugged country by Europeans was by people from Cartagena. They advanced across steep slopes and dense forests seeking gold, which was thought to be plentiful in the stream gravels. They did discover a wealth of precious metals, and in 1541 they founded the town of Antioquia, near the Río Cauca, but for permanent settlement the region was handicapped by a lack of Indian workers. By the end of the colonial period, Antioquia was a poor region. Despite its gold, there was little prosperity because of the high cost of food.

For many years Medellín and adjacent settlements remained isolated from the rest of Colombia. Between Medellín and Cartagena is a deeply dissected country which is so difficult to cross that not until 1955 was a highway completed all the way to the Caribbean. A trail, used mostly by mules, connected Medellín with Cartago at the northern end of the Cauca Valley, from which communication with the Pacific port of Buenaventura by way of Cali was relatively easy. The most common route used for the small trade of Medellín was an oxcart road, 120 miles long, to Puerto Berrío on the Magdalena, crossing the Cordillera Central by a pass only 5800 feet above sea level. Although Medellín soon replaced the older town of Antioquia as the chief center of the region, the people remained isolated and largely self-sufficient by exporting a little gold and silver.

The Antioqueños   Antioqueños, the people of Antioquia, are known throughout Colombia for their shrewd business sense, their willingness to work hard, their aggressiveness, and their ability to colonize new lands. Racially, they are similar to the mixed populations of other parts of Co-

---

[2] James J. Parsons, "Antioqueño Colonization in Western Colombia," *Ibero-Americana* 32 (1949).

lombia. In their protected valleys the Antioqueños developed a strong sense of unity. They were, and still are, predominantly Catholic and politically conservative.

Without their extraordinary birth rate, this small group of isolated settlers might not have achieved more than local significance. Large families have long been the rule. Since the original occupation in the sixteenth and seventeenth centuries, the Antioquia region has not received many new immigrants, yet settlement has expanded rapidly. The tendency to establish new colonies beyond the original nucleus first appeared in about 1800, long before coffee entered the economic life of Colombia. This was a time when the only exports from the region were gold and silver, and when the agriculture of Antioquia consisted of the shifting cultivation of maize, sugarcane, bananas, and beans for local subsistence.

Medellín was settled in 1675, mostly by European refugees. The settlers did not have black slaves, nor did they rely on the Indians as agricultural workers, for they were amply supplied with family labor. Expansion occurred chiefly south of Medellín, following the forested intermediate slopes on the western side of the Cordillera Central. When highly productive soils were discovered, the movement of settlers southward went on more rapidly.

The population increase in the Antioquia region is remarkable considering that it has not been supported by immigration. In 1778 there were 46,000 people in Antioquia. By 1883 the number was over 500,000. In 1918 the two departments of Antioquia and Caldas, into which the original political unit was divided, numbered 1.2 million, and there were more than 170,000 Antioqueños in the neighboring departments of Tolima and Valle. In 1954 the two departments had more than 2.8 million people. Medellín, which had 88,000 people in 1924, had reached 1.5 million by 1980. Manizales, 75 miles south, little more than 130 years old, is now a city of 250,000; it was founded by farmers from Medellín.

Transformation of Antioquia   Coffee appeared as a new crop in Santander soon after 1850, but it was slow to reach remote Antioquia. Even as late as 1880, the Antioqueños still practiced a shifting cultivation of food crops for local consumption, but during World War I and the following decade, they suddenly embraced a new economic life. In the short span of two decades, the small farmers began to plant coffee on their *fincas.* Land devoted to the new crop was shaded by the cover of taller trees, forming a thick cover of vegetation over the steep slopes. The entire Antioquia area began planting coffee, and the frontier of new settlement continued southward all the more rapidly. Three long fingers of pioneer settlement now extend southward on either side of the Cauca Valley and in Tolima east of the Cordillera Central, overlooking the Magdalena Valley.

Change in economic activities has altered the routes of circulation. At one time Antioquia could be reached only by oxcart road from Puerto Berrio to Medellín, or by trail from Ibague to Armenia over the Quindío Pass (11,434 feet). When Antioquia began to produce coffee for export, transportation facilities had to be provided. One of the earliest of these was a cable line over the mountains from Manizales to a point on the railroad near Honda. A railroad from the Pacific port of Buenaventura to Cali was extended to Cartago, and then to Manizales. For many years Medellín was reached by a single-track, narrow-gauge line that climbed to the pass from Puerto Berrío and almost, but not quite, joined another line from Medellín to the western side of the pass. The gap was covered by oxcart. Today there is a tunnel through this pass, and the railroad connects Puerto Berrío and Medellín.

A modern transport investment program to encourage interregional and foreign trade is progressing rapidly. Main highways connecting the three largest cities of Bogotá, Medellín, and Cali are being improved, as are feeder roads connecting rural areas with the highway network. Urban mass transportation, as well as rail, river, and air improvements, is also being undertaken. A major project is the Bogotá-Medellín Highway, but high priority has also been given to construction of the last incomplete stretch of the Pan American Highway across the Darién Gap. A 100-mile highway between the port of Turbo and Medellín will be reconstructed and paved. Medellín is no longer isolated, having become a major industrial and commercial center with airline, highway, and railroad connections to the rest of the nation.

The Cauca Valley

The Cauca Valley is that part of the structural depression drained by the Cauca River between Popayán in the south and Cartago in the north. The cluster of population that forms this region remains distinct from that of the Antioquia region, which occupies the slopes on either side of the valley but does not descend into it. The Cauca Valley area of concentrated settlement is in Riseralda, Valle del Cauca, and Cauca departments.

The deep rift valley that separates the Cordillera Occidental and the Cordillera Central southward from Cartago is a major surface feature of both Colombia and Ecuador. This structural valley is deeply filled with ash in the southern part of Colombia. The rivers that drain it have cut deeply into the ash, leaving the original surface as high terraces on either side. Popayán is located on such a terrace, 5500 feet above sea level and about 1000 feet above the Río Cauca which here emerges from the Cordillera Central. From Popayán northward the Cauca flows through a narrow gorge, cut deeply into the valley floor, but just upstream from Cali it emerges onto a floodplain some 15 miles wide, which extends northward for about 125 miles to Cartago. Downstream from Cartago the Cauca again enters a narrow gorge. At Cali the floodplain of the Cauca is in the *tierra caliente*, about 2300 feet above sea level. Cali is built on a low terrace on the western side of the floodplain at an elevation of 3140 feet, while Palmira occupies a similar site

to the east of the floodplain at an elevation of 3500 feet.

When the Spaniards entered the Cauca Valley, they found mostly a wooded savanna. Along the floodplain the swamps were filled with bamboo, interspersed with dense evergreen broadleaf forest. Away from the floodplain, the terraces and alluvial fans were covered with tall, rank grass and scattered low trees. The rainfall that supports this vegetation is abundant, coming in two seasons: from March to May and from September to November.

The Spanish invaders who entered the Cauca Valley from both the north and south found little gold, but they did find land well suited for that new wealth-bringing commercial crop, sugarcane. The Indians, however, proved inadequate as a supply of labor, and were soon replaced by African slaves.

Cali and Popayán have played contrasting roles in the social and economic life of the Cauca region. Popayán became an aristocratic town, occupied largely by people of unmixed European ancestry. Even the owners of the placer gold mines of Antioquia came to live in Popayán, leaving their mines, as the sugarcane planters left their estates, in the care of overseers. Yet, Popayán is not a national capital like Caracas. It is still a predominantly European place, the home of aristocratic families, a town rich with colonial tradition. This city of about 106,000 people was founded in 1536 and has been the birthplace of seven Colombian presidents.

When slavery ended, the old sugarcane economy collapsed. Sugarcane plantations were reestablished in parts of the Cauca Valley during the latter part of the nineteenth century, and cane is now grown on the low terraces along the eastern side of the valley. Around Cartago there are also plantations of cacao, and the Palmira area is known for its plantations of tobacco.

Even as late as 1960, the Cauca Valley was used mostly for the pasture of beef cattle. Some 8 percent of the owners of land in the valley held more than two-thirds of the area, most of it divided into ranches exceeding 125 acres. The pastures were not cultivated. The animals fed on wild grasses, moving onto the river floodplain at low water and to higher land on either side during times of flood. The result was minimal production of food per acre from the potentially most productive land. Meanwhile, most subsistence farmers of the region raised food crops on the bordering mountain slopes.

The Cauca Valley became an area seriously disturbed by sporadic and uncoordinated peasant revolt. In Colombia this is known as *la violencia.* Bands of armed men would rob and kill, or occupy unused private land and defy the authorities to remove them. Clearly, in the Cauca Valley the stage was set for an explosion. Yet, the landowners resisted any real redistribution of property, and landowners held the political power. The turmoil of this era lasted from 1948 to 1962 and resulted in the loss of 200,000 lives.

The Upper Cauca Valley is now the focus of Colombia's prospering sugar industry. Sugarcane plantings cover about a third of the valley's total farm area, and yields are among the highest in the world. Sugarcane bagasse is the basis for pulp and paper production, and it is expected that sugar here, as in Brazil, will be important for the manufacture of gasohol. Cacao, maize, and soybeans also are grown in the Cauca Valley, but cacao is gradually being displaced by sugarcane. The Upper Cauca Valley is also important for the grazing of cattle.

Cali   Cali at an early stage became the focus of travel routes for the entire Cauca region. By 1918, the opening of the Panama Canal stimulated economic development along the Pacific Coast, and Cali had grown to a population of 45,000. The city's modern expansion began in the late 1950s, with Cali becoming a thriving commercial center with a population well over 1 million.

Industrial growth proceeded rapidly in the 1960's, as large manufacturing plants owned by foreign firms were established. There were branches of well-known North American com-

panies, such as Colgate-Palmolive, Gillette, Goodyear, Quaker Oats, and Squibb, and there were factories owned by the French, Germans, Swiss, Swedes, and Lebanese.

The government of Colombia has sought to develop the Cauca Valley in a comprehensive, integrated manner, much like the Tennesse Valley Authority (TVA) in the United States. Dams built on tributary streams now generate electric power that is carried to the entire valley in an interconnected grid of power lines. The industries at Cali are major users of this electric power. Schools have been inaugurated, and a major university has been developed in Cali. Large numbers of peasants have migrated to the city; hence, the provision of housing and public services remains acute.

The Pasto Region

The southern part of Colombia is composed of two mountain ranges, the Cordillera Central and the Cordillera Occidental. Where the folded Cordillera Oriental comes to an end, about 2° north of the equator, the Cordillera Central becomes the easternmost of the Colombian Andes. Between the Cordillera Central and the Cordillera Occidental lies the southward continuation of the great rift valley, drained, in this region, by the Río Patía. Here, the rift valley resembles the upper part of the Cauca Valley around Popayán, for its ash fill is deeply dissected by streams and the initial surface is represented only by isolated terrace remnants. The elevation of these terraces is considerably higher than that around Popayán: 8510 feet at Pasto. The Río Patía escapes from the rift valley through a gorge that crosses the Cordillera Occidental to the Pacific Coast, so narrow and deep that it is useless as a travel route. The entire region receives abundant moisture, with two rainy and two dry seasons.

Pasto, the capital of Nariño Department, was once an important commercial center. People were involved in the gathering of rubber and cinchona bark in the valleys of the Río Caquetá and the Putumayo. Despite their isolation, the natives could export their products by using the rough mountain trails that led to Pasto. Transportation was partly on muleback, and partly by Indian porters. A small trade in food products still descends from Pasto to supply the communities of African-descended gold miners in the coastal zone, but Pasto is primarily a trade center for the adjacent agricultural region.

Penetration roads have been built to the pioneer zone of forests and oilfields of Putumayo Department and the eastern plains. Yet, immigration from the rural areas continues, bringing unskilled workers to Pasto and other urban communities. Numerous handicraft industries utilize their talents. To develop the port of Tumaco into a more diverisifid terminal, the 170-mile highway from Pasto is being rebuilt and paved.

The Caribbean Coastal Lowlands

The sixth group of settlements is on the Caribbean coastal lowlands. Three clusters of people occupy this region: one centers on the port of Cartagena, in the departments of Bolívar, Córdoba, Sucre, and Antioquia; another centers on Santa Marta, in the departments of Magdalena and César; and the third centers on Barranquilla, in the Department of Atlántico.

Where the Río Magdalena emerges from its structural valley between the Cordillera Oriental and the Cordillera Central, it enters a lowland plain built by deposits of river alluvium. This land is subject to frequent inundations, and much of it is swampy even at low water. The maze of swamps, oxbow lakes, and rivers is fed by the waters of four streams: the Magdalena,

THE PORT CITY OF SANTA MARTA, COLOMBIA. THE LARGE SILOS ARE USED FOR STORING GRAIN.

Cauca, San Jorge, and César. Most of the surface east of the Magdalena is permanently covered with water, including vast shallow lakes or reed-filled marshes with fluctuating outlines, known as *ciénagas*. West of the Magdalena large areas disappear under water during the flood season and dry out during the single dry season from October to March, at which time cattle may be pastured. The main outlet for all this water is through the mouth of the Magdalena, which discharges into the virtually tideless Caribbean. Until recently, the river mouth was so clogged with sandbars that ocean ships could not enter and riverboats could not leave. The whole area is covered with dry scrub woodland and wet savanna.

The Río Magdalena, unsatisfactory as it is for navigation, has remained a major route between the coast and the centers of population in the highlands. At the northern end of the Magdalena route, Cartagena, Santa Marta, and Barranquilla have competed for traffic, and as the fortunes of each has risen or fallen, the clusters of people in the districts around them have increased or decreased.

Cartagena   Cartagena was founded in 1533 on a protected harbor along the Caribbean Coast. The tiny village prospered as merchants, adventurers, Spanish soldiers, and even pirates swarmed ashore in search of gold and other treasure. After a disastrous fire in 1552, the Governor ordered the city rebuilt with stones, bricks, and tiles. With the labor of African slaves, a series of forts was built to protect the approaches from the sea, but even such safeguards did not prevent attacks on the city.

In 1811 Cartagena declared its independence from Spain, at which time a force of 10,000 sailed across the Atlantic in an attempt to recapture the city. After a siege of more than 100 days, the assailants broke in and found only corpses. The heroic men had chosen to die of starvation and disease rather than surrender. It was not until 1819 that Simón Bolívar declared Colombian independence from Spain.

Cartagena is now a dynamic city of more than 450,000 people. Still a major port, it is continually expanding. Bordering the Caribbean are tourist complexes, elegant homes, and wide avenues. The Cartagena Free and Industrial Trade Zone and the Cartagena Industrial Export Processing Zone have been developed in an area designed to accommodate 30 export-oriented businesses.

In the hinterland of Cartagena, three IN-CORA projects have been established. One is southeast of the port, where an irrigation system has made nearly 70,000 acres available for the resettlement of 2500 families and for the production of a variety of food crops, including rice and maize. A second is along the lower Sinú, including its delta, where land once used for cattle-grazing has been divided into farms for settlers who grow food crops and cotton. The third project is along the eastern side of the Gulf of Urabá, where a pioneer zone has been made available through the construction of highways. A zone of banana plantations has developed along the highway from Medellín to Turbo. More than 80 percent of Colombia's export bananas now come from the Urabá region.

Cartagena is the main outlet for products of the Atrato Valley. Connection with this remote part of Colombia is by riverboat and airplane. Products brought to Cartagena for export include gold and platinum, for this is the most productive gold mining area of South America. Colombia has long been among the leading producers of platinum in the world. Both gold and platinum are taken from the stream gravels near the headwaters of the Atrato and San Juan rivers; but placer mining requires few workers, and when modern mechanical dredges are used, the number of jobs in mining is greatly reduced. Hence, there has been a steady migration of black miners into the forests of the west coast south of Buenaventura, where they have proved better able to maintain themselves than have the Indian inhabitants. Some blacks have also moved northward along the Pacific Coast into Panama. The Atrato region has a small nucleus of population around Quibdó, but is otherwise sparsely settled. Quibdó enjoys the dubious distinction of having an average annual rainfall of 415 inches, which makes it one of the rainiest places on earth.

The Caribbean Coast near Cartagena includes terminals for two oil pipelines, one from the Barrancabermeja field in the Magdalena Valley and the other from the Colombian part of the Maracaibo lowlands. Nearby Mamonal is a thriving industrial complex of 18 major oil companies, six of which are owned and staffed by Colombians.

The economic importance of Cartagena has changed greatly since World War II. Completion of an all-weather highway from Medellín to Cartagena has given this port an advantage over its rivals in handling commerce of the Antioquia region. Much of the coffee that once went to the Magdalena and thence to Barranquilla now moves by truck to Cartagena. Coffee exports are about equally divided between these two places. Furthermore, El Dique, the old silted up channel of the Magdalena, has been dredged deep enough to permit the largest river steamers to reach Cartagena. The railroad to Calamar has been abandoned. The population of Cartagena, which was only about 75,000 after World War II, exceeded 450,000 in 1982.

Santa Marta   On the eastern side of the Caribbean coastal lowlands is the city of Santa Marta. This city, like Cartagena, has a good harbor, and through the *ciénagas* on the eastern side of the Río Magdalena, shallow-draught boats and canoes have been able to reach Santa Marta as easily as they can reach Cartagena. When Cartagena's connection with the main river was no longer navigable, Santa Marta became for a time the leading commercial center of the region. Both of these rivals, however, were placed at a disadvantage when steamboats began to operate on the Magdalena, for neither Santa Marta nor Cartagena could be reached by these larger vessels until El Dique had been dredged.

Santa Marta was well known as a banana port until 1940, when Panama disease all but destroyed the plantations. Now the United Fruit Company leaves banana production to independent growers. The company buys bananas and ships them through special banana docks at Santa Marta. The city has also become a sea resort for the people of Bogotá, but Santa Marta has also become famous for quite another industry. Whole neighborhoods of new houses are said to be occupied by "drug kingpins" whose business is conducted in Ríohacha, capital of the Department of Guajira. This is the hub of the country's biggest illegal business, the growing and shipping of marijuana to the United States.

Guajira is largely a flat piece of land, the northernmost point of South America, that juts out into the Atlantic Ocean. Colombian police report that traffickers have built 150 clandestine airstrips that are used mostly at night to fly tons of marijuana and cocaine out of the country. Authorities estimate that 240,000 acres of marijuana have been planted that will produce 60,000 tons a year, which would amount to $6 billion at wholesale prices. This is almost double the estimated total legal exports of products from Colombia in 1980, which amounted to about $3.1 billion. At the present time, Colombia is also the world's largest producer of cocaine. Aerial surveys have shown that marijuana is no longer restricted to the Guajira, having spread to Magdalena, Antioquia, Meta, the Chocó, and even the Amazon River Basin.[3] It is estimated that 150,000 Colombians are employed in this agroindustry.

Guajira has the equivalent of 650 million barrels of oil in its natural gas reserves on the Atlantic Coast. This will be used for private consumption in large cities. Gas pipelines connect the fields in La Guajira with Barranquilla and Cartagena. Additional pipelines from the coast will go to Medellín, Cali, and Bogotá. Salt, gypsum, cop-

---

[3] *Miami Herald,* November 9, 1980, p. 7E.

per, nickel, and coal are the other major natural resources of the region.

Barranquilla   The third commercial center of the Caribbean coastal lowland, and the one that has most controlled the trade of the Magdalena Valley, is Barranquilla. This town achieved a leading position in about the middle of the nineteenth century, when steamboats began to navigate the river as far as the rapids at Honda. In 1862 a short railroad was built from Barranquilla to a place just west of the river mouth, where a pier was constructed far enough into the shallow waters of the Caribbean to accommodate ocean ships. Later, this port was abandoned because of rapid silting, and Puerto Colombia was built. This settlement is now connected with Barranquilla by both railroad and highway.

Barranquilla is a large urban industrial community with a population of almost 900,000. It is the nation's leading Caribbean port and a major international air terminal. Barranquilla also has a free-trade zone with a labor force of about 38,000. The zone specializes in the manufacture of lightweight goods that are easily transported by air.

The cluster of people in the Department of Atlántico also includes some agricultural settlers. West of the Magdalena there is a small but densely populated area devoted to cotton and sugar production — an area of small properties and a predominantly black population. The natural levees on either side of the Magdalena are similarly occupied by farmers, who raise vegetables and fruit and bring them to the Barranquilla market by canoe.

Transportation

The rugged terrain in which these six regions of concentrated settlement are embedded has greatly complicated the problem of communication and transportation. The physical barriers that separate one region from another are greater than those that separate Colombia as a whole from its neighbors. Despite the obstacles to navigation, each of the highland centers has established an outlet through a river port. Ocaña and Bucaramanga have their own small river ports. The thriving communities of Antioquia around Medellín are connected with Puerto Berrío. Manizales is connected by a cable line over the Cordillera Central to a station on the railroad near Honda. The settlements of Boyacá and Cundinamarca are reached by several roads that descend to the river banks between Honda and Girardot.

Honda was once a commercial center of major importance. Because of its location at the end of navigation on the lower river, it was the point where goods destined for the coast were gathered from all the surrounding country. A road from Honda winds over the Quindío Pass from Ibagué to Armenia and continues by way of Cali, Popayán, and Pasto to Quito, Ecuador. The importance of Honda was undermined by the construction, in 1884, of a railroad that provided transportation around the rapids. This railroad was eventually extended up-river to Girardot and Neiva. A second railroad was built directly from Bogotá to navigable water downstream from Honda, bypassing both it and Girardot. In 1961 this line was extended to Santa Marta (418 miles), giving Bogotá an all-rail connection with the Caribbean for the first time. Construction of a modern 125-mile railroad between Bogotá and the coal basins will make possible the transportation of heavy loads of coal to the Caribbean port of Santa Marta. The trip upstream from the Caribbean ports to Bogotá once took from eight days to a month. Now, passengers and mail are flown from Barranquilla to Bogotá in about an hour.

Colombia had the first commercial airline in the Western Hemisphere. Established in 1919, a German line was placed in regular operation along the Magdalena Valley between Barranquilla and Girardot. The airlines are now nationally owned, and regular flights reach all major urban centers of the country. Planes carry sea-food from both coasts to the interior, collect bags of coffee from rural areas, and transport salt, sugar, fresh vegetables, and passengers to jungle outpost or modern city.

Colombia as a Political Unit

Colombia is unique in that no other political unit in Latin America contains so many differing elements within a comparable area. The territory east of the Andes includes vast areas that have only sparse settlement. The gateway to the Amazon, Leticia, is a small town 2000 miles from the mouth of the river. With the completion of an airport, passenger and cargo traffic is now possible. The Amazonian region is rich in iron, tin, manganese, bauxite, gold, diamonds, gas, and petroleum. The Putumayo area contributes greatly to Colombia's output of oil, while the Amazon River Basin itself has considerable hydroelectric potential.

The land of the Colombian Llanos is devoted to cattle raising, but within the western third of the national territory, where most of the Colombians live, there is an extraordinary diversity of physical conditions. This diversity results not only from the rugged surface, but also from the range of vertical climatic zones.

The variety of the Colombian scene, however, is more than a matter of physical setting. The varied regions of concentrated settlement exhibit contrasts in racial composition, economic life, level of living, and political attitudes. It is not surprising that even the genius of Bolívar was unable to hold Gran Colombia together. Rather, it is notable that he was able to hold together in one country the six regions of settlement closest to Bogotá.

The Economic Situation

As in all other Latin American countries, the population of Colombia is increasing, but its rate of increase has fallen from 3.2 percent a year to 2.1 percent. Yet, by 1985 the country had more than 30 million people. Major shifts of population occurred as large numbers of rural peasants moved into all of the big cities, partly to seek employment, partly to escape the social disorder and disruption of community life resulting from *la violencia.* Some urban areas grew at fantastic rates, and the number of unemployed city people increased. Government and regional agencies have been active in funding programs of apprenticeships, technology, and management training. The latest efforts are designed to support government attempts to decentralize industry, promote exports, and expand small and medium-sized enterprises. To stimulate the economy in general, the government has favored the establishment of companies to produce food from raw materials based on agriculture. President Turbay Ayala declared in 1978 that "the problems of the rural population derive from unproductive small holdings, and not from large-scale modern enterprises which generate employment, industrial inputs and foreign exchange."[4]

Colombia's mining sector is rapidly assuming a more dynamic role in the economy. Among the new mining ventures are major coal and nickel developments, as well as exploratory drilling for natural gas and oil. The most important nickel deposit is at Cerromatoso, which contains about half of the nickel ore reserves of Colombia. A plant is being built to produce 42 million pounds per year for 26 years. The deposit is relatively near a main highway by which nickel ore will be hauled by truck to the Río Cauca and then shipped by barge northward via the Magdalena to Cartagena and Barranquilla for export.

One of the world's most valuable commercial deposits of coal is located at El Cerrejón, on the Guajira Peninsula, close to the Caribbean Coast. It is a clean coal, near the surface, and can be mined by open-pit or strip-mining methods. Furthermore, it is low in sulphur content and can be burned without creating pollution problems.

---

[4] *Comercio Exterior de México* 26, No. 4 (1980): 125.

Production goals indicate that coal may become Colombia's second most important export commodity after coffee.[5] It is estimated that Colombia has 60 percent of South America's coal reserves.

Economically, the future of Colombia will also depend substantially on developments within the petroleum industry. Colombian oil production began in 1922, in the Middle Magdalena Valley, and expanded rapidly. Much of the nation's crude oil is still derived from this area, supporting a major refinery complex at Barrancabermeja and accounting for most of the cargo traffic on the river. Colombia exported petroleum until 1975, after which it became a net importer. Since then major new discoveries have occurred in the Llanos of eastern Colombia, raising expectations that Colombia may regain self-sufficiency in the near future.

Tourism in Colombia has expanded rapidly since the early 1970s. About 75 percent of the visitors arrive by air, totaling more than half a million annually. Tourist promotion stresses the contrasts that Colombia has to offer. There are modern luxury hotels in all large cities, unspoiled islands like San Andrés with fine beaches, clear mountain lakes, and beautifully preserved colonial towns throughout much of the country.

The Political Situation

Amid grave social tensions, Colombia has maintained one of the few civilian-controlled political systems in Latin America. The very nature of the country, its internal arrangement, and resulting transportation problems have led to a regional development of political parties.

Long after Bolívar was forced to give up his efforts, Colombia was torn by civil war. The economy suffered severely from the devastation. Only the Antioqueños were spared the most serious effects of the disorders because of their isolation. Then for a time it seemed as if a political balance had been achieved and democratic procedures could be developed. This occurred partly because the six regions of settlement were more or less balanced, and no one of them was powerful enough to subdue the others.

The rise of Antioquia after World War I changed all this. When Antioquia turned to the commercial production of coffee, and especially when Medellín developed as a leading industrial center, this one region gained an economic advantage over the others. The Antioqueños saved their money and invested it in new capital, new factories, and new means of production. The Antioqueños were the business people of Colombia, and as they played a greater role in the economy, they also began to play a larger part in the political life. Yet, they continued to think of themselves as Antioqueños first and Colombians second. A military dictatorship gained control of the country, as the conservatives of Antioquia sought to keep the liberals of Bogotá from recapturing the government. Civil liberties disappeared, newspapers were tightly controlled, and political order was imposed by force.

No small part of *la violencia* was politically inspired by the bitter rivalry of the two political parties, each occupying a separate part of the national territory. Nor was this something new in the Colombian experience, but only a return to the violence of the nineteenth century. Outlaw bands could act in the name of one or another political party, making life in rural villages highly insecure.

In 1957 a bold effort was made to find a compromise between the extremes of political opinion. The conservative and liberal parties agreed to a coalition in which liberal and conservative leaders would assume the presidency alternately every 4 years for 16 years, and in which the various government offices would be staffed equally by people of both parties.

After a 16-year period the National Front had served its purpose, and the return to competition between the dominant political parties was relatively peaceful. Economic prosperity, together with favorable export prices for Colombian commodities during the 1970s, aided the smooth

---

[5] *Colombia Today,* 14, No. 2 (1979).

transition. Still, there was much dissatisfaction with the government. Guerrilla forces operated in some parts of the country during the following decade. Unemployment, civil unrest, and a severe economic recession further tested the viability of democracy in Colombia.

As a country in many ways typical of all Latin America, Colombia merits close observation. Its resources are varied and extensive; its prospects for the future could be bright. Much will depend on economic development and the political capability to bring the benefits of that development to the majority of Colombians throughout this diverse country.

ECUADOR

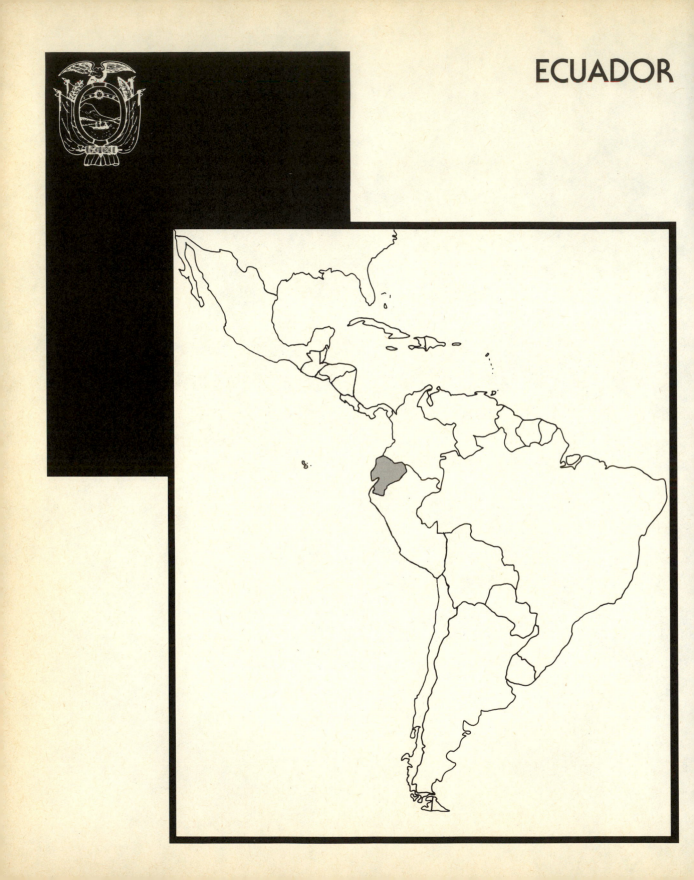

22

REPÚBLICA DEL ECUADOR

Land area   108,627 square miles

Population   Estimate (1985) 9,100,000
Latest census (1982) 8,072,702

Capital city   Quito 918,900
Largest city: Guayaquil 1,278,900

Percent urban   45

Birthrate per 1000   35

Death rate per 1000   8

Infant mortality rate   70

Percent of population under 15 years of age   42

Annual percent of increase   2.7

Percent literate   84

Percent labor force in agriculture   50

Gross national product per capita $1430

Unit of currency   Sucre

Leading crops in acreage   rice, potatoes, coffee, grains

Physical Quality of Life Index (PQLI)   70

COMMERCE (expressed in percentage of values)

Exports

crude oil	67
fish products	8
coffee	7
bananas	7
petroleum products	5

Exports to

United States	57
Panama	10
Colombia	6

Imports from

United States	34
Japan	9
West Germany	9
Italy	5
Brazil	5

Data mainly from the 1985 World Population Data Sheet of the Population Reference Bureau, the 1985 Britannica Book of the Year, and the 1985 World Almanac.

During much of its history Ecuador has functioned almost as two separate countries. One is the country of Indians; the other is the country of mestizos and those of Spanish descent. The cultures of each has been markedly different, especially as the latter turned their attention increasingly to industry, commerce, and government. Now Ecuador is experiencing rapid change and is gradually becoming a more unified nation.

Ecuador is much simpler in its geographic arrangement than is its northern neighbor, Colombia. The backbone of the country is formed by high ranges of the Andes. There are two main

cordilleras, surmounted by towering volcanoes and separated by a series of 10 intermont basins, all of them lying within the *tierra fría*. To the west of the Andes is the coastal region, composed of swampy lowlands and low hills. The coastal region is covered in the north by a dense rain forest, but toward the south, as the rainfall decreases, the vegetation changes from wet forest to dry scrub woodland and savanna. Some of the world's leading banana plantations are located here, but most of Ecuador's bananas are produced on small farms. The third part of Ecuador lies east of the Andes. This is the *Oriente*, the

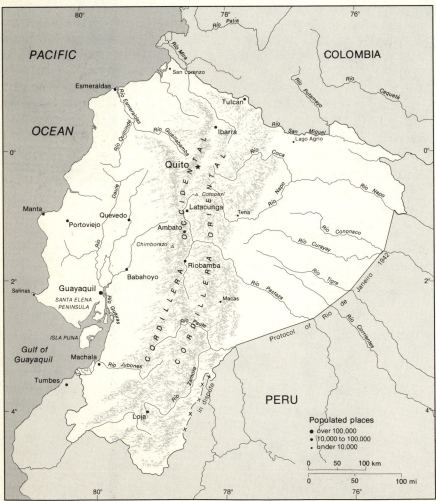

FIGURE 22-1

rainy, forested eastern slopes and piedmont of the Andes, which descend toward the Amazon Plains.

The size of the national territory is uncertain, because of a boundary dispute with Peru. The dispute is the longest unresolved boundary controversy among the Latin American states. At one time Ecuador claimed, but did not effectively occupy, large areas east of the Andes which are now recognized as Brazilian, Colombian, or Peruvian territory. In 1942 Ecuador accepted a treaty that gave Peru ownership of the left-bank tributaries of the Río Marañón as far upstream as they were navigable by launches. However, in the late 1950s when it became evident that the arrangement of surface features and drainage lines was not as expected, the Ecuadorians proclaimed the treaty void. Nevertheless, Peruvians occupy and administer all of the disputed area, most of which constitutes the hinterland of Iquitos, a Peruvian city at the head of navigation for ocean-going vessels on the Amazon River.

Armed conflict erupted in 1981, along a 50-mile strip of territory in the Condor mountain region where no boundary has yet been defined. Ecuador continues to press for direct access to the Amazon, while Peru seeks to maintain control over the territory it has occupied. Ecuador and Peru signed a Brazilian-sponsored Treaty of Amazon Cooperation in 1978, thus recognizing Ecuador as an Amazonian state, a long-desired goal that had been blocked by Peru.[1]

Population and Settlement

The first national census of Ecuador was taken in 1950, and the official count registered slightly more than 3 million people. In 1985 it was estimated that the population had exceeded 9 million. Estimates concerning ethnic origins vary widely. Official sources report 40 percent Indian, 40 percent mestizos, 10 percent Spanish, and 10 percent black. Indians of pure stock predominate throughout the rural areas of the highlands, while people of European origin are concentrated in the cities of Quito, Riobamba, Latacunga, Cuenca, and Guayaquil. Mestizos also settle in the cities as they acquire a Spanish lifestyle. People of African origin are concentrated mostly in the northern part of the coastal lowland.

Course of Settlement

A dense population of sedentary Indians occupied the high basins of Ecuador before the Spanish conquest. The Incas had extended their empire northward to include Quito before the Europeans arrived. However, the distance separating this northernmost outpost from the capital, Cuzco, was so great that communication was difficult. Central authority was maintained only because the Inca king spent much of his time in Quito.

In 1534 a Spanish expedition advanced northward from Peru and occupied the Indian town of Quito. From there the Spaniards continued northward into the Cauca Valley of Colombia, founding the towns of Pasto and Popayán and in 1538 Guayaquil was established where hilly ground borders the Guayas River.

Poverty and isolation, rather than strong sentiments of nationality, gave Ecuador its independence in 1809. When the Wars of Independence freed the colonies from Spain, Ecuador was included with Nueva Granada (Colombia) and Venezuela in the Republic of Gran Colombia, which Bolívar attempted to form and administer from Bogotá. The collapse of Bolívar's state left the more remote parts, Venezuela and Ecuador, free to announce their independence.

Formation of the Northern Boundary

Historical events have left certain curious features in the political geography of this part of

---

[1] "Ecuador-Peru," *International Boundary Study* (Washington, D. C.: Office of the Geographer, U.S. Department of State, 1980).

South America. Included is the location of the northern boundary of Ecuador with reference to areas of concentrated settlement, for this is one of but three examples of the entire continent where an international boundary bisects a cluster of population. The territory around Pasto, Colombia, originally settled from Quito, might have become part of Ecuador had not a revolt in the Ecuadorian capital given the armies of Bogotá the advantage at a critical moment. This they accepted by pushing southward into territory formerly administered from the rival center. Since the treaty negotiators lacked detailed information concerning the intermont basins and Indian communities, the clear separation of the Basin of Tulcán from the Basin of Ibarra by the Páramo de Boliche was ignored. The boundary was established on an obvious natural feature, the Río Tulcán, which cuts through the middle of the intermont basin.

The development of national sentiment which made the drawing of a boundary necessary occurred among the Spaniards and the mestizos who adopted Spanish attitudes and objectives. In the political centers, Bogotá and Quito, the desire for freedom was strongest, and the position of the dividing line that separated the territory adhering to one or the other of these centers was determined more by conditions in the capitals than by local arrangements along the border. The people separated by the boundary are mostly pure-blooded Indians to whom the distinctions of nationality even to this day are of little importance.

The Regions of Settlement

The population of Ecuador has traditionally been concentrated in the highlands. However, in recent decades there has been a steady migration toward the coastal lowlands. Now the population is about equally divided between the Andes and the coast. Settlement in the mountains follows a north-south pattern corresponding to the series of 10 intermont basins. In the coastal low-

lands the greatest population density is near Guayaquil and upstream along the Río Guayas and its tributaries, also near the Pacific Coast in the Province of Manabí. Only about 3 percent of the total population is located in the *Oriente*, which comprises half of the total national territory. Throughout Ecuador there is a strong current of migration to the cities. Consequently, about 45 percent of the total population is now urban. Guayaquil, the largest city and principal seaport, has a population of about 1.3 million; Quito, the national capital, has about 900,000.

The Highland Region

From the northern border of Ecuador to the Basin of Cuenca, about 250 miles southward, the Andes are similar in geologic structure to the Cordillera Occidental and Cordillera Central of Colombia. Together these two ranges form an arch of crystalline rocks, with a section along the axis that has collapsed to form a continuous north-south rift, but in Ecuador the volcanic activity that borders this rift is much greater than in Colombia. Standing on either side of the structural depression in Ecuador are about 30 volcanoes, including Cotopaxi (19,347 feet), the highest active volcano in the world. Even higher is Mount Chimborazo (20,577 feet), whose majestic cone rises far into the zone of permanent snow. Stream dissection in the loose volcanic ash is intensive; hence, the slopes of inactive volcanoes are cut into rugged "badlands." Where rivers extend their headwaters into the ash-filled intermont basins, deep canyons have been excavated. Beyond the reach of these streams gentle slopes of the initial basin floors remain.

The natural vegetation of highland Ecuador is related not only to climate, but also to the peculiarities of soil material. Whereas the rainfall, which is concentrated in one rainy season from November to May, is enough to support forests where temperatures permit, so porous is the unconsolidated ash that moisture is quickly absorbed. Forests cover the slopes of mountains to about 10,000 feet above sea level. Much of the

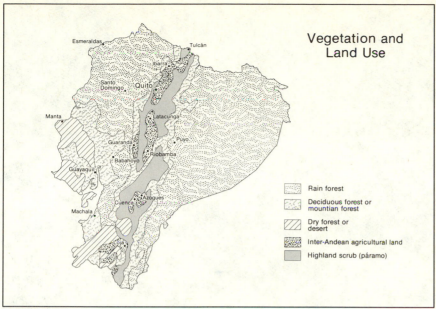

Vegetation and
Land Use

- ⊡ Rain forest
- ⊡ Deciduous forest or mountian forest
- ⊘ Dry forest or desert
- ⊟ Inter-Andean agricultural land
- ▦ Highland scrub (páramo)

FIGURE 22-2

brush has been cut to make charcoal, or burned to provide pasture, so that today grassy *páramos* cover about half of the total area of the highlands. Altitudinal limits define the distribution of crops.

Intermont Basins   The people of highland Ecuador are concentrated in distinct clusters occupying the various intermont basins. Within the zone between the two cordilleras, the 10 large basins are arranged in a line from north to south. Each basin is occupied by a dense population, mostly Indian and mestizo, which uses the land for subsistence crops or the grazing of cattle.

The three northernmost basins in Ecuador are drained by streams that descend westward to the Pacific. The Basin of Tulcán, bisected by the boundary between Ecuador and Colombia, lies at an elevation of 9500 feet and is, therefore, too high for most crops except potatoes. The Páramo de Boliche, south of the Basin of Tulcán, is used as common pasture by nearby Indian communities but contains few fixed settlements. Just south of this *páramo* lies the Basin of Ibarra, formed as an amphitheater of high terrace remnants around the head of a deeply incised valley. On the remnants of the basin floor, 7000 or 8000 feet above sea level, are numerous Indian villages, and around them farmers raise grains and potatoes. In the bottom of the valley, only 2500 feet above sea level, are small plantations of sugarcane and cotton, owned by the people of Ibarra. Here, too, are settlements of blacks who have migrated upstream along the valley of the Río Chota.

Next, to the south, is the Basin of Pichincha. Quito is built on a rim of this basin at an elevation of 9350 feet. The Río Guayllabamba, which drains westward to join the Río Esmeraldas, has cut its gorge through the Cordillera Occidental. When it reached the loose ash deposits in the Basin of Pichincha, it excavated the central part of the basin, forming an inner basin with a generally level floor 7500 feet above sea level. This inner basin is divided into large estates owned by affluent people who live in Quito. Indian tenants on these properties raise maize and pasture dairy cattle. On the higher outer rim and on the bordering mountain slopes, the population is comprised almost entirely of Indians. They are sub-

TIME TO RELAX AT THE PLAZA INDEPENDENCIA, IN QUITO, ECUADOR.

sistence farmers who raise wheat, barley, and potatoes, and herd sheep. Commercial dairy farming is also important.

Quito has been listed as one of the 13 wonders of the world. Laws, therefore have been enacted to preserve its rich heritage. The stately colonial section of the city has been protected and restored, while growth of the newer, modern sections has spread northward. The many squares and parks are alive with flowers, and the city's relaxed pace is a delight to residents and tourists alike. Few cities in Latin America have experi-

enced such rapid urban growth, yet maintained their colonial charm.

South of the Basin of Pichincha, high *páramos* form a narrow connection between the eastern and western cordilleras. The Basin of Latacunga includes the towns of Latacunga and Ambato, which were almost entirely destroyed by an earthquake in 1949 but have been rebuilt. This is a dry and relatively poor area, deeply dissected by the headwaters of east-flowing rivers. The Basin of Latacunga and the Basin of Riobamba, farther southward, are almost separated by the massive pedestal of Mount Chimborazo. The area around Riobamba, like that around Ambato and Latacunga, is composed of such porous material that only small parts of the district are suitable for crops. Passing over this difficult terrain, avoiding the deeply dissected valleys and winding around the lower slopes of Chimborazo, is the railroad which was built at great cost to connect Quito with Guayaquil.

South of the Basin of Riobamba is the Basin of Alausí. This is not a structural basin, like the others, but a widening along the valley of a river that drains westward to the coastal region. The railroad to Quito followed this valley in its difficult ascent from the lowlands and is no longer operational between Guayaquil and Riobamba. The Basin of Alausí, lying below 8000 feet and having alluvial soil, is intensively utilized for subsistence crops of wheat, barley, and potatoes.

The southernmost structural basin is a large one in which the city of Cuenca is located. Cuenca, with a population exceeding 105,000, is the only city in the highlands, other than Quito, that has a predominantly urban function. It was founded in 1557 and is filled with historical churches and museums that are part of a magnif-

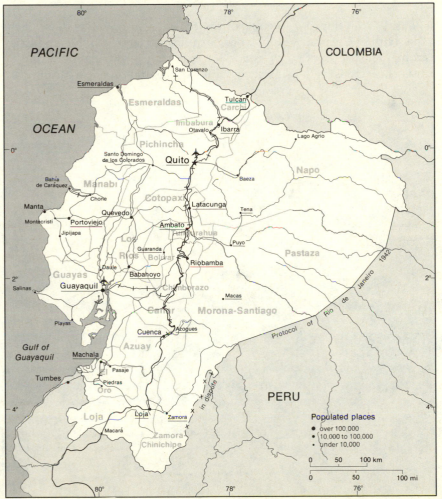

FIGURE 22-3

icent past. The basins of Cuenca and Pichincha are the two most productive areas of the highland region. Conditions near Cuenca are suitable for the cultivation of maize and for dairy cattle. More of the basin floor is devoted to the production of food for humans rather than feed for animals, in contrast to the other high basins. Cuenca is a center for the weaving of Panama hats, which makes use of fiber from the coastal regions. Since World War II various other industries have located in Cuenca to take advantage of low-cost labor. Manufactures include ceramics, sewing machines, television sets, radios, watches, automobile springs, and tires. Cuenca is also a major handicraft center.

Three other intermont basins, developed along stream valleys, lie to the south of Cuenca. These are Oña, Loja, and Zaruma. The people of these districts cultivate subsistence crops and raise cattle. Loja and Zaruma are the only highland settlements that do not send their few surplus products to Guayaquil exclusively. They supply some cattle to the oases of the Peruvian coast.

The highland region suffers from too great a density of rural people to be supported by farming. Hence, there has been a steady current of emigrants from the Andes to the coastal lowlands. In 1870 more than 90 percent of Ecuador's population was in the highlands. By the 1980s the proportion had decreased to 45 percent.

The Coastal Region

People of the coastal region, unlike those of the highlands, are engaged largely in the production of goods for export. The population is mostly mestizo, with a concentration of people of Span-

HOUSES ON STILTS ABOVE THE WATER IN GUAYAQUIL.

ish ancestry in Guayaquil. In the north, including Esmeraldas, there is a considerable mixture of black and mulatto. To the south, along the coast, there are communities that are almost pure Indian.

The Land

One of the narrowest zones of climatic transition in all of South America appears along the coast of Ecuador. Within a few degrees of latitude, one passes from tropical rain forest in the north to desert in the south. In the northern coastal region, Ecuador shares with Colombia the double maximum of rainfall, with enough annual precipitation to support a selva. Farther southward the double rainy season gives way to a single rainy season from December to June, and at Guayaquil from January to May. The Santa Elena Peninsula receives little rainfall, and arid conditions reach the shore south of the Gulf of Guayaquil at about the border between Ecuador and Peru. A great dry belt stretches southward from there, first on the western side of the Andes and then on the eastern side, almost to the Strait of Magellan.

Distinctive features of the coastal lowland are the huge alluvial fans along the western base of the Andes, sloping gently westward from the mountain valleys across the plains for 30 or 40 miles. The fans are largest in the north, where the volcanic ash accumulation in the highlands is greatest. Beyond the fans the coastal lowland is relatively flat. Only west of Guayaquil is there a belt of low hills that rise above the plains.

The Guayas Basin, which includes the most extensive area of level land in coastal Ecuador, is drained by a network of rivers that converge to form the Río Guayas just upstream from Guayaquil. Both rainfall and subsurface water are abundant, so that most of the basin was originally forested. Soils, largely comprised of alluvial deposits, are generally fertile and support most of the country's export crops. As the rivers of the Guayas Basin helped to focus commerce on the city of Guayaquil, so also did the railroad between Quito and Guayaquil. Development of a national highway network, construction of deep-water port facilities, and completion of a bridge over the Río Guayas just upstream from the city have helped to expand the hinterland of Guayaquil to include a substantial portion of the total national territory.

Settlement Along the Coast

Land use and population in coastal Ecuador change from north to south. In the tropical rain forests, south of the Colombian border, black subsistence farmers have long conducted a shifting cultivation. As they migrated southward along the Pacific Coast of Colombia, these migrants proved to be more successful in occupying the rain forests than were the native Indians. Only in isolated places have Indians survived in the forests. The black communities raise maize, manioc, and other food crops, and periodically pan for gold in the stream gravels. They also cut balsa wood and export it through the port of Esmeraldas. After World War II, when a new all-weather highway was built into the forests southeast of Esmeraldas, black farmers spread away from the coast along this road and started to plant bananas. Bananas are brought by boat or truck to Esmeraldas and are then transported by barge to the river mouth where ocean-going banana ships ride at anchor.

Esmeraldas is a city of about 60,000 inhabitants. Its chief exports are crude oil from the 300-mile Trans-Andean pipeline from the *Oriente* which ends nearby, plus coffee, cacao, bananas, and seafood. There is an oil refinery at Esmeraldas, and a petrochemical complex is planned. The city has also become a popular tourist resort.

South of the equator, in the Province of Manabí, the Indians conduct a sedentary subsistence type of farming. There are small plantations of cacao and bananas, and cotton has become a major commercial crop. From the woodlands comes the tagua nut, a vegetable ivory, formerly used in the manufacture of buttons. This district also produces Panama hats. Since 1950 the port of Manta has had a new surge of economic activity. Included is the marketing of frozen fish. Vast shrimp beds lie about 10 miles off the coast and extend southward to Peru. A fleet of small boats puts out to sea twice a year to catch tuna, bonito, and shrimp. Tuna fishing is among the finest in the world, and there is some fishing for lobster.

Land along the divide between the Esmeraldas drainage and the Guayas drainage was for a long time isolated from the major settlement centers of Ecuador. This area was occupied by the Colorado Indians, who were the only Indians to survive in the rain forests and managed to retain possession of their lands. In 1947 the isolation of the region was ended by the construction of a truck road from Quito down the western slopes to the apex of the alluvial fan at Santo Domingo de los Colorados. Santo Domingo became a boom town overnight, as pioneer settlers moved down from the highlands to clear the forests and plant commercial crops, especially bananas. Since then, the community has acquired paved streets, public utilities, and other urban amenities. It now serves as a regional trade center for an extensive rural area.

The Guayas Lowland

North and east of Guayaquil, and along the eastern shore of the Gulf of Guayaquil, is a region of great commercial importance. It is the most productive part of Ecuador agriculturally and a primary source of exports. Yet, its population remains less dense than that of the highland basins. Early in the present century, Ecuador was a major producer of cacao, and until 1941 this was the principal export crop. During World War II rice exports were more valuable, and in the 1950s Ecuador became the world's leading supplier of bananas.

Cacao and Bananas   For many years cacao was the chief source of income for the large landowners of lowland Ecuador. Tenants and managers were expected only to provide income, not to care for the land or the trees. After 1961, Ecuador's cacao plantations were heavily damaged by a succession of plant diseases. Cacao, however still occupies a large acreage in the Guayas lowland. It is grown mostly along the riverbanks where high rainfall and humidity, and the absence of winds, offer great natural advantages. About a third of this premium grade cacao goes to the United States.

Bananas are native to many areas of lowland Ecuador and have long been produced for the domestic market, but not until the late 1940s did shipments reach a significant volume. A spectacular expansion of banana production then occurred, and by 1953 Ecuador had become the world's leading supplier. In 1972 this fruit constituted about 62 percent of Ecuador's total export trade, but thereafter the value of all agricultural exports was far exceeded by petroleum. The United States is still Ecuador's best customer for bananas, taking more than half of its total sales.

Banana production in Ecuador, which has always been predominantly by small-scale plantations, developed in five different areas. The first to appear was in the vicinity of Esmeraldas, with shipments sent through that port. Others were in coastal Manabí, along the lower Río Daule, and, later, around Santo Domingo de los Colorados. A fifth zone, which included plantations of the United Fruit Company, is in the southernmost lowlands, bordering the Gulf of Guayaquil.

Major factors in the rapid expansion of banana cultivation include a road-building program by the Ecuadorian government which opened the exceptionally fertile lands of the Río Guayas and Río Esmeraldas basins, and connected them with the highland communities and Pacific ports. Yet, Ecuadorian producers have also experienced marketing problems. There are no specialized banana-loading docks here, such as are found at various Caribbean ports, and there are no railroads from most producing areas to the coast. Consequently, transportation from field to refrigerator ship has been by trucks, rafts, and boats, resulting in high costs in transshipment, labor, and fruit damage. Near Santo Domingo extensive acreages have been diverted

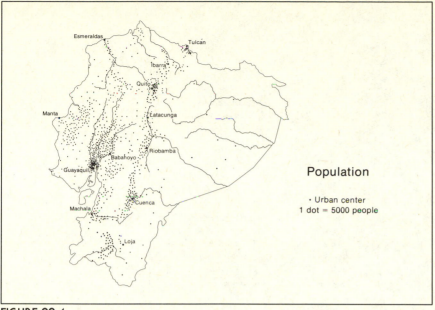

Population

• Urban center
1 dot = 5000 people

FIGURE 22-4

from bananas to African oil palm, food crops, and pasture.

Ecuador has attempted to solve the shipping problem by building an ocean port near Guayaquil. Before 1963 banana ships anchored in the Gulf of Guayaquil and the fruit was brought to them by barge. Then a new deep-water port, Puerto Neuvo, was opened on a side-channel about five miles south of Guayaquil. A canal has been cut from the Río Guayas to this channel, so that barges carrying bananas can reach the new port. Here the ships can tie up at wharves. As a result, since 1964 some 70 percent of the banana shipments have moved through Puerto Nuevo, and only about 10 percent have been shipped through Esmeraldas. Another result is that banana cultivation has been intensified in regions with easy access to this port and has declined in areas that are less accessible.

Rice, Coffee, and Abacá   Rice, a staple food crop in Ecuador, grows well in the Guayas lowland and Andean foothills. Production is concen-

trated in a relatively small area, but there are two harvests each year. On the large Ecuadorian estates, tenant workers first clear the forests on previously unused land. The planted crop is given almost no care; harvesting is by hand, with the machete. After the shocks are dried in the sun, the rice is threshed by beating it on the ground. Since rice is in short supply, additional amounts must be imported from Colombia, Venezuela, or other countries.

While coffee exports have decreased in recent years, coffee remains a significant source of foreign exchange. Most of it is grown at altitudes of less than 3000 feet; hence, its flavor is inferior to that of Venezuelan or Colombian coffee.

Ecuador is also a leading producer, along with the Philippines, of abacá. This plant which was introduced in the late 1930s produces a hard vegetable fiber used in the manufacture of twine, mats, high-quality paper, surgical clothing, and disposable diapers. Ninety percent of the harvest is shipped to the United States, the world's largest abacá market. Ecuador's entire output of abacá is grown on about 37,000 acres.

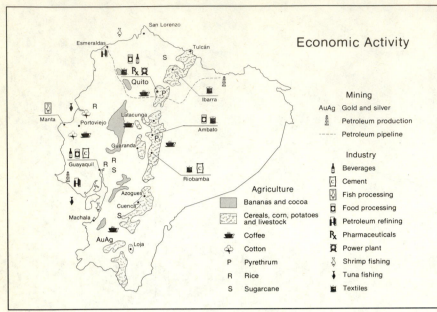

Economic Activity

Mining

AuAg Gold and silver

Petroleum production

- - - - Petroleum pipeline

Industry

Beverages

c Cement

Fish processing

Food processing

Petroleum refining

Rx Pharmaceuticals

Power plant

Shrimp fishing

Tuna fishing

Textiles

Agriculture

Bananas and cocoa

Cereals, corn, potatoes and livestock

Coffee

Cotton

P Pyrethrum

R Rice

S Sugarcane

FIGURE 22-5

The Galápagos Islands

Almost 150 years after Charles Darwin visited the Galápagos Islands, scientists still study them as a unique "laboratory of evolution."[2] These isolated islands, belonging to Ecuador and located about 650 miles west of the mainland, rise some 5000 feet above the surface of the Pacific Ocean. Bathed by the Peru Current, they are characterized by cool temperatures, and along the shores there is little rainfall. The Galápagos include a half-dozen large islands and many smaller ones, all of volcanic origin. The last eruption occurred in 1979, as the volcano Pequeño erupted on the main island of Santa Isabela. The total population of the islands has increased to about 5000, as people from the mainland have moved there in search of employment. The number of tourists arriving annually has risen to more than 14,000. In 1965 the whole Archipelago de Colón was made a national park, and a list of

endangered species was drawn up, including the iguana, the giant anteater, Baird's tapir, the Galápagos penguin and tortoise, the lava lizard, and a variety of plants.

Ecuador is dedicated to preserving the habitats of this amazing variety of flora and fauna. A major problem is in the eradication of feral animals, brought to the islands by whalers, sailors, and colonists. Animals such as cats, rats, burros, and dogs run wild and are a threat to thousands of plant and animal species. Worst of all are the goats, which have multiplied rapidly. On one island alone there are more than 40,000 of these destructive creatures, which could become a source of meat.

When the Panama Canal opened, the Galápagos Islands became of strategic importance for its defense. Offers by the United States to purchase the islands from Ecuador met with vigorous opposition. During World War II, Ecuador permitted the United States to establish an air base on the island of Baltra, but it was returned to Ecuador in 1946. Since 1967, the islands have been the site of a satellite tracking station.

[2] M. G. Scully, "Darwin's Galápagos Islands: The Survival of a Laboratory of Evolution," *Chronicle of Higher Education* 20, No. 13 (1980).

Ecuador as an Economic Unit

Historically, the economic life of Ecuador has been domestically oriented. Throughout the colonial period and early years of independence, the population was concentrated primarily in the high basins of the Andes, and the crops produced were largely for subsistence or for local urban markets. Even today only a few commodities of the highlands, such as pyrethrum, extracted from flowers as raw material for insecticides, find their way into international commerce.

Once transportation facilities were developed in the Guayas Valley and along the Pacific Coast, especially after World War II, Ecuador became relatively important as an exporter of cacao, rice, coffee, balsa wood, and a variety of other products. Guayaquil was by far the leading commercial and industrial center, while Quito developed as a secondary manufacturing center, and even Cuenca established a successful industrial zone.

Ecuador enjoys certain special privileges, along with Bolivia, as one of the less developed nations of the Andean Pact. Heavy industry, such as iron and steel manufacture, automobile assembly and petrochemicals, is contemplated, but the leading manufactures to date are processed food, textiles, wood products, and pharmaceuticals. Most enterprises are labor-intensive, small-scale operations. Future developments will depend largely on the extent to which petroleum and natural gas production can support the national economy.

Ecuador as a Political Unit

The existence of two distinct groups of people, concentrated in different parts of the national territory, makes Ecuador's efforts to establish order and stability a difficult assignment. The country has a long history of insecure governments and uncertain policies, but with its oil discoveries, Ecuador has acquired a greater potential to implement whatever national objectives may be formulated.

During its first 95 years of independence, Ecuador experienced a succession of 40 presidents, dictators, and *juntas*. From 1925 to 1948, some 22 presidents or chiefs of state came to power, but none was able to complete the four-year term of office. José María Velasco Ibarra was elected President five times but completed only one full term, 1952–56. To make the challenge even greater, the Constitution of 1979 has extended the term of office to five years.

A basic problem in the governance of Ecuador is that only a small portion of the population is politically represented, and even this segment is severely divided. A major division is between the conservative highlands and liberal coast, but within each of these there are still further divisions. In 1981 there were no fewer than 12 legal political parties, none of which could claim to represent anywhere near a majority of Ecuadorian opinion.

One unifying factor is the desire to reclaim the Ecuadorian territory lost to Peru, an objective that does not appear likely to materialize soon. Meanwhile, the government strives to incorporate into the effective national territory that part of the *Oriente* that is under Ecuadorian control. This effort was substantially aided by the discovery of petroleum in the northern margin of the *Oriente* in 1967, after which Ecuador became an OPEC member country and gained third rank among Latin American nations in petroleum exports.

The government is diligently forging ahead with plans to improve the national economy. With foreign assistance, major efforts are being devoted to hydroelectric power development, both in the *sierra* and in the Guayas Valley. A new civilian government has attracted foreign oil interests so that exploration will continue, especially in the Gulf of Guayaquil, where natural gas already has been discovered. Other projects in the offing include the improvement of roads and highways, the development of heavy industry, and the strengthening of the fishing sector. Yet to be achieved are the internal unity and organizational effectiveness needed to generate confidence in the nation's destiny.

PERU

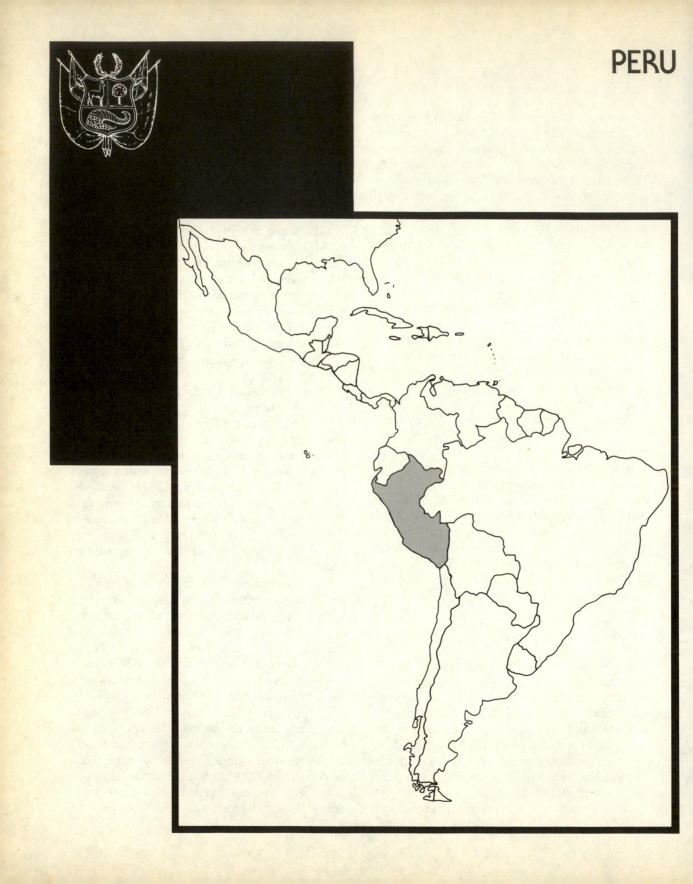

23

REPÚBLICA DEL PERÚ

Land area   496,222 square miles

Population   Estimate (1985) 19,500,000
              Latest census (1981) 17,005,210

Capital city   Lima 4,601,900

Percent urban   65

Birthrate per 1000   35

Death rate per 1000   10

Infant mortality rate   99

Percent of population under 15 years of age   41

Annual percent of increase   2.5

Percent literate   78.7

Percent labor force in agriculture   40

Gross national product per capita $1040

Unit of currency   Sol 9056 per U.S. $1

Physical Quality of Life Index (PQLI)   65

COMMERCE (expressed in percentage of values)

Exports

petroleum	18	silver	13
copper	15	lead	10
zinc	10		

Exports to

United States	33
Japan	15

Imports from

United States	33
Japan	9
West Germany	7
Brazil	6

Data mainly from the 1985 World Population Data Sheet of the Population Reference Bureau, the 1985 Britannica Book of the Year, and the 1985 World Almanac.

Peru is the third largest country in South America, after Brazil and Argentina, and is also one of the poorest. Like Ecuador, it has struggled for more than 400 years with problems resulting from the existence of two contrasting cultures within the same national territory. Spanish colonial settlement was directed toward Peru, and Lima, like Mexico City, became a primary settlement center. Cultural patterns developed by the Spaniards were superimposed on the Indians, but the two cultures have only recently begun to blend. Even today, Peru remains as much Indian as it is Spanish.

The country includes a wide variety of landscapes and climates. Bordering the Pacific Ocean is a desert coast, less than 50 miles in width, which rises to the towering ranges of the Andes Mountains. The eastern slopes of the Andes are rainy and densely forested, as are the plains of the Amazon Basin. A further change has come to Peru through the growth of an industrial society, which differs sharply from either the traditional Spanish or Indian way of life.

The Empire of the Incas

The ancient civilizations of China, India, Egypt, and Mesopotamia all began in highly productive river valleys, where closely knit and coherent societies were drawn together, in part, through the necessity for cooperative use of water for irrigation. The Incas, however, built their civilization in a small intermont basin of the high Andes, which seems unsuited to the development of an advanced society. From there the lines of authority were extended to the north and south over a land of profound canyons and forbidding snow-capped ranges.

The Andes of Peru, Bolivia, and Ecuador were occupied by well-developed cultures for thousands of years before the appearance of any unifying structure to tie them together. Along the Peruvian coast there were expert weavers of textiles as early as 8500 B.C. In fact, these Indians had discovered every known technique of weaving textiles before the Europeans arrived. Around Lake Titicaca an ancient civilization, much older than the Inca, had domesticated the potato and other wild plants of the area.[1]

The first Inca conquest was the densely populated Basin of Titicaca, where earlier Indian civilizations had flourished and declined. Subsequent conquest of the Peruvian coast was facilitated by the fact that coastal communities were dependent on water for irrigation, which was readily controlled from the highlands. The borders of the empire were eventually extended southward to middle Chile, southeastward to the edge of the Argentine plains near present-day Tucumán, and northward to what is now southern Colombia. It is difficult to estimate the population of the Inca Empire, but one leading authority has placed the figure between 16 and 32 million. Recent estimates of the present population of Ecuador, Peru, and Bolivia total just over 34 million. This means that it has taken more than 400 years for the population of this mountainous area to return to densities comparable to those of the Inca era.

Despite the earlier diversity of people and ways of living in the conquered territory, the benevolent and paternalistic rule from Cuzco[2] minimized the major differences and placed a uniform stamp over all but the most recently invaded communities. The entire empire was divided into four parts, each served by one of the great Inca roads that focused on the capital city. With reference to Cuzco, these parts lay to the north, northwest, southeast, and south. Within each of these four parts, the subjects of the Sapa Inca were arranged systematically in groups of standard sizes.

---

[1] E. P. Lanning, "Early Man in Peru," *Scientific American* 213 (1965): 68–76; David R. Harris, "New Light on Plant Domestication and the Origins of Agriculture: A Review," *Geographical Review* 57 (1967): 90–107.
[2] *Cuzco* is the Quechua word for navel, referring to the outline and sharply defined borders of the intermont basin.

THE HISTORIC INCA RUINS OF MACHU PICCHU, PERU.

The Incas recognized the family, rather than the individual, as the basic unit of the empire. Groups of families were administered as communities by appointed officials: groups of 10, 50, 100, 500, 1000, 10,000 and 40,000 families. The groups were kept at these approximate sizes by the resettlement of families in new colonies. This logical and systematic organization was responsible to and derived its authority from the Sapa Inca, who held complete power over the lives of his subjects and in whom was vested the ownership of land.

Inca engineers increased the arable area by control of the water supply and by construction of terraces on the valley sides. Just as wheat, barley, and rice formed the food bases of the ancient civilizations of Asia and Africa, maize was the basic food of the Incas, but it was supplemented by other foods. At the higher altitudes the hardy grain *quinoa* and potatoes were important. *Quinoa*, combined with peppers, was used in a kind of soup. Potatoes were used to produce *chuñu*, a potato meal that is still one of the principal foods of the Andean Indians. Dried llama meat was also included in the diet of ordinary people, whereas fresh meat was generally restricted to the Inca and his family. In addition, the common people drank an alcoholic beverage made from maize—a kind of beer known as *chicha*—and were addicted to the chewing of coca leaves. The highland Indians from Ecuador to northern Chile still consume both chicha and coca.[3]

The problem of communications within the empire was formidable. The roads, which led in four directions out of Cuzco, were paved or in places cut out of solid rock. Suspension bridges spanned deep canyons. Since there were no wheeled vehicles or domestic animals that people could ride, travel was entirely on foot and the roads were designed accordingly. Communication was maintained by relays; men were especially trained from boyhood for this particular service.

The domestication of some native Andean animals was also a major achievement. Indians of the Inca Empire had little poultry, and no horses, cattle, sheep, hogs, or cats. They used dogs mostly as pets but in some cases for hunting. They did, however, domesticate three closely related animals: the llama, alpaca, and vicuña. The llama remains one of the most important beasts of burden in the highlands of Peru and Bolivia, although as a carrier it is by no means as efficient as the mule. Normally gentle and easily handled, it has a stubborn disposition when tired or overloaded and may then lie down and refuse to move, or may even resort to spitting, camel-like, at those who come within range. It can carry only about 100 pounds and must be driven at a leisurely pace, in a herd, grazing along the way. A llama herd can cover little more than 10 miles a day. The smaller and less sturdy alpaca, a relative of the llama, is still raised for its wool.

The Incas' artistic and technical achievements were of a high order. The design and weaving of textiles from alpaca wool and cotton were among their most notable skills. Their pottery was colorful and finely modeled, yet was made without the use of a wheel. Inca engineers never made

use of wheels, never built towers, columns, or keystone arches. Yet, they were able to build suspension bridges, long irrigation ditches across rugged surfaces, terraces on mountain slopes, and massive walls with stones so closely fit that, without mortar, they have resisted the forces of destruction to the present day. The Incas were skilled metal workers, and, although they had no knowledge of iron, they did know how to make bronze from copper and tin. They also had a written language which, unfortunately, has been lost, and only now is their unique plaintive music becoming fully appreciated. Accounts were kept in the form of *quipos:* colored strings with differently spaced knots, but these were destroyed by Spanish priests who did not comprehend the system. The Inca decimal system was much less cumbersome than the Roman system used by the Spaniards.

The European Impact

The amazing story of the conquest of Peru by Francisco Pizarro and his little band of fewer than 200 men, equipped with 27 horses, is well known. Taking advantage of dissension among the rulers of the empire, the Spaniards overthrew the Incas within the comparatively brief period between January 1531 and November 1533, when the conquerors made their victorious entry into Cuzco. Once the leadership of the ruling group was removed, most of the inhabitants, long accustomed to unquestioning obedience to central authority, easily accepted the new rulers.

Profound differences between the Spanish and Indian ways of living soon became apparent. The Spaniards were interested in their overseas connection, and, the once remote coastal region therefore became the center of political and economic activity. Short transverse roads leading to Pacific ports replaced the longitudinal roads of the Incas. In 1535 Lima was founded and, with its port Callao, became the primary settlement center from which Spanish culture spread over most of western South America. It also, became a

---

[3] Not to be confused with cacao, the source of chocolate. Coca is a small shrub from which cocaine is extracted.

center of political power, social life, and commerce, a city of fabulous wealth.

The European conquest caused a serious decrease in food supply. The introduction of cattle and sheep did make extensive areas of mountain grassland habitable, but great numbers of Indians were removed from the coastal oases and highland agricultural centers and sent to work in the mines. The irrigation systems built by Inca engineers and maintained by the Indian communities were allowed to break down. The subsequent decrease in food supply was accompanied by an enormous decline in the Indian population. Great numbers of Indians died from overwork in the mines and from epidemics of smallpox and measles. By 1580, about 50 years after the arrival of the Spaniards, the number of Indians in the territory of the Inca Empire had been reduced by about a third.

The population of Peru is still predominantly Indian, however. Some 46 percent of the inhabitants are now of unmixed Indian ancestry, about 40 percent are mestizo, 1 percent is Oriental, and the rest are of European descent.

The Highlands

About 70 percent of the population in the highland provinces is composed of Indians. Here the minorities are white or mestizo landowners, government and army officials, storekeepers, and priests. Some communities, however, are supported by mining and have a larger proportion of Europeans, and in areas of commercial agriculture, the proportion of pure Indian drops to about 60 percent.

Surface Features

Three fundamental surface elements predominate in highland Peru. The first is a high-level surface of gentle slopes. The second, at still higher elevations, is composed of towering ranges with snow-capped peaks. Below this is the third element: deep canyons with steep, vertical sides. The westernmost rim of high ranges, which overlooks the Pacific Coast, forms the continental divide. The tributaries of the Amazon rise in the highest areas, cross the high-level surface through broad valleys with low gradients, and descend turbulently through canyons toward the eastern plains.

The ranges that form the second element of the Peruvian highlands stand on the high-level surface as on a platform. Most conspicuous is the Cordillera Blanca, the "white range," within which Mount Huascarán (22,205 feet) rises to the highest elevation in all of Peru. This range includes about 30 peaks with elevations exceeding 19,000 feet and extends as a massive wall for a distance exceeding 100 miles. It is here, too, that some of the world's greatest natural disasters have occurred, in the form of earthquakes and landslides. In 1962 a massive piece of ice and snow broke loose from the northern summit of Mount Huascarán and careened down the Callejón de Huaylas to the valley of the Río Santa. In addition to great physical destruction, more than 50,000 lives were lost and additional thousands of people were injured.

The canyons that have been cut below the high-level surface of gentle gradient are of tremendous size. Some are nearly twice as deep as the Grand Canyon of the Colorado in southwestern United States. In many places, the rivers pass through narrow gorges with vertical rock walls. In a few spots along the canyon bottoms, there are narrow ribbons of terrace land or small alluvial cones where tributary streams reach the main valley. Yet, because the rivers can be followed neither upstream nor downstream, and because of the difficulty of climbing the canyon walls to the high-level surface far above, these little valley flats are among the most isolated places in the nation.

Volcanic activity is limited to the southern part of the highlands. Here a fourth element is added to the complex of surface forms. South of latitude 14°S, great cone-shaped volcanoes appear and continue southward along the western side of Lake Titicaca and along the border of Chile and

Bolivia. The best known is El Misti, which overlooks the city of Arequipa. In this part of the Peruvian highlands, the high-level surface is deeply covered with ash falls and lava flows.

Climate and Vegetation

Climatic conditions within the highlands of Peru are as varied as the surface features. The upper limit of crops is roughly 14,000 feet above sea level between Cuzco and Arequipa. The annual range of temperature is somewhat greater than it is nearer the equator, and the diurnal range is greater than the seasonal range. At higher altitudes freezing temperatures may occur during any month of the year.

High altitude exerts a limiting effect on human energy. Whether because of decreased pressure, lack of oxygen, or other factors, physical labor at altitudes over 12,000 feet is difficult. Mountain sickness, known in Peru as *soroche*, produces a feeling of nausea and dizziness and may have serious effects. Respiratory diseases at these altitudes can be fatal. Since so much of Peru lies near or above 15,000 feet, altitude is of importance in understanding the human distribution.

The vegetation reflects not only vertical zoning, but also the increasing width of the dry zone from north to south. The forested eastern slopes and the plains beyond are called the *montaña*. In Peru forests occur only in the east and north. The selva climbs the lower east-facing slopes of the Andes, but higher up it gives way to a "mountain forest" composed of a dense growth of smaller trees. The Peruvians and Bolivians call this formation the *ceja de la montaña*, or "eyebrow of the forest." In drier spots, the desert vegetation known as "xerophytic shrub" makes its appearance. The grasses and shrubs of the highlands become increasingly xerophytic toward the south. The vegetation known as *puna* appears first on the western side of the highlands a little north of Huarás and extends over most of the width of the highlands in the south. In this formation the bunch grasses are more widely spaced than in the mountain grassland of the north, and other plants associated with the grasses are low, without stalks, and with hairy leaves. The *puna* becomes even more xerophytic in the zone of volcanic activity in the south, where the tola bush, a hardy resinous plant, maintains itself in a land that suffers from low rainfall, porous ash soil, and continuously low temperatures.

Population and Settlement

The highlands include more than half of the total population of Peru. Quechua-speaking tribes formed areas of dense population before the Inca conquest, and the same population pattern has survived the centuries of both Inca and Spanish rule. Wherever arable land is found, mostly along river valleys, it is fully occupied and intensively used. The cities are relatively small and serve mainly as markets and service centers for the neighboring rural areas and mining districts.

Agricultural practices in the highlands have changed little through the centuries. In fact, perhaps crops are planted and harvested less effectively now than when agricultural activities were supervised by Inca administrators. Following the Spanish conquest, most of the land was held in large estates, and the Indian communities served as the source of labor. Not until 1968, when a military government came to power, was major attention paid to the nation's agricultural structure and related problems of productivity. An agrarian reform law was then enacted to eliminate the *latifundia* and to promote the development of cooperatives by the former *hacienda* employees, who now had access to land of their own.[4] Between 1967 and 1974, the government expropriated 16 million acres, and about 216,000 farm families received parcels of land.

Despite efforts at agrarian reform, extreme inequality of landholding among the peasantry remains. The vast majority remain landless or hold properties too small to provide even the mini-

---

[4] P. J. Ferre, "Peru's Agricultural Reform Program Picks Up Headway," *Foreign Agriculture* 14, No. 3 (1976): 10–12.

FIGURE 23-1

mum harvests needed for subsistence farming. For many, the only alternative is to abandon the land and migrate to the cities of the highlands or, more likely, to metropolitan Lima or other cities along the coast.

Mining Centers of the Central Andes

In the central Andes, inland from Lima, there is a large concentration of mining settlements. The oldest and most prominent of these, Cerro de Pasco, is located about 110 miles northeast of Lima. Silver ores near Cerro de Pasco were discovered in 1630, and about 30 years later Peru become one of the world's leading silver producers. For hundreds of years these ores were reduced in crude smelters near the mines, and silver bars were hauled over a steep trail to the city of Lima. When the mines neared exhaustion, Cerro de Pasco declined in population, for at this altitude there was no other support for so large a concentration of people.

In 1902 a mining company was organized to take over 941 mineral claims in the vicinity of Cerro de Pasco. The venture was highly speculative, for the installation of modern machinery, construction of a railroad from La Oroya, and other necessary developments cost about $25 million. The corporation prospered, however, because it could exploit a variety of minerals that previously had not been of much value, and Cerro de Pasco soon became the primary mining center of Peru. Indian laborers were trained to perform the skilled work required in the mines and smelters. The most important mineral of this zone is now copper, but silver, gold, lead, zinc, and bismuth production is also significant. The Peruvian government nationalized the Cerro de Pasco mines and related facilities in 1974.

Cerro de Pasco is perched on the edge of a mine where ores are extracted by the open-pit method. The ores continue under the city; hence the Peruvian Congress passed a law approving removal of the city to a new location. A new city, named San Juan, was built to house the thousands of people displaced, and mining continues.

The principal smelter for the ores produced in this district is located at La Oroya.

Other minerals have been produced in various quantities in this part of Peru. A significant supply of vanadium has been mined periodically at Mina Ragra, and in colonial times the mercury mines at Huancavelica were of great importance. Coal is mined at Goyallarisquizga, and tungsten is produced at San Cristóbal, in Junín Department. In general, the zone of metallic minerals extends from beyond Cajamarca in the northwest to Huancavelica in southcentral Peru. However, the nation's largest mining operations, based chiefly on copper and iron ore deposits, have been developed in recent decades near the Pacific Coast in southern Peru.

Settlement in the Southern Andes

The old Inca capital of Cuzco is the urban center for one of the most densely populated parts of the Peruvian Andes. The city has more than 100,000 inhabitants and occupies a small basin approximately 11,000 feet above sea level. Just to the north is the Urubamba Valley, and to the west is the partly isolated Basin of Anta, an old lake plain with exceptionally productive soils. These three basins, Cuzco, Anta, and the Urubamba, have been occupied since the earliest recorded time by a relatively dense population of Indian farmers. Today, most of the Indians are small-scale farmers who raise crops such as maize, wheat, barley, quinoa, and potatoes, or they are herders of cattle, sheep, or llamas and alpacas. Cuzco also has become a leading tourist attraction, along with the famous Inca ruins at Machu Picchu which are reached from Cuzco by rail.

Still another major concentration of people is located around the shores of Lake Titicaca, extending without interruption across the international boundary into Bolivia. Here, as in the Basin of Tulcán on the border of Colombia and Ecuador, the international boundary was drawn without reference to the cluster of Indian popula-

FIGURE 23-2

STATELY LLAMAS MARCH DOWN THE STREET IN SICUANI, PERU AFTER DELIVERING PRODUCE TO THE MARKET.

tion already on the land. The Indians around Lake Titicaca speak Aymara, not Quechua, and are the descendants of people who formed the pre-Inca civilization. Indian farmers of the area still grow potatoes, barley, quinoa, and maize.

At higher altitudes, domestic animals provide the basis of human existence. Most important for meat, wool, and transportation is the llama, but the alpaca is also common. Alpaca wool and low-grade sheep wool are gathered from the entire highland region, in part for the export market. Where roads and rail lines have penetrated the mountains, minerals and a few agricultural and pastoral commodities have been brought down to the coast. The wool of southern Peru is sent out through Arequipa and Matarani, while wheat from the Basin of Huancayo is sent to Lima.

The Eastern Border Valleys and Eastern Plains

Quite different from the relatively stable Indian communities of the highlands are the small settlements of the eastern border valleys of the Andes and of the eastern plains, both of which are included in the forested region known as the *montaña*. Few of these settlements have exhibited stability, although the *montaña* has sometimes been described as a utopia in terms of climate and productivity of soils. Isolation defeated most efforts to bring this region into the mainstream of Peru's economy until the arrival of the airplane. automobile, and truck.

The *montaña* contains so many obstacles to travel that it, rather than the higher country to the southwest, forms the principal barrier to trans-Andean communication. The 60 percent of Peru's total national territory that lies east of the mountains therefore is utilized effectively in only a few scattered places. By 1968, however, highways were constructed across the mountains, and a highway along the eastern side of the mountains, known as the *Carretera Marginal de la Selva,* was undertaken to connect with other highways in Bolivia and Brazil. This project opened millions of acres in the Amazon Basin for colonization by the land-poor farmers of Peru. At times, of course, the roads are impassable because of weather conditions, landslides, and poor repair.

In the lowlands travel is especially difficult, except on the rivers. Small ocean-going vessels can navigate the Amazon as far as Iquitos, about 2500 miles from the Atlantic Ocean but only 350

feet above sea level. Upstream from Iquitos, the rivers are navigated mostly by shallow-draught launches and canoes.

Settlement

Penetration of the eastern border valleys has been achieved in only three districts. Most important is the central district, which is connected with the highland towns of La Oroya and Cerro de Pasco. Settlers have also gone into this region from Cuzco to the south and Cajamarca in the north.

The Central District   The settlements of Huánuco, Tarma, Pozuzo, and Tingo María illustrate the vital importance of transportation. Huánuco, some 100 miles north of Cerro de Pasco, was founded in 1539 in a small valley basin along the Río Huallaga, about 6300 feet above sea level. Its settlers were enthusiastic about the mild climate and productivity of the soil. Yet, the colony languished because of the enormous cost of transportation. Today, Huánuco has about 45,000

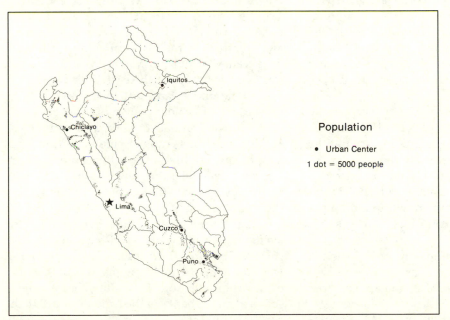

FIGURE 23-3

people. It is surrounded by fields of sugarcane, wheat, maize, barley, and sorghum. With an abundance of ruins, some of which are undergoing restoration, the city also supports a modest tourist industry. Tarma, just 45 miles northeast of La Oroya, was another colonial settlement in a pleasant and productive location. So isolated was this town that only products of great value per unit of weight could be sent out to market. The chief exports of these two places were coca leaves and *aguardiente* (sugar brandy), although maize, fruit, and vegetables were also grown. Pozuzo was founded by experienced German farmers from Bavaria, but there was no prosperity. For a time coca leaves were used to produce cocaine. When this was prohibited, the settlers were forced to rely on unprocessed coca leaves and brandy.

When a road reached Huánuco, that community suddenly came alive, and Tingo María, 135 miles to the northeast, became a small boom town. Because of this connection to market, settlement progressed rapidly. The leading product sent over the highway to Lima was lumber, but coffee, bananas, sugarcane and tea, grown by Japanese farmers, were also produced on local farms. Two other highways leave La Oroya. One descends to Tarma, and the other extends southeastward through Huancayo, Ayacucho, and Abancay to Cuzco, giving adjacent areas easy access to markets.

Of greatest importance to settlement of the central district in recent years has been the exploitation of petroleum, particularly around the river port of Pucallpa. With the rapid growth of this community have come a highway extension from Tingo María, modern airport facilities, and an increased flow of traffic downstream on the Río Ucayali to the Amazon River at Iquitos.

The Southern District Settlement of the eastern border valleys in southern Peru is restricted almost entirely to the lower Urubamba. The upper part of the Urubamba Valley ends abruptly a short distance northwest of Cuzco at the beginning of the Gorge of Torontoy. In the next 20 miles, the river drops 3000 feet, from over 11,000 to about 8000 feet. Then, at elevations between 1000 and 8000 feet, a series of small basins along the lower river is occupied by settlers. A railroad was built from Cuzco toward the isolated settlements below the Gorge of Torontoy early in this century, but of the proposed 112 miles of line from Cuzco to Santa Ana, only 50 were completed before the collapse of rubber production stopped all such costly projects. Settlers of the lower Urubamba remained in economic stagnation, unable to send their products to market, until a highway was built after World War II.

At present, there is gold mining in southeastern Peru, centered near Puerto Maldonado in the Department of Madre de Dios and in northern Puno. Thousands pan for gold in these forested areas, and the government has given encouragement to investors and miners through exemption from taxation and duty-free imports of equipment, vehicles, and supplies.

The Northern District The northern district became the newest zone of pioneer settlement after it was made accessible to the Pacific Coast in 1968. A highway starts at Olmos, 50 miles north of Lambayeque on the edge of the coastal desert, and climbs over the mountains, down into the deep valley of the Río Marañón, and up again over the next range to the east. Then it traverses rugged mountain and valley country passing close to the old colonial town of Moyobamba. It continues via Tarapoto to Yurimaguas, on the Río Huallaga. A trans-Andean pipeline 530 miles in length now roughly parallels this route, connecting extensive oilfields in northeastern Peru with the shipping and refining center of Bayovar on the Pacific Coast.

Iquitos is the chief commercial center of the *montaña*. Founded in 1863 by Peruvians, this hot, steamy city thrived during the rubber boom. It was able to import luxury items from abroad and even imported a hotel from Paris that was reassembled on the banks of the Amazon. In 1918 the rubber boom burst, and for some time the town was dormant. Iquitos is now a city of 115,000 people. It is a commercial center that exports lumber, petroleum products, furniture, tropical fish, and plants. Iquitos also includes a floating shantytown of some 15,000 people.

Tarapoto lies in a fertile farming region. Its population has more than doubled since 1966, when it had 20,000 inhabitants. For some years it has had all-weather roads, but it is the *Carretera Marginal de la Selva* that will open millions of acres of land for colonization and agriculture. Eventually, it will link Venezuela, Colombia, Ecuador, Peru, and Bolivia, and will facilitate connections between the Pacific Coast and the Atlantic.[5] This ambitious undertaking was engendered by Peru's architect President Fernando Belaúnde Terry during his first term in office (1963–68). When completed, the road will extend about 3720 miles. Political and economic conditions, plus the declining flow of foreign aid for *La Marginal*, resulted in a slowdown of construction activities in 1968, but by 1976 the Peruvian section was almost completed. The Colombian segment has gone beyond the planning stage, Ecuador has had the benefit of oil revenues to aid its construction of *La Marginal*, whereas Bolivia and Peru have used army troops and engineers in building their share of the highway. The date set for completion is 1995. The strategic importance of this highway to the countries involved cannot be overestimated.

The Coast

The third of the great natural divisions of Peru is the coast, which includes the political and economic heart of the country. Between 20° and 30° of latitude on the west coasts of all continents, there is a combination of cold ocean water and dry land, but in no other continent does this desert condition extend so far equatorward as in South America along the coast of Peru. The first signs of increasing aridity are noted on the coast

of Ecuador, only about 1°S of the equator. The transition from tropical rain forest to desert is unusually abrupt: within about four degrees of latitude, one passes through scrub woodland, wooded savanna, and scattered xerophytic shrub, to barren desert. In Peru, between Tumbes and Chiclayo, where the gravelly surface is not covered with live sand dunes, there are scattered desert plants, but south of Chiclayo most of the land is completely barren.

Surface Features

Along the coast of Peru the areas of flat land are small and disconnected. In the north the coastline bends in a broad arc toward the west, leaving, between Piura and Chiclayo, a wide belt of lowland. Most of this area is covered with moving sand dunes. Between Chiclayo and the mouth of the Río Santa, the coastline and base of the mountains come gradually together, and the coastal lowland is pinched out. From the Río Santa to Pativilca, the precipitous rocky slopes of the Andes rise directly from the Pacific. Along this stretch of coast are protected harbors in tiny rock-encircled bays, such as that of Chimbote. South of Pativilca as far as Pisco, alluvial fans built by the rivers are so large that they have almost merged along the coast, producing a narrow lowland, interrupted here and there by rocky spurs from the Andes. From Pisco southward, the coast is bordered by a range of low mountains which, back of Mollendo, reaches an altitude of 3000 feet. Behind the coastal range, separating it from the base of the Andes, it is a bleak, rocky surface, deeply and intricately dissected by streams that remain dry throughout most of the year. The places where agricultural settlements can be established on this coast are limited not only by aridity, but also by ruggedness of the surface.

Climate and Fauna

The basic cause of the dryness and coolness along the coast is the presence of cold water off-

[5] H. Horna, "South America's Marginal Highway," *Journal of Developing Areas* 4 (1976): 409–424.

shore, the so-called Peru Current.[6] Along all continental west coasts, except for that of Australia, there are equatorward flowing currents of cold water, accompanied along the immediate shore by upwelling water with especially low temperatures. However, in no other part of the world are these phenomena as strongly developed as off the west coast of Peru and northern Chile. This cold-water current is composed of two distinct parts. The *Peru Oceanic Current* is found at an irregular distance offshore: it is cold, but not as cold as is water nearer the land. It is poor in marine organisms and is deep indigo in color. This current, influenced by the earth's rotation, tends to swing westward away from the land. The other division of the coastal waters is the *Peru Coastal Current,* which is 50 to 100 miles in width and fed by the upwelling water. It moves northward faster than the oceanic current. Its water is colder, and it is remarkably uniform in temperature all the way from Chile to northern Peru. This current contains a tremendous quantity of marine organisms and is of greenish color.

The coast normally receives very little rainfall. The south and southwest winds that prevail along the Peruvian coast cross water that is progressively cooler toward the land.[7] Crossing cold water, air is cooled near the earth's surface. With cool, heavy air below and lighter air above, the atmosphere remains stable, and there is little or no rain. Nevertheless, because the temperature of the air is continuously lowered, it may be brought to the point of condensation, and sea fogs may form along certain parts of the coast.

From the Chilean border northward to Lima, and beyond to about 11°S, the coastal cloud cover is thick and persistent from June to October. Despite gray skies, more than a few drops of rain seldom fall, and the land below remains parched and barren. Yet, where the cloud rests against the slopes of the coastal range or the lower foothills of the Andes, a heavy mist, known as *garúa,* supplies soaking moisture to the land. The dense growth of quick-flowering plants and grasses that appear with the *garúa* is described in Peru as *loma.* Near Lima, the zone of the *lomas* begins about 2600 feet above sea level and extends to approximately 4600 feet. The *lomas* supply pasturage for animals during the cloudy season, from June to October, at just the opposite season from the rainy period in the highlands, which comes in the summer, from October to April.

The climatic peculiarities of the coastal region are matched by other unusual elements. In contrast to the barren water of the Peru Oceanic current and the barren surface of the land is the water of the Peru Coastal Current. The chemical composition of this water, together with protection from too much sunlight under the coastal cloud, provides the necessary environment for a rich organic life. Microscopic organisms, in turn, provide food for an amazing number of fish. Immense schools of small fish are preyed on from below by larger fish, and from above by a bird

---

[6] This current was formerly called the Humboldt Current by some writers because the great German geographer, Alexander von Humboldt, was the first to measure its temperature. Oceanographers have adopted the policy of using geographic names for all major currents, hence the agreement to use the name "Peru Current."

[7] According to classical descriptions of the world's wind systems, this part of the Southern Hemisphere should lie in the zone of the southeast trades. On a rotating earth, however, all moving bodies tend to advance along curving lines. In the Northern Hemisphere deflection is to the right and in the Southern Hemisphere to the left. A more realistic generalization of the world's wind systems, therefore recognizes the existence of great whirls of air around the permanent areas of high pressure over the eastern parts of the ocean basins about latitude 30°. Clockwise whirls in the Northern Hemisphere and counterclockwise whirls in the Southern Hemisphere produce, in the low latitudes, generally easterly winds and in the middle latitudes, westerly winds. However, along the continental west coasts the winds are more nearly from a poleward direction, and along the east coasts from an equatorward direction. In fact, the winds on the coast of Brazil south of latitude 8°S are predominantly from the northeast, and those on the west coast are from the south and southwest between middle Chile and the equator.

population that is one of the most extraordinary spectacles the world has to offer. There are long lines of pelicans flying close to the water; there are flocks of cormorants that hover above the schools of fish; and there are places where the water is broken into spray by flights of gannets diving for their food. The excrement of these birds, deposited in part on the islands and promontories where they nest, and preserved in the arid climate, forms one of Peru's notable resources — *guano*, a fertilizer containing from 14 to 17 percent nitrogen.

The bird colonies on the islands and promontories of the coast occur in almost unbelievable numbers. A study of one of the Chincha Islands, off Pisco, estimated that in one relatively small area there were 5.6 million birds. To feed such a colony requires at least 1,000 tons of fish per day. A million birds provide about 10,000 tons of guano per year. Here, indeed, is one of the most remarkable examples of an intricately balanced complex of the organic and inorganic.

Periodically, the harmony is utterly destroyed. Every year along the north coast, as far south as Chimbote, winds from across the equator bring southward a back eddy of warm water, which spreads over the surface of the cold water. Because this eddy appears about Christmastime or during January, February, and March, it is called El Niño, after the Christ Child. In normal years it forms only a thin surface over the cold water and is soon dissipated. At irregular intervals, however, El Niño is more vigorous. When El Niño arrived in 1982–83, it had awesome effects as the equatorial current reversed direction across the entire Pacific. Temperatures of the sea waters rose about 14°F above normal as warm waters stretched 8000 miles along the equator and brought low-pressure systems that generated torrential rains and consequent flooding. Fish departed from the coast of Peru, following the colder water with its food supply; floods of water destroyed irrigation systems; houses crumbled and collapsed; and roads and bridges were washed out. The loss of human life and property

was incalculable. Only slowly was the balance of nature restored.

Agriculture

Life in this unusual region must adapt to the prevailing aridity. The Incas, who practiced agriculture here before the arrival of the Spaniards, recognized that they had to build and maintain terraces and irrigation works on the lower slopes. Of the 52 streams that descend from the western slopes of the Andes to the coastal regions of Peru, only 10 have sufficient volume to continue across the desert. From August to October most of the channels are dry. It was the Spaniards who neglected and abandoned the terraces and irrigation systems. Not until 1884 did foreign-owned corporations begin to improve these systems and plant crops, such as sugarcane and cotton, for export. By the early 1940s more than a million acres of irrigated land along the northern coast were planted to rice, vineyards, cotton, sugarcane, and vegetables.

Since then, many irrigation projects have been created, but agricultural production in Peru remains disappointing. The reasons are many: price controls, a restrictive marketing system, credit shortages, a shortage of machinery and storage facilities, an out-of-date research system, problems arising from the old agrarian reform system, and, of course, the prevailing aridity.[8]

With the election of a civilian government in 1980, significant changes in agricultural policy occurred. International agricultural development aid expanded, and measures were taken to establish an effective research and extension program.

The Northern Oases

The northernmost of the major irrigated areas of the Peruvian coast include the group of oases around Piura, Sullana, and San Lorenzo. This

---

[8] "Peru: Agricultural Situation Report," 1980, Attaché Report, Foreign Agricultural Service, U.S. Department of Agriculture, PR-0003.

group is served by the port of Paita. Water is supplied by two rivers: the Río Chira and Río Piura. The Chira carries the largest annual flow of water among all the Peruvian west coast rivers. The Piura carries much less water and is less dependable. In 1532, when Pizarro started his march southward toward Cuzco, he founded the town of Piura near its present site, which makes it the oldest Spanish settlement in Peru. Irrigation was implemented with a minimum of river control. At high water the river spread out among the sand dunes south and west of the town, and the area thus moistened was used to grow food crops and cotton. In modern times, both Piura and Sullana grow mostly an extra-long staple variety of cotton known as Pima, which was introduced to this area from the United States. Cotton acreage and production have increased as a result chiefly of changes in the official pricing policy.

In 1967 a large irrigation project was completed, adding more than 160,000 acres of irrigated acres to the Piura-Sullana Oasis. In a development known as the San Lorenzo project, an entire new town was built.

New irrigation projects continue. The first stage of the Chira-Piura project was completed in 1979. This included a large reservoir, the Piura irrigation canal and secondary drainage canals in the lower Piura Valley. The Jequetepeque-Zana irrigation project began in 1979 and is expectd to assure that the traditional rice-producing area it serves will no longer be threatened by water shortages. The largest irrigation project ever undertaken in Peru, that of Olmos, is financed by the Soviet Union. The Olmos Oasis, once used only by nomadic goat herders, was created out of desert land beginning in 1962. Water is brought to Olmos via a tunnel drilled through the continental divide. When completed, in about the year 2000, the project will supply water for 280,000 acres.

The Chiclayo Oases   About 50 miles south of Olmos is one of the older irrigated areas of the Peruvian coast. Three distinct irrigated areas are served by the ports of Pimentel, Eten, and Pacasmayo. In the oases of the Chiclayo district the higher parts of the alluvial fans, where water is more abundant, are used by a few large sugarcane estates. If there is enough water, small farmers downstream can use it to raise rice and other food crops. In fact, the Chiclayo Oases produce a large proportion of Peru's rice crop, but this proportion may be reduced when the Olmos district gets into full production. Even in the Chiclayo Oases, projects for an increase of the water supply have been undertaken. A considerable expansion of the irrigated area has been gained along the northern border of the district, where additional water is provided by tapping some of the Amazon drainage.

The Trujillo Oases   South of the Chiclayo Oases are the irrigated lands around Trujillo. These oases are among the most productive of the coastal region. Here, the latest agricultural techniques and most modern machinery are employed. Reflecting the healthy state of the oases is the commercial and industrial city of Trujillo. Its many industries include food products, beer, tractors, candles, soap, leather, and cocaine. Connected by rail to Trujillo is the port of Salaverry. A breakwater provides shelter for ocean vessels and a pipeline allows ships to load molasses without coming to the docks. Other products exported are coca, cotton, fish, sugar, and rice.

Chimbote and the Río Santa   From Trujillo southward, the slopes of the Andes reach the edge of the ocean, and, although the Río Santa is second only to the Río Chira in annual flow, there is almost no flat land on which to develop an oasis. Yet, Chimbote has become one of the most important cities along the coast, and water from the Río Santa is fully utilized. This water is used to generate hydroelectric power and is also sent by long canals northward to enlarge the Trujillo Oases, and even beyond to the valley of the Río Chicama.

The Río Santa rises in the mountains north of Cerro de Pasco and, after crossing stretches of the high-level surface, plunges abruptly into a deep valley on the Pacific side of the continental divide. The highest mountains in Peru, the Cordillera Blanca, overlook the Santa Valley from the northeast. Tributaries bring to the Santa large volumes of water throughout the year from the glaciers and snowfields of these mountains. In the Cañon del Pato, downstream from Huarás, the Río Santa drops 1400 feet in six miles. Here, hydroelectric plants have been built inside the canyon walls, and much of the power generated is made available to Chimbote by transmission lines.

Peru's largest steel plant is in Chimbote where electricity is used in the smelting process. Carbon is provided by pulverized anthracite coal which is mined only a short distance inland, while limestone is quarried along the same railroad that serves the coal mines. Iron ore comes from the Marcona mines, inland from San Juan, 500 miles to the south. Chimbote is also a major fishing port and center for boatbuilding. It was, in fact, the world's leading fishing port in 1970, but a severe earthquake in that year caused extensive damage to the city, and the decline of Peru's fishing industry has taken an even greater toll. Thirty years ago this city was booming from the development of anchovy meal processing. Today, due to overfishing, many of its more than 240,000 inhabitants are unemployed.

The Middle and Southern Oases

The middle oases are those adjacent to rivers south of the Río Santa as far as, and including, the Río Ica. Along this part of the Peruvian coast there are few large areas on which irrigation can be developed. The largest is that supported by the Río Rimac, on which Lima and Callao are built. The important crop here is cotton. Around Lima, much of the irrigated area is devoted to truck crops which supply the metropolitan area. South of Lima, the areas around Pisco and Ica specialize in vineyards. The port of Pisco gives its name to a popular brandy sold throughout the Americas.

South of Ica, most of the oases are used only for subsistence crops. However, two of the oases south of Ica are commercially important. The larger of these is the oasis of Arequipa, in a valley at the base of the volcano El Misti. Water is brought from several small streams, including the Río Chili and its tributaries. The irrigated land is used for intensive cultivation of food and feed, which are consumed locally. Animals fattened on alfalfa are exported. Arequipa, a city of 300,000 inhabitants, is a major wool market, serving the entire southern part of the highlands. The city also has a modern industrial park and is noted for the manufacture of electronic equipment, cement, glass, plastics, and alcoholic beverages. A railroad and road extend inland to Cuzco and to Puno on the shores of Lake Titicaca.

The port that has served Arequipa since colonial days is Mollendo. This place, however, is one of the spots along this harborless coast where landing operations are most difficult. Mollendo faces southward and lies open to the sweep of waves driven by the prevailing winds. Consequently, the new port of Matarani was opened about seven miles northwest of Mollendo in 1941.

The other commercial oasis of the southern group, Moquegua, lying south of Arequipa and served by the port of Ilo, is noted for its vineyards and olive plantations. Tacna, the southernmost Peruvian oasis, grows crops primarily for local consumption.

Mineral Production on the West Coast

Peru became the first commercial producer of petroleum in Latin America when oil was discovered in 1863 along the northern Pacific Coast, at Zorritos, just south of Tumbes. Most of the country's production subsequently has come from the nearby Lobitos and Negritos fields, in a zone extending about 40 miles north and south of Talara. From 1930 to 1960, Peru was a significant net exporter, but local consumption increased while production declined. The opening of oilfields in

FARMERS IRRIGATE THE DESERT NEAR AREQUIPA, PERU TO CULTIVATE ALFALFA.

the *montaña* of northeastern Peru in 1976 increased total production threefold and again made the country a net exporter. In 1979 an off-shore field with reserves of 80 to 100 million barrels of light crude, was discovered after which petroleum became the nation's leading export commodity. Reserves are now estimated to be 63 percent in the *montaña*, 28 percent on the coast, and 9 percent offshore.

It has been known for many years that good-quality coal exists in Peru. Included are the anthracite deposits of the Río Santa Valley, but these did not attain international significance until a railroad was constructed to the mines and docks were built at Chimbote. Copper ore is also shipped from Chimbote, being transported from the Aguila Mine, north of Lima, by a slurry pipe-line to Chorro and the remaining distance by truck. Other mineral resources include phosphate, potash, and uranium. A major find of phosphate in the Sechura Desert of northern Peru could eventually make Peru an important world supplier. The deposits are located at Bayovar and are estimated to include about 15 percent of the world's reserves.

During recent decades mining in Peru has been influenced primarily by large-scale reserves of copper. At Toquepala, in southern Peru, there are about 1 billion tons of low-grade copper ore. To handle such reserves, a railroad and highway were built to the port of Ilo, where a breakwater and loading dock were constructed. Among the world's largest copper facilities are those of adjacent Cuajone. This mine's output is reduced to "blister" copper at the Ilo smelter. Also in southern Peru, at Quechua, copper deposits of about 80 million tons have been discovered.

Important bodies of iron ore were developed during the 1950s at Marcona, inland from the port of San Juan. This ore supplies the steel industry at Chimbote, plus a smaller iron and steel mill at Lima. A pelletizing plant at San Juan, facilitates exports of iron to the United States, Europe, and Japan.

Lima and Callao

The political, social, and economic focus for all of Peru is the capital city of Lima and its adjacent seaport, Callao. During the colonial period most of western South America came under the dominance of Lima, an influence that has subsequently waned. In 1535 Pizarro selected this site for his capital because it combined two special advantages. First, here was one of the larger irrigable areas of the coastal region, for the Río Rimac provides a dependable supply of water to a broad alluvial plain. Second, the offshore island of San Lorenzo and a long gravelly promontory of the mainland gave protection from waves of the open Pacific.

Lima, eight miles inland from Callao, was laid out in characteristic Spanish manner with a rectangular pattern around a central plaza. It incorporated all of the features that characterized Spanish cities from California to the Strait of Magellan, although this city was built before the plan was prescribed by law.

Lima remains a city of ancient traditions, but has become a rapidly growing modern industrial center. With a population exceeding 4.6 million, it performs a variety of functions. It is the chief commercial center of Peru as well as the center of art and education. The University of San Marcos, the oldest university in South America, is located here and attracts students not only from all of Peru but also from many foreign countries. Government provides employment for a great num-

ber of workers, as does industry. The factories of Lima and Callao produce cotton, woolen and synthetic textiles, cigarettes, foodstuffs, and metalware. About 70 percent of the nation's total economic activity is located within the confines of metropolitan Lima. Callao remains the nation's leading seaport, and also serves as a major fishing port and naval base.

In the midst of Lima's historic buildings and narrow streets, new styles of architecture are bringing changes that are making this city look increasingly like all other modern urban centers. Industrial slums, and the problems that accompany them, have been added. Squatter settlements surround Lima and house thousands of poor people. It has been estimated that more than half of Lima's residents live in *barrios marginales.* These are places of hope for migrants who seek a home and education for their children, instead of remaining in the highlands where only a meager living can be gained by farming. Many of the inhabitants are not newly arrived immigrants but are members of families that have moved out from the inner city. Even migrants from rural areas are often from fairly well-to-do families in their home areas and are among the better educated.

Peru as a Political Unit

Peru has far to go in solving its social and economic problems. This impoverished country includes vast areas that are too dry, too mountainous, or too wet to be agriculturally productive. Marine resources are abundant but have been overexploited, and mining remains a mainstay of the economy. Unfortunately, mining is based on the exploitation of nonrenewable resources, as is the related development of petroleum resources. Extensive areas of the country remain to be opened for settlement, but this requires the investment of vast amounts of capital.

Social stratification, as elsewhere in Latin America, constitutes a major problem. Despite the rhetoric of revolutionary change, the rich remain rich and the poor remain poor. The rural Indian areas of Peru contrast sharply with the urbanized, Westernized communities of the Pacific Coast or the regional trade centers elsewhere in the country. Meanwhile, a continuous flow of migrants gravitates from the farmlands and pasturelands to urbanized areas in search of employment, education, medical care, and a variety of other attractions.

The coast and the Andes have been settled for many centuries, and population there is relatively dense in relation to arable land. Hence, the vast eastern plains, although largely covered with selva, are considered by many to be the "land of the future." This Amazonian region comprises more than 60 percent of the total national territory, but is occupied by only 7 percent of Peru's population. New roads and an expanding airline network are reducing the region's isolation, yet many years will pass before most of eastern Peru is fully integrated as part of the effective national territory. Meanwhile, the region does contribute a substantial flow of petroleum exports, by pipeline over the Andes and by barge down the Amazon, and is a significant exporter of hardwood lumber. Tourism has also expanded due in part to direct airline connections between Iquitos and Miami.

Peru as a political unit, then, is large, diverse and complex. The achievement of social and political harmony, and the maintenance of progress toward common objectives, are most difficult tasks. This government is faced with a formidable challenge.

The Economic Situation

The commercial dimension of the Peruvian economy focused on mineral production throughout the Spanish colonial period, and mining has retained its prominence in terms of foreign trade. It was not until the late nineteenth century that sugarcane and cotton began to rival mineral exports, and the coastal areas became a primary focus of the national economy. Then,

petroleum production expanded rapidly along the north Pacific Coast, and copper and iron mining was developed in southern Peru.

The most spectacular change in the economic sector during recent decades was the sudden rise of Peru to the position of world's leading fishing nation in terms of tonnage caught. From 31,000 tons in 1947, Peru's total fish catch rose to 6.9 million tons in 1963. During this period and until 1971, the rapid expansion of fisheries was based on the exploitation of anchovies, which were processed into fishmeal and fish oil largely for the export market. Thousands of highland Indians migrated with their families to the coastal ports, where they were employed on fishing trawlers or in fishmeal factories. The fishing industry employed half a million people, and Chimbote became the world's leading fishing port.

AN ANCHOVY CATCH ENTERING A FISHMEAL PLANT IN CALLAO, PERU.

On Christmas Eve 1971, disaster struck. El Niño swept southward and killed much plankton, which anchovies require to survive. The anchovy catch plummeted 57 percent in one year, from 10.3 million metric tons in 1971 to 4.4 million in 1972. By 1980, Peru had to import fish oil to meet its own domestic demand for edible oils. The number of fishmeal plants was reduced from 100 in 1973 to 33 in 1980, of which only 7 were in operation, and the number of fishing boats was reduced from 1500 to 200. Fishmeal production declined from 2.25 million tons in 1970 to less than 0.5 million tons by 1980. Reconstruction of the fishing industry is now being pursued, based especially on the canning and export of tuna and sardines, plus expansion of the domestic market for fish products. The sea still provides a livelihood for about 650,000 persons, and seafood ranks third in export earnings after minerals and agriculture.

Agriculture in relation to other sectors of the Peruvian economy has declined steadily in recent years. In the early 1950s agriculture employed almost 60 percent of the economically active population, a figure reduced to 40 percent in 1984. Its share of the gross domestic product declined accordingly. In 1980 Peru imported 91 percent of its wheat, 97 percent of its soybeans, 35 percent of its maize, 30 percent of its dairy products, 20 percent of its rice, and 6 percent of its sugar. The government's 10-year economic program, 1979–89, for very practical reasons, seeks to provide additional support to the agricultural sector, mineral production, and export industries generally.

The Political Situation

Until 1962 political control of Peru remained firmly in the hands of the traditional elite, namely, the large landowners and the officers of the army. In 1924 a political party known as the *Alianza Popular Revolucionaria Americana* (APRA) was formed. The *Apristas,* or members of APRA, urged a land-reform program that would return land to the Indian communities, a system of universal education that would reduce illiteracy, and an economic program that would include Indians in the national system of production and consumption. When they tried to initiate revolutionary change based on the Mexican pattern, they were resisted by the traditional authorities, and the leaders were forced to seek asylum in foreign embassies. Yet, the APRA program gained enthusiastic support from a majority of the literate people of Peru and was prevented from being accepted only by the authority of a dictator backed by the army. Because of this movement, communism in Peru remained weak.

In 1962 Peru held an election in which people who could qualify as literate were permitted to vote. In this election the conservatives formed an alliance with the *Apristas,* which had the effect of shocking the officers of the army into a break with the conservatives for the first time in Peruvian history. The army remained solid in its refusal to permit APRA to come to power, even when APRA received more votes than any other party. The election was annulled and a new election called. In 1963 many people realized that APRA would never be permitted to take office and that a vote for this party would be wasted. An alternative was presented by the Popular Action Party led by Fernando Belaúnde Terry.

From 1963 to 1968 Belaúnde gained support from an overwhelming majority of the Peruvians, so that it became increasingly difficult for the army to interfere. In 1964 his administration enacted a program of agrarian reform that drew protests from the landowners. Many of the traditional elite, whose privileges were threatened, insisted that Belaúnde was a Communist.

Belaúnde was dramatically overthrown by a military coup in 1968. The new military government, under General Juan Velasco Alvarado, sought to profoundly change the country within a short period of time. It nationalized the petroleum industry, eliminated the traditional *latifundia,* and emphasized industrialization and economic unification with the other Andean countries. Mines, sugar plantations, factories, fisheries, and banks, especially those that were foreign owned, were brought under state control. Major educational reforms were also undertaken. Still, the country did not prosper.

In 1975 Velasco's radical military regime was replaced by a more conservative one under General Francisco Morales Bermúdez, who began to dismantle the revolutionary structures of his predecessor and diminish the authority of the central government in favor of private enterprise. Having come full-circle after 12 years of military rule, the presidency was again transferred to Fernando Belaúnde Terry, by election, in 1980. Also transferred to the new civilian government were major national problems such as massive unemployment, uncontrolled inflation, a huge public debt, and widespread poverty and malnutrition. A program of austerity in public spending was instituted, yet there was renewed emphasis on agricultural self-sufficiency, road building, and settlement of the *montaña*. A major hope was for political stability within a general framework of representative democracy.

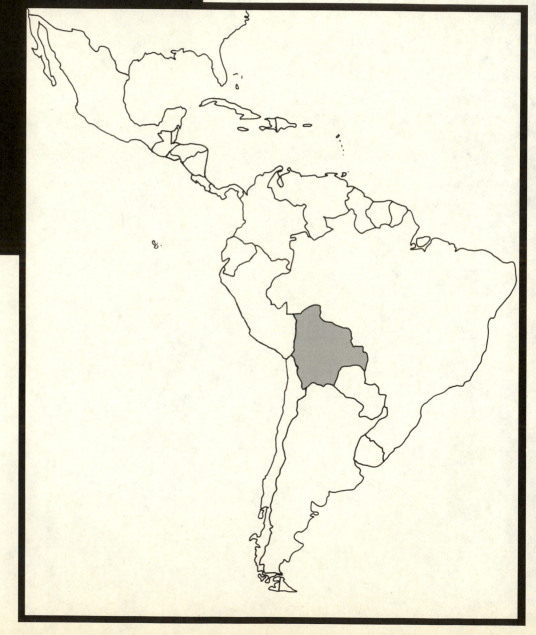

24

REPÚBLICA DE BOLIVIA

Land area 424,162 square miles

Population Estimate (1985) 6,200,000
Latest census (1976) 4,613,486

Capital city La Paz 953,634
(Legal Capital) Sucre 84,505

Percent urban 46

Birthrate per 1000 42

Death rate per 1000 16

Infant mortality rate 124

Percent of population under 15 years of age 44

Annual percent of increase 2.7

Percent literate 75

Percent labor force in agriculture 47

Gross national product per capita $510

Unit of currency Peso Boliviano

Physical Quality of Life Index (PQLI) 39

COMMERCE (expressed in percentage of values)

Exports

mineral fuels and lubricants	44
natural gas	43
tin	31
coffee	2

Exports to		Imports from	
Argentina	52	United States	29
United States	26	Argentina	15
Netherlands	4	Japan	11
West Germany	4	Brazil	10
United Kingdom	3	West Germany	7
Peru	3	Chile	3
		Peru	3

Data mainly from the 1985 World Population Data sheet of the Population Reference Bureau, the 1985 Britannica Book of the Year, and the 1985 World Almanac.

The establishment of a united nation with a stable government has proved to be especially difficult in Bolivia. In 1952 a social revolution eliminated the traditional large landowning class and resulted in nationalization of the principal source of wealth, the tin mines. Hopes were expressed that the status of the Indian could be enhanced and that steps would be taken to attack the prevailing widespread poverty. In subsequent years, however, a mixture of social idealism and political opportunism diverted the revolution from its original goals.

The problems of Bolivia can be clarified by examining the nation's geographic setting, including the internal arrangement of land and people within the national territory. To a greater extent than in any other Latin American country, the population of Bolivia is arranged in small, scattered clusters isolated from each other and from the outside world.

Before the War of the Pacific, 1879–83, Bolivian territory was far more extensive than it is today. As a result of that war, Bolivia lost its outlet to the Pacific Ocean, and Chile acquired an extensive area of the Atacama Desert with its valuable deposits of nitrate and copper. Next, Argentina seized a section of the eastern Chaco, and in 1903 Brazil took possession of the rubber-rich territory of Acre. In 1932 Bolivia and Paraguay went to war over what was thought to be rich oil-bearing land in the Chaco region, by which time Bolivia had lost two-thirds of its original territory and had become a land-locked country. In this war, Paraguay gained about 100,000 square miles of land from Bolivia, and an equal number of lives were lost in battle.

The physical conditions of the habitat in which the Bolivians are struggling to establish a coherent state are marked by extremes of altitude and aridity. The Bolivian people are located mostly on the high Altiplano south of Lake Titicaca and in valleys of the Eastern Cordillera. Separating these clusters of people from the west coast are high, bleak, wind-swept plateaus, towering ranges of volcanic peaks with passes above 13,000 feet, one of the driest deserts in the world,

and an escarpment more than 2000 feet high which descends abruptly to a harborless coast. In western and southern Bolivia the great dry belt of South America which extends from southern Ecuador almost to the Strait of Magellan crosses the Andes diagonally from northwest to southeast. Between latitudes 20°S and 30°S, within that zone which is dry on the western sides of all continents, South America is so deficient in moisture that the land is almost totally barren. Even in the high Western Cordillera of the Andes, there is so little rainfall that few streams emerge from the mountains. Unlike coastal Peru, where numerous streams flow across the desert to the ocean, in what is now northern Chile only one river, the Río Loa, rises in the Andes and flows all the way across the desert to the Pacific.

The internal force by which a state might be able to expand across such physical barriers is lacking in Bolivia. Like Ecuador and Peru, this state is composed of people of very different cultural traditions, the Indian and the Spaniard, but in Bolivia the problems of cultural diversity are compounded by the arrangement of the population in small, isolated clusters and ribbons of settlement. Furthermore, the Indians of the Titicaca Basin and the Atliplano between Lake Titicaca and Lake Poopó are Aymaras, descendants of people whose civilization flourished in this region a century before the time of Christ. Conquered first by the Incas and then by the Spaniards, they have nevertheless maintained their language and customs with little change and have clung stubbornly to their ancestral lands. Among all these divergent elements, the Bolivian state has been unable to develop any strong sense of political coherence. In Bolivia as a whole, people of unmixed Indian ancestry comprise about 55 percent of the population, mestizos 35 percent, and people of unmixed European ancestry, about 10 percent.

Around the shores of Lake Titicaca, the people are almost exclusively Indian. Even in the capital city of La Paz, Indians form the majority of inhabitants. In the Department of La Paz, Indians comprise about 75 percent of the total. In the lower valleys and basins of the Eastern Cordillera there are more mestizos and Europeans. In the Basin of Cochabamba, for example, which is densely populated, 70 percent are mestizos or Europeans.

Although the Aymaras, mostly illiterate and unable to speak Spanish, have had little to say about Bolivia's political and economic policies until recently, they nevertheless comprise the overwhelming majority of farmers, herders, and miners of the country. Repeatedly, their traditional communal system of land tenure has been officially abolished, yet the Indian village con-tinues to operate on a communal basis as it has for thousands of years.

Physical Diversity

Bolivia, like Colombia, is about one-third mountainous and two-thirds lowland, and most of the lowlands lie outside of the effective national territory. The plains of eastern Bolivia are drained partly by the headwaters of the Amazon and partly by the Río Paraguay and its tributaries. The plains in the southeast, south of Santa Cruz,

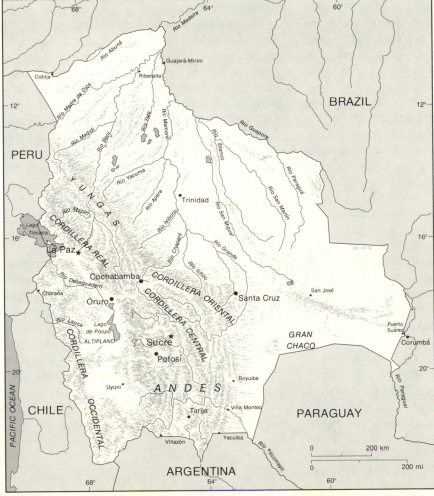

FIGURE 24-1

are part of the great lowland region known as the Gran Chaco. Most Bolivians live in the one-third of the national territory that is mountainous.

It is in Bolivia that the highlands of the Andes reach their greatest width. About latitude 18°S, between Arica on the Pacific Coast and Santa Cruz on the eastern plains, the mountain zone is approximately 400 miles wide. The highlands here are made up of three distinct parts: the Western Cordillera; a string of high intermont basins known collectively as the Altiplano; and the Eastern Cordillera.

The Western Cordillera is a southward extension of the volcanic region which begins north of Arequipa in Peru. Along its crest are numerous active volcanoes, some of which, in the symmetry of their cones, rival the famous El Misti. East of the Western Cordillera are the basins of the Altiplano, the northernmost of which is occupied by Lake Titicaca. Spurs extending eastward from the Western cordillera all but separate these high basins, but on the eastern side of the Altiplano, along the west-facing front of the Eastern Cordillera, there is a continuous passageway of gentle gradient which extends from the shores of Lake Titicaca southward across Bolivia.

The northern part of the Eastern Cordillera is a southeastward continuation of the high ranges of Peru. The mountains that stand northeast of the Titicaca Basin and the city of La Paz are very high, some of them more than 21,000 feet above sea level, and the descent from these great elevations to the eastern plains is remarkably abrupt. This rainy and heavily forested northeastern slope of the Eastern Cordillera, which is the equivalent of the eastern border valleys region of Peru, is known in Bolivia as the *Yungas.*

A line drawn east-west from near Santa Cruz to the edge of the Altiplano, passing just north of Cochabamba, marks the approximate southern end of the Yungas. South of this line the Eastern Cordillera is composed of a great block of the earth's crust which has been tilted eastward, its upper surface sloping gently toward the eastern plains and its western edge forming a sharply defined escarpment overlooking the basins of the

Altiplano. The top of this block, like the highlands of Peru, is composed of a high-level surface of slight local relief, standing, near the western border, between 12,000 and 14,000 feet above sea level. Above this surface there are irregularly placed and discontinuous ranges of high peaks, and below it are deeply excavated valleys and basins. This high-level surface in the Eastern Cordillera of Bolivia is known as the *Puna.*[1] The Eastern Cordillera south of latitude 17°S can be divided into a western high part where the Puna surface is only slightly dissected, and an eastern lower part where the streams have cut the Puna into long fingerlike remnants standing high between the valleys.

Highland Bolivia is crossed diagonally by the zone of aridity. From about latitude 20°S almost to the Strait of Magellan, the eastern base of the Andes is dry. As a result of this climatic arrangement, each of the major surface divisions of Bolivia can be subdivided into a northern wetter part and a southern drier part. Most of the people occupy the northern basins of the Altiplano and the basins and valleys of the Eastern Cordillera along the line of transition between the very wet Yungas and the very dry south.

The Western Cordillera and the Altiplano

For rural people, the habitability of western Bolivia decreases from northeast to southwest as aridity increases. The Western Cordillera has the smallest population of the highland regions, for here aridity and altitude are combined. In the northern part, several small rivers rise among the volcanic peaks, some draining westward toward the Atacama of northern Chile and some eastward toward the Altiplano of Bolivia. In little valleys between 11,500 and 15,000 feet above sea level, there are narrow ribbons of irrigated

---

[1] The word *Puna,* designating a region, should not be confused with the same word used to designate a vegetation type.

land on which meager crops of potatoes and hay are raised. Toward the south, where few sources of water are to be found, the land remains almost uninhabited except by seminomadic shepherds.

The Titicaca Basin

In the Titicaca Basin environmental conditions are not as extreme as in the Western Cordillera. The zone of deficient moisture crosses the Altiplano south of Lake Titicaca, and to the north rainfall is adequate for the cultivation of crops without irrigation. Around the shores of the lake, temperatures are moderated by the presence of open water. Because the lake is very deep, 918 feet at the maximum, the water temperature remains nearly constant throughout the year, at about 51°F. As a result, air temperatures around the lake do not drop as low at night or in winter as

they do at similar altitudes farther from the water. The surface of Titicaca is 12,507 feet above sea level, but maize and wheat ripen in the Titicaca Basin to an elevation of 12,800 feet and as many as 70 different varieties of potatoes are grown.

These advantages for an agricultural people have been reflected since earliest recorded history by a relatively dense population around the shores of Lake Titicaca. This area of concentrated settlement formed the core of the ancient civilization whose ruined temples can still be seen on promontories and islands of the lake. Just east of Guaqui are the ruins of Tiahuanaco, once the chief market city of the highlands. It was in this region that the potato was probably first cultivated, and it continues to be the staple food of the Aymaras. They also grow quinoa and supplement these foods with dried llama meat and fish.

BOATS MADE OF WOVEN REEDS ARE USED FOR TRANSPORTATION AND FISHING ON LAKE TITICACA.

The only items regularly brought to the region from elsewhere are coca leaves from the Yungas, which the Indians chew to deaden hunger pangs, and maize from which they make *chicha*. Sheep are raised, and a new breed of cattle is helping to bring prosperity to this poverty-stricken area. By crossing *criollo* cattle, a type that thrives in the rarified Andean altitude, with the Holstein, an exotic animal has been created, one that can produce 20 quarts of milk daily. From this milk, Indian farmers produce cheese which can be sold at a reasonable profit.

After the Spanish occupation, the Aymaras were forced to labor in the silver mines of Potosí and other places. Where hard work was to be done, the Aymaras were the ones who did it. The Titicaca Basin was divided into large private estates owned by people of Spanish descent, but the Indians continued to occupy their own lands on a communal basis, paying the landowners rent. The owners lived elsewhere, some in Cochabamba, some in La Paz, and some even in Europe, leaving overseers to collect the rents.

The fact that the dense conglomeration of people in the Titicaca Basin is bisected by the boundary between Peru and Bolivia illustrates the extent to which Spanish culture was superimposed on that of the Indian. Political boundaries and all they connote are Spanish importations, and until the Indian can see a difference between being a Peruvian or a Bolivian, the boundary has little meaning. Twice each year, a great fair is held at Copacabana, on the peninsula that almost bisects the lake near the boundary between the two nations. Indians arrive on foot or by raft from all around the Titicaca Basin. They come from as far away as Cuzco and from Tucumán in Argentina, not only to Copacabana but also to other fairs in various parts of the highlands.

COPACABANA, ON LAKE TITICACA, WAS SETTLED BEFORE THE RISE OF THE INCAS.

Indian Settlements South of Lake Titicaca

The number of agricultural communities decreases south of the Titicaca Basin. Indian settlement extends southward along the valley of the Río Desaguadero about as far as the margin of sufficient moisture. As aridity increases, the Desaguadero takes on the braided, shifting channel of a typical dry-land river, and the settlements along its banks are spaced at wider intervals.

The Desaguadero drains into the salty Lake Poopó. Although the waters of Lake Titicaca are only slightly brackish, the Río Desaguadero on its way south crosses saline beds from which a certain amount of salt is carried away in solution. The saltiness of Poopó is also due to the fact that water escapes from it mostly by evaporation. Lake Poopó averages 12,120 feet above sea level, and only 10 feet in depth, but in times of flood rises rapidly and may overflow into the Salar de Coipasa, a salt flat to the southwest. Farther to the south, the Salar de Uyuni is a great windswept salt flat which is now totally arid.

East of Lake Poopó there is a more or less continuous band of Indian settlements along the west-facing piedmont of the Eastern Cordillera.

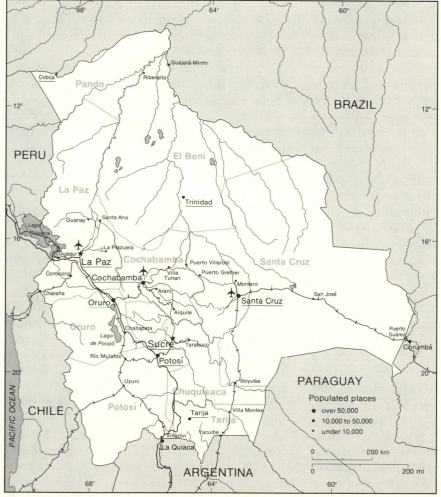

FIGURE 24-2

Each stream that emerges from the mountains has built an alluvial fan, and on these fans Indian communities have become established. The pattern of settlement differs from that of the Titicaca Basin, however, in that the villages are located high on the fan slopes where water is available, rather than close to the lake. Cultivated fields are located on the upper parts of the fans; the lower slopes where water is less certain are used for pasture.

Mining Communities of the Altiplano

In addition to the agricultural Indian communities of the Altiplano are two clusters of population which originated as mining centers. One of the oldest mining communities in this part of South America is located just east of the Río Desaguadero and south of the main railroad line between La Paz and Arica. This is Corocoro. Since earliest history, a mining population has been located here, for this is one of the two sources of native copper in the Americas (the other being the Keweenaw Peninsula of northern Michigan). Because copper was found here in pure form, not as an ore to be smelted, native peoples used it long before knowledge of metallurgy made possible the utilization of other sources of copper. Corocoro is still exlusively a mining community, supplying about 90 percent of all copper production in Bolivia.

Oruro is the center of another mining district of the Altiplano. A range of low hills, about four square miles in area and rising 1200 feet above the general level of the Altiplano, contains ores of silver and tin. During the colonial period, this district was one of the chief sources of silver, and Oruro, at the eastern base of the hills, attained considerable importance. However, mining communities are generally subject to major fluctuations of prosperity and population because of the speculative character of the business of seeking and selling the ores, especially in a land that is too high and dry for inhabitants to find any other means of livelihood. With the decline of silver production during the nineteenth century, Indian workers moved away and Oruro was largely abandoned. With the expansion of tin mining in more recent years, the community has regained its former prestige. Oruro is the site of Bolivia's first tin smelter. Its population of about 110,000 is composed mostly of Indians, many of whom work the tin, tungsten, and silver mines.

The Altiplano is composed of a number of more or less separate basins. Along the eastern border of the Altiplano the several sand and clay-filled basins are connected, and a level route of travel is available from the southern end of the Titicaca Basin far to the south. This route is followed by one of Bolivia's main railroad lines. Two major branches join this line in the vicinity of Oruro: one serves the tin mining district around Uncía in the Eastern Cordillera; the other reaches the agricultural Basin of Cochabamba, lower on the eastern slopes. Oruro, as a rail center and the focus of highway routes as well, has become a supply center connecting Bolivia's most productive agricultural district with its most productive mining district.

La Paz

La Paz, the main political and commercial center of Bolivia, is the world's highest large city. Located 12,000 feet above sea level, this city of 950,000 people is wedged into a narrow valley in the Altiplano, where a tolerable compromise exists between comfort and accessibility.

The remarkable site occupied by La Paz is the result of river cutting. The Río La Paz, a tributary of the Río Beni and the Amazon, has cut headward from the rainy eastern slopes of the Cordillera Real through the heart of the range until its headwaters now include some streams descending westward toward the Altiplano. The gorge which the river has cut through the Cordillera is one of the most spectacular features in a continent where spectacular gorges are commonplace. From summit to summit along the crest of the range, the gorge is only about 12 miles wide, but

LA PAZ, BOLIVIA, OCCUPIES A DEEP CANYON. ADJACENT TO THE ALTIPLANO AND HIGH PEAKS OF THE CORDILLERA REAL.

nearly 11,500 feet below the crest the river plunges through a channel bordered by almost vertical cliffs, an opening impassable for modern transport. Where the headwaters of the Río La Paz have reached the loose lake-bed deposits of the Altiplano a deep chasm has been excavated parallel to the mountain front.

The Spaniards, not the Indians, chose the site for La Paz. The native people regarded the deep trench along the front of the mountains as uninhabitable. For the Spaniards this site combined accessibility to the principal colonial route of travel with a measure of protection from the cold winds of the Altiplano. The main road from Lima to the silver mines of highland Bolivia passed Lake Titicaca on the southwest side, where the

terrain is less rugged than on the northeast side. Emerging from the Titicaca Basin, the road followed the eastern side of the Altiplano to the edge of the chasm where the way to the south was open. Once established, La Paz controlled the passage of goods to and from Lima. Later, when railroads were built, they were all brought to a focus on this city.

The site of La Paz has proved anything but favorable for the development of a large urban center. The cramped space in which the buildings are wedged has handicapped growth. Yet, the city continues to expand, not only within the chasm but also up the slopes and onto the Altiplano. The climate, even in this protected spot, is not easy to endure. The rapid changes in temperature make respiratory diseases common, and the rarified atmosphere at these altitudes causes physical exhaustion and nervous strain. In almost every respect La Paz is unique as a national capital and as a major metropolitan center.

Settlement in the Eastern Cordillera

The valleys of the Eastern Cordillera south of the Yungas have been cut into the Puna surface by two main river systems: the Río Grande, a tributary of the Río Mamoré, the Madeira, and the Amazon; and the Río Pilcomayo, a tributary of the Paraguay, Paraña, and Plata. The valleys vary in width. There are places where they are narrowly constricted, but upstream from such places they generally broaden out to form long ribbons of flat land or, in a few instances, fairly wide basins. The streams generally follow winding courses, even where they are deeply entrenched in the Puna surface, which the physiographers describe as "entrenched meanders." The valleys and basins have been excavated several thousand feet below the Puna, forming irregularly shaped belts where the temperature is considerably higher than that of the Puna country on either side. Since the streams carry an abundant supply of water, irrigation is possible even where local rainfall is sparse.

·La Paz, then, was established along one of the major trade routes of colonial South America, a route that connected the primary settlement center on the west coast with the rich mines of the Bolivian highlands, and led southeastward to the plains of Argentina. At one end of the road was Lima, the center of Spanish colonial life. The major objective of the road in what is now Bolivia was to serve the silver mining community of Potosí.

Mining Communities

In 1544 the Spanish conquerors of Peru in their restless search for El Dorado discovered Cerro Rico, a conical mountain that stands above the Puna surface of the Eastern Cordillera and reaches 15,680 feet above sea level. Within this mountain is one of the richest ore bodies known anywhere in the world, an ore containing not only silver, but also tin, bismuth, and tungsten. In 1554, however, the Spaniards wanted silver, for tin was then much more cheaply supplied to Europe from local sources, and bismuth and tungsten had no uses. Out of this one mountain, between its discovery and the beginning of the seventeenth century, came about half of all the silver produced in the world during those 56 years. The "royal fifth" which poured into the Spanish treasury played a vital role in shaping the course of European history.

The desire for wealth can lead people to endure the most severe hardships and, at least temporarily, to overcome the most severe handicaps. The mining town of Potosí at one time had a population of 160,000 despite the fact that the climatic conditions were so harsh that La Paz must have seemed blissfully comfortable by comparison. The average winter day ranges in temperature from about 3° to 45°F, and snow is not uncommon. Moreover, colonial engineers faced a difficult problem in providing power for the mines and ore crushers. Fuel on the highlands was scarce and remains so even today. Consequently, a water-power system was devised by building more than 30 small reservoirs in the vicinity, which provided enough water to turn the wheels of more than a hundred mills.

Potosí, after its dazzling rise to wealth, suffered the fate of most mining communities. Exhaustion of the readily available ores, advances in technology that made possible the use of poorer ores in more accessible locations, and a series of natural disasters at Potosí — including the flood of 1626 when one of the dams above the city broke — combined to terminate the preeminence of the area. First Peru and then Mexico took the lead in silver production. Potosí began to decline in population, since no other means of support could be found in such a locality, and for more than two centuries Potosí remained a ghost town, occupied by only a few hundred people.

In the late nineteenth century the exploitation of Bolivia's tin ores began. Increasing demand resulting from new industrial uses, as well as exhaustion of the more accessible European sources, led to a rise in the market price of tin. The result was a reawakening of mining activity around Potosí. Besides tin, there were substantial deposits of lead and copper, and small amounts of silver to be mined.

Nearly 100,000 people live in Potosí today, but time has taken its toll. Recognizing that the artistic and historic preservation of Potosí is urgent, the Organization of American States has proclaimed Potosí an American monument, with "its innumerable treasures that should be preserved as testimony to the creative power and identity of the American people.[2] Other mining communities in the Eastern Cordillera of Bolivia are now more important than Potosí. The rich tin ores between Oruro and Uncía were discovered at the end of the nineteenth century. By the beginning of World War II, three big mining corporations — Patiño, Hochschild, and Aramayo — controlled 80 percent of Bolivian tin production, and the Bolivian treasury derived 70

---

[2] *Américas* 32, No. 2 (1980): 160

THE CATAVI MINE IS THE WORLD'S LARGEST SOURCE OF TIN.

percent of its income from tin exports. The Bolivian Simón Patiño (1860—1947) owned the mines at Uncía, from which he built one of the world's largest fortunes. From his home in France, Patiño for many years virtually dictated Bolivian government policy.

Frequent armed conflicts developed in the mining districts. The workers were Aymara Indians, recruited from communities of the Altiplano. They were paid such low wages that all suffered from an inadequate diet. When the Indians' deep resentment was expressed in strikes, order was restored by armed force. One of the first steps of the revolutionary government that came into power in 1952 was to nationalize the properties of the three mining interests, which since that date have been operated by the government.

Bolivia remains dependent on mineral production for approximately 70 percent of its foreign exchange earnings. While the country also produces bismuth, lead, zinc, wolfram, copper, gold, and antimony, tin and natural gas have a preeminent position. Deposits of tin ore are confined mostly to the mountainous, high-altitude areas where access is difficult, and mineralization usually occurs in narrow veins embedded in hard rock which requires extensive tunneling. Furthermore, the deposits are often of low quality. Were it not for strategic factors in the defense of the Western Hemisphere, many Bolivian mining enterprises would soon cease to be economically viable.

Agricultural Communities

The most densely populated clusters of rural population in Bolivia are found in the intermont basins and valleys at lower altitudes in the Eastern Cordillera. Communities in the larger basins of Cochabamba, Sucre, and Tarija, together with many smaller communities in this general region, form the heart of the Bolivian state, for here are to be found most of the people for whom nationality has any real meaning.

Cochabamba, Sucre, and Tarija The largest concentration of rural settlement in Bolivia is in the Basin of Cochabamba, a geographically constricted area surrounded by high, almost uninhabited Puna. The city of Cochabamba, with its mild climate, is a comfortable place to live, and fertile soils in the basin produce excellent crops of maize, barley, alfalfa, and fruit. Cochabamba is the second-largest city of Bolivia, with a population exceeding 200,000.

A notable feature in the political geography of Bolivia is that the country supports two national capitals: La Paz and Sucre. In 1825, soon after independence was declared from Spain, Sucre (then Chuquisaca) became the seat of government. It was named in honor of Antonio José de Sucre, who had served under Simón Bolívar and became first President of the new nation. As the legal capital, it continues to be the seat of the Supreme Court of Bolivia and archbishopric of the Church. La Paz, the *de facto* capital since 1898, retains the executive and legislative branches of the government.

Like Cochabamba, Sucre is the center of a dense agricultural population, but many of the city's 90,000 inhabitants are employed in cement manufacturing, oil refining, and other industries. Known as the "White City" because most of its buildings must be painted white by law, it exhibits picture-postcard beauty with its formal gardens and impressive colonial homes.

Tarija (50,000 population) one of the oldest cities in Bolivia, also has a mild climate and is situated in the Guadalquivir Valley. It serves as a market for many miles of ribbonlike farms that produce olives, pears, apples, peaches, and grapes.

Problem of Accessibility

People who practice the European way of living do not prosper in isolation. Although these eastern valleys are comfortable and productive, providing transportation is costly. Products such as coca, cacao, and coffee, which can stand high transportation costs because of high value per unit of weight, cannot be grown in the country south of the Yungas because of increasing aridity. For such bulky products as grain or perishable produce as fruit, a cheap means of transportation is essential.

The main obstacles to transportation development are not only the rugged terrain and steep grades necessary to reach the Altiplano, but also the geographical arrangement of the settlements themselves. The focus of economic and political power is not sharp, because people of this core of the Bolivian state are scattered in numerous relatively small communities, each anxious for government support of local projects, but each ready to block the projects that bring advantages only to other communities. The largest cluster of people is in the Cochabamba Basin, but the area of this basin is small. Other communities are still smaller, many of them strung in narrow bands along the valley bottoms. The difficulty of providing these scattered communities with a common political and economic focus constitutes a problem that is not easily solved.

All-weather highways, here as elsewhere, provide a better solution to the problem of accessibility than do railroads. Since the 1950s, roads have been built to connect with the Peruvian highways on either side of Lake Titicaca. From near La Paz a main highway has been built along the old colonial road through Oruro to Potosí and on through Tarija to connect with Argentine highways along the eastern piedmont of the Andes. A road has been built from La Paz over a high pass in the Cordillera Real and down through the Yungas to a river port on the Río Beni. Another road connects Oruro with Cochabamba, and this has now been extended to Santa Cruz on the eastern plains. Both Oruro and Cochabamba are connected with Sucre, and from Sucre a road descends to the eastern piedmont on the edge of the Chaco. The highway development program has made the eastern plains accessible to the highland communities and Santa Cruz is the main transportation center. Construc-

tion of a marginal highway along the front of the mountains will connect Sant Cruz northward with the Peruvian *Carretera Marginal de la Selva* and southward with the Argentine highways at Yacuiba. The opportunity exists to develop an enlarged domestic market for agricultural products from various parts of Bolivia.

Settlement of the Eastern Plains

The part of Bolivia lying east of La Paz, Cochabamba, and Tarija can be divided roughly into a northeastern region that is rainy and a southeastern region that is relatively dry. In both divisions, pioneer settlements within the mountains and along the mountain piedmont appear as soon as the highway provides access, but much of the land that extends eastward to the borders of Brazil and Paraguay remains outside of Bolivia's effective national territory.

The Bolivian Yungas forms a distinct natural division of the country. It includes the northeastern slopes of the Cordillera Real northeast of La Paz and north of Cochabamba and is part of a region that extends unbroken along the eastern Andes from Colombia, across Ecuador, and Peru, as far south as Santa Cruz. In Peru this rainy, densely forested land is known as the eastern border valleys and in Ecuador as the *Oriente*. The much narrower fringe of vegetation that continues south of Santa Cruz along the mountains well into Argentina is composed of a more open type of forest.

A belt of sandstone ridges through which the Río Beni and its tributaries pass in a series of water gaps marks the foothills of the Andes. Beyond lie the eastern plains, covered partly with selva and partly with wet savanna. These plains of northeastern Bolivia and neighboring parts of southeastern Peru suffer from extreme isolation, not only because of the steep mountain barrier to the west, but also because the Río Madeira, which gathers the water of the Madre de Dios, the Beni, and the Mamoré, is interrupted by a long stretch of rapids as it flows northeast to the Amazon.

Since the arrival of the Spaniards, three kinds of products have attracted settlers to the Yungas and the plains beyond. Among the earliest sources of wealth was gold. The gravels of the Río Tipuani, about 60 miles east of La Paz, in the zone of the Front Ranges, proved to contain large quantities of this metal. Until 1800, while Bolivia accounted for about 10 percent of the total gold production of South America, the Yungas was the principal source. Even today, placer works along the Río Tipuani remain active. All workers must be recruited from the Indian communities of the highlands.

Settlers also have been attracted to the valleys of the Yungas at intermediate altitudes, from about 2000 to 6000 feet, by the possibilities of producing coca, coffee, and sugar. These products, like gold, can more readily support the high costs of transportation. Coca leaves continue to

find a steady demand among the highland Indians, as they did before the arrival of the Spaniards. Coca production, for the most part, is confined to the Yungas and Chaparé areas which are frost free and have plenty of rain. It is not illegal to raise coca or sell coca leaves in Bolivia, but it is definitely illegal to produce derivatives of coca, such as cocaine. Yet, Bolivia is a major source of cocaine being smuggled northward to the United States. Sugar, converted into *aguardiente*, also finds sufficient demand to permit shipment from the Yungas, and meat is regularly flown from cattle ranches in the Beni region to the urban market of La Paz.

Cinchona bark and rubber were once sources of wealth for the settlers of northeastern Bolivia. Highland Indians were called upon to collect these wild products from the selva, and many communities at the intermediate altitudes in the Yungas were all but abandoned in the rush for profits. However, first cinchona and then rubber

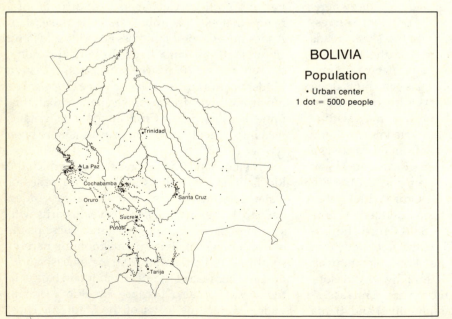

BOLIVIA

Population

• Urban center
1 dot = 5000 people

FIGURE 24-3

were virtually eliminated as South American exports by the cheaper plantation methods of Malaya, Sumatra, and Java.

The Bolivian Chaco

Most of the piedmont of the Andes, like the plains that stretch eastward to the Río Paraguay, is covered with dry scrub woodland and patches of dry savanna. The Front Ranges are composed of parallel ridges or cuestas of brilliant red sandstone, with broad longitudinal valleys 2 to 15 miles in width. A trellis pattern of drainage has developed, with the main streams crossing the ridges in water gaps so narrow that access to the neighboring valleys must be gained by climbing over the intervening ridges rather than by following the rivers. The eastern plains form a part of that large area, shared by Bolivia, Paraguay, Brazil, and Argentina, which is known as the Gran Chaco.

Founded in 1560 by settlers from Asunción, Santa Cruz has become the focus of settlement in this region. Once a drowsy outpost, it is now an oil-boom city with a population of about 140,000. In this once isolated spot, a few of the Spanish settlers received large grants of land which they turned into self-sufficient estates devoted mainly to cattle-raising. The first break in the isolation of the Santa Cruz area came after World War I. In 1920 a concession was granted to a North American petroleum company to explore for oil along the Andean piedmont. Several oilfields were discovered along the front of the mountains south of Santa Cruz, near the Argentine border, and it appeared that much larger reserves would eventually be discovered. Bolivia was not enthusiastic either about permitting the development of these fields by foreign-owned corporations, or about marketing the petroleum in oil-poor Argentina and Brazil.

After the loss of much of the Chaco, in its war with Paraguay, the Bolivians were jarred into action. The concessions of the North American company, which had been held in reserve and had not been further explored, were expropriated in 1937. A government-owned corporation, the *Yacimientos Petrolíferos Fiscales Bolivianos* (YPFB), was formed to assume the responsibility for developing the oilfields. The actual exploration, drilling of wells, and building of refineries and pipelines were done by foreign-owned corporations under contract with YPFB. Hence, considerable development of oil production from the fields south of Santa Cruz took place. In 1965 about 75 percent of the total Bolivian oil production came from these fields, but by then new and promising oilfields had been developed north of Santa Cruz. Southeast of Santa Cruz a large gas field was discovered. To get all this oil and gas to market, pipelines were built. The first line connected southward to Argentina, and another led into the mountains to Sucre, Cochabamba, and La Paz. In 1966 a line was extended to Arica, and the first shipment of Bolivian crude oil left that port in northern Chile. In 1982 heavy oil was discovered in Santa Cruz Department. A new gas pipeline is planned from Santa Cruz to São Paulo, Brazil. Another will link Cochabamba, Santa Cruz, Oruro, and La Paz. Refineries were built in Cochabamba, Sucre, Santa Cruz, and in the oilfields of southern Bolivia. Today, there are additional oilfields at Colpa and Caranda and a large gas field at Río Grande. The most recent finds are in eastern Santa Cruz and in Chuquisaca. Oil production, however, has declined steadily since 1975, so that Bolivia became a net importer by 1980.

In the 1960s, Santa Cruz became a major center of pioneer agricultural settlement. The government made great efforts to convince the highland Indian farmers that life in the eastern lowlands was promising. A small movement of colonists began after the road to Santa Cruz was completed in 1953, and by 1967 the number of migrants had increased greatly. By the 1970s various ethnic groups came with entire families to engage in a myriad of agricultural enterprises. Mennonites were joined by Indians from the highlands, Japanese from Okinawa, and German immigrants from Europe.

Bolivia as a Political Unit

Until recent decades, Bolivia's continued possession of the Chaco seemed most unlikely. The isolation of this region from the rest of the country was a matter of grave danger for international peace when it became clear that the Chaco would some day emerge as an important source of petroleum. The Chaco was easily accessible to Brazil and Argentina, both of which lacked sufficient oil, and the stage was set for conflict.

The Chaco War and resultant loss of territory forced the Bolivian government to shift its attention eastward, with major emphasis on road building, air transport, and agricultural colonization. The region tributary to Santa Cruz is now more dynamic in its growth and development than are the highland communities of the Altiplano. Yet, many people in many parts of the country feel little sense of national unity. Hence, the political leaders must not only continue to seek the development of modern infrastructure for the nation, but also formulate a state-idea powerful enough to unite a majority of the people against the forces of disintegration.

The Economic Situation

Bolivia is the poorest nation on the continent of South America, with a gross national product of only $510 per capita. This means that most of its people live in abject poverty. More than a fourth are illiterate, and about half are engaged in agriculture, generally at a primitive level. Bolivia has been called a country of contrasts, contradictions, and frustrations. Yet, it is twice the size of France, with only half the population of New York City, and possesses a relative abundance of natural resources. During brief periods of political stability, it has shown evidence of rapid progress toward modernization.

About 5 percent of the population controls more than 90 percent of Bolivia's wealth and nearly all of its political and military power. The mestizos have made some gain within the military and in small-scale commerce, while the Indians have yet to achieve any real influence. Before 1952, the minority of wealthy landowners and mine owners, like the higher officers of the army, lived in a world apart. The great majority of Bolivians were Indian farmers who remained outside of the modern economic system. Since that date, an effort has been made to improve the living conditions of these people and to bring them into the national economy.

Bolivia's immediate economic problem is its $3 billion foreign debt and $400 million balance of payments deficit. The petroleum-based boom is now over, and Bolivia's oil production has declined steadily. A persistent problem has been the high cost of tin production, since the country's "hard rock" mines cannot compete economically with the open-pit mining of alluvial deposits in Southeast Asia. Consequently Bolivia dropped from second place in world tin production in 1977 to fourth place in 1980. Vast iron and manganese deposits are known to exist at Mutún, in easternmost Bolivia, but exploitation awaits major capital investments, regional development, and access to external markets.

Only about half of Bolivia's 25,000 square miles of fertile land has been exploited, and agricultural production has not expanded rapidly. Subsistence agriculture prevails in the Altiplano. Coffee is the leading commercial crop in the Yungas and is produced in sufficient quantities to supply the domestic market. Amazonian Bolivia is devoted to cattle-raising and the gathering of products from the tropical forest, to the extent that it is utilized at all. Most agriculture of a modern character has been developed in the area around Santa Cruz, where corn, sorghum, sugarcane, cotton, and cattle are raised in relative abundance. Agricultural productivity, plus revenues from petroleum and natural gas sales, make this the most dynamic region within the Bolivian nation.

Experts on agricultural and energy policy have criticized the neglect of agricultural development which, for example, requires Bolivia to import 250,000 tons of wheat per year. The nation cannot feed itself despite vast stretches of arable land, and it is thus forced to import $100 million worth of food products annually. Meanwhile, a major generator of foreign exchange is the illicit trade in cocaine, with a value estimated at $600 million per year.

The Political Situation

Government in Bolivia had long been dominated by the military forces. The three large tin mining corporations once held the political power, and policy decisions were made in Paris or New York, not in La Paz. The army kept order in Bolivia and did not hesitate to use force when needed, but the war with Paraguay dealt the army a humiliating defeat. Soldiers recruited among the Indians not only experienced the shock of losing a war, but also for the first time ventured beyond the limited horizons of their ancestral communities. After the war, many drifted to the cities or the mines, where they became second-class citizens. The national election of 1951 revealed the nature of the problem: out of a population of 3.5 million, fewer than 200,000 were permitted to vote.

In 1952 a revolutionary party seized control of the government. The leader was Víctor Paz Estenssoro, a member of the landowning class and an intellecutal with a strong sense of social idealism and nationalism. The party was the *Movimiento Nacionalista Revolucionario* (MNR), which issued a statement of objectives similar to those of APRA in Peru. The MNR had the support of most politically active people: the intellectuals, students, city workers, and miners. Opposed were the landowners, army officers, and some of the illiterate tenant farmers who had little idea of what issues were involved.

The first step was expropriation of the tin mines, in 1952. The decree was enormously popular because many Bolivians felt that the corporations were exploiting national resources without adequate return investments. The Patiño interests produced 60 to 65 percent of all tin exports, by value, and the other two, combined, 10 or 15 percent. The housing and medical services provided by the mine owners could not outweigh the well-known fact that almost everyone employed in the mines was in debt to the company stores.

In 1978 the country had gone 11 years without an elected President. When elections were finally held, there was an outcry of fraud, and again there was a military coup. After years of jockeying by various candidates, Wálter Guevara became interim President for one year, and this provoked yet another coup. A short time later, the Congress appointed Lydia Gueiler as the first woman President in Bolivia's history. She, too, was toppled by the military and was replaced by General Luis García Meza.

It has been properly stated that "Bolivia has lived under the permanent threat of a *coup d'état* led by forces of the ultra-right, the agroindustrial oligarchy, the mine-owners, the transnational corporations and the military and civilian sectors linked to drug and arms smuggling."[3] Within a three-year period from 1978, there were three general elections, four revolutions, five interim governments, about 80 political parties, and more than 1000 strikes. In 1980 the 195th *coup d'état* occurred within the 155-year history of the Bolivian Republic, indicating that the average political administration remains in power for a period considerably less than one year. Political stability, which is a key to social and economic development in any country, is therefore conspicuously absent.

Internally, the short-term goal of the Bolivian government is to bring about economic growth and political stability. In a longer term perspective, the hope is to achieve a reasonable standard of living for all Bolivians and their integration into a modern industrial society. However, the latter will require a heavy investment of external capital.

In terms of foreign relations, greatest emphasis has been placed on two major objectives: advantageous economic integration into the group of Andean Pact nations, and recovery of an outlet

---

[3] "Bolivia: Return to Democracy Frustrated by the Military," *Comercio Exterior de México* 26, No. 8 (August 1980).

to the Pacific Ocean. Bolivia and Ecuador have been granted special advantages within the regional alliance, yet neither has achieved major gains toward industrialization. An outlet to the Pacific depends on negotiations with the government of Chile, which has thus far insisted that a grant of land to Bolivia for such an outlet must be matched by a grant to Chile of comparable Bolivian territory. The international port of Arica, or a spot nearby, is where the Bolivian government wishes to establish its own sovereign port.[4] It is understood, however, that both Chile and Peru must agree to any such plan since much of what is now northern Chile was acquired from Peru and Bolivia in the War of the Pacific. Meanwhile, Bolivia remains one of but two land-locked countries in all of Latin America.

[4] R. B. St. John, ''Hacia el Mar: Bolivia's Quest for a Pacific Port,'' *Inter-American Economic Affairs* 31, No. 3 (1977): 73.

CHILE

25

REPÚBLICA DE CHILE

Land area   292,257 square miles

Population   Estimate (1985) 12,000,000
Latest census (1982) 11,275,440

Capital city   Santiago 4,225,300

Percent urban   83

Birthrate per 1000   24

Death rate per 1000   6

Infant mortality rate   23.6

Percent of population under 15 years of age   32

Annual percent of increase   1.8

Percent literate   95.6

Percent labor force in agriculture   9

Gross national product per capita $1870

Unit of currency   Chilean Peso

Physical Quality of Life Index (PQLI)   79

COMMERCE (expressed in percentage of values)

Exports

copper	47	meat and fish meal fodder	7
fruits and vegetables	8	vanadium and molybdenum ores	7
		wood pulp	6

Exports to		Imports from	
United States	21	United States	24
Japan	12	Venezuela	7
West Germany	11	Brazil	7
Brazil	8	Japan	6
United Kingdom	5	West Germany	6

Data mainly from the 1985 World Population Data Sheet of the Population Reference Bureau, the 1985 Britannica Book of the Year, and the 1985 World Almanac.

Chile is like no other country in the world. It stretches 2630 miles from Arica in the north to Cape Horn at the southernmost tip of the South American continent. Yet, at no place is it as much as 250 miles wide. Beyond its traditional boundaries, claims have been made over extensive areas of the ocean and a pie-shaped sector of Antarctica. Island possessions in the South Pacific include Rapa Nui (Easter Island), Sala y Gómez, San Félix, San Ambrosio, and Juan Fernández.

The core of the Chilean state is composed of just one area of concentrated settlement. Southern Chile is one of the rainiest parts of South America, where glaciers descend from snow-covered mountains to a deeply fiorded coast. In contrast, northern Chile is one of the driest places on earth. It has one of the few weather stations in the world where no rainfall has ever been recorded. These two extremes are Chilean only in the sense of possession. The real Chile is the land between: a narrow strip between high mountains and the sea; a land of crops and pastures, bordered by graceful rows of Lombardy poplars, eucalyptus, or weeping willows; a land of dense population.

The geographical unity of Chile's one central cluster of people gave this country a distinct advantage in the development of a coherent society. This advantage was supported by a degree of racial homogeneity that is uncommon among the countries of Latin America. As elsewhere, social distinctions appeared between the minority of landowning aristocrats and the majority of landless tenants, known as *inquilinos* or *peones*. Yet, this social gap was bridged in Chile by the strong paternalistic attitude of the landowners, who tended to live on their estates and participate in the life of the rural communities. In this respect, the development of Chile was notably different from that of most other countries.

Even in Chile, the traditional landowning elite gradually lost influence to a rising middle class whose wealth derived from the expansion of mining, business, and industry. The landowners had lost much of their political power by 1925, when a new constitution provided for a highly centralized national government and made Santiago the unrivaled center of influence. By the 1960s they had begun to lose their properties, as a result of land reform. Rapid urbanization hastened the process. The population of Chile was more than 67 percent urban by 1960, a figure that increased to 83 percent by 1985. From 1971 to 1973, under the regime of President Salvador Allende, more than a fourth of the total agricultural land was expropriated.

The People

Chile is a mestizo country, with none of the profound racial diversities found in the countries farther north. Only about 5 percent of the people are pure-blooded Indians. People of Spanish descent whose ancestry remains unmixed total approximately 25 percent, and 66 percent are mixtures of Spanish and Indian. This mestizo group is more Spanish than are the mestizo groups of Peru, Bolivia, and Ecuador. In the more prominent families of Chile there have been no Indian ancestors for many generations.

The Indians

The Chilean Indians were not like the Quechua and Aymara who were dominated by Inca rule. At the time of the Spanish conquest there were perhaps 500,000 natives in the middle part of Chile, between present-day Valparaíso and Puerto Montt. In the forests south of the Río Bío-Bío, the warlike Araucanians held the Incas at bay, using the woods to good advantage in fighting armies whose previous experience had been in open country. The Araucanians achieved a culture level like that of the Iroquois of North America: they were hunters and fishers, but they derived much of their food supply from a shifting cultivation of maize.

Araucanian tribes occupying the more open country north of the Río Bío-Bío were brought under the influence of the Incas. The Incas extended their conquest south to the Río Maule, but since the road to Cuzco was long and difficult, their hold on this remote frontier was weak. Just inland from Valparaíso lies a basin drained by the Río Aconcagua in which the Incas established their southernmost settlement. Slightly farther south, near the present city of Santiago, the Araucanians had either been taught or discovered how to construct simple irrigation ditches and raise crops repeatedly on the same land. Thus established, they formed a kind of "march" to protect the Inca settlements to the north from attacks by the still warlike tribes from the southern forests. These sedentary Araucanians, rather than the more primitive tribes farther south, made the greatest contribution to the racial composition of the Chilean mestizo.

The Spanish Conquest and Race Mixture

The first Spanish explorers to penetrate the formidable barriers that isolated middle Chile were motivated by the same desires that led Pizarro to conquer Peru, but Chile had little to offer. Its stream gravels contained little gold, and its Indians were much less docile than those of Peru. When Pedro de Valdivia tried to organize an expedition, he found few volunteers ready to face the hardships of the overland march to a land of reported poverty. Nevertheless, with a small army, he reached middle Chile and founded Santiago in 1541 and Concepción in 1550. Almost at once the newcomers were involved in warfare with the Araucanians, a situation which continued with few interruuptions for 300 years. Despite the absence of precious metals and the warlike character of the Indians, the Spaniards were attracted to middle Chile because of its productive soils and pleasant climate. The surface of middle Chile was soon marked off in huge agricultural or pastoral estates, the ownership of which created a new aristocracy.

Race mixture was as free and without prejudice in middle Chile as in other parts of Latin America. There were few white women in Chile during the early years of the Spanish conquest, and each soldier was soon attended by several native women. It is said that there were four women to every man in the frontier posts established to protect the settlements against Araucanian raids. During the year 1580, there were 60 children born in one week at a post where 160 men were stationed. The territory was soon swarming with mestizo children produced by the mating of two exceedingly virile and enterprising racial types.

Social differences soon began to appear in this new society. The contrast between landowners and the landless was increasing, despite the fact that no racial differences separated the two classes. Among mestizo children, girls had the better chance to marry well; boys usually were less fortunate. Indian and Spanish blood, however, mixed freely whether among the landowning aristocracy or among the tenants who became increasingly attached to the large estates.

Concentration of People in Middle Chile

The central region, focusing on Santiago, is what might be termed the "cradle of Chilean nationality." In the Aconcagua Valley, and in the Central Valley between Santiago and Concepción, a relatively dense population has developed. In the rural areas around Santiago, the density is well over 250 per square mile. Santiago itself is a city of more than 4 million; Valparaíso, together with Viña del Mar, has a population exceeding 1 million; Concepción's metropolitan area includes about 750,000.

About three-fourths of the total population is in middle Chile, the area between Coquimbo and the Río Bío-Bío near Concepción. If to this core area is added the northernmost part of southern Chile, between the Río Bío-Bío and Puerto Montt, about 90 percent of the total population is included. Concentrated in middle Chile are the preeminant activities of Chilean life; the political, economic, social, and artistic center of the nation is Santiago.

In 1974 the Chilean government established a new administrative division of the country to encourage local and regional development. This plan called for 12 official "regions", plus a special metropolitan district for greater Santiago. Each region is comprised of a number of provinces, and each province has its respective "*comunas*" (counties).

Middle Chile

The core of the Chilean nation is complex in its internal arrangement. Although this small ribbon of territory has a natural unity because of its climate, the surface features divide it into various smaller units.

Surface Features

From one-third to one-half of the width of middle Chile is occupied by high ranges of the Andes. South of about latitude 27°S the Andes narrow to one dominant cordillera, on the crest of which stand some of the highest peaks of the continent. At the head of the Río Aconcagua, in Argentina, Mount Aconcagua reaches an elevation of 22,834 feet, the highest peak in the Western Hemisphere. As far as latitude 33°S, all the passes over the cordillera are above 10,000 feet. South of Talca a string of active volcanoes stands prominently to the west of the main range. The summits of the higher mountains are snow-covered; the lower slopes are densely forested; and between the trees and permanent snow the zone of alpine pastures becomes narrower toward the south, until it disappears entirely south of about 40°S.

Another third of the width of middle Chile is occupied by a zone of coastal plateaus and terraces. The highest elevations are reached in the north, where the flat-topped surfaces stand some 7000 feet above sea level. In the south, near Concepción, the highest elevations are between 1000 and 2000 feet. The plateaus are deeply dissected by small streams that have cut steep-sided, V-shaped ravines and are interrupted by larger valleys where the Andean streams flow through to the ocean. Cliffs rise abruptly along most of the

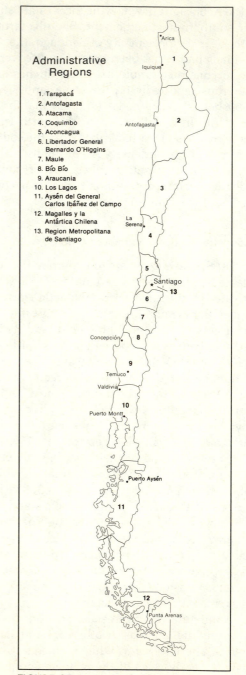

Administrative Regions

1. Tarapacá
2. Antofagasta
3. Atacama
4. Coquimbo
5. Aconcagua
6. Libertador General Bernardo O'Higgins
7. Maule
8. Bío Bío
9. Araucania
10. Los Lagos
11. Aysén del General Carlos Ibáñez del Campo
12. Magalles y la Antártica Chilena
13. Region Metropolitana de Santiago

FIGURE 25-1

THE MICROWAVE TOWER IN DOWNTOWN SANTIAGO ENHANCES CHILE'S TELEPHONE SERVICE.

coast, and there are no harbors except where promontories give partial protection from the prevailing southerly winds.

Between the Andes and the coastal plateaus is a structural depression of varying width known as the Central Valley. Streams that descend from the Andes cross this depression at right angles and plunge through canyons in the plateaus to reach the sea. Only along the southernmost of the rivers, the Río Bío-Bío, does a broad, flat-floored valley extend all the way to the Pacific. The Central Valley is divided by spurs of the Andes into individual basins, in some cases completely separated from others to the north and south. Between Santiago and Concepción, these basins are continuous. The Aconcagua Valley is isolated by a mountain spur from the valley of the Río Mapocho, in which Santiago is located,

and between the Aconcagua Valley and the beginning of the desert at Coquimbo, the spurs of the Andes extend so far westward that the basins of the Central Valley are pinched out. The floors of the basins that form the Central Valley are not level, since the Andean streams have built great alluvial fans sloping from east to west.

Climate and Vegetation

The most distinctive feature of middle Chile is its climate. Between Coquimbo at latitude 30°S and Concepción south of latitude 36°S, there is a transition between the desert north and the continuously rainy lands to the south. This is a climate of hot, dry summers and cool, moist winters, to which the term *mediterranean* is commonly applied. A similar type of climate is found on the west coasts of all the continents between 30° and 40° of latitude.

The temperatures are never extreme. At Valparaíso the coldest months (June and July) average 53°F, whereas the warmest month (January) averages 64° — temperatures slightly lower than those of San Diego, California. In the Central Valley, where protection from the chilling effect of the sea is just about compensated by increased elevation, averages are similar to those of the coast, although the extremes are somewhat greater. Temperatures at Santiago average 46°F in the coldest month and 69° in the warmest month. Frost is sometimes experienced in the Central Valley, but snow is a rare occurence.

Rainfall increases steadily from the desert margin toward the south. The length of the summer dry season becomes shorter until, south of the latitude of Concepción, there is no season that is essentially rainless. Throughout middle Chile, however, summer is either entirely dry or receives an average of less than half an inch in the driest month.

The winter rains of this region are produced, as in other mediterranean regions of the world, by the interaction of cold and warm air masses, known as cyclonic storms.[1] The normal air movement over middle Chile in both summer and winter is from the southwest, circulating around the permanent center of high pressure at about latitude 30°S in the South Pacific. Throughout the year this stream of relatively warm, humid, and light air is penetrated periodically by masses of heavy, cold air from the Antarctic. As a cold air mass advances northward, light oceanic air forms a whirl along its front which circulates in a clockwise direction. The advancing cold air, being relatively heavy, forces the whirling air ahead of it to rise, and this causes rain. The ap-

proach of a cold air mass, therefore, is heralded by a shift of wind from the usual southwesterly direction to the northwest and north. Since the cyclonic whirls in winter sometimes develop considerable velocity, and since the storm winds come from a northerly direction, Chilean harbors, protected from the south but open to the north, were dangerous for ships caught in them in the period before breakwaters were built for protection. Because these cold air masses never extend beyond Concepción, in summer, this season in middle Chile is very dry, and because only the strongest winter storms affect the region north of Concepción, rainfall brought at this season decreases toward the north.

A mediterranean evergreen broadleaf woodland appears on slopes of the coastal plateaus a little south of Coquimbo, and becomes denser and more extensive toward the south. Throughout the region, however, there are many areas where this forest is replaced by a growth of low evergreen bushes of a type similar to the *maquis* of Southern Europe and the *chaparral* of California. This typical mediterranean vegetation ends abruptly along the Río Bío-Bío.

The Chilean Hacienda

The qualities of land and people that distinguish this core of the Chilean state are only variations on the common Latin American theme. The institution around which the process of settlement took place was the hacienda. As elsewhere in lands newly conquered by Spain, the Spanish Crown divided the area into private holdings granted to members of the army. The size of grant varied according to the position of the person to whom the grant was made. There were town lots for those who wished to live in the urban centers; small farms for the soldiers of lower rank; and vast estates, measured in square leagues, for officers of higher rank. Large landowners and their descendants assumed their places, in accordance with the Spanish agrarian tradition, as members of the aristocracy, as leaders of the political, economic, and social life.

---

[1] The term *cyclone* should not be confused with the popular word cyclone which refers to a tornado. Meteorologists use this term to refer to any whirling storm. In most parts of the middle latitudes of the world, the alternation of warm and cold air masses is associated with the passage of cyclones.

The census of 1925 reflected four centuries of settlement. In that year middle Chile, outside of the cities, was divided into 82,000 rural properties. Of these, 77,000 were classed as "small," since they were less than 200 hectares in size.[2] Only 5000 properties, about 7 percent of the total number, were classified as haciendas. However, this relatively small number included 89 percent of all farmland in middle Chile. In the Aconcagua Valley 98 percent of all farmland was included in 3 percent of the properties. These figures remind one of the concentratioin of landownership in Mexico before 1910.

Large landowners in Chile, like those in other Spanish-American countries, were interested primarily in raising cattle and horses. Consequently, they used the small area of good land in middle Chile to raise feed crops under irrigation, chiefly alfalfa, oats, clover, and vetch. Animals were driven into high mountain pastures above the tree line in summer, and irrigated feed crops in the Central Valley kept the animals fat during the winter, a characteristic form of *transhumance.* In the Central Valley only about 1 field in 10 in any one year was used to grow food crops. The chief food was, and still is, wheat. Chile is the only country in Latin America in which wheat occupies more area than maize. Bread made from wheat rather than from maize is eaten by all classes. To supplement their diet, workers on the haciendas were permitted to use small plots of land to raise potatoes, beans, peas, lentils, onions, artichokes, and peppers.

Some small areas were also used for vineyards, which were found in Chile from Arica southward to the Río Bío-Bío. Throughout middle Chile they occupied some land on each hacienda, but the major concentration of this land use was in the Aconcagua Valley near Talca. Less than half of the area in vineyards was irrigated;

the vines were usually planted on bordering slopes too steep for other uses.

The *inquilino,* or tenant worker, was the traditional victim of this agricultural system. Tenant families occupied the same haciendas for centuries, and the hacienda owner, the *patrón,* was accepted as the hereditary master. The owner would not think of letting his *inquilinos* starve, or go without legal advice, or do without the services of midwives; but nothing could persuade the owners to improve the miserable standards of living that tradition accorded to the tenants. The *inquilino* family was given a house with a mud floor and thatch roof. There was no provision for cooking, which was done outside even in winter. The houses were not heated. There were no toilet facilities, and the water supply came from contaminated irrigation ditches. It is little wonder that a fourth of all children born in Chile would die within the first year after birth.

Population Pressure

Population pressure occurred in middle Chile as early as 1870. The number of rural people in the area between Coquimbo and Concepción was about 1,425,000 according to the census of 1885; in 1925 the figure was 1,497,000. Meanwhile, the total population of Chile nearly doubled, and after 1925 the number of rural people in middle Chile began to increase rapidly. Something had to change.

Before World War II the increasing population was absorbed in six different ways. First, army recruitment for the war with Peru and Bolivia (1879–83) jarred many *inquilinos* out of their traditional attachment to the haciendas. Second, the rise of the nitrate industry in northern Chile attracted workers, who came in a steady stream from middle Chile. Third, the rise of other mining industries, chiefly copper and coal, absorbed many workers from middle Chile. Fourth, a steady current of emigration carried excess population from middle Chile across the Andes into the Argentine oases to the east. Fifth, many people from rural middle Chile became pioneers

---

[2] One hectare equals 2.47 acres; 200 hectares equal 494 acres. There are 640 acres in a square mile. These "small farms" of about 500 acres are much larger than those considered to be small farms in the United States or in the Old World Mediterranean regions.

who pushed the frontier of settlement southward into the forests of southern Chile. Sixth, and of greatest importance, within middle Chile itself, large industrial cities, whose rise began during World War I, absorbed a major part of the excess population from the rural districts. Partly by expansion, partly by emigration, and partly by internal rearrangement, Chile thus accommodated its growing numbers without increasing the rural population density of the nucleus.

Land Redistribution

By 1950, the pressure of people on the traditional economy was becoming explosive. To be sure, the economy was expanding more rapidly in Chile than in most other Latin American countries. Consumption of electric power per capita was the highest in Latin America, and Chilean landowners were not unwilling to invest money in business. Members of the aristocracy did not lose prestige by managing industries or commercial firms. After 1950 an increasing number of landowners actually divided their properties into small lots and sold them to invest the proceeds in manufacturing. When land was redistributed in this way, there was a decrease in feed crops and an increase in area devoted to orchards — oranges, apples, pears, peaches, and lemons. Most buyers of these small country estates were white-collar workers in the cities.

In 1967 the government passed a new agrarian reform law that made quite drastic provisions for land redistribution. By 1972 some 15 million acres of privately owned land were expropriated and made available in small farms to more than a million landless farmworkers. For many decades, Chile had to import wheat, butter, potatoes, and vegetables because of the failure of its agricultural system to produce these items. Aside from ending the institution of tenantry, a basic purpose of the new law was to increase production of these basic foods. Yet, in every Latin American country where land redistribution programs have been implemented, the first result has been a decrease in food production. In Chile, food

shortages were a primary factor in the overthrow of the Allende regime in 1973.

Northern Chile

Northern Chile is generally considered to be that part of the country north of latitude 30°S, at Coquimbo, which is an arid region known as the Atacama Desert. The Atacama is one of the most distinct natural divisions of South America. Here the two principal groups of people who have occupied the desert have done so in sharply contrasting ways. The Indian settlements, closely attached to sources of water, have remained isolated, static, unchanging; the Europeans, after neglecting the region for centuries, suddenly acquired a strong interest in the desert because of its unique resource, sodium nitrate. Whereas the Indian villages survived many centuries almost unchanged, the Europeans, during the past century, "conquered" the desert in a spectacular wave of settlement and exploitation, only to find that their occupation of the region was precarious. Disputes and conflicts over landownership led to warfare in which Chile was victorious over Peru and Bolivia. More recently, Europeans in the Atacama have had to face the problem of markets, which arises largely from economic and political conditions outside the region and beyond the control of the Chileans.

The Coast

No part of the west coast of South America is more forbidding, more utterly desert-like in aspect, than the stretch of about 600 miles between Arica and Caldera. From the water's edge, the cliffed escarpment of the coastal plateau rises as an unbroken wall 2000 to 3000 feet above the sea. There are no harbors, no protected anchorages. Along the lower slopes of the coastal escarpment are narrow wave-cut terraces, now lifted above the sea by gradual emergence of the land; and on these narrow shelves, clinging with apparent insecurity, for there are severe earth-

FIGURE 25-2

quakes in this region, are such towns as Pisagua, Iquique, Tocopilla, Mejillones, Antofagasta, Taltal and Caldera.

The Desert

The wealth of the Atacama, for a people of European culture, lies back of the coastal plateau. Each port is dependent on its connections with mining districts of the interior, and these connections are maintained by railway, roads, or cable line. The railroads zigzag up the steep escarpment that faces the Pacific, passing over the crest through shallow, dry ravines about 2000 feet above sea level. East of the plateau a series of dry basins, or *bolsones,* 50 miles in width and about 2000 feet in elevation, separate the coastal plateau from the base of the Western Cordillera of the Andes. The *bolsones* are invaded on the east by enormous alluvial fans that spread out from the mouths of Andean valleys and extend into the basins as far as 40 miles west of the mountain front. The lower parts of the *bolsones,* therefore, are near their western sides, near the rim of hills at the crest of the coastal escarpment. These basins were once filled with lakes, which have completely dried, leaving deposits of salts. A series of strata is recognized, one of them containing valuable *caliche* which is composed of sodium chloride, sodium nitrate, and a variety of other substances including iodine salts. The *caliche* varies in thickness from a few inches to many feet, with an average of perhaps one foot.

The Atacama is one of the driest places on earth. For years at a time no rain falls, so that average figures of precipitation are deceptive. Over a period of 20 years, for example, 14 years passed at Iquique without a drop of rain, and during the 6 years in which some rain fell the total was only 1.1 inches. Back of the coastal plateau, at Calama, no rain has ever been recorded.

Although both the coast and interior are rainless, in other ways they are different. The coast has a much higher relative humidity. Iquique averages about 81 percent, while Calama aver-

ages only 48 percent. The coast is usually cloudy, while the interior is almost cloudless. Temperatures on the coast are more uniform than those of the interior, while the interior has a greater diurnal and seasonal range. Under the clear skies in the interior, the rapid loss of heat at night and in winter often brings the temperature close to freezing. Low fog banks then hang over the desert, soaking the surface with dew.

Permanent settlement in a desert is dependent on water, and supplies of surface water in the Atacama are scarce. Tacna is the southernmost of the Peruvian oases, and just accross the border in northern Chile is its twin, Arica. Between Arica and Copiapó only one river, the Río Loa, gathers sufficient volume in its headwaters to flow across the desert. The oasis of Calama is located where the Loa emerges from the Western Cordillera. Immediately downstream from Calama, the Loa enters a deep, narrow canyon from which it emerges to enter the Pacific. In this canyon the water is not available to support oasis settlement. Even the mouth of the Loa is not used as the site of a port, for the gap through the coastal plateau is too narrow to serve as a route to the interior. The first surface water to reach the ocean south of the Loa is the Río Copiapó; the oasis of Copiapó, a ribbon of cultivation almost 90 miles long, is generally regarded as marking the southern limit of the Atacama.

The Atacama as a Region of Transit

Throughout its history until 1825, the Atacama remained an area of sparse population, a barrier with no intrinsic value. The Indians of the Inca Empire wished to cross it from north to south because it separated settlements in the highland basins of Cuzco and Titicaca from the frontier region of middle Chile. The Spaniards wished to cross it from east to west because it lay between the mines of Bolivia and the sea. Each arrived at a workable arrangement that made it possible to cross this inhospitable region with a minimum of hardship.

The Indians established a north-south road

THE SMOOTH, ROLLING ATACAMA DESERT AT IQUIQUE, CHILE.

through Calama and Copiapó and on into middle Chile. To cross the desert between its principal oases, however, required numerous smaller supply stations where food, water, and shelter could be found. The little valley oases of the Western Cordillera served this function. In their mountain valleys they were sheltered from the winds and dust of the desert—minute habitable spots in a barren land which were so well hidden that in some cases lighthouses were built on nearby ridges to guide travelers to them.

The Spanish conquest shifted the emphasis from north-south communications along the Andean front to east-west lines of travel. The Spaniards soon recognized the importance of the Indian settlement of Calama and made it a major focus of communications. Calama established its connections with the Pacific at several places in turn. At first, the chief port was Cobija; later, Mejillones and Antofagasta shared the trade of Calama. None of these ports was easily reached, and none possesses outstanding advantages over the others.

When the Wars of Independence established the existence of Peru, Bolivia, and Chile, the political boundaries in this region were not carefully defined. The important oases came definitely under one or another of these countries, but the exact boundary in the uninhabited country between was not a matter of concern. Tacna was definitely Peruvian; Calama belonged to Bolivia; and the oasis of Copiapó was in Chile.

Rise of the Nitrate Industry

Although a main problem in the Atacama until the early nineteenth century was the maintenance of a route of travel across it, there were some attempts to explore the desert for minerals. Prospectors entered the region from Tacna and from Copiapó, and the first discoveries were silver and copper ores. Between 1832 and 1845 silver mining near Copiapó resulted in the attraction of so many people that they could not all be fed, a situation Charles Darwin noted when he visited the area on his voyage around the world. As late as 1880, the development of silver mines east of Taltal led to the establishment of that port. Copper was found and mined in the Western Cordillera, but copper mining was of little importance except during the years 1850 to 1875.

Prospectors who discovered silver and copper in the mountains bordering the desert also found layers of sodium nitrate under the sands and gravel of the desert floor, but sodium nitrate was only a curiosity until someone found a use for it. According to legend, a German living in Chile in 1809 put some handfuls of sodium nitrate on his garden and was amazed at the luxuriant growth of his plants. For the world's leached or heavily overworked soils, one of the pressing needs was for a fertilizer that would replace the nitrogen. Sodium nitrate from the Atacama began to find markets in the Cotton Belt of the United States, in parts of Europe with soils of low natural fertility, and in Egypt for use on its exhausted soils. In 1831 a shipment of 110 tons of nitrate was sent to England, where it quickly found favor. By 1860 a thriving mining industry had been established in the Atacama. With abundant nitrate just below the desert surface and no competing source of natural nitrate anywhere in the world, an unlimited future in the fertilizer market seemed assured. By 1860 exports exceeded 50,000 tons per year.

The real mining boom in the Atacama began when a use was found for sodium nitrate in the manufacture of smokeless powder. While black powder was in use, sodium nitrate had no value, because this salt has the property of absorbing moisture from the air and going into solution. Smokeless powder came into use in explosives around 1860, and sodium nitrate found a new and rapidly expanding market. By 1895 exports from Chile had gone well beyond the million mark.

When, early in 1914, speculative overproduction of nitrate brought such financial difficulties that the whole economic structure of the industry was threatened, thousands of people had to flee this region, which offered no other means of support. Then came World War I, and exports at times exceeded 3 million tons a year. Workers streamed back to the Atacama, and mining was conducted with renewed confidence. At this time about 65,000 workers were employed in the nitrate operations, while the entire region, including its ports, had a population of 270,000.

Political Changes in the Atacama

Meanwhile, political control of this region was rearranged. Even before 1860 vague boundaries in the Atacama caused disputes over jurisdiction. In 1879 about 59 percent of the capital invested in nitrate mining was Peruvian, whereas 19 percent was Chilean, 14 percent British, and 8 percent German. Some of the richest fields were in Bolivian territory, but the Bolivians had invested little capital and sent few laborers to this remote part of their country. In 1876, the Peruvian government, finding itself in serious financial difficulties, sought to expropriate private nitrate enterprises within its territory and set up a government monopoly. Amid growing confusion, Chile landed troops at Antofagasta, declaring war on both Peru and Bolivia.

The War of the Pacific lasted from 1879 to 1883. Chile won, and its forces even occupied Lima. The Treaty of Ancón (1884) resulted in transfer of the Peruvian section of the Atacama to Chile, reserving for later decision the fate of the oasis of Tacna and its port Arica. Bolivia was deprived of its part of the Atacama and its access to the sea. Chile, in complete control of the nitrate country, further developed the industry and collected expanding revenues, thereby offering the world the rare example in modern times of a profitable war.

From 1884 to 1929 the final disposal of Tacna and Arica remained unsolved. They were beyond the northernmost of the nitrate fields, and ownership was not directly related to the exploitation of minerals.Chile perhaps sought to protect the fields from this nearest likely base of attack; Bolivia, having lost Antofagasta, pressed claims for Arica; and Peru claimed both Tacna and Arica as sacred territory occupied by a conqueror. Finally, the territory was divided between Peru and Chile. Tacna was awarded to Peru, Arica to Chile, and Bolivia was given duty-free entry through Arica and use of the railroad from there to La Paz.

Decline of the Nitrate Industry

During the period of controversy, exports of nitrate brought wealth to Chile, since it controlled the only important source of this mineral. Into the Chilean treasury there poured a golden stream exceeding $30 million a year, and foreign capital investment increased. By 1901, British investments represented 55 percent of the total; Chilean, 15 percent; German, 14 percent; and Spanish, 10 percent. Most of the workers were Chileans.

Chile lost its monopoly and its speculative profits after World War I. In 1928 there was a sudden collapse of the world nitrate market, as the product of new plants extracting nitrogen from the air became available. Since then, Chile has continued to export nitrate, but at a marginal profit. Iodine, formerly considered a byproduct, now has an export value equal to that of nitrates. Most important to Chile, however, is the mining of copper.

The "Conquest" of the Atacama

During these chaotic developments in the Atacama involving warfare, transfer of territory, economic boom, and collapse, Indian villages along the Andean piedmont remained undisturbed. Each community, nestled in its little valley, is essentially an independent unit, producing little for export and buying little from the outside. Water remains the primary need, and the population of each oasis is closely adjusted to the amount available.

Near the villages the valley sides are terraced and irrigated, and on these fields maize is raised for human consumption, and alfalfa for the sheep, llama, and goats. The food supply consists of maize, meat, goat's milk, and cheese. The women make pottery, and spin and weave wool for clothing. Inhabitants of the piedmont towns gain some support from activities outside the immediate locality. They own sheep, llamas, and goats, which they pasture during part of the year on sparse grasses of the higher Andes. In years of heavier rainfall, pasturage there may be so abundant that animals are imported from a distance, but such periods occur infrequently.Of the contrasting kinds of settlement, the unchanging "out-of-the-world" communities of Indians appear to have formed a more permanent working relationship with the land. All mining communities face two ultimate dangers: that the supply of ore will be exhausted, or that technological innovation will make the ore no longer useful.

Present-Day Atacama

Most of the Atacama is as barren as the surface of the moon. Yet, it has been thoroughly tramped over, scratched, and dug. Abandoned diggings and deserted settlements abound where mines and mills were once active. Far from total abandonment, however, the Atacama remains one of the most productive deserts in the world.

Copper is by far the leading export from Chile, and three of the four large-scale mines (Chuquicamata, El Salvador, and Andina) are in the Atacama. Chuquicamata is one of the world's largest open-pit mines and alone accounts for about half of the nation's copper production. Chile possesses the world's largest copper reserves, of which a large part lies within this desert. Nitrate mining continues, along with the production of iodine, and 40 percent of the world's metallic lithium reserves lie in or near the Salar de Atacama. One of the world's highest mines produces sulphur, at 19,700 feet, on the upper slopes of the volcano Aucanquilcha near the Bolivian border northeast of Calama.

Settlements along the Pacific Coast of the Atacama have long depended on the export of minerals from their hinterland. Largest of these is the port city of Antofagasta, which also functions as a fishing, refining, and regional supply center. Arica, under Chilean administration since 1929, was given free-port status in the early 1950s. Soon thereafter, automobile assembly began, largely on the basis of imported parts, some of which continues. Arica, however, is far from national sources of supply and from the primary markets. Hence, this attempt at regional decentralization failed, and most of the automotive industry now centers around Santiago. Arica continues to serve as a frontier post and as a transit point for the international trade of Bolivia.

Large-scale iron ore deposits have been developed near Copiapó, Vallenar, and Coquimbo, and an iron pellet plant has been constructed on the coast at Huasco to supply both domestic and foreign needs. Manganese is one byproduct of these ores, and vanadium is extracted from blast furnace slag at the steel mills. Most of the coastal settlements have benefited from a major expansion of the fishing industry. Chile has now surpassed Peru as the world's leading exporter of fish meal and ranks among the five leading nations in total fish tonnage caught.

Southern Chile

Beyond the Río Bío-Bío, the open landscapes of middle Chile suddenly disappear, and a once-forested terrain comes into view. Instead of irrigated fields on sloping alluvial fans, farms have been created on cleared lands. Instead of whitewashed mud buildings, with thatch or red-tiled roofs, there are frame houses with shingles.

Long, stately rows of Lombardy poplars and graceful weeping willows are entirely lacking. Here, instead of the large estates of a landed aristocracy, medium-sized farms have long prevailed. The contrast between the region north of the Bío-Bío and that south of it is more than superficial; it is a contrast of social systems, a contrast of mental attitudes. For a forest-bred people, this region is potentially as rich as the desert of the north was for a people who best understood the exploitation of mineral wealth.

The northern part of southern Chile, formerly Araucanian territory, was first successfully developed for farming after German colonists arrived in 1840. After the war with Peru and Bolivia, pioneer small farmers entered the forested region, and this movement continued until almost all the arable land was occupied. In addition to farming, coal mines were developed on the Lebú Peninsula just south of Concepción. The Germans remained as large estate holders, but most of the Chilean pioneers were bought out by wealthy families, and the largest estates in Chile, until the agrarian reform, were located in this part of the country.

South of Puerto Montt, the coast of Chile is deeply embayed, like the coast of British Columbia and southern Alaska. This is one of the stormiest parts of the world, where high winds, storm clouds, rain and sleet, and a roaring ocean pound the continent. Except for sheep pastures east of the Andes, some timber enterprises, and the oilfields of Tierra del Fuego, this part of Chile is largely empty country.

Surface Features

The fundamental elements of the land in middle Chile are continued southward. To the east, the Andes dominate the scene; on the west, the coastal plateaus and terraces border the sea, both on the mainland and on Chiloé Island. Between the Andes and the plateaus is the Central Valley. The Andes are not as high in southern Chile as to the north, and passes through them are much lower. During the Ice Age glacial erosion, which

THE RAMUNTCHO CLIFFS MARK THE RAGGED COAST AT THE MOUTH OF THE BÍO-BÍO RIVER, NEAR CONCEPCIÓN.

affected only higher parts of the cordillera at the latitude of Santiago, extended to lower altitudes with increasing distance south of the equator. Just south of Temuco, the valley glaciers at one time emerged from the mountains into the Central Valley in great tongues of ice. They scraped and broadened the mountain valleys into U-shaped troughs, and around their lower ends they piled the coarse gravel torn from the heart of the cordillera into a horseshoe pattern of knobby moraines. Behind each moraine, south of latitude 39°S, is a valley lake, similar in origin to the

famous lakes of the Alpine piedmont in Italy. Above these lakes the Andean range has been heavily sculptured by ice, featuring sharp horns, high cirques, and deep valley troughs. Now, only a few small relict glaciers remain high in the mountains in sheltered spots facing to the south, but the limit of permanent snow is lower than it is near Santiago.

South of latitude 35°S, a row of enormous cone-shaped volcanoes, snow-covered at their summits, stands to the west of the cordillera, helping to make the Chilean lake district one of the most spectacular scenic attractions of the world. Many of the volcanoes remain active, and earthquakes are common. During the present century alone, there were earthquake disasters in 1906, 1928, 1939, and 1960.

Coastal plateaus border the Pacific as in middle Chile. Beyond the gap at Concepción, the coastal range reappears in typical form—that of a dissected plateau—in the Lebú Peninsula. Its maximum elevation, about 5000 feet, occurs south of Valdivia, and along the backbone of Chiloé Island the range is approximately 2600 feet in altitude. Drowned river mouths at Talcahuano and Valdivia provide protected harbors.

The Central Valley also continues into southern Chile. Its eastern margin, however, is no longer smoothed by broad alluvial fans. Instead, its Andean border is festooned with moraines and dotted with marginal lakes. The rivers cross it at right angles, and between them mountain spurs separate the valley into a series of compartments. The valley floor continues southward from the Bío-Bío at about 300 to 400 feet above sea level. It ends abruptly at Puerto Montt, descending to sea level in a series of terraces.

Climate and Forest

Chile and the Pacific Coast of the United States are very similar in climate and landscape. Traveling southward in Chile, or northward in California, one leaves a climate of hot, dry summers and cool, moist winters for one of stormy winters and cool summers. The summers, although less rainy than the winters, are nevertheless not dry. Many days feature gray skies, violent storms, and heavy rainfall. These are the characteristics that distinguish the west coast climates of all continents poleward of 40°. Valdivia has about the same average temperature in its warmest month as Tacoma, Washington (62°F), but its coldest month (45°F) is 7 degrees warmer. Valdivia, however, receives almost three times as much rainfall (105 inches per year) as does Tacoma. Storminess increases toward the higher latitudes, even more so than in North America, making

southern Chile one of the stormiest parts of the world.

The division between a climate with dry summers and one with wet summers is remarkably sharp, just south of the Río Bío-Bío. The vegetation reflects this change, as dense forest replaces the open scrub woodland and brush. South of Osorno, the forest is so dense and wet that it cannot even be burned off. Here, the mild winters and heavy rainfall support an evergreen, broadleaf forest that is more difficult to penetrate than the tropical selva because of the dense growth of underbrush.

First Contacts with the Araucanians

In both the Inca and Spanish colonial periods, the sharp climatic and vegetation boundary along the Río Bío-Bío was reflected in a sharp cultural boundary. Yet, the significance of the physical contrast between the regions separated by the Bío-Bío has been different for each of the peoples who have occupied this part of Chile. The Araucanians knew how to live in the forest and support themselves by hunting game, which they supplemented by incidental farming. They could scarcely have withstood the all-conquering armies of the Incas and the Spaniards without the protection afforded by the forest.

The Spaniards who conquered the Inca Empire and marched across deserts and high plateaus to seek wealth and position in Chile were not the kind of people to stop on the borders of unfamiliar country. Under the leadership of Pedro de Valdivia, these fearless adventurers plunged southward into the heart of Araucanian country. Before 1560 they placed settlements on navigable waters at Arauco, Imperial, Valdivia, and on the island of Chiloé at Castro. They also established settlements inland wherever meadowlands offered the opportunity, including Cañete, Angol, Villarica, and Osorno.

For more than 200 years after the Spanish conquest, southern Chile remained Indian country. The seminomadic Araucanians gradually established fixed settlements as they adopted ideas that crossed the frontier. They practiced a shifting cultivation of maize and potatoes, chiefly in the meadows and glades where the heavy labor of clearing the forest was unnecessary. From the Spaniards they adopted cattle and poultry. Along the frontier and around the isolated Spanish settlements of the coast, warfare and raiding continued until 1885.

Settlement of "Araucania"

The Araucanians, along with the Seminoles of Florida, have the distinction among all Indian tribes of the Americas of never having been conquered by force of arms. They were subdued, little by little through contacts with the rival civilization. Even today this sturdy group of Indians, although incorporated in the Chilean nation, persists in its occupation of a part of the southern region. Approximately 400,000 pure-blooded Indians, mostly of Araucanian ancestry, live in the southcentral area around Temuco.[3] These Indians refer to themselves and are known to others as *Mapuche,* a term meaning "people of the land." The Chilean government has attempted to convert their communal holdings into private farms, but the Indians resist such efforts and continue to own and cultivate their lands in common.

The first serious steps leading to settlement of the interior were taken by foreign immigrants in the decades after 1850. Small groups of Germans, each containing only a few hundred people, were established between 1850 and 1854 near Valdivia, at Puerto Montt, and at Puerto Varas on the shores of Lake Llanquihue. Pioneering was not easy in this wet country where travel through the forest was almost impossible during the winter months, but the German colonists showed what could be done. Until 1864, when the arrival of settlers from Germany practically ceased, the total number that had entered this part of Chile was only 3367. Even today descen-

dants of these settlers form an insignificant proportion of the total population, yet the southern provinces of Osorno, Valdivia, and Llanquihue reflect a strong German influence. The Germans soon became loyal citizens of Chile, and they implanted a distinctly German civilization in the new land—German in the architecture of the buildings, the agricultural methods, the cleanliness, the emphasis on schooling, and social manners and customs through various generations.

While these foreigners led the way, and still play an economic role out of proportion to their numbers, the great flood of pioneer settlers came from middle Chile, recruited largely from the *inquilino* class. After the war with Bolivia and Peru in 1883, the returning army was sent to the southern frontier to finally establish Chilean control of the Araucanians. Chile might have settled many of its soldiers on this frontier, but this was resisted by the ruling class, which sought to assure an abundance of labor in the older parts of Chile. Officially, about 6000 new colonists were settled in the south, and probably many more became squatters on Indian communal lands. The war had broken the strong traditional ties of the *inquilino* with the hacienda. The population of provinces carved from the old frontier was nearly 2 million in 1960. Relatively few upper-class Chileans were involved in this movement. Hence, Chilean literature, which is almost exclusively the product of the upper classes, shows little of the effect of the frontier. Yet, the migration of people from middle Chile, continued over the decades, has been one of the most potent forces in the transformation of Chilean life.

The farms of southern Chile are devoted chiefly to livestock and, to a lesser degree, the production of food crops. A large proportion of the cattle in Chile are raised on pastures south of the Bío-Bío. Most of these animals are used for meat rather than for dairy products, although the climate is ideally suited to dairying. Owing to the absence of pastures in the Andes, transhumance is not practiced in southern Chile, but there is some movement of herds through the relatively

---

[3] "Chile," *Background Notes,* Publication 7998, (Washington, D.C.: U.S. Department of State, 1983).

PERU

BOLIVIA

Tacna

Arica

Iquique

PACIFIC
OCEAN

Tocopilla

Chuquicamata

Calama

Mejillones

Antofagasta

24°

Chañaral

Caldera

Copiapó

Huasco

La Serena
Coquimbo

Ovalle

ARGENTINA

30°

Illapel

San
Felipe

Valparaíso

Los Andes

San Antonio

Santiago

Rancagua

San Fernando

Curicó

Talca

Linares

Cauquenes

Chillán

72°

18°

Pan American Highway

Talcahuano

Concepción

Chillán

Lebu

Los
Angeles

Angol

Temuco

Villarrica

Valdivia

Osorno

Puerto
Montt

San Carlos de Bariloche

Ancud

Castro

ISLA DE
CHILOE

42°

ARGENTINA

Puerto Aysén

48°

Populated places

● over 100,000

● 25,000 to 100,000

● under 25,000

ATLANTIC
OCEAN

Strait of Magellan

Punta Arenas

Porvenir

54°

Ushuaia

72°

66°

0   200 km

0   200 mi

FIGURE 25-3

low passes to the oases of northern Patagonia in Argentina, especially Neuquén.

Less than 20 percent of the total area of southern Chile is devoted to food crops. On this area, however, is produced an important share of Chile's wheat crop, mostly soft wheat adapted to the rainy climate. Most of the wheat lands are just north of Temuco. Other crops include sugar beets, potatoes, oats, apples, and hay. There is also an abundance of trees, some of which help to support the rapid expansion of forestry.

The nation's extensive forest resources are concentrated in the southcentral section, from Santiago southward to the Province of Llanquihue. Past exploitation has been limited, but in recent years forestry has become one of the most dynamic sectors of the economy. In 1981 the value of forest-product exports exceeded that of agricultural exports, the primary items shipped being wood pulp, sawn timber, newsprint, and logs. Places such as Valdivia have become centers for both lumbering and the manufacture of furniture.

Settlement of Chiloé

The island of Chiloé occupies a geographical position similar to that of Vancouver Island in North America and is part of the frontier region of southern Chile, but settlement of this island has been more difficult than that of the mainland. It is a chilly, foggy place with evergreen forests so dense that clearing the land for plowing has proved most difficult. The island's 120,000 people engage mainly in ranching, lumbering, and fishing, but sugar beets, potatoes, wheat, and fruit are grown. The fine stands of evergreens and hardwood provide material for a small boatbuilding industry.

The Far South

The remainder of southern Chile, southward from the island of Chiloé, comprises a third of the total national territory, yet is occupied by barely 1 percent of the population. Chile's "Far South" is a region of high winds and heavy rains, steep rocky slopes, and storm-tossed waters. Glaciation in the Andes, during the Ice Age, was increasingly vigorous toward the south. The glacial troughs inland from Valdivia are small compared with the deep excavations made farther to the south, which completely cross the Andes, so that many streams draining to the Pacific have their headwaters on the eastern side of the mountains. These troughs are drowned along the coast, forming an amazingly intricate pattern of channels and islands. Between Puerto Montt and latitude 44°S, structural conditions similar to those in middle Chile continue southward, partly under water. Farther southward, the Central Valley disappears, and there is only a labyrinth of fiords with steep rocky margins and a dense tangle of forest wherever trees can fix their roots. This landscape is almost constantly shrouded in cloud and driving rain, where the outermost islands are shaken by the pounding surf as the world's stormiest ocean dashes against a continent.

The snow line descends rapidly toward the south. On the volcano Osorno, northeast of Puerto Montt, it is about 5000 feet above sea level. In Tierra del Fuego the zone of permanent snow is at only 2300 feet. Many mountains are permanently ice-capped, and at several places glaciers still discharge icebergs into the coastal fiords.

It would be difficult to find a more unpleasant climate. There are few hours of sunshine, few hours when the wind is not blowing. Rainfall exceeds 200 inches per year in places. Snow and sleet are common throughout the winter. Only protected places, such as Punta Arenas, near the eastern border of the mountains on the Strait of Magellan, offer somewhat better conditions. There, the average temperature of the warmest month is 53°F, and the annual rainfall is only 19 inches. Some of the world's most primitive tribes once occupied this southern coastland of Chile, and a few of their descendants remain.

Pastures and Oil Fields

Tierra del Fuego and Chilean Patagonia were discovered by the Portuguese Ferdinand Magellan in 1520. Despite this early discovery, the southernmost part of South America was largely neglected until the nineteenth century. As long as ships were moved by sails, the Strait of Magellan was not very useful, but when steamships were developed it became possible to navigate the narrow waters and to sail against the wind to the west. Punta Arenas was founded in 1847 because Chile wanted to maintain its claim to the Strait. A short distance inland a small coalfield was discovered, and Punta Arenas became an important coaling station on a major shipping route. After the opening of the Panama Canal, the Strait was visited only by local coasting ships or tourist ships. When these were converted to oil, Punta Arenas declined in importance.

Meanwhile, it became apparent that the climate was ideal for wool and mutton production. Sheep were brought in through Punta Arenas in 1878 and pastured on the unfenced short-grass steppe along the eastern front of the Andes. They did well in this area, producing good-quality wool and mutton. Punta Arenas became a port of call for ships exporting these products. Today, the city is a major regional trade center for all of southernmost Chile, with a population of about 65,000.

In 1945 oil and gas were discovered on Tierra del Fuego across the Strait of Magellan from Punta Arenas. Further exploration has enlarged the known reserves, and pipelines have been built to the ports. Oil and gas also have been discovered north of the Strait and under it, to the east of Punta Arenas. The oil is sent by tankers to refineries in middle Chile. The gas is liquified at plants on Tierra del Fuego and is sent northward to supply the cities between Santiago and Concepción. Now, an agreement has been made with Argentina to build a gas pipeline from Argentina's Neuquén Province to Santiago.

Thus, in the past several decades, this windswept land has been developed. In 1980 about 60 percent of Chile's 12 million barrels of oil came from Tierra del Fuego, and in the following year Chile was able to supply about 45 percent of its domestic consumption. Settlements in the oilfields and at the shipping ports are modern and comfortable, with shops, theaters, recreational facilities, and automobiles. Although agriculture is little developed, experiments have shown that vegetables such as potatoes, lettuce, cabbage, cauliflower, beets, turnips, and carrots do well under local soil and climatic conditions.

The Mines of Copper, Iron, and Coal

Punta Arenas is 1200 miles by ship from middle Chile. Even its rapid development in recent years does not mean that any large number of people could emigrate from Chile's core area to the far south. Neither the north of Chile nor the south can provide an outlet for pioneer settlement, or do more than provide some additional support for the national economy through the production of oil, copper, nitrate, chemical fertilizers, and additional minerals. Yet, these products are critically important to a country in which mining generates 60 percent of the total foreign exchange.

The Copper Mines

The Andes of Chile contain vast amounts of low-grade copper ore. In the nineteenth century, when copper mining began, only high-grade ores could be used. However, improvements in technology during the first decades of the twentieth century made possible the use of low-grade ores, found in such enormous quantity that the huge investment needed to exploit them is amply justified.

The northernmost of the Chilean ore bodies is at and around Chuquicamata, where mining began in 1915 by the Anaconda Copper Company. The large mines of Chuquicamata, El Teniente, El Salvador, and Andina account for about 85 percent of annual output and are under control of the state *Corporación Nacional del Cobre de Chile* (CODELCO). These ore bodies are mined largely by open-pit methods. A notable exception is El Teniente, southeast of Santiago, the world's largest underground copper mine.

Significant problems for Chile include a widely fluctuating demand for copper on the world market and the progressive exhaustion of high-grade ores. Chile, however, remains the world's leading copper producer. Byproducts of copper have shown substantial growth in recent years, as has direct production of gold, silver, and molybdenum. Other byproducts, such as uranium and selenium, offer significant potential for the future.

Iron Ore, Coal, and Steel

Chile also possesses the raw materials for a steel industry. The mining of iron began in 1908 at El Tofo, five miles inland from Cruz Grande in northern Chile. From 1922 to 1952 the Bethlehem Steel Company, which then owned El Tofo, extracted 50 million tons of ore. El Tofo is now almost exhausted, but since 1956 iron ore mining has moved to newly discovered sites. Iron ore has been found in three locations: (1) a body of high-quality ore in the high mountains east of Antofagasta near the Argentine border; (2) a belt parallel to the coast, about 50 miles inland, from southeast of La Serena to just north of Chañaral; and (3) in southern Chile, south of Concepción, where there is a large amount of low-grade (30 percent) iron ore. Next to copper, iron ore is Chile's most valuable mineral, although it is a distant second. The industry employs only about 5000 workers. Chile is the third largest producer of iron ore in South America, and it exports about 85 percent of the total production to Japan.

Most of Chile's coal production is concentrated in the central part of the country. About two-thirds comes from the Lota mines just south of Concepción and slightly southward, on the Lebú Peninsula, is an area of low-grade coal. A further 25 percent is located in the Osorno-Chiloé area. A coal gasification plant is planned for

Talcahuano to reduce domestic petroleum consumption, and it may satisfy all of the needs for electricity in Concepción Province when completed.

Chile's only large iron and steel mill is located at Huachipato, near Concepción. It has two blast furnaces and is operated by the *Compañía de Acero del Pacífico* (CAP), a state-owned corporation that also operates the nation's largest iron mines. About 60 percent of the plant's needs for coal are supplied from domestic sources nearby, the remainder being imported from the United States and Canada. CAP also operates a smaller plant at Rancagua, south of Santiago.

Like most other countries in Latin America that are heavily dependent on petroleum, Chile must plan to use alternate surces of energy. Hydroelectric potential is abundant and far from being fully utilized. Most of this form of energy is generated at sites in middle Chile, near the primary centers of consumption. Chile has abundant uranium reserves and expects to have a nuclear plant in operation soon. Chile's geothermal resources are considerable, but there has been little development in this sector. However, a new geothermal power station is expected to supply all power requirements of the Chuquicamata copper mine.

The Modern Industrial Cities

By 1985, 83 percent of Chile's population lived in cities. Largest of the conurbations is Santiago, which is the home of a third of all Chileans. In addition to its function as national capital, it is the industrial, commercial, and cultural hub of the country.

Santiago was founded in 1541 on a typical rectangular plan. As the focus of political life in the thriving new colony, Santiago soon established its position as Chile's largest and most attractive city. Since 1865, when the city had 115,000 inhabitants, it has grown enormously and reached a population exceeding 4 million by 1985. The original rectangular nucleus is surrounded by a wide zone of industrial suburbs and *poblaciones* (shantytowns), where the city has expanded into the countryside. Santiago lies adjacent to some of the nation's best farmlands, while the towering Andes border the city to the east. Colonial buildings are intermingled with modern structures, none of which are skyscrapers because Chile is part of an extensive earthquake zone and the height of buildings must be carefully controlled.

Like most other primate cities of the world, Santiago reflects the best and worst features that society has to offer. Among its attractions are a spectacular scenic location, parks, playgrounds, and museums. Urban renewal has provided a pleasant urban mall in the heart of the commercial district, and Santiago possesses one of the finest subway systems in the world. The nation's three largest universities are located in this city, as are most government offices, banks, insurance companies, and industries. Manufacturing is a major source of employment, including textiles, foodstuffs and beverages, pharmaceuticals, electrical goods, rubber and plastic products, metalware, wood products and furniture, and printing and publishing. Its markets overflow with seafood, fruit, and vegetables. Such attractions stimulate intensive rural-to-urban migration so that problems such as unemployment, traffic congestion, pollution, inadequate housing and social services are never satisfactorily resolved.

Chile's second city is the seaport of Valparaíso, near the mouth of the Río Aconcagua. A highway and railroad descend to the coast through a narrow valley cut through the coastal plateau, and reaches the sea at the beautiful resort city of Viña del Mar, immediately north of the port. Valparaíso itself is built on a north-facing bay which provides shelter from the southerly winds. The business district is built on a narrow terrace near the edge of the water. Above it are residential areas, and still higher hilltop slums overlook the entire scene. A breakwater gives the bay protection from the northern storm winds, making it possible for ships to tie up at the docks. Neighboring Viña del Mar is one of the

FIGURE 25-4

A SERPENTINE SECTION OF THE TRANS-ANDEAN HIGHWAY IN CHILE.

most famous resort cities in Latin America. With its extensive beaches, race tracks, and casino, it serves as the hub of the so-called Chilean Riviera. Together, Valparaíso and Viña del Mar have a population well in excess of 1 million. Chile's third-largest metropolitan center is Concepción, with the adjoining steel city of Huachipato and ports of Talcahuano and San Vicente. This is the focus of the nation's heavy industry. Its population exceeds 700,000.

There are numerous other large and rapidly growing cities. These include Antofagasta (205,000) in the north and Valdivia (110,000) toward the south. Valdivia is a center for fishing, forestry, and cattle-raising. Its factories produce textiles, wood products, leather goods, and foodstuffs. Among the residents of Valdivia and Concepción are numerous descendants of German immigrants; Temuco has the largest Indian population. Puerto Montt is a tourism center in the Chilean lake district and serves as a transportation gateway to Argentina and the far south of Chile.

Chile as a Political Unit

Chile is one of the more privileged nations in Latin America. It has an abundance of mineral, forest, marine, and agricultural resources in relation to its modest total population. The Chileans are racially and culturally homogeneous and enjoy a relatively high standard of living. There is a well-developed infrastructure of transportation, communication, and electrical power, and an advanced educational system. Traditionally, the government has been highly centralized in Santiago, but increasing attention is now being given to regional development.

Chile has some tradition as a democratic nation, but during recent decades drastic changes in political orientation have taken place. In 1970 a Marxist government came to power, with characteristic emphasis on land redistribution, expropriation of private enterprise, and centralization of political power. Since the military coup of 1973, such trends have been abruptly reversed. The ultimate objective of most Chileans is economic prosperity and political democracy. Hence, the primary challenge in the years ahead will be the achievement of these objectives, in which the degree of wisdom exercised by the nation's military forces will be a critical issue in Chile's long-term welfare.

The Economic Situation

Chile's economy is in a state of transition, perhaps more so than that of any other country in Latin America.[4] Under the current administration, improved economic efficiency has been given highest priority. Tariff barriers have been reduced, the tax structure has been changed, and land expropriation has ended. By mid-1978 most of the enterprises taken over by the Allende government were returned to private ownership, and individual titles had been issued to about 37,000 peasant farmers. Programs to reach the needy, while eliminating subsidies to the middle- and upper-income groups, were initiated and promoted. Employment in government positions was sharply reduced.

Adherence to the principles of free enterprise, unprotected by government interference and subsidies, helped to achieve rapid economic recovery after near-bankruptcy of the Chilean economy under the Allende regime.[5] It also resulted in Chile's withdrawal from the so-called Andean Pact in 1976. The most negative aspects of current policies have been in the areas of human rights, including the legitimate right of dissent.

A basic economic principle of the current regime is that Chile must promote and develop those economic activities in which it has some natural competitive advantage. Thus, a massive reforestation program has been undertaken which is expected to quadruple Chile's long-fiber resources by the late 1990s. Another initiative will move the country toward a privately oriented social security system. A continuing problem is that is has one of the world's highest burdens of foreign debt.

The Chilean economy has been, and remains, highly dependent on world copper prices. The quality of this ore is declining, however, and heavy emphasis has been placed on the need for export diversification. Foreign investment laws have been liberalized, and Chile has become one of the most attractive countries in the world for transnational corporations.[6] Major efforts are also being made to develop coal, hydroelectric, and thermal power to make the nation less dependent on foreign energy resources. Lines of communication are being expanded to all parts of the national territory, including new highway construction to further develop the southern frontier between Puerto Montt and Puerto Aysén and onward toward Tierra del Fuego.

Agricultural productivity has been a continuing problem, despite a relative abundance of arable land. Wheat and barley are well suited to the climate of middle Chile, yet these grains must be imported on a large scale. Although Chile is Latin America's only significant producer of sugar beets, it cannot supply its own needs for sugar. Products in which Chile does excel relate especially to horticulture (grapes and wine, apples, peaches, pears, cherries, plums, and vegetables), plus fisheries and forestry.

---

[4] Chile: An Economy in Transition, *A World Bank Country Study* (Washington, D.C.: World Bank, 1980).

[5] *Wall Street Journal,* August 15, 1980.

[6] *Wall Street Journal,* January 28, 1980.

Few countries of the world are more dependent on internatonal trade than is Chile. During the Great Depression of 1929–32 Chile was hit hardest, since nitrate and copper prices on the world market bottomed out and these commodities constituted 80 percent of the nation's foreign trade. The results included high unemployment, poverty, and political instability. Fifteen political administrations, both civilian and military, came to power within a period of 18 months. During the early 1980s Chile was hurt by the worst economic recession in the past 50 years, but political stability was maintained by a military dictatorship that seemed capable of dealing with the crisis.

The Political Situation

In the years before 1920, and again from 1932 to 1972, Chile experienced a high degree of institutional stability and political pluralism. The rural population was very poor, but the paternalistic nature of the hacienda did not make poverty so burdensome as to cause a popular revolution. Furthermore, the War of the Pacific generated a heightened sense of nationalism, and after the war, pressures in middle Chile were relieved by the existence of the southern frontier of pioneer settlement. However, there was a growing demand for land reform and a steady trend toward state direction of national development. The movement of *inquilinos* into the cities and the rise of an urban-industrial complex have brought new pressures to bear on the pre-industrial institutions and have inevitably led to unsettled conditions.

The election of President Salvador Allende in 1970 was accompanied by a manifesto promising state control over mining, banking, credit, and external trade, along with agrarian reform and expansion of central planning. Large wage increases were granted without a corresponding increase in productivity. Nationwide food shortages occurred, as did a rampant black market trade in consumer goods. A breakdown of the Chilean economy was a quick result, followed by a military coup in 1973.

The incoming military regime, under General Augusto Pinochet, suspended all activity by political parties and labor unions. Planning, economic development, and the nation's educational system were decentralized. Price inflation was reduced from an annual rate exceeding 500 percent in 1973 to less than 10 percent in 1981. The government divested itself of control over industry, banking, commerce, and transportation, and essentially eliminated most forms of tariff protection. It assured political stability by institutionalizing military rule until at least 1997, at the cost of personal liberties and traditional democratic processes.

External affairs have not been harmonious for Chile. Since the War of the Pacific, relations with Peru and Bolivia have ranged from tense to hostile. The same is true concerning relations with Argentina, with which Chile shares an international boundary that in places is ill-defined. In the far south, the ownership of three small islands in the Beagle Channel was long contested, and these two nations have maintained widely overlapping claims to territory in Antarctica. Curiously, the area Chile claims in Antarctica exceeds its total national territory on the continent of South America. Perhaps more than any other Latin American nation, Chile is externally oriented and is likely to remain so in at least the immediate future.

PARAGUAY

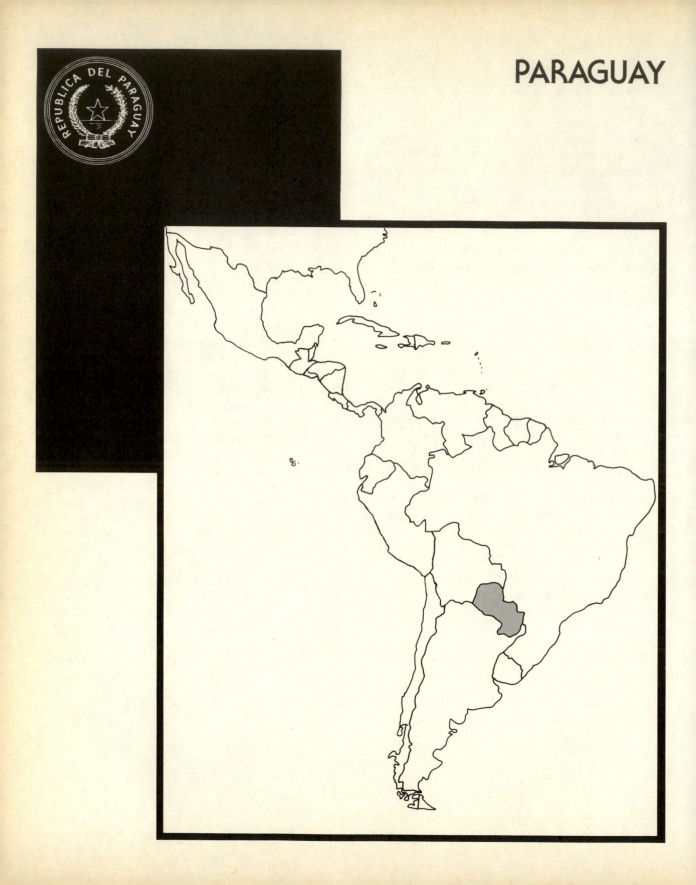

26

REPÚBLICA DEL PARAGUAY

Land area   157,047 square miles

Population   Estimate (1985) 3,600,000
Latest census (1982) 3,015,670

Capital city   Asunción 780,000

Percent urban   39

Birthrate per 1000   35

Death rate per 1000   7

Infant mortality rate   45

Percent of population under 15 years of age   42

Annual percent of increase   2.8

Percent literate   86

Percent labor force in agriculture   44

Gross national product per capita $1410

Unit of currency   Guarani

Physical Quality of Life Index (PQLI)   75

COMMERCE (expressed in percentage of values)

Exports

cotton	33
soybeans	31
timber	7
animal fodder	5
tung oil	5
tobacco	4
cowhide	3

Exports to		Imports from	
Brazil	21	Brazil	28
Netherlands	14	Argentina	19
West Germany	12	Algeria	14
Argentina	12	West Germany	7
United States	9		

Data mainly from the 1985 World Population Data Sheet of the Population Reference Bureau, the 1985 Britannica Book of the Year, and the 1985 World Almanac.

Paraguay is one of the least known countries of Latin America. About the size of California, it is located in the interior of the South American continent and is land-locked except for an outlet to the Atlantic Ocean via the Paraguay-Paraná-Plata River system. Although once the center of Spanish exploration and settlement in eastern South America, Paraguay has been a hermit nation during much of its history since independence. Its traditional isolation appears likely to end within the near future as the country develops a transportation system, its agricultural and forestry potential, and its hydroelectric resources.

The Paraguayan state is geographically uncomplicated. It is composed of one nucleus of clustered population, and its boundaries are drawn through sparsely occupied territory. The one cluster of people, moreover, is racially homogeneous. Yet Paraguay, like the other nations of Latin America, illustrates in its own peculiar way the dominant theme of diversity. Not only are the Paraguayan people quite different in physical appearance and racial composition from other population groups of the region, but within the country there are profound differences between the minority of landowners, army officers, and government officials on the one hand, and the majority of landless workers on the other. The Paraguayans are a gentle people who seem anything but warlike and belligerent, yet the destiny of this inland state has been warped by two disastrous wars within a century. Now, Paraguay has entered a new era, one in which the internal characteristics of the nation will change and in which foreign relations will be of increasing significance.

The People

The Paraguayans are mostly mestizo. Some 97 percent have at least one Spanish ancestor, but the predominant racial strain is Guaraní Indian. Only 3 percent are pure Indian, and people of pure European ancestry are even fewer. Some are the descendants of Germans and other Europeans who settled in Paraguay after 1870. There are very few blacks and Orientals.

The Indians of Paraguay belong to the linguistic family known as Tupi-Guaraní. This group was at one time concentrated in the Basin of the Río Paraguay and from there spread over a large part of South America east of the Andes. The Guaraní had even invaded the Quechua country in the Front Ranges of eastern Bolivia and established themselves there before the Spanish conquest. They comprised most of the coastal people of Brazil, and tribes that speak this language still inhabit the interior of the Amazon Basin. The Tupi-Guaraní tribes practiced a shifting cultivation of maize and manioc, supplementing their diet with fish and game.

The Spanish Conquest

The establishment of a Spanish primary settlement center on the eastern side of South America had quite different results from those that followed the founding of Lima. In 1536 an expedition under Pedro de Mendoza landed on the shore of the Plata and established a settlement called Buenos Aires. To the sixteenth-century Spaniards the Argentine grassy plains were low in potential productivity. Despite its strategic location near the mouth of the Río de la Plata, Buenos Aires lacked the qualifications to become a primary settlement center, so, shortly after its foundation, the colony was abandoned.

Meanwhile, the Spaniards had pushed far up the Paraná, hoping to find a short route to Peru. In 1537 they advanced far enough upstream to get beyond the savage Pampa tribes, and on the first high ground that bordered the river within Guaraní country, they founded the town of Asunción. The town was isolated, but in a territory that has since become the nucleus of Paraguay and a primary settlement center. Lacking any source of wealth comparable with that which made Lima one of the wonders of the sixteenth-century world, this inland town nevertheless became a settlement center from which the Spanish occupation of southeastern South America radiated. Colonists advanced northwestward across the Chaco to found the town of Santa Cruz not far from the eastern base of the Andes. Settlements also spread eastward, and the final successful establishment of Spanish colonies on the margins of the Argentine Pampas was accomplished by people who descended the Paraguay-Paraná-Plata from Asunción. Santa Fé was one of these colonies, founded in 1573, and in 1580 the site of Buenos Aires was reoccupied.

Present Population

Few Spanish reinforcements came to Paraguay after the early expeditions of the sixteenth century. As a result, the Guaraní contribution to the Paraguayan mixture is relatively large, not only of blood, but also of language and ways of living. The Guaraní language is still common in Paraguay, and many place names throughout this part of South America, including the southern part of Brazil, are Tupi or Guaraní words.

A third element of the Paraguayan population is composed of European immigrants who entered the country after 1870. The number is small, for most of the Europeans who came to South America during this period settled in Argentina or Brazil. A few Italians, French, Spaniards, English, and Germans found their way to Asunción, however, and intermarried with the Paraguayans. At present, the influence of this group in the country's economic, political, and social life is of much greater importance than their numbers would suggest. In 1936 an experimental pioneer colony of Japanese farmers was settled about 80 miles southeast of Asunción. After World War II, several other Japanese colonies were established, most of them just north of Encarnación. Until recently, the Japanese have mixed but little with their Paraguayan or European neighbors. Internal migration has been extensive in recent decades and is mostly rural-to-rural in nature, rather than from rural areas to the primate city. More than 20,000 immigrants entered the country from 1960 to 1970. Of these, 22

percent were Argentine, 17 percent Brazilian, 15 percent Japanese, and 3 percent Korean.

Most of the population of Paraguay is concentrated in a pie-shaped triangle extending about 100 miles eastward from Asunción toward Puerto Presidente Stroessner and southeastward toward Encarnación, in each case about half the distance to the Río Paraná. Within this area the land is densely occupied and characterized by minifundia, a myriad of tiny subsistence plots. The soil is approaching exhaustion from lack of fertilizers, and the surplus population forms a stream of migration to Asunción and the eastern border region. About 97 percent of the total population occupies that part of the country lying east of the Paraguay River, whereas the remaining 3 percent inhabits the vast Chaco area to the west.

The Land

Paraguay possesses many natural advantages. The climate is mild, yet provides the stimulating effects of changeable weather. There is an adequate supply of good soil for agricultural use, although only 4 percent of the nation's total land area is under cultivation. Forests and grasslands are extensive. The country is neither monotonously flat, except for parts of the Chaco, nor sharply separated into mountains and plains. The eastern third of Paraguay is an elevated plateau varying from 1000 to 2000 feet in altitude. This is the western part of the great Paraná Plateau, a land composed of successive flows of dark-colored lava interbedded with layers of red sandstone. From central Río Grande do Sul state in southern Brazil, the southern and western edge of the plateau forms a commanding scarp or cuesta, cliffed at the top. The scarp continues northward across Paraguay and the Brazilian state of Mato Grosso. The Paraná, which flows southward through the center of the plateau, forms the great Guaíra Falls where the northern border of Paraguay reaches the river. From the falls to where the Paraná emerges from the pla-

FIGURE 26-1

teau near Posadas and Encarnación, the river occupies a deep valley cut through the lava flows. This valley forms the eastern border of Paraguay, and much of it will be flooded by the construction of three huge dams on the Río Paraná.

From the edge of the Paraná Plateau westward to the Río Paraguay there are low plains, generally inundated at high water, interrupted in two places by "peninsulas" of crystalline hills. The ancient crystalline rocks that form the basement complex of the Brazilian highlands are deeply buried beneath the lava flows of the Paraná Plateau, but they emerge to the west to pro-

duce a country strikingly different from the plateau. Instead of the sharp features of the plateau, with its tabular profiles and steep-sided canyons, the crystalline uplands to the west are gently rounded and of relatively slight relief. One of the peninsulas of crystalline hills extends westward to the Río Paraguay north of Concepción, even forming a few isolated mounds west of the river. The other reaches the left bank of the Paraguay at the site of Asunción. The central area of concentrated settlement in Paraguay is located on the belt of crystalline hills between Encarnación and Asunción.

The remainder of the territory east of the Río Paraguay is composed of a lowland plain, much of it subject to annual floods. Between the two peninsulas of crystalline hills two great bays of lowland extend far to the east of the river. One of these lies south of Concepción; the other forms a triangle in the extreme southwest, bordered to the northeast by the hilly country of central Paraguay, and on the other two sides by the Paraguay and Paraná rivers. Along the immediate riverbanks are natural levees sufficiently elevated to stand above all but the highest floods. Back of these narrow strips are marshes that are filled with stagnant water during the rainy season and are only partially dry during the remainder of the year.

West of the Río Paraguay lies that vast alluvial plain known as the Gran Chaco, a region composed of unconsolidated sands and clays eroded from the Andes. The Río Pilcomayo, which forms the northern border of Argentina, and the Río Bermejo cross this plain with winding, shifting courses. Their braided channels change after each flood season, leaving an intricate pattern of crescent-shaped swamps and abandoned levees. The remainder of the Chaco is an almost featureless plain sloping gently from the base of the Andes to the Río Paraguay.

Vegetation and Climate

The natural vegetation of Paraguay shows a general correspondence with climatic conditions but conforms to the pattern of surface and soil. Since rainfall is greatest on the Paraná Plateau and diminishes toward the west, the vegetation is most dense in the east and thins westward. Asunción has an average annual rainfall of 53 inches, which in these latitudes is moderate. Semideciduous forests, reflecting an abundance of rain, cover the eastern part of Paraguay. They are composed of tall broadleaf trees, some evergreen and some deciduous. The forest is dense in the moist valleys of the plateau and thinner on the red sandy soils of the crystalline hilly belts. The

lava soils, too, support a relatively dense forest, but sandy soils on the plateau, where red sandstones come to the surface, are marked by patches of scrub woodland and palm.

Between the semideciduous forest and the Río Paraguay, the vegetation is mostly savanna. In the wet spots there are areas covered with tall, coarse grass, but the savannas generally are mixed with scattered palms. Each tributary stream is bordered by a dense galeria forest.

Along most of its course the Río Paraguay is a major vegetation boundary. To the west are deciduous scrub woodlands. In a zone along the river these woodlands grow luxuriantly, and among the species of scrubby trees is the quebracho (*Quebrachia lorentzii*), a tree valuable as a source of tannin. As rainfall decreases toward the west, however, the woodland becomes increasingly xerophytic. There are thickets of thorny, deciduous trees and brush, interrupted in places where the soil is sandy by irregular openings covered with coarse savanna grasses. Over much of the Chaco there are no surface streams, but the water table is only a few feet below the surface. During the long dry season patches of alkali appear, and in many places the groundwater is salty.

These features of surface, soil, drainage, and vegetation are combined in a land near the margins of the low latitudes. Temperatures in both summer and winter vary considerably with the passing weather. Cold air masses from the south alternate with warm air from the north. During the passage of a cold front, temperatures may drop as much as 30°F within a half hour.

The Course of Settlement

After a nucleus of Spanish colonization was established around Asunción, the penetration of eastern Paraguay began in 1608 with the arrival of Jesuit missionaries. Scattered, migratory tribes of Guaraní were gathered around the missions and taught a sedentary way of living. Thirty-two Jesuit missions were established in Paraguay east of the Río Paraguay. For the Indians, the new life based on farming, cattle-raising, and the collection of forest products meant a more adequate and varied diet and greater security from famine. Unfortunately, the Jesuits could not maintain their isolation. They began to produce goods for sale outside of their small communities, eventually even selling wines and tobacco in distant settlements along the eastern front of the Andes. This economic expansion brought the Jesuits into conflict with the large landowners who wanted both the profits of commercial enterprise and the labor of the Indians. In 1767 the Jesuits were expelled, and the mission communities disintegrated. Missions in outlying sections of the country were entirely abandoned as the Indians drifted toward the central area around Asunción.

Independence and War

In 1810, the first independent Argentine government was established in Buenos Aires, and jurisdiction was claimed over the entire area included in the viceroyalty of La Plata. The Paraguayans, however, had no desire to be ruled from Buenos Aires. Since the only connection with Spain was downstream via Buenos Aires, the Paraguayans soon had to set up their own independent government.

A major concern of the Paraguayan rulers was to gain an outlet to the sea that would free them from Argentine control. They were negotiating with Uruguay for such an outlet in 1864 when Brazil intervened to protect its southern boundary. The Paraguayans then crossed the arm of Argentine territory east and south of the Río Paraná, and they invaded southern Brazil. This brought Brazil, Uruguay, and Argentina into an alliance against them and started a war that lasted from 1865 to 1870. It took five years for the armies of the Triple Alliance to defeat the Paraguayans, but at the end of the war Paraguay was devastated. The prewar Paraguayan population of about 1.3 million was reduced to less than 250,000, of whom fewer than 30,000 were males.

From this crushing disaster Paraguay struggled slowly back, aided by the European immigrants who brought new hope to a tired people. By 1912 the population reached about a million, and the ratio between the sexes was almost normal again. Economic development was slow however, largely because of high transportation costs. In 1913 a railroad was completed from Buenos Aires to Asunción, using ferries across the Paraná near Buenos Aires and again near Posadas. Yet, the volume of traffic was so low that high freight rates were required. Before World War II, the cost of shipping a cargo from Buenos Aires to Asunción was about equal to that of shipping the same cargo to Yokohama.

The Río Paraná has never offered an easy outlet for Paraguay. Its channel winds about to such a degree that many miles of travel are required to cover a short direct distance. The main channel touches the base of higher ground not subject to flood at only a few places, such as Rosario and Santa Fé in Argentina, as does that of the Río Paraguay at Asunción. Modern, ocean-going vessels can ascend the Paraná only to Santa Fé, and they encounter much difficulty above Rosario. Despite the fact that the Paraguay-Paraná-Plata is such a poor river system for navigation, until the present century it provided the only connection between Asunción and the outside world.

The Eastern Border Region

Except for the Jesuit missions prior to 1767, the eastern border region of Paraguay was occupied only by dense forest. The defeat of Paraguay in the War of the Triple Alliance (1865–70), however, was followed by the extensive sale of public land in this region to provide the government with badly needed revenue. This led, in turn, to the expansion of latifundia and the exploitation of yerba maté (Paraguay tea). An Anglo-Argentine company, *La Industrial Paraguaya*, became the largest landholder in eastern Paraguay and

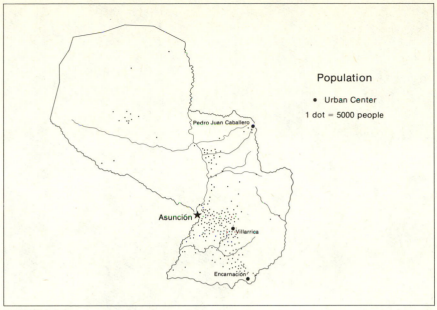

Population

- Urban Center
1 dot = 5000 people

FIGURE 26-2

the largest single employer in the entire country. In 1911 this company employed 5000 workers in its *yerba maté*, logging, and cattle-ranching operations, and owned 17 percent of all Paraguayan territory east of the Río Paraguay. Other European, Argentine, and Brazilian companies also bought land in Paraguay and engaged in similar types of activities, using the river system to export their products via Argentina. After 1940, the demand for *yerba maté* declined rapidly, and much of the land was abandoned.

Since 1960 the entire eastern border region, from Pedro Juan Caballero to Encarnación, has been the scene of rapid transformation. A primary task was to transfer the surplus agricultural population to the eastern departments of the country and to establish colonies of foreign immigrants in the area. In part, this action also was designed to help maintain Paraguay's sovereignty along its eastern border. Of particular concern is the growing wave of Brazilian settlers who migrated westward from Paraná and Mato Grosso states since 1960, buying land, clearing the forests, and planting commercial crops such

as soybeans.[1] Land on the Paraguayan side of the border is far less expensive than in Brazil, and the soybeans are transported by truck to the Brazilian port of Paranaguá where products from Paraguay can be exported duty-free.

The population of Paraguay's easternmost departments increased dramatically after 1970 as a result of internal migration and colonization by foreigners. In 1972 no less than 43 percent of the population in the newly created Department of Canendiyú, bordering the Río Paraná, was composed of Brazilians. Facilitating such settlement is a growing transportation network, including a paved highway from Asunción and the central core area to Puerto Presidente Stroessner and a bridge over the Río Paraná at that location. Accompanying the influx of population is a massive exploitation of natural resources, much of it destructive in nature. Particularly alarming is the

---

[1] R. Andrew Nickson, "Brazilian Colonization of the Eastern Border Region of Paraguay," *Journal of Latin American Studies* 13, No. 1 (1981): 111–131.

THE ACARAY DAM IS THE SITE OF A HYDROELECTRIC PROJECT IN PARAGUAY.

wholesale destruction of the region's extensive hardwood forests.

Hydroelectric Development

The vast hydroelectric potential of the Río Paraná has been contemplated for many years but only recently developed. Expansion of the Acaray dam project, just north of Puerto Presidente Stroessner, increased the power output from 94,000 to 240,000 kilowatts by 1976. Meanwhile, Brazil and Paraguay, signed the Treaty of Itaipú in Brasília in 1973 which provided for the construction of the world's largest dam and hydroelectric complex.

Work began on Itaipú Dam in 1975, just 10 miles upstream from Puerto Presidente Stroessner on the Río Paraná. This dam is almost a mile long, 650 feet high, and six times larger than the Aswan high dam of Egypt in terms of generating capacity. More than 40,000 workers were employed at peak periods of construction. The dam was inaugurated in 1983, and its full power capacity of 12.6 million kilowatts was scheduled for 1990, multiplying Paraguay's power capacity 40 times and making this country the world's largest exporter of electricity. The dam is owned jointly by Brazil and Paraguay, and Paraguay's share of the construction cost is to be paid by selling to Brazil most of its half of the electricity generated. Long-term prospects include the expansion of electric service throughout Paraguay, a major expansion of industry, continued large-scale export of electricity, and the growth of tourism.

By treaty with Argentina in 1975 another joint venture was undertaken on the Río Paraná, 200 miles downstream from Itaipú, near Encarnación. Here, the Yacyretá-Apipé Dam will be constructed to generate 5.4 million kilowatts of electric power, and to facilitate navigation, flood control, and irrigation. It could provide a fifth of Argentina's total electric power by 1990 and will further enhance Paraguay's economic growth and its role as an exporter of electricity. Yet another giant Argentine-Paraguayan project is at Corpus, midway between Itaipú and Yacyretá-Apipé. It is expected to add 5 million kilowatts to the total power production from the Río Paraná and to contribute further to regional development. No longer will the Río Paraná flow uncontrolled through largely empty country.

The Chaco

Most Paraguayans are concentrated on the eastern side of the Río Paraguay. The Chaco lies west of the river. This great alluvial plain between the river and the base of the Andes bears a striking resemblance to the Ganges Valley of northern India. The climate of the two regions is similar; the scrub forests of each represent the same general category of natural vegetation; and both regions are traversed by great sprawling rivers. Whereas the Ganges Valley is densely populated (more than 1000 rice- and wheat-growing farmers per square mile), the Chaco is one of the larger areas of sparse population in Latin America. By 1900 about 90 percent of the Chaco had been acquired by foreigners.[2] Among the owners, for a short time, was the North American Percival Farquar, who had taken possession of about 14 million acres, which amounted to one-

fifth of the Chaco. The Paraguayans were not pleased by this "takeover," but the truth is that none of those who came to exploit the Chaco introduced any important changes in the area. The Chaco is divided among four states: Argentina, Paraguay, Bolivia, and Brazil; yet, only a small portion of it is within the effective national territory of any of them.

This is the wilderness over which the Paraguayans and Bolivians fought a war beginning in 1932. Each government set forth abundantly documented legal arguments to support its claims to the territory north of the Argentine and west of the Brazilian border. Between 1926 and 1931 the Paraguayans established about 35 villages of Canadian Mennonites some 125 miles west of the Río Paraguay on land that the Bolivians also claimed. Meanwhile, the Bolivians placed army detachments far to the east in territory the Paraguayans claimed. The Bolivians were led to believe that the extension of political territory to the banks of the Paraguay would in some miraculous fashion solve their problem of isolation. The Paraguayans were no doubt motivated in part by the hope that oil would be discovered not only along the Andean front, but also in the plains to the east. From 1932 to 1935 the two countries were locked in a death struggle. Each of them, already burdened with debt, contracted new loans to pay the huge cost of armaments. The war ended in 1935 when both sides were literally exhausted, but with Paraguay the victor. A new boundary was drawn along the approximate line of battle at the time of the armistice. It included a considerable gain of territory for Paraguay but left the known oilfields as part of Bolivia.

Paraguay as a Political Unit

Paraguay, like Bolivia and Uruguay, constitutes a "buffer state" in the political geography of South America. Surrounded by larger nations, it has served as a battleground and has had to defend

---

[2] J. H. Williams, "Paraguay's Unchanging Chaco," *Américas* 34, No. 4 (1982): 14–19.

its national territory with tenacity. In modern times, military power must be replaced by international cooperation as the nation seeks to preserve its territorial integrity. Such conditions help to support a centralized government, and in this regard Paraguay is a classic example.

The Economic Situation

In 1985 Paraguay had a population of 3.6 million, concentrated in the hilly belt between Asunción and Encarnación. About 61 percent of the people are rural, and almost half of the working force is employed in agriculture. The estimated rate of growth is 2.8 percent per year, but Paraguay's wars have reduced the population drastically. Not more than 3 percent of the Paraguayans own land, and many of the farmers are squatters on public or private property. Land has always been abundant in Paraguay in relation to the total population; however, because little prestige is attached to landownership, squatter occupancy is accepted as normal. Most of the farmers raise subsistence crops of maize and manioc, with lesser acreages of rice, beans, oranges, and vegetables.

Paraguay has taken the initiative to improve its economy. In 1954 the only Latin American country with a lower gross national product per capita than Paraguay was Haiti, and in 1984 the GNP per capita of $1610 placed Paraguay in fifteenth place. One step in moving the economy forward has been the construction of all-weather highways. It is now possible to travel by automobile at any time of the year from Asunción to southern Brazil or to northeastern Argentina. A bridge over the Río Paraguay at Villa Hayes connects Asunción and all of eastern Paraguay with the Trans-Chaco Highway, which extends from the bridge northwestward more than 460 miles to the Bolivian border.

A highway system has enhanced the viability of outlying farm colonies and facilitated the resettlement of landless farm families from the crowded central area. The Mennonite colonies in the Chaco have expanded, and other colonies have multiplied in eastern Paraguay. The most productive of the Mennonite colonies centers on Filadelfia, a town 270 miles northwest of Asunción, which was connected by road to the central area in 1961. Since that date, the farmers have been able to market their cotton, dairy products, meat, and poultry in Asunción. There are now about 18,000 people in these Mennonite communities, operating farms that cover more than 2 million acres.

Industry is little developed in Paraguay. Such industries as do exist are engaged largely in the processing of raw materials from agriculture and ranching, such as meat, flour, sugar, tobacco, candy, leather goods, and textiles. An oil refinery operates at Asunción, as do factories for the manufacture of clothing, furniture, clay products, soap, matches, beverages, and metalware. Cement production has expanded greatly with the construction of giant dams on the Río Paraná and its tributary, the Acaray. A new steel mill began production in 1982 at Villa Hayes, on the Río Paraguay, using iron ore from Brazil and electricity from Paraguay. With the vast surplus of electric power available, a major expansion of industry is anticipated.

Agriculture, ranching, and forestry remain the primary bases of the Paraguayan economy. Agricultural production has increased steadily, in large part related to rapid settlement of the eastern border region. There, Brazilian immigrants have introduced soybeans, coffee, mint, and upland rice, all produced on a commercial basis. Forestry has likewise expanded, but agriculture and lumbering are based largely on the exploitation of previously unused land and forest resources. An urgent need exists for the employment of modern conservation practices if sustained yields are to be achieved.

Throughout much of its modern history Paraguay has been an exporter of people. Unable to find employment opportunities in their own country, thousands migrated to Buenos Aires and other parts of Argentina. During the 1970s the flow of migration was reversed, as Paraguay experienced one of the highest economic growth rates in Latin America. Economically, few countries of the region had equal cause for optimism.

The Political Situation

The Paraguayans have developed a greater sense of national unity than most of their neighbors. This may be due in part to Paraguay's inland position and the geographic unity of its one nucleus of concentrated settlement. Ever since 1810 Paraguay has sought freedom from outside interference, but this objective has been largely frustrated. Although Paraguay emerged victorious from the Chaco War with Bolivia, the military effort left the country even more impoverished than before. Moreover, the returning soldiers were less ready to accept poverty and restrictions against political expression than they had been before the war.

In 1954, when the government in power sought to nationalize some of the new industries, General Alfredo Stroessner, with backing by the conservatives, took over the presidency. Since that time his authoritarian administration has remained in power. His primary sources of support have been the armed forces and the Colorado Party. Elections have been held on a regular basis every five years, with the result that President Stroessner has been returned to power in 1958, 1963, 1968, 1973, 1978, and 1983.

The Stroessner government has provided Paraguay with more than 30 years of political stability. In the process, freedom of the press has been restricted, civil liberties have been curtailed, and foreign sources have repeatedly charged the government with violations of human rights. In reality, Stroessner has not felt the need to resort to excessive police action. The great majority of Paraguayans are outside the range of political affairs, and most of those who hold political convictions give their support to the dictator. They value personal security and the near-absence of crime, the lack of political turmoil, and the insulation Paraguay experiences from conflicts between the world's major military powers. There is widespread support, too for administration policies designed to bring long-term economic prosperity to the nation, especially with regard to increased employment in construction and industry as a result of harnessing the hydroelectric power of the Río Paraná.

If current expectations are not raised excessively, the Paraguayan people should enjoy continued peace and increasing prosperity. Of considerable concern in the years ahead, however, will be the need for a peaceful transition of power from the present administration to some unidentified successor. Also of concern is the problem which "buffer states" throughout the world experience: how to benefit from the country's association with larger and more powerful neighbors while preserving territorial integrity and political independence.

ARGENTINA

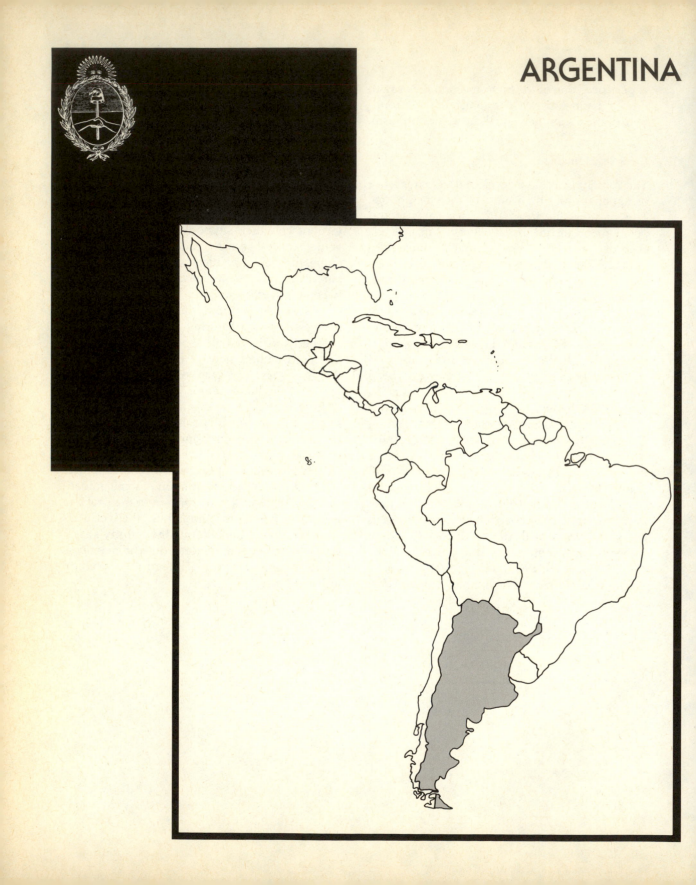

27

REPÚBLICA ARGENTINA

Land area   1,073,394 square miles

Population   Estimate (1985) 30,600,000
Latest census (1980) 27,862,771

Capital city   Buenos Aires 9,766,000

Percent urban   83

Birthrate per 1000   24

Death rate per 1000   8

Infant mortality rate   35.3

Percent of population under 15 years of age   30

Annual percent of increase   1.6

Percent literate   95

Percent labor force in agriculture   19

Gross national product per capita $2030

Unit of currency   Austral

Physical Quality of Life Index (PQLI)   85

COMMERCE (expressed in percentage of values)

Exports

wheat	9	meat and meat products	6
maize	8	animal fodder	6
petroleum products	7	soybeans	5
machinery	7	oils	5

Exports to		Imports from	
U.S.S.R.	21	United States	22
United States	13	Brazil	13
Netherlands	8	West Germany	9
Brazil	7	Japan	8
West Germany	4	Bolivia	7

Data mainly from the 1985 World Population Data Sheet of the Population Reference Bureau, the 1985 Britannica Book of the Year, and the 1985 World Almanac.

Argentina has a reputation of being one of the most advanced nations in Latin America. It is the eighth largest country of the world and, in relation to its population, possesses abundant natural resources. The process of modern economic growth and development in Latin America began in Argentina, and by 1930 this country had the highest gross national product per capita and the highest value of foreign trade. It also had the highest percentage of literacy, the highest rate of urbanization, the largest area of farmland per person, and the smallest proportion of the labor force employed in agriculture. Among the nations of the modern world, only a few produce a surplus of agricultural commodities. Argentina is one of these, and with its favorable ratio of people to land it is virtually assured of a high standard of living for many years into the future. Prosperity depends not only on natural endowment, however, but also on effective organization and management.

Argentina's rapid economic growth before 1930 was aided by a substantial investment of British capital and the export of primary products such as meat, wheat, maize, and flax. When the Great Depression led to a collapse of international trade, however, Argentina was a principal victim. The dominance of primary exports caused Buenos Aires not only to become a great port city, but also to dominate almost every aspect of Argentine life. The nation is composed of contrasting regions, each with its own social, economic, and political characteristics; yet, each region is more closely attached to Buenos Aires than to neighboring parts of the country.

By 1930, Argentina had enjoyed almost 70 years of uninterrupted constitutional government. Its internationally famous newspaper, *La Prensa,* had long stood for equality before the law, widespread public education, free access to knowledge, and open discussion of political issues. Since 1930, Argentina's economic growth has been sporadic, military intervention and dictatorship have been commonplace, and international affairs have been characterized by conflict. The Argentinians are racially homogeneous and

have achieved an advanced economic and cultural status, yet they remain sharply divided on basic issues of public policy. The future of Argentina depends largely on the achievement of national cohesion and constructive, dynamic leadership.

The People

Argentina occupies a vast territory, ranging from sea level to the highest point in the Western Hemisphere. The distribution of population, however, is highly uneven, inasmuch as more than a third live within the metropolitan area of Buenos Aires. Another zone of relatively dense population includes the oasis settlements along the eastern piedmont of the Andes, focusing chiefly on Tucumán, Mendoza, and Córdoba.

The degree to which Argentine national life is concentrated in the immediate hinterland of Buenos Aires is extraordinary, especially considering that this concentration is a product of only the past century. In 1871 the city had just 170,000 people, but an influx of European immigrants brought the population to about 1 million by 1905. Buenos Aires soon became the largest city in Latin America, the largest Spanish-speaking city in the world, and the most elegant city in the Southern Hemisphere.

About two-thirds of the Argentine population resides in the Humid Pampa, which comprises about 22 percent of the national territory, and it is here that Argentina's productive capacity is concentrated. In relation to the concentration of economic activity in this region, the density of rural population is surprisingly low. The other clusters of population in Argentina are quite different from those around Buenos Aires: the oasis settlements of the Andean piedmont represent a much older civilization which came from the north and west, rather than from the east across the plains. Aside from the piedmont oases and some smaller concentrations in the North, the rest of Argentina is sparsely settled. Almost half of the national territory, in fact, is occupied by less than 8 per-

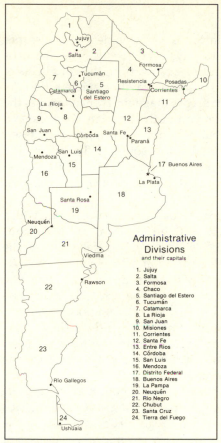

FIGURE 27-1

cent of the people. Over vast areas the population density is scarcely two people per square mile.

Racial Character and Origins

The great majority of the Argentines are of unmixed European descent. In 1930 it was estimated that 74 percent were Argentine-born of European parents, and 24 percent were recent immigrants born in Europe. Since 1930, the current of immigration has decreased sharply. Official estimates place the proportion of people of unmixed European ancestry at 97 percent, but this figure conceals the presence of many mestizos in outlying parts of the country. In the Humid Pampa the population is exclusively Eu-

ropean and is composed mostly of families that arrived in Argentina after 1850. Less than 3 percent of the population is pure Indian, and the proportion of blacks is negligible.

The Argentine territory was originally settled from Peru, Paraguay, or Chile. Asunción was the primary settlement center from which the Spaniards spread over much of the surrounding territory, especially along the Paraguay-Paraná-Plata River system. Meanwhile, northwestern Argentina was occupied by people who came directly or indirectly from the other great Spanish culture center, Lima. The main route of settlement followed the old Inca road to Tucumán. Because the early route to Chile avoided the Atacama and the high Puna country via a long circuit to the east, then crossing the single range of the Andes south of latitude 28°S, this eastern piedmont of the highlands was intimately connected with the settlement of Chile during the sixteenth century. The first Spanish stronghold and center from which other colonies in this region were established was Santiago del Estero, founded by people who returned from middle Chile in 1551 and 1553. From this center other settlements were made at Tucumán (1565), Córdoba (1573), Salta (1582), La Rioja (1591), and Jujuy (1593). The eastern piedmont settlements farther south were also settled by people who came across the mountains from Chile: Mendoza (1561–62); San Juan (1562); and San Luis (1596). A strong current of immigrants from Chile in more recent times has supplemented the population of the Argentine oases from Mendoza to Neuquén, and people of Chilean origin are the principal settlers of the eastern Andean border in southern Patagonia.

All of the early settlements had trouble with the nomadic Indians of the Argentine plains. The Abipones, Puelche, and other tribes of the Pampas and Patagonia, while not numerous, were independent and warlike. They resisted the Spanish invasion as they had that of the Incas. These migratory hunters of the guanaco and rhea could not be tamed for agricultural labor as were the Guaraní. They lacked the shelter of the forests that the Araucanians employed, but they were more than a match for the Spaniards in arid and semiarid areas where knowledge of water sources was of primary importance. The adoption of horses and firearms had much the same effect on these Indians as it had on the Indians of the Great Plains of North America. Greatly increased mobility and capacity to kill wild game was a temporary advantage, but the decreasing number of animals made the Indians even more warlike in their struggle for wider hunting grounds. As late as 1876, it was estimated that 40,000 head of cattle were stolen every year in Indian raids, many being sold in Chile. The Indian days have passed, and only on reservations in the more remote regions can pure Indians be seen today.

During the colonial period the piedmont settlements of the northwest belonged to a different sphere. They were connected economically with the west coast. At Salta a fair was held each year where mules and cattle from the grassy plains were sold to the mining people of the Andes. Meanwhile, the seminomadic *gauchos* of the plains, or residents of the small Plata ports, had little real contact with the inhabitants of the piedmont oases.

The rapid increase of European immigration after the middle of the nineteenth century, the spread of railroads and agriculture, and the growth of Buenos Aires occurred primarily within the Humid Pampa. The result was a new and different kind of population and environment from that of the older piedmont settlements. The latter have been reoriented economically and now definitely come within the influence of Buenos Aires. They maintain little commerce with countries on the other side of the mountains. Tucumán and Mendoza, too, have received a considerable current of immigration, in part composed of people from other Argentine provinces. Today, the long struggle for economic domination between these contrasting portions of Argentina is at an end. Buenos Aires is the focus of everything Argentine.

The Land

Argentina includes a wide variety of landscapes in four major physical divisions. The first division, the Andes, extends from the cordilleras of the dry north to the heavily glaciated, ice-covered mountains of Patagonia. It encompasses the arid southern part of the Bolivian Altiplano and the lower but also dry basin and range desert west of Córdoba and south of Tucumán. The eastern piedmont of the Andes with its succession of oasis settlements is included in this region.

The second major division is the North, which is composed of three principal types of landscape. There is the vast alluvial plain of the Chaco, with its tropical scrub woodland cover. East and south of the Río Paraná is the Argentine Mesopotamia, the land between the rivers (the Paraná and the Uruguay), composed partly of floodplain and partly of gently rolling, well-drained interfluves. In the far northeast is an arm of Argentine territory which extends onto the Paraná Plateau.

The third major division of the country is the Pampas, the great plains lying south of the Chaco and east of the Andean piedmont. Most of the Pampas was originally covered with a growth of low scrubby trees and grasses, a vegetation type known as *monte*. Toward the southeast, however, where the rainfall is heavier and summers remain cool, tall prairie grasses were more common. It is customary to divide the Pampas into a wetter eastern part and a drier western part, designated as the humid Pampa and the Dry Pampa. When the Argentines speak of La Pampa, they are referring to the province of that name which lies mostly within the Dry Pampa.

Patagonia, the region south of the Río Colorado, is the fourth major physical division of Argentina. This is the land of arid, wind-swept plateaus, crossed at wide intervals by strips of green vegetation along the valley bottoms. In the far south of Patagonia, Argentina shares with Chile the land of continuously cool and stormy

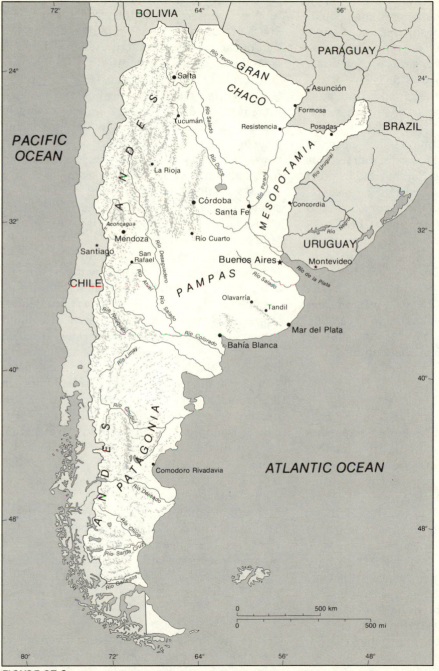

FIGURE 27-2

weather, where winters are never severe but where there is never any summer.

Various currents of settlement brought changes not only in the character of the population, but also in the relationship between people and the land. These settlements have helped to develop and diversify the regions of Argentina. Some of the most distinctive divisions of the country have been given their marked individuality by the selective process of human settlement.

The regions of Argentina as developed by their human occupants all focus on that nucleus of the country, the Humid Pampa, and its great urban center, Buenos aires. Until 1853, however, the Humid Pampa played a subordinate role, and the outlying portions of the country (the Northwest, Mesopotamia, the Chaco, and Patagonia) had most of their connections with foreign countries. The domination of Buenos Aires has been extended gradually to even the most remote places. Contact with this growing urban center meant the possibility of selling in an expanding market, but such contact also led to economic vulnerability in times of financial depression.

The Northwest

No physical change distinguishes southern Bolivia from the Northwest of Argentina. From west to east are: the Altiplano, with its dry intermont basins; the undissected Puna; the eastern margin of the Puna, where the east-flowing streams have cut deep valleys across the zone of the Front Ranges; and, finally, the Chaco. The Puna and Altiplano together in northern Argentina are approximately 250 miles wide and stand between 11,000 and 13,000 feet above sea level. Above the Puna surface are isolated ranges, with some elevations exceeding 19,000 feet. The commanding east-facing front of the Puna is breached in places by broad valleys that offer relatively easy routes of access to the high country beyond. Between the eastern front of the Puna and the western margin of the Chaco is a zone of Front Ranges, composed of parallel cuestas and hogback ridges, separated by roughly north-south structural depressions, or *valles*. At about the latitude of Tucumán this country comes to an end. South of Tucumán the mountain landscape and that of the eastern piedmont are quite different from the landscape farther north.

The natural vegetation is also a southward extension of the types found in Bolivia. The high Puna is covered with a sparse growth of widely spaced xerophytic shrubs. On the east-facing edge of the Puna and on the tops of some prominent Front Ranges is a narrow belt of mountain forest, the *ceja de la montaña*, composed of a dense growth of broadleaf trees. A deciduous scrub woodland covers the Chaco. All these types reach a southern end at approximately the latitude of Tucumán.

Decreasing precipitation along the eastern piedmont is reflected not only by the xerophytic character of the natural vegetation, but also by the type of agriculture. South of Tucumán no crops can be raised along the piedmont without irrigation. From Tucumán northward, irrigated and unirrigated crops are mixed. Some sugarcane near Tucumán is unirrigated, as are some fields of maize in the *valles* around Salta and Jujuy. In this zone there are also many irrigated spots, especially where the alluvial soils are porous. Tucumán is abundantly supplied with water from streams that descend from the Sierra de Aconquija, a commanding range more than 17,000 feet high, which here forms the eastern border of the mountain zone.

South of latitude 27°S, the territory between Tucumán, Mendoza, and Córdoba is dry land, composed of elongated north-south ranges and broad depressions with salt flats or salt lakes along their bottoms. This is a typical basin and range landscape, mostly without exterior drainage.

Colonial Settlement in the Northwest Even before the Spanish conquest, permanent settlements in this part of South America were made by people from the highlands. The Incas established a fortress near Tucumán and maintained control over Indians of the neighboring valleys. Proceeding along the Inca road, the Spaniards, en route to Chile, passed through what is now the Argentine Northwest, and all important towns were established between 1551 and the end of the sixteenth century.

During the Spanish colonial period, these settlements were attached economically to the highlands and the west coast. The Argentine plains produced just one item of importance to the highland centers. It was the mule, an animal that made possible the transportation of goods over rugged mountain trails at very high altitudes. The Argentine plains became the main source of this patient, sturdy, but sterile, offspring of a mare and a donkey.

The mule trade developed distinct routes and centers of commerce. Mule breeding focused on the grassy basins between Córdoba, Rosario, and Santa Fé. In this district there were seminomadic herders of cattle and mules who, like the *llaneros* of Venezuela, established no permanent settlements. These were the picturesque *gauchos*. Cattle were valued chiefly for their hides, but the demand for young mules was apparently unlimited. An annual fair was held at Salta, where three-year-old mules were traded for silver from Peru and Bolivia, or for expensive products imported through Lima and Panama from Spain. During the colonial period Salta, through its annual fair, was the principal commercial center of this entire region. Here, as many as 60,000 mules were traded annually.

Salta and Jujuy in the Modern Era   Salta (260,000) is no longer the focus of economic life in northern Argentina but maintains the appearance of a colonial town with its many historic buildings and monuments. Activities center around the cattle industry, lumbering, farming, and mining. Its mineral wealth includes copper, iron, and petroleum. Salta Province is in the process of industrialization, with electric energy from the Cabra

Corral dam and reservoir. Borax, lead, brimstone, and oil from Campo Durán contribute to the economy of this regional trade center, as do the nearby sugarcane fields.

Jujuy is a small (60,000) provincial capital and serves as a cultural and economic center. Sugarcane, tobacco, citrus, and vegetables are produced within the surrounding area, and the mining of lead, zinc, silver, and iron ore provides additional employment. Only minor commercial contacts are maintained with communities of the *valles*, which produce wheat and maize, and communities of the highlands, which produce wool and salt. Some large estates include many square miles of mountain pasture, still used primarily for the production of low-grade beef animals. In winter, when the lowlands are dry, cattle are driven to the highlands; in the summer they return to lowland pastures. Arable land in the lower areas is used mostly for the food crop maize and the feed crop alfalfa, on which the animals are fattened for local markets.

Tucumán   The growth of Tucumán to a population of 322,000 and the concentration of more than 300 people per square mile in Tucumán Province result from an interesting combination of factors. During the colonial period, Tucumán enjoyed a strategic position between the Argentine plains and Salta. The roads from Rosario, Córdoba, and Santa Fé converged on Tucumán because here the crossing of the dry belt was relatively easy. Two rivers, the Salado and Dulce, provide an ample supply of surface water along this route. North of the Salado no road crossed the Chaco from the Paraná to the Andean piedmont. Tucumán was the center through which almost all of the colonial trade of Argentina passed. It was an important supply point not only for people starting northward toward Salta, but also for those entering the grassy plains to the southeast. In the course of more than two centuries considerable wealth was accumulated by the people of Tucumán, through the manufacture of wagons and harness. The rise of Tucumán as a sugarcane district came late in the nineteenth century. Physical and economic conditions were favorably combined in just this area for the successful planting of sugarcane to supply the Buenos Aires market. Tucumán was hardly within range of the metropolis, and transportation costs were too high for the shipment of sugar from places farther north. Tucumán also lies just within the climatic range of sugarcane, since frosts are too severe for this crop farther south and east.

The Tucumán district possesses several physical advantages for sugarcane cultivation which are not duplicated elsewhere in northwestern Argentina. The zone of Front Ranges ends just north of the city, and the eastern border of the Puna is surmounted, just west of Tucumán, by the high Sierra de Aconquija. No obstruction stands between this range and the warm, moist winds from the east. The result is a zone of abundant rainfall on the eastern slopes, supporting at intermediate altitudes the southernmost end of the tropical semideciduous forest. From this belt of abundant rain and from snowfields at higher elevations, several permanent streams descend to the piedmont. They cross alluvial fans to join the Río Salí, which flows by the site of Tucumán itself. Rainfall at Tucumán totals about 37 inches a year, and between the Río Salí and the base of the Aconquija there is not only an abundance of water in the streams, but also enough precipitation to render irrigation unnecessary in most years. Rainfall is subject to irregularities, however, and irrigation is practiced to guard against drought. East of the Salí there are no tributary streams to provide water for irrigation, and rainfall decreases rapidly away from the mountains.

This island of ample rainfall produced by the slopes of the Sierra de Aconquija is also an island of relatively mild winter temperatures. Cloud banks reduce night radiation so that minimum winter temperatures remain higher than those under clear skies even somewhat farther north. Near the base of the mountains the Tucumán area is frost free, because of rapid air drainage near the tops of the alluvial fans. The frost-free area extends about 35 miles east of the mountain front, providing excellent conditions for citrus, especially lemons, as well as for sugarcane.

Sugarcane Plantations   Sugarcane cultivation requires an abundance of labor. As this industry developed, it was necessary to recruit gangs of laborers in communities such as Santiago del Estero, Catamarca, and even Córdoba. The population of the Tucumán area was thus increased by the arrival of people from other parts of northern Argentina who settled permanently on the cane plantations or near the *ingenios*. Few came directly from foreign countries, and for this reason Tucumán retains a character distinct from that of major cities in Argentina. At present, the Province of Tucumán has a population exceeding 900,000 and a much higher rural density than that of any other part of Argentina.

During recent decades, the focus of Argentine sugar production has shifted northward toward Salta, Jujuy, and the border with Bolivia. The plantations and mills of this district are larger and more modern, rendering those of Tucumán relatively obsolete. Consequently, the economy of Tucumán Province has become somewhat decadent, resulting in high rates of unemployment, political unrest, and attempts at economic diversification. Production of gasahol from sugarcane in Tucumán was begun in 1979, both to use a surplus agricultural commodity and to help the nation achieve self-sufficiency in energy resources. Within the city of Tucumán, new industries such as automotive assembly and newsprint manufacture have been established. Yet, what was once known as the "pampered province," as a result of government subsidies to the sugarcane industry, remains a "problem province" because of prevailing economic conditions.

Mendoza and Other Vineyard Oases   Agriculture changes notably south of Tucumán, where alfalfa rather than maize covers the largest area, and the vineyards assume first place in commercial production. There are three vineyard oases: Mendoza in the middle, San Juan about 100 miles to the north, and San Rafael equidistant to the south. In addition there are smaller irrigated areas around La Rioja and Catamarca, and between San Juan and Tucumán.

Wherever streams tap snowfields in the high Andes and emerge onto the eastern piedmont, oases are established. The oasis of San Juan is supported by the Río San Juan and is a major center of viticulture. For many miles southward no permanent streams break through the eastern rampart of the Andes. Then, two important rivers, the Río Mendoza and the Río Tunuyán, emerge within a short distance of each other and support the large irrigated tract around Mendoza. Still farther south two other rivers, the Diamante and the Atuel, supply the oasis of San Rafael. The region as a whole is one of interior drainage. Streams descend from the mountains over broad alluvial fans which stretch eastward with gradually decreasing slopes. Only large streams persist in their flow across these fans. All unite in a zone of sloughs or playas, from which water only occasionally makes its way to the Río Colorado.

The rural landscape of the vineyard oases is distinctive. Always in the background are the naked, rocky slopes of the easternmost ranges which obstruct the view of higher peaks, such as Aconcagua. In the foreground on irrigated land are straight rows of vines, some festooned on trellises, some pruned low on wires, but all threaded with small irrigation ditches. A large area is devoted to alfalfa, which is used for fattening range cattle. Between the fields and along dusty roads are long rows of tall, slender poplars. Houses are of one-story construction with whitewashed adobe walls and red-tiled roofs.

Mendoza preserves less atmosphere and tradition of colonial Argentina than does Tucumán. A disastrous earthquake in 1861 destroyed the colonial city, and modern Mendoza dates from that catastrophe. The population of Mendoza Province grew rapidly during the early years of the present century, owing largely to immigration from Italy. In 1914 the population included 31 percent who were foreign born. Today, Mendoza is western Argentina's largest city, with a population of almost a half million, and is a center of petroleum exploration and production.

Vineyards, alfalfa, and deciduous fruits are of primary importance in the oases south of Tucumán. Almost 1 million acres of irrigated land are included in the oases of Catamarca, La Rioja, San Juan, Mendoza, and San Rafael, and more than half of this area is devoted to vineyards. The remainder is mostly in alfalfa. The principal concentration of wine manufacture is in the Mendoza area, where 40 percent of the irrigated land is used for grape cultivation.

In addition to the agricultural communities of the northwest, there are a few clusters of people engaged in the exploitation of minerals. Mining is not of great importance in Argentina, but the country does possess a wide variety of mineral deposits and substantial reserves. Only iron ore, coal, petroleum, and natural gas are produced on an economically significant scale, and mineral exports are negligible. Iron ore is mined in Salta and Jujuy provinces to supply a small government-owned steel mill at Zapla, near the city of Jujuy. More than one-fourth of Argentina's crude oil is produced in the Northwest, and local energy resources are supplemented via a natural gas pipeline from southern Bolivia.

Argentine Mesopotamia

In contrast with the mountainous and arid northwest, the northeastern part of Argentina is a land of abundant precipitation, luxuriant forests, and rolling grassy plains. Not only does it differ physically from the Northwest, but it also differs in its economy and historical background. This region is known as the Argentine Mesopotamia.

Mesopotamia, the land between the Paraná and Uruguay rivers, is composed of gently rounded, grass-covered interfluves and swampy, forested valleys. It is a region of hot, rainy summers and mild winters. The arm of Argentina that bends around the southern and eastern part

PACIFIC OCEAN

BOLIVIA

La Quiaca

Jujuy

Salta

Tucumán

PARAGUAY

Asunción

Formosa

Santiago del Estero

Resistencia

Encarnación

Corrientes

BRAZIL

Posadas

Catamarca

La Rioja

Paso de los Libres

Uruguaiana

San Juan

Córdoba

Santa Fe

Concordia

Aconcagua

Villa María

Paraná

Salto

Colón

Río Cuarto

Rosario

San Nicolás

Paysandú

Mendoza

San Luis

Pergamino

Ibicuy

Zárate

URUGUAY

Santiago

San Rafael

Junín

Colonia

CHILE

Buenos Aires

Montevideo

La Plata

Trenque Lauquen

Pehuajó

Santa Rosa

Olavarría

Azul

Tandil

Tres Arroyos

Mar del Plata

Neuquén

Bahía Blanca

Zapala

Choele-Choel

San Antonio Oeste

Viedma

San Carlos de Bariloche

ATLANTIC OCEAN

Trelew

Rawson

Sarmiento

Comodoro Rivadavia

Colonia Las Heras

Puerto Deseado

Populated places

● over 100,000

• 25,000 to 100,000

· under 25,000

FALKLAND ISLANDS
(ISLAS MALVINAS)

Río Gallegos

Río Grande

Ushuaia

0   500 km

0   500 mi

FIGURE 27-3

of Paraguay, known as Misiones, includes a portion of the Paraná Plateau, a region of ample rainfall and dense mixed forests of pine (*araucaria*) and broadleaf species. Here, the Paraná and its tributaries have cut deep canyons in the flat-topped plateau. Where these rivers drop over the edges of lava formations there are spectacular falls, including Guaíra Falls on the Paraná and Iguazú Falls on the Río Iguazú.

The Río Paraná has been a barrier, isolating the people on either side of it, throughout most of its history. With its winding and shifting course, annual floods that cover a wide area, and shallow channel frequently clogged with sandbars, the Paraguay-Paraná-Plata is a poor route of travel. The small city of Corrientes would seem to occupy a position of extraordinary strategic importance, for just upstream the two great branches, the Alto Paraná and the Paraguay, join. Far from being a focus for routes of travel, this city has occupied only a remote and isolated corner of Mesopotamia. The district around Corrientes has remained chiefly pastoral in its economic activities, although tobacco has been of some importance since the colonial period. The city and the district it serves have benefited little until recently from their location along the river.

Mesopotamia is now mainly livestock country, but fruit, wheat, maize, rice, and sunflowers are grown toward the south. Since the 1930s the southern part of the region also has been a leading flax producer in Argentina. In Corrientes Province about 9 million head of cattle and sheep are produced each year, the sheep supplying the export market with superior grades of wool.

Settlement of Misiones

The settlement of Misiones has been supported since colonial times by the production of *yerba maté*, or Paraguay tea. *Maté* is a beverage made from leaves of a tree native to the Paraná pine forests of this part of South America, the *Ilex paraguayensis*. The first European settlements in this area were established by Jesuit Fathers, who built missions and brought the native Indians together in fixed agricultural communities. The planting of *yerba maté* was first attempted around these missions. When the Jesuits were expelled, the plantations were abandoned, and *maté* production was limited to the collection of leaves from trees growing wild in the forests.

After World War I the plantation system was again introduced, this time in the area just east of Posadas. Small clusters of people are grouped permanently on these plantations, but since labor requirements are heavy only at harvest time, the groups of permanent workers are small.

Maté is little used outside of Argentina, Paraguay, Uruguay, and southern Brazil. Argentine planters sell most of their product in the domestic market, although since World War II there has been an attempt to export *maté* to North America and Europe. Posadas, the capital of Misiones, is a small city (44,000) in the midst of the *yerba maté* and tobacco-growing area. Its major advantage is a deep natural harbor where ships can anchor. Tobacco is exported from here to France, Germany, and the United States.

In recent decades, it has been found that Misiones is well suited to the cultivation of tea (*Thea sinensis*) and other subtropical commodities. After World War II, Argentina became the principal producer of tea in the American Hemisphere, mostly from plantations located on the high ground of southcentral Misiones midway between the Paraná and Uruguay rivers. Tung nuts are likewise harvested from plantations in Misiones, their oil being a valuable ingredient in fast-drying paints and varnishes. Wood pulp and paper industries have expanded in the region and help to make Argentina 60 percent self-sufficient in these materials.

Misiones remains one of the least developed provinces of Argentina. Yet, it has great potential for forestry and tree crops, as does the remainder of Mesopotamia for agriculture and agroindustrial activities. All of Mesopotamia appears destined to experience rapid change within the near future, as huge dams are constructed on the adjacent rivers and a network of highways, tunnels, and bridges helps to integrate this region with Uruguay and with other regions of Argentina.

The Chaco

Between the Río Paraguay-Paraná and the piedmont settlements of the Northwest lies another large division of the Argentine North. This is the Gran Chaco, the Argentine part of a great lowland region which extends northward from about latitude 30°S into Paraguay, eastern Bolivia, and western Brazil. The Argentine Chaco, like the country farther north, is a region of deciduous scrub woodland interspersed with patches of savanna. There are places where the thorny, deciduous trees grow in veritable thickets. Elsewhere, especially near the rivers, taller trees form bands of dense, semideciduous forest. There are places, too, where the scrub trees are widely spaced, like apple trees in an old orchard, and where the forest floor itself is grass-covered. Finally, there are places where the forest is interrupted by extensive grasslands, perhaps the result of repeated burnings, or perhaps the result of edaphic conditions.

Some of the highest temperatures recorded anywhere in South America occur in the Chaco. This region is located on the margins of the tropics, in a climatic position similar to that of the Ganges Plain of India or the Gulf Coast of Texas and Louisiana. Summers in these latitudes are hot. Winters are relatively mild and dry, with occasional frosts in the southern part of the region.

The entire Chaco is a great lowland plain, interrupted in only a few places by prominent surface features. It is composed mostly of alluvium deposited by rivers from the erosion of the Andes. During the summer rainy season, vast areas near the streams are inundated. The drier western side of the Chaco is flooded only along the courses of the few streams: the Pilcomayo,

Bermejo, Salado, and Dulce. These rivers follow braided and shifting courses, changing position each year during the period of high water, and in some years radically changing the pattern of their channels. As a political boundary, therefore, the Pilcomayo is less than ideal.

The Province of Chaco, which includes only a fraction of the total Chaco region, is a land of vast cotton plantations that have replaced the quebracho forests of a century ago. Cassava, sugarcane, and sunflowers are also important to the agricultural economy. Food-processing plants, cotton gins, spinning mills, and regional packinghouses are scattered throughout the area. Resistencia, the province capital, serves as the primary trade center and a shipping point on the Paraná for cotton, quebracho, and animal products.

Settlement of the Southern Chaco

Most of the Chaco region remains sparsely populated. The principal areas of permanent settlement are along its southern and eastern margins, focusing on Santiago del Estero, Sante Fé, Resistencia, and Formosa. The oldest settlements on the margins of the Chaco are a string of agricultural communities along the Río Salado and Río Dulce. Long before the Spanish conquest, Indians made sporadic use of the floodplains of these rivers, for after the annual floods there was enough moisture to support crops of maize. In the shallow depressions of the floodplain, after the floods recede, good yields can be gained from the planting of maize, wheat, flax, and cotton. Croplands irrigated in this manner by annual floods are known in Argentina as *bañados*.

Santiago del Estero, the oldest city in Argentina, is located near the southern margin of the Chaco and has derived its economic base from local irrigation, the provision of labor to the piedmont oases, and the exploitation of regional resources. The first groups to enter the Chaco were interested primarily in exploiting the forest itself, not in the development of agriculture. Among the scrub trees of the Chaco is a species known as *quebracho* (literally "break-ax," because of its very hard wood), a tree which contains a high percentage of tannin, used in the tanning of leather. Nowhere else in the world is there a similar forest from which this valuable substance can so easily be obtained. Industrialization of quebracho for its tannin began in 1889.

True quebracho, known botanically as *Quebrachia lorentzii,* grows under special conditions. It is found chiefly in the eastern Chaco just west of the Paraguay-Paraná, and it apparently occurs in dense growth only where the groundwater is strongly impregnated with salt.

In the western Chaco is another species of quebracho which has a dark red wood and contains much less tannin. The *Quebracho santiagueño,* or *colorado,* the red quebracho, contains only about 10 percent of this substance, whereas the true quebracho, known in Argentina as the *Quebracho chaqueño,* contains as much as 30 percent. Red quebracho is used mainly for its wood: for telephone poles, fence posts, railroad ties, firewood, and the manufacture of charcoal. In the treeless plains of Argentina red quebracho is a valuable commodity.

Settlement of the Eastern Chaco

Large-scale agricultural development in the eastern Chaco occurred during the 1930s and was associated with the rapid increase of cotton production in Argentina. There are two chief axes of pioneer settlement. One is along the railroad northwestward from Resistencia, and the other along a railroad extending in a similar direction from Formosa. Cotton can be grown without irrigation as far west as the 32-inch rainfall line, about 160 miles west of the Paraguay-Paraná.

The location of specialized cotton districts in the northern part of the Chaco rather than in the south, near Santa Fé, is due to the distribution of private properties. Wherever the quebracho cutters leave clearings in the forest, land is available for agricultural or pastoral use, but in the southern part of the region the land is already divided into large private estates, used chiefly for the grazing of beef cattle.

The majority of colonists are of European origin and have come from the Humid Pampa. Many nationalities are represented.

Most of the cotton grown in Argentina comes from settlements in the provinces of Chaco and Formosa. It is largely hand picked and is shipped in bales by railroad to Resistencia and Formosa. From there it is transported by riverboat to Buenos Aires for distribution to the textile factories of Argentina and, on a smaller scale, for export to Japan and other countries of the Far East. Other crops harvested in the eastern Chaco include sunflowers and sorghum which, like cotton, yield well in areas of limited rainfall.

Patagonia

The hot, forested areas of northern Argentina present a sharp contrast to the cool, dry, windswept plateaus of the south. South of the Río Colorado lies that part of Argentina known as Patagonia. This vast area, which comprises more than 25 percent of the national territory, is occupied by less than 3 percent of the Argentine people. The population density is generally less than 1 per square mile, and most of the 600,000 people live in urban communities.

In Patagonia the gusting wind seldom ceases. It is a boisterous, stormy wind that carries rolls of cloud with it and frequently changes direction as different air masses sweep by. The haze of dust makes objects in the typical Patagonian landscape indistinct even at short distances. In the west are some of the world's most spectacular mountains, carved by glaciers, and even now mantled in certain parts with glacial ice which is the nearest approach to an inland icecap outside of the Polar regions. Along the piedmont of the Andes is a succession of marginal lakes, their upper ends deep in the mountain canyons where the water laps against the ice cliffs of descending glaciers. The lower ends of these lakes are contained by knobby moraines, dumped by the formerly more extensive ice tongues of the glacial period. Here the green lake waters, churned into whitecaps by violent winds, dash noisily on shingle beaches.

Despite its relatively high latitude, Patagonia is not a land of extreme temperatures. This part of South America lies between 38° and 55° south latitude, equivalent in North America to the territory between Chesapeake Bay and Labrador. Severe winter temperatures, like high summer temperatures, however, are the result not only of latitude but also of protection from the moderating influence of open oceans. In South America the tapering of the land toward the south means that increasing latitude brings decreasing distance from the sea.

No part of Patagonia receives much rainfall except the mountainous area along the western border. Furthermore, a notable contrast exists between the summer maximum of rainfall at Mendoza and the winter maximum from Neuquén southward. As far south as the Strait of Magellan, moisture is brought chiefly by winter storms, but the total amount of precipitation is minimal. A moisture deficiency persists along the east coast from near the mouth of the Río Negro to just north of the Strait.

The desert of Patagonia is the only example in the world of an arid east coast in latitudes poleward of 40°. This aridity is only partly due to the rain barrier of the Andes. As cold air masses cross the southern part of the continent, cyclonic whirls precede the cold fronts as in the Northern Hemisphere, but with a clockwise, rather than counterclockwise, rotation. This brings air onshore from the Atlantic. Moisture would be precipitated by these cyclonic storms if the air from the east had not crossed a wide zone of cold water: the Falkland Current, which bathes eastern South America as far north as the southeastern edge of the Humid Pampa. This cold water is as important as the Andes in accounting for the aridity of Patagonia. Heavy sea fogs, like the

garúas, of Peru, are common, especially in the far south.

Surface Features and Water Supply

The Patagonian highland is composed of two surface elements. There are vast areas of level-topped plateaus, rising like a series of gigantic steps toward the west, where the highest tablelands are well over 5000 feet above sea level. These great plateaus are formed of horizontal strata, some of sedimentary origin, some of dark-colored lava. The second surface element stands above the plateaus: areas of hilly land, composed of resistant crystalline rocks.

Deep canyons cross the Patagonian plateaus from west to east, most of them containing no surface water at any time of the year. A few carry permanent streams, and some have surface water intermittently. Along the canyon bottoms, wells sunk in the gravel fill can tap a good supply of groundwater. These canyons, with their water supply and cliffed sides, shelter most of the ranches of the country and offer the only safe routes of travel across the desert.

Plateaus and crystalline hills come abruptly to an end where the Andean structures begin. In many places the westernmost plateaus stand with cliffed sides facing toward the Andes, but separated from the steep mountain slopes by a narrow belt of lowland. This lowland forms a discontinuous series of basins between 1000 and 2000 feet above sea level which, taken together, is called the Pre-Andean Depression. From Punta Arenas on the Strait of Magellan, the Depression offers a continuous passage northward as far as Lake Argentino. North of this lake the Depression is interrupted at intervals by spurs of the Andes. The northernmost basin which forms a part of the Pre-Andean Depression lies north of Lake Nahuel Huapí.

Even before the glacial period, rivers on the western side of the mountains had cut headward across the cordillera until they had shifted the continental divide to the western rim of the plateaus. The Pre-Andean Depression was drained mostly to the Pacific by rivers that plunged westward through narrow canyons. The ice of the glacial period invaded these canyons and gouged them into broad U-shaped troughs.

The relation between the drainage divides and the crest of the cordillera gave rise in the late nineteenth century to what might have become a serious boundary dispute between Chile and Argentina. The Chileans claimed the drainage divide as the boundary, while the Argentines claimed the crestline of the mountains. Instead of settling the dispute by battle, the two countries requested Great Britain to make an award. The award granted Argentina much of the lake country in the north, but gave Chile a wide stretch of country east of the mountains in the south, including both sides of the Strait of Magellan.

A dispute remained until 1984 over three islands near the Beagle Channel, off Tierra del Fuego: Picton, Nueva, and Lennox. The islands themselves are of little value, but how the boundary is drawn at this point could vastly influence the claims of Argentina and Chile in the South Atlantic Ocean and Antarctica. Resolution of the conflict was achieved largely through the papacy at Rome.

European Colonization

European colonization in Patagonia is a relatively recent development. Until the nineteenth century the only settlement along the east coast was at Carmen de Patagones, near the mouth of the Río Negro. Because this community was a source of salt, shipped by boat to Buenos Aires, it enjoyed a certain degree of permanence. The remainder of the country was left to the nomadic tribes of Puelche and Tehuelche who hunted the fleet-footed guanaco and the rhea, a type of ostrich. When the tribes adopted horses and firearms from the Europeans, they became formidable opponents of European settlement. The sparse population of natives was finally all but wiped out, however, in a series of vigorous military campaigns between 1879 and 1883.

As soon as the threat of Indian attack was eliminated, new colonists began to move into the region. The pioneers included people of many nationalities, especially Welsh, Scottish, and English. Settlement spread into Patagonia along three chief routes. From the Humid Pampa at Bahía Blanca, one stream of settlement spread southward along the coast and inland up the canyons. Meanwhile, another stream was advancing northward from the Chilean port of Punta Arenas and eastward along the Strait and into Tierra del Fuego. A third route of colonization led southward from Neuquén into the northern part of the Pre-Andean Depression, past Lake Nahuel Huapí, in part supported by people who came from Chile. In addition, the Welsh settlement at Trelew, established in 1865, sent out new colonies to the south and west. The discovery of oil in 1907 led to the development of a small settlement at Comodoro Rivadavia, which is still Argentina's leading petroleum center. European nationalities, as well as Argentine and Chilean, are now represented in the population of Patagonia.

Stock-Raising

Most of Patagonia is devoted to sheep-raising, a way of life that supports meager population. The land is divided into ranches occupying hundreds of square miles. On these properties the few people are clustered around the ranch headquarters, which are usually located in the shelter of a canyon where water is available. Little irrigation is practiced in these places, and where small bits of land are cultivated, it is usually to provide feed for horses.

The uplands south of the Río Negro support herds of sheep, goats, and cattle. Large-scale ranches exist with 40,000 to 100,000 head of sheep, and even the smallest operate with several hundred sheep and goats. Overgrazing is a serious problem, and everywhere the pastures are eroded. However, grazing continues throughout Patagonia, and a total of about 20 million sheep are raised. The only major area for cattle ranching is in the northern Pre-Andean Depression.

The River Oases

The only cropland of importance in Patagonia is along the exotic rivers, mainly the Colorado, the Negro, and the lower Chubut. Simple irrigation works make cultivation possible along the valley floors. Being 100 to 300 feet below the plateaus, these areas are sufficiently protected from violent winds, where alfalfa is the chief crop. Along the Río Negro between Neuquén and Viedma, and between Rawson and Puerto Madryn, there are small concentrations of cattle.

About 148,000 acres have been made available for crops along the Río Negro. These have been laid out in small properties of 250 acres each. Alfalfa is still the chief crop of this area in terms of acreage, but vineyards and fruit orchards have been added, creating an oasis similar to that of Mendoza. The Río Negro Valley is now Argentina's chief producer of apples, peaches, pears, and other deciduous fruits, and supplies a substantial export market.

In recent decades the Argentine government has invested heavily in the development of energy resources. Large dams have been constructed on the headwaters of Patagonian rivers, and hydroelectric power has become abundant in provinces such as Neuquén. In the same area, petroleum and natural gas have been discovered, gold and copper are mined, and forestry is of some importance. Irrigation is being extended throughout the northern valleys. Tourism, once focused chiefly on the world-famous winter resort of San Carlos de Bariloche, has also been expanded.

The Coast

The coast of Patagonia is rugged and barren. Cliffs at the water's edge mark the beginning of the plateaus. To the south there are a few embayments of river mouths, such as those on which Santa Cruz and Río Gallegos are situated, but even here the building of port facilities is made difficult by a great tidal range. The spring tide at the mouth of the Río Santa Cruz, for example, is as much as 48 feet.

A conspicuous feature of the Patagonian Coast is the Valdes Peninsula, attached to the mainland by a narrow isthmus. Adjacent waters support an extraordinary variety of wildlife, including great numbers of whales, dolphins, seals, penguins, and birds.[1] The Magellan penguins breed here and can be seen in small schools in pursuit of the fish and crustaceans that comprise their food. The whale population is greater than anywhere else in the world, and the general abundance of wildlife in this region requires that special attention be given to the problems of conservation.

Most ports along the coast of Patagonia are used primarily for the shipment of wool. Usually, there is little activity at these places and little trade with the interior. After the shearing season, when wool begins moving to the coast, there are a few months when the shipping facilities are kept busy. From the ranch headquarters the wool is transported by truck or rail to the nearest port, from which small coasting vessels carry it to the big wool depot at Buenos Aires.

Some of the port cities have meat-packing plants called *frigorificos*, and others have become modern industrial centers. In the far south, Río Gallegos has a small plant for the slaughter of sheep as well as port facilities for the shipment of coal. Low-grade bituminous coal is mined near the Chilean border at Río Turbio, where almost all of Argentina's known coal deposits are located, and is tranported by rail to the coast. Farther northward is the petroleum center at Comodoro Rividavia, and beyond that a modern aluminum complex at Puerto Madryn. Bauxite for the aluminum smelter must be imported, but abundant electric power (440 megawatts) is pro-

---

[1] R. Ellis, "Argentina's Valdés Peninsula," *Américas* 32, No. 9 (1980): 13–21.

vided by the Futaleufú Dam on the upper Río Chubut.

The history of settlement in Patagonia is similar to that of other parts of Argentina in terms of remoteness from Buenos Aires. Today, however, the destiny of Patagonia is as closely tied to the metropolis as is that of the North and Northwest, and the mineral resources of Patagonia are beginning to play an increasingly important role in the national economy. Petroleum has been discovered in many places along the coast and at offshore locations in the Atlantic, and pipelines now supply the coastal cities and Buenos Aires with natural gas from Tierra del Fuego and Neuquén. Recent discovery of extensive iron deposits in the Sierra Grande of southeastern Chubut Province has provided Argentina with its first significant domestic supply. The magnetite ore averages 55 percent iron and is produced from underground mines less than 10 miles from the coast. There, a pellet plant upgrades the ore for shipment to steel mills along the Río Plata-Paraná.

The Humid Pampa

A vast grass-covered plain can, in its own way, be as spectacular as snow-capped mountains or wind-ruffled lakes. For days the early Spanish colonists could travel without encountering features distinctive enough to vary the scene. Repeated endlessly were the views of tall, plumed grasses, or of coarse rushes in the marshy spots, rustling in the shifting winds which were almost never still. Through the haze of wind-borne dust, the sunset colors were splashed brilliantly across the skies or reflected in the tawny waters of the Río de la Plata.

Although the Spaniards found this region covered with prairie grass, the landscape may already have undergone an important modification of its natural condition. To the west the land was covered with *monte*, an impoverished continuation of the tropical scrub woodlands of the Chaco. The *monte* is essentially a tree-steppe composed of deciduous broadleaf scrub trees and bushes with a ground cover of short grass. The tall bunch grasses of the prairie could survive the fires set by the Indians, which seem to have pushed the *monte* border farther toward the dry west.

Climate

Rainfall over much of the Humid Pampa is sufficient to support trees. Indeed, the many trees planted around the *estancia* headquarters since the agricultural conquest have survived without difficulty. At Buenos Aires and throughout the entire northeastern part of this region the average annual rainfall is about 37 inches. Moreover, it is quite evenly distributed throughout the year. Rainfall decreases toward the southwest, totaling 21 inches at Bahía Blanca and 16 inches along the dry margin of the Humid Pampa. The seasonal distribution shows more of a concentration in the summer months (December through February) toward the west.

The dependability of rainfall has been a critical matter since the spread of agriculture. The area around Rosario and Pergamino rarely suffers from drought. This "island" of dependable rainfall, however, is replaced by more arid conditions in any direction from Rosario. The dependability of summer rain, which is a critical matter because of the maize crop, also decreases toward the north. The greatest rainfall uncertainty, however, is in the dry lands to the west.

The Humid Pampa is, in general, a region of mild winters and hot summers. The growing season between killing frosts varies from about 300 days along the Paraná-Plata shore to 140 days south of Bahía Blanca. Despite the annual frosts, winters are mild, and the absense of continuous snow cover makes agricultural and pastoral operations possible throughout the year. Summers are especially hot in the Northwest, but toward the southeast they are cooler. The region around Mar del Plata and Tandil is distinctly cool. Hence, grains do not thrive in this section, and dairying is of greater importance. Tandil

(46,000) is a regional trade center and one of the major cheese-producing centers of Argentina.

A comparison of average temperatures at Buenos Aires and cities of the eastern United States is instructive. Buenos Aires has about the same temperature in January as does New York City in July. During winter, however, the average temperature of the coldest month at Buenos Aires is about equal to that of the coldest month at Charleston, South Carolina. Cold air masses from the south cross the Argentine plains toward the northeast or north, meeting warm, humid, and relatively light air of tropical origin. This interaction of air masses is similar to that which occurs in North America where cold air from Canada moves southeastward toward the Gulf of Mexico or Atlantic Ocean. Along the front of advancing cold air masses, the light tropical air forms whirls or eddies which appear on the weather maps as "lows" or cyclones, rotating in North America in a counterclockwise manner and in South America in a clockwise direction. Along the immediate cold front abrupt local up-currents of warm, moist air produce thunderstorms. The Buenos Aires area is noted for the violence of its thunderstorms and magnificent displays of lightning. During the passage of these air masses, the weather changes from cloudy, muggy, and depressing to clear, dry and bracing. Occasionally in winter, severe cold spells do reach Buenos Aires. The people of the Humid Pampa distinguish between the *norte*, or sultry north wind, and the *pampero*, or invigorating wind that comes from across the Pampa.

Soil and Surface

Wind is an important element in the physical character of the Humid Pampa. Although less violent and blustering than the wind of Patagonia, there are times when the *pampero* may blow with considerable force and when the air is filled with dust whirled aloft from the dry surfaces to the west and south. This burden of fine rock particles carried by the wind, picked up in the arid west and dropped in the more humid

east, is responsible in large measure for the soil characteristics and surface of the Humid Pampa.

The Argentine Plains are composed of deep accumulations of loose material resting on a hilly surface of granite and other ancient crystalline rock that continues southward from Uruguay. South of the Plata, the hilly surface is deeply buried. At Buenos Aires, for example, bedrock lies beneath 985 feet of river alluvium and wind-blown dust. The more prominent features of buried surface still protrude above the cover: the Sierra del Tandil (1600 feet) and the Sierra de la Ventana (4200 feet) form conspicuous hills rising above the general level of the plains. Around the base of each of these Pampa sierras is a gently sloping apron of alluvium. A similar, but larger, apron of alluvial material spreads out from the easternmost margin of the Sierra de Córdoba. Elsewhere, the surface of the Humid Pampa is so level that only the marshy spots, or surveyor's instruments, reveal the slight irregularities. Except near the Pampa sierras and Sierra de Córdoba, the soil is entirely free of stones or pebbles. Here is a land where gravel is unknown.

The Northeast Pampa Rim merges gradually with a zone of marshes southwest of Rosario and Buenos Aires. This part of the Humid Pampa is drained by the Río Salado.[2] During wet periods a rising water table converts much of the surface into shallow reed-filled lakes, inhabited by innumerable water birds. More commonly, the Salado Valley is composed of marshes with the water table at, or slightly below, the surface, and the sluggish current is lost among the coarse grasses.

Beyond the Salado, extending to the southern limit of the Humid Pampa, is an area that has been called the Southern Pampa. This surface is also bordered along the sea by an abrupt slope

---

[2] At least three rivers in Argentina are named *Salado*. There is the Río Salado which crosses the southern Chaco from near Tucumán to Santa Fe; the Río Salado which drains the zone of playas east of Mendoza, and occasionally enters the Río Colorado; and the Río Salado on the Humid Pampa, to which reference is made here.

which is especially prominent along the southern shore, east of Bahía Blanca. The surface, except for alluvial aprons around the Pampa sierras, is composed of almost imperceptible swells of high ground, interspersed with shallow, marshy depressions. In a few places sand dunes stand above the surface of the plain, but in most areas the soil is developed on a powdery, wind-blown material known as *loess*.

The entire western part of the Humid Pampa has been invaded by relatively coarse, wind-blown material from the Dry Pampa to the west. Here the soil material is sandy, rather than powdery, and in places there are strings of sand dunes. This is the dry margin of the Humid Pampa, where soils similar to those of the North American Great Plains would develop if it were not for the continued deposition of new soil material blown in from arid lands by the prevailing westerly and southwesterly winds. Throughout the Western Pampa the water table remains close to the surface so that it is easily reached by the roots of alfalfa.

Settlement Before 1853

When Spanish colonists again settled along the Paraná-Plata shore, they came downstream from Asunción. Santa Fé was founded in 1573, and the site of Buenos Aires was reoccupied in 1580. The first colony at Buenos Aires failed largely because people at Asunción refused to permit sufficient support. When the site was reoccupied, the chief purpose was to provide a place where ships, after the long voyage from Spain, could find fresh provisions before beginning the long and difficult trip up-river to Paraguay. Moreover, Buenos Aires could guard Spanish interests at the mouth of the Plata. A century later, in 1680, when the Portuguese founded Colonia on the opposite shore of the Plata, Buenos Aires became an important outpost in the Spanish and Portuguese struggle for control of the Plata estuary.

Within the Northeast Pampa Rim where pastoral settlement was concentrated, the Humid Pampa underwent a rapid transformation. For the native guanaco and rhea, the Europeans substituted herds of domestic animals and, perhaps without intention, the seeds of European grasses. The new species of grass spread rapidly, and soon the native bunch grass was replaced by a dense lawnlike growth of European grasses that formed a thick sod. A sharp vegetation boundary thus developed along the Salado where no such boundary existed before. Travelers of the early nineteenth century commented on the great difference between the tall bunch grasses of the Indians' hunting ground southwest of the Salado and the rich green carpet of well-cropped grasses in the pastoral zone.

The *estancia* owners, who hired *gauchos* to do the work on the estates, were not an agricultural people. When wheat farming was introduced near the end of the eighteenth century, it was done by tenants and was scorned by the vast majority of settlers. Wheat was raised near the Paraná and the Plata, and on the Pampa in a semicircle some 25 miles in width around Buenos Aires. Late in the eighteenth century, flour was actually exported to the purely pastoral region of Mesopotamia.

Meanwhile, the Río Salado remained a sharp cultural boundary separating the Europeans from the Indians, and the European grasses from the unchanged prairie grass. Convoys of wagons crossed the Humid Pampa to the southwest in search of salt, but there was little interest in laying out *estancias* in the vast territory between the Salado and Bahía Blanca. This lack of attention to one of the world's most favorable natural pasturelands cannot be laid to the hostility of the Indians, for the *gauchos* were not the kind of people to be stopped by a few scattered tribes of natives. Rather, the owners of the cattle estates placed little value on this Indian country, since there was no opportunity to develop its agricultural possibilities. The world was not ready for such development, and for the pastoralist whose mules were driven to the market at Salta there was plenty of good grazing land west of Rosario.

Following independence from Spain (1810–

16) to the drafting of a federal constitution (1853), the frontiers of occupied country remained little changed while the great issue of the day, centralism versus federalism, was fought out. During the long domination of the dictator Juan Manuel de Rosas (1829–52), strong central authority from Buenos Aires was imposed on the outlying parts of the country. Rosas was overthrown in 1852, and in 1853 a federal system was adopted. Until 1861 the Province of Buenos Aires tried to remain separate from the federation, and during this time Paraná was the Argentine capital. When, in 1862, Argentina was united under President Bartolome Mitre and Buenos Aires again became the capital, changes began which have produced this modern nation. Argentina advanced toward the modern period with four basic characteristics: (1) a sparse population—only about 900,000 in 1800 and 1.2 million in 1852; (2) a people almost exclusively interested in livestock rather than in agriculture; (3) an abundance of high-quality land for grazing and grain farming; and (4) the tradition of large private estates.

Beginning of the Modern Period

The Argentine grasslands, like the other mid-latitude grasslands of the world, entered a period of spectacular development. The advance of frontier settlement across first-class agricultural lands was accompanied by widespread prosperity, owing to expanding markets and increasing land values. This phenomenon was characteristic of one period of economic history and one kind of place. Techniques developed during the past century not only made possible the agricultural use of the grasslands, but also tied these remote places with urban markets by cheaper forms of transportation. Agricultural machinery for the first time in history made possible the cultivation of large areas from which yield per acre was small. The new wire fences made possible the separation of agricultural land from pastures and one pasture from another so that the breeding of animals could be controlled. Well-drilling machines and inexpensive windmills helped to secure water where it was difficult to find at the surface, and railroads and steamships facilitated the transport of farm products to the expanding city markets. A fundamental characteristic of this period was the rapid increase in population growth rate throughout the world. This formed the basic support for expanding markets, which made possible the system of credits and investments. Credits and investments, in turn, led to continuously expanding production.

Transportation   Transformation of the Humid Pampa began with the development of transportation: the construction of roads and railroads. Roads were remarkably difficult to maintain. As long as travel was free in any direction, as it was before fences were built, the problem was not serious. When ruts were worn too deep into this loose soil, carts or horses could follow a new route to one side of the old one. When fences were placed along property lines and roads could not be shifted, wagon wheels soon cut through the grass cover and exposed the fine Pampa soil. With each rain the fine material turned into a quagmire, and when the mud dried out, the wind picked up the powdery dust and whirled it away, with the result that, in time, the roads were several feet below the general level of the plain.

The construction of railroads was fairly simple. On the nearly level surface there was no need for cuts, fills, bridges, tunnels, or even curves. In few parts of the world are construction costs lower. The first railroad began operations in 1856 and ran for six miles straight southwestward from Buenos Aires. Soon thereafter a railroad was built along the old colonial mule route from Rosario to Córdoba and Tucumán. During the following decades railroads were extended from all the leading ports of the Humid Pampa, usually in strait lines toward indefinite and distant objectives. By 1910 Argentina's central region was crisscrossed by a series of overlapping fans, converging on Buenos Aires, Rosario, Santa Fé, and Bahía Blanca.

During more recent decades, an extensive network of roads and highways has been constructed to connect all cities and towns throughout the Humid Pampa. Air transport also has expanded, with the result that railroads in Argentina, as in most other parts of the world, are in a state of decadence.

The Pastoral Base   Life on the *estancias* of Argentina has grown out of a pastoral tradition. The owners were mostly Argentine creoles, whose main occupation was the herding of cattle, sheep, and horses. At first, the herds were comprised of the descendants of scrub animals from the colonial period. These were permitted to run wild on the vast, unfenced range and to breed without thought to quality. Such animals were good for the production of hides, tallow, and salt beef, but the meat was lean and had a strong taste. In 1877 the first refrigerator ship made possible the shipment of frozen meat to Great Britain, but British

taste would not accept Argentine meat. The result was the importation of high-grade beef cattle from Britain and the careful breeding of these animals on fenced-in pastures. This shift from scrub cattle to carefully selected animals occurred between 1880 and 1900.

The change in pastoral technique created many other changes in the relation of people to the land and in productivity of the land. Scrub cattle fared well on the poor grasses of the *monte* and thrived on the untamed prairie. They could endure the insect pests of the Chaco or the long overland marches with little feed and water. The larger beef animals bred from British stock lacked these advantages and could not survive the ravages of Texas fever. For the first time, the southern limit of the area infested with ticks became a significant geographic boundary. In addition to requiring tick-free pastures, the new animals required a better source of feed than the unculti-

CATTLE RANCHING IN PATAGONIA, NEAR RÍO GALLEGOS.

vated pastures could supply. After 1890, the *estancieros* realized the need to shift from a livestock industry based on uncultivated pastures to one based on a cultivated crop. Alfalfa was the crop they adopted, and on the Humid Pampa alfalfa did exceptionally well. Alfalfa had to be planted on plowed land, however, and then cut and fed to the animals. This required the services of many more workers than had been needed to care for the wild scrub animals. At last the *estancieros* wanted immigrants.

Immigrants   In 1856 the first group of agricultural immigrants arrived from Europe. It consisted of 208 Swiss families. Because the more accessible parts of the Humid Pampa were already in private hands, this colony was established on land granted by the Province of Santa Fé. The first colony, Esperanza, was soon bordered by other colonies of European immigrants. After 1882 the district around Santa Fé became one of the first major sources of wheat.

The tide of imigration increased rapidly. The total population of the country, which was 1.2 million in 1852, increased to 2.5 million in 1880, including 173,000 people born in Europe. The first peak of immigration was reached in 1889, when about 220,000 passengers came to Buenos Aires. In every year, however, there was a considerable return migration. Immigration figures for 1889 must be balanced against 40,000 emigrants. The depression of 1890–91 resulted in a net loss of population in Argentina, after which the tide swelled rapidly again, until the peak year of 1913 when 302,000 entered and 157,000 left. Between 1857 and 1900 approximately 2 million immigrants arried in Buenos Aires and 800,000 departed, for a net increase of 1.2 million. After World War I, another peak was reached in 1929, with 427,000 arrivals and 348,000 departures. During the 73 years between 1858 and 1930, total immigration amounted to 6.3 million people. After 1930 immigration was reduced to a mere trickle.

The racial character of the Argentine population was profoundly altered by this stream of European immigration. The immigrants whose arrival so radically changed the character of the Argentine people were largely of Italian and Spanish nationality. These two groups totaled almost 80 percent of the newcomers. Between 1857 and 1924, of those who remained in the country, 1.3 million were Italians and 1 million were Spaniards. Also represented in the stream of immigration by substantial numbers were French, Germans, Austrians, Russians, British, and Swiss. After 1930 most of the immigrants came from Eastern Europe, especially Poland.

Tenants and the Rise of Agriculture   The immigrants brought agriculture to Argentina, and agriculture was encouraged by the landlords as a byproduct of the expansion and improvement of grazing. The most effective way to prepare land for alfalfa was to rent it for a period of four or five years to tenants and permit them, for a share of the crop, to raise wheat. The *estancieros* and their hired hands were neither numerous enough to undertake this work nor willing to do so. They were eager, however, to secure tenants for their estates, having found that in addition to increasing their alfalfa acreage they could derive considerable profit from a share of the crops. The contracts obliged tenants to plant alfalfa and to move away after a specifid number of years. Alfalfa yielded well, giving as many as three cuttings a year, for five to ten years, after which new tenants were secured, and the cycle repeated. It was the tenant group that made the Humid Pampa one of the world's leading surplus grain and meat regions.

With the development of commercial agriculture, between 1880 and 1900 the present outlines of the Humid Pampa came sharply into focus. As the new alfalfa-fed beef animals of the fenced pastures replaced the half-wild animals, and as cultivated grains and alfalfa replaced the prairie grass and *monte*, a new regional boundary began to appear. Grain farming and alfalfa pushed westward as long as yields were profitable in the Humid Pampa; farther west there was not enough moisture to support adequate yields.

A SHEEP RANCH IN PATAGONIA.

These crops were cultivated about as far as the 16-inch rainfall line west of Bahía Blanca, and to the 23-inch rainfall line west of Santa Fé. The Humid Pampa, as distinct from the Dry Pampa, made its appearance at this time. Development of the Humid Pampa is reflected in export statistics: in 1894 the three chief pastoral products (wool, meat, and hides) comprised 63 percent of the total value of exports; in 1903, for the first time, the combined value of agricultural products (maize, wheat, and linseed) exceeded that of animal products. Immigrant farmers supplied the manpower to reshape the economic destiny of the region.

Land-Use Districts

The transformation of the Humid Pampa from an open range country of low productivity to a productive grazing and farming region resulted in a new pattern of land-use districts. These subdivisions of the Humid Pampa, which appeared before 1930, are four in number. First is the pastoral district, in the southeast, between Mar Del

Plata and Tandil. Second is the alfalfa-wheat district, in which alfalfa is the chief crop in terms of acreage. This is also the principal area of wheat cultivation. The third division is the maize district around Rosario where maize is more important than wheat. The fourth division is the truck, dairy, and fruit district around the margins of Buenos Aires.

Pastoral District   The pastoral district is that part of the Humid Pampa where more than 80 percent of the productive area is used for grazing. Alfalfa and wheat cannot be grown, except at high cost, because of poor drainage and the unfavorable moist, cool summers. In this district livestock ranching without agriculture is the predominant form of land use. The pastoral district is now a zone of high-grade mutton and wool sheep, butter production, and the breeding of beef cattle.

Because agriculture is so little developed in the pastoral district, this area has received few immigrants. Few people are needed to care for herds of sheep or cattle, although the spread of dairying requires more workers. At present, this district has the lowest rural population density of any part of the Humid Pampa, ranging from 10 to 25 people per square mile.

The Alfalfa-Wheat District   Most of the Humid Pampa is included in the alfalfa-wheat district. In a broad crescent, extending 600 miles along the western side of the Humid Pampa from Santa Fé in the north to Bahía Blanca in the south, the chief commercial crop is wheat.

The base of the rural economy of the alfalfa-wheat district is still pastoral. Owners of the *estancias* are more interested in producing high-grade animals than in farming. Wheat and alfalfa are grown by tenants as a byproduct to maintain the productivity of the pastures. Because of the temporary relationship between the tenants and the lands they cultivate, the region can register sharp differences in crop and pasture acreages from year to year without severe economic dislocations. When the price of grain is down, the landowners dismiss their tenants. On the former cultivated land the tenants plant grass before moving away, and the pastures can be maintained by not overgrazing them until a new cycle of alfalfa and wheat is needed.

There has been a gradual decrease in the number of large estates as the family *estancia* has become subdivided. However, the *estancia* is still the chief form of land tenure, and there are small farms on which the owners raise vegetables or pasture dairy cattle. In the alfalfa-wheat district, tenant farms range from 370 acres in the north to more than 1000 acres in the south. There is a relationship between increase of farm size and the increased use of machinery. About a fourth of the wheat crop is now raised on owner-operated mixed grain and livestock farms, on which the workers, hired for wages, are more permanently attached to the land than are the renters.

Wheat was the first grain crop to achieve permanence against the pastoral tradition of the Argentine Humid Pampa. Its rapid spread was coincident with settlement south of the Río Salado after 1880. Wheat also first brought prosperity to the colonies of European immigrants northwest of Santa Fé. In time, wheat was planted in almost all parts of the Humid Pampa and proved successful in all but the southeast.

The best yields are in an area southwest of Rosario, where an average of 17 to 20 bushels per acre is reported. Yields between 13 and 17 bushels occur in the central part of Buenos Aires Province and southward to the coast. They decline rapidly, however, to the southwest, to between six and nine bushels along the dry margin west of Bahía Blanca. Little actual production comes from the area of highest yields.

The Maize District   The exclusion of wheat from the area of highest yields is due to the competition of other grains. Since 1895, wheat has been challenged successfully by maize, which also has its highest yields in the central district around Rosario. In this district an average of 30 to 40 bushels per acre is to be expected. Maize predominated over wheat because its yield was far greater, while both crops brought about the same returns to the growers per bushel of grain. Beyond the Rosario district the yield of maize declines, indicating the relatively small area in which a favorable combination of fertile, well-drained soil, moderate temperatures, and dependable rainfall is found.

The third major agricultural division of the Humid Pampa was developed after 1895 when maize proved more profitable than wheat in the central district of high grain yields. Today, the maize district centers on the river port of Rosario. With the start of World War II, when the acreage of all grains declined rapidly, maize was reduced more than the others. Maize now occupies about 7.5 million acres, considerably less than the figure for 1912.

The Argentine maize district produces largely for export, not for the fattening of hogs and cattle as in the United States. Most of the maize is of the "flint" variety, characterized by small, hard grains, low in water content. This type is easily shipped, and for a long time it held a preeminent position in Europe as a poultry feed. The "dent" varieties used in North America chiefly to fatten animals are not popular in Argentina, and even in the maize district cattle are fattened on alfalfa.

Flax appeared on the Humid Pampa in about 1900 and was planted extensively in the maize area. At first, it was a favorite crop in the preparation of new land for other crops or pasture. It, too, yields better in the Rosario district than elsewhere, but because of its greater resistance to heat and drought, its yields do not decline as rapidly as those of maize, especially to the north and east.

The Truck, Dairy, and Fruit District   The maize district is bordered on the east by a zone of intensive farming, devoted to supplying the market of metropolitan Buenos Aires. This district around the great city offers two major advantages for intensive farming: easy access to the urban market and land suited to the production of garden crops. The soil is fine-grained, deep, and well drained; rainfall is abundant and evenly distrib-

uted throughout the year; and winters are so mild that there is no season when fresh vegetables cannot be provided.

Dairying is now concentrated in the zone southeast of Buenos Aires, extending into the eastern part of the Humid Pampa. Because of the large proportion of swampy land, cool summers, and abundant rains, rich pastures furnish good grazing for high-grade dairy animals. Alfalfa and maize do not prosper. The population, therefore, is composed largely of pastoralists rather than farmers. Most of the land is in large estates, subdivided and rented to tenant dairymen.

Vegetables for the Buenos Aires market are produced near the city. The farms range from garden plots to 25 to 30 acres. Truck farms are clustered along the railroad lines and on land within the urban limits not otherwise used. As the Argentine diet changes to include more fresh vegetables, the prosperity of this fringe of intensive truck farming increases.

Along the floodplain of the Río Paraná, the presence of warm water from the north permits a long southward extension of tropical or subtropical kinds of vegetation. Fruits such as apples, pears, plums, and peaches give high yields on islands near the junction of the Río Paraná and the Uruguay. Much of the land on the islands is owned by large fruit companies and is worked by hired laborers. In addition to the orchards are plantations of willow and poplar, which supply materials for the construction of baskets and boxes in which the fruit is sent to market. Within the relatively small area of this district of intensive dairying, truck farming, and fruit-raising, the density of rural population is much higher than anywhere else in the Humid Pampa.

Buenos Aires

Buenos Aires constitutes another division of the Humid Pampa. Here, in an area of 1400 square miles, about 35 percent of the nation's population is concentrated. Buenos Aires is the third-largest urban center in Latin America, after Mexico City and São Paulo, and the largest

ONE OF THE BUSIEST SHOPPING AREAS IN ARGENTINA: AVENIDA FLORIDA IN BUENOS AIRES.

Spanish-speaking city in the Southern Hemisphere.

Like all the great cities of the Occidental world, Buenos Aires preserves in its patterns and forms a record of its history. Signs still remain of the functions it has performed during its advance from a frontier post of the Spanish colonial empire to one of the world's great commercial and industrial metropolises. Most of the city's growth, however, occurred during a time of architectural planning when such great avenues lined with buildings of uniform front as the Champs Élysées of Paris, the Avenida Río Branco of Río de Janeiro, the Paseo de la Reforma of Mexico City, or Commonwealth Avenue of Boston were built. The central avenue in Buenos Aires is the famous Avenida de Mayo. The modern period of architecture is also finding expression in store fronts and in skyscrapers that now diversify the Buenos Aires skyline.

Founding and Growth   The place first selected for the establishment of a colony in 1536, and again in 1580 when the Spaniards came downstream from Asunción, possessed one important advantage. It was the only spot along this stretch of the

Plata shore where the water was deep enough to permit ships to reach a dry landing place at the base of the Northeast Pampa Rim. The southern shore of the Plata is low and marshy, with wide mud flats exposed when the wind blows from the south, but at this one place a small tributary stream known as the Riachuelo provided an anchorage for the shallow-draught ships of the sixteenth century near the higher ground of the Pampa.

From 1580 to 1853 Buenos Aires was of minor importance. After 1680, when the Portuguese established a fort on the opposite shore of the Plata at Colonia, Buenos Aires became a defense post of great importance in the struggle for control of the Paraná-Plata. When the British at-

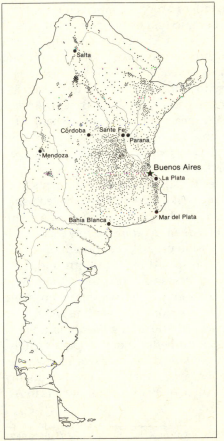

FIGURE 27-4

tempted to occupy Buenos Aires in 1806 and 1807, the fact that this little town held control over a wide and potentially rich hinterland was beginning to be understood.

The development of agriculture in the Humid Pampa was accompanied by a spectacular growth of the urban nucleus on which this development focused. In 1778, when Buenos Aires was opened as a port, the population was only 24,000. Buenos Aires passed the million mark in 1909, and by the census of 1914 it had 1.5 million people, including more than half of the total urban population of Argentina. After World War I the growth of the city continued, reaching 2,197,000 in 1932. At present the population of the city proper is estimated at almost 10 million. Associated with this rapid growth has been all of the social and economic phenomena characteristic of modern Occidental cities: the rapid rise of land values in the center; the development of "blighted areas" in the old residential zone near the center and in the suburbs; and a rapid expansion of the city along lines of travel, including the establishment of detached suburbs and satellite towns.

The Urban Pattern   The urban pattern of Buenos Aires has developed around the original rectangular nucleus just north of the Riachuelo. At the center is the Plaza de Mayo, from which the Avenida de Mayo extends westward to the capitol building. The main artery which continues to the west and southwest is the Avenida Rivadavia. As the city has grown in this direction, various subdivisions have been built along the Avenida, their right-angle streets conforming to the orientation of the central thoroughfare. Extensions of the city to the northwest have been attached similarly to the main avenue that leads through San Isidro and San Fernando to Tigre.

In the center of Buenos Aires, traffic congestion led to modification of the original rectangular plan. From the Plaza de Mayo diagonal avenues were cut to the northwest and southwest. The northern diagonal, the Avenida Sáenz Peña, has become the axis of the commercial core.

Buenos Aires also has built two systems of underground electric railways which converge on the Plaza de Mayo from the west and northwest.

Buenos Aires has spread far beyond its original boundaries and those of the Federal District. Population figures for the political city have little meaning. All the people in communities that are functionally related to the central city and that form an almost uninterrupted built-up area must be included in the Buenos Aires conurbation.

In the late nineteenth century, Buenos Aires faced the need to build an artificial port. The original Riachuelo, where small sailing ships of the sixteenth century reached dry ground, was totally inadequate for steamships. A deep channel was dredged to Buenos Aires, and to keep the channel open, dredging operations were continuous. Four connected basins were built at Buenos Aires where ships could tie up at docks. Even by the time of World War I, however, these docks had become entirely inadequate. Many of Argentina's exports were consequently diverted to the outport of La Plata, 35 miles to the southeast. This community became a major industrial center, including meat packing and grain milling, and its expansion has been enhanced in recent years by completion of a gas pipeline from Tierra del Fuego.

Argentina as a Political Unit

Argentina started along the road to economic development earlier than any other Latin American country and entered the "take-off" stage as early as World War I.[3] By 1930 it led all the other Latin American nations in gross national product per capita, in adequacy of diet, in the proportion of people living in cities, in percentage of literacy, and in the stability of constitutional government. Then, development ceased, and constitutional processes were abandoned. Many Argentines found themselves confused, with no clear course of action or common purpose.[4]

Political stability was enforced during the regimes of Juan Domingo Perón, 1946–55 and 1973–74, but the country entered into a period of catastrophic economic conditions. Foreign debt rose dramatically, unemployment became widespread, and rates of inflation were among the highest in the world. Guerrilla warfare, with terrorism and kidnapping, was especially rampant during the mid-1970s. The greatest problem for modern Argentina is the apparent inability to achieve representative, democratic government, despite a relatively high standard of living, a high rate of literacy, and racial homogeneity.

The Economic Situation

During the latter part of the nineteenth century, Argentina became one of the great commercial nations of the world. The Civil War of the United States, which followed shortly after Rosas' long dictatorship offered Argentina an unforeseen opportunity to gain a foothold in the European markets and especially in Great Britain. Argentina supplied meat and wheat in exchange for coal and manufactured articles from Great Britain. British investments in railroads and packing plants, purchases of Argentine foods, sales of manufactured goods, and coal shipments that comprised the bulk cargoes of British steamship lines all formed links connecting Argentina with Great Britain. Argentina could supply food at relatively low cost to urban Great Britain, and British industry prospered by the return trade. For owners of capital and land, the system was highly satisfactory. Even tenants who were doing most of the work in Argentina were, for a time, satisfied with the arrangements, since they were living much better than in Spain or Italy. The system contained the seeds of its own destruction, however, for there was great inequality in the distribution of income. Landowners in Ar-

---

[3] W. W. Rostow, *The Stages of Economic Growth* (Cambridge, England, 1960), p. 38.

[4] Aldo Ferrer, *The Argentine Economy* (Berkeley, Calif., 1967), p. 209.

gentina were, as a group, among the world's wealthiest people.

When Argentina was cut off from the usual sources of manufactured goods, it turned to the development of domestic manufacturing. Machinery was imported, factories were built, and tariffs were established to protect infant enterprises. By 1920, shoes made from local leather could compete, with imported shoes for the ordinary market. Similarly, woolen textiles made in Buenos Aires could meet the competition of all but the highest grade textiles from Great Britain. Argentina's exports made possible the purchase of industrial machinery and fuels.

For its coal supplies Argentina was dependent on Great Britain, which provided high-quality coal at low cost. Because it formed bulk cargo on ships sailing from Britain and returning with wheat, transport costs also were low. When World War I suddenly reduced coal shipments, Argentina faced a serious fuel crisis. During World War II this happened again. To keep the railroads running, wood, and even maize and other grains soaked in linseed oil, were used for fuel. By the end of the war, British coal had become more expensive, partly because the British mines were nearing exhaustion. Within Argentina small coal seams of poor quality were found along the Andean piedmont between San Juan, Mendoza, and San Rafael. The largest domestic source, however, was found at Río Turbio, near the Chilean border in southern Patagonia. This coal has a high sulphur content and cannot be made into coke, but it is useful in some types of industry.

For more than 75 years Argentina has been searching and drilling for oil and gas. In 1907 oil was discovered at Comodoro Rivadavia, and since then this field has accounted for the greatest share of Argentine production. Almost half of all the oil produced in the country has come from the San Jorge Basin in this area. Recently, Argentina has built a gas pipeline from Tierra del Fuego that extends 22 miles under the Strait of Magellan. This line connects with another from Neuquén en route to Buenos Aires. Additional gas is produced in the Mendoza area of Northwest Argentina, and another pipeline brings gas to Buenos Aires from southern Bolivia. Consequently, natural gas now supplies about a fourth of Argentina's total energy needs.

Oil and gas exploration requires a huge supply of risk capital as well as technical skill. Once an oilfield has been developed, however, many people become resentful of the profits that flow out of the country if foreign oil companies are involved. Government ownership of national resources is then advocated. As early as 1936, even before Mexico nationalized its oil, the Argentine government placed all oil and gas under the exclusive control of a government agency, *Yacimientos Petrolíferos Fiscales* (YPF). Oil production declined, and Argentina had to import 60 percent of its supply. When the private companies were again permitted to operate in the 1960s, a major expansion in production occurred. Political pressure on the government to draw up new contracts with the companies then reversed the trend. Argentina now seeks to attract foreign capital and expertise to expand both exploration and production, and the country has become about 92 percent self-sufficient in petroleum.

The Economy Since World War II   The world never returned to the economic patterns of the pre-1930 era. Since World-War II, no country has been willing to accept the role of supplier of raw materials in exchange for manufactured products from industrialized societies. The policy now is to build an integrated industrial structure, based on such fundamental necessities as steel. It has been demonstrated that modern economics are vastly strengthened, and the levels of living raised, when countries with diversified, modern industries are free to exchange products over a wide range of the international market.

The dictator Juan Domingo Perón, who came to power in 1946, established a fascist-type state in Argentina. The government assumed complete control of the national economy. In 1947 the Perón government announced its first five year plan, which included large public works and industrialization designed to make Argentina economically independent of the rest of the world. Roads were built, irrigation systems were improved, slums were cleared, and cities were given a modern appearance. The number of factories increased from about 40,000 in 1935 to more than twice that number in 1950, and the number of workers employed in manufacturing rose from 577,000 to 1,108,000. In addition to the food industries already in existence, there was a considerable increase in leather goods and textiles, metal and machinery, petrochemicals, paper, and plastics. To compete with imported steel, construction of an integrated iron steel mill was initiated at San Nicolás on the Río Paraná downstream from Rosario.

These giant steps in economic development soon exhausted the capital available to Argentina and undermined the productivity of the meat and grain producers, on whose efforts the country's economy previously was based. Perón was asking the large landowners to finance the nation's economic development. The result was a decrease in the acreage of grains and, eventually, even a shortage of beef for domestic uses. Perón did succeed in attracting private investors from the United States. However, when he endorsed a contract with the Standard Oil Company of California for oil exploration, which was submitted to the Argentine Congress in 1955, the armed services revolted and forced Perón's resignation.

Since 1955 Argentina has struggled to rebuild its shattered economy. Not only did Perón flee the country with a substantial portion of the national treasury, but he also left a mountain of debts. Perhaps the most serious problem was the large number of workers who had "never had it so good" and who were ready to vote for a return of Perón's policies, or even of Perón himself.

In 1960 the San Nicolás steel plant began production. The iron ore came from Brazil, Chile, and Peru; coal came from the United States; and gas was supplied from Bolivia and northern Argentina by pipeline. A smaller steel plant was built at Rosario, and a huge complex of petrochemical industries was begun at San Lorenzo. A nationally integrated industrial system appeared likely. An upturn in the economy occurred in 1964. Unemployment was reduced, industrial production increased, and agricultural products became more abundant. By 1966 Argentina had a gross national product per capita second only to that of Venezuela in Latin America.

Much remained to be done. A major problem in 1967 was a virtual collapse of the railroads. Owned and operated by the government, they were losing a million dollars a day and contributed about half of the total budget deficit. Of the 173,000 railroad workers, perhaps 50,000 were superfluous, but the powerful railroad unions resisted all efforts to reorganize the workforce. Of the 3570 steam locomotives the railroads owned in 1966, about 2500 were more than 50 years old.

By the early 1970s, serious political and economic problems resulted in a rash of demonstrations, strikes, and riots. The cost of living rose 60 percent, and there was discontent and disillusionment in Argentina. After seven years of military government, Juan Perón was again elected President in 1973. Even as this government came to power, the situation worsened. Perón died in 1974, and his third wife, María Estela Martínez de Perón, became the first woman Chief of State in South America. She was replaced by a military *junta* in 1976, when unemployment had become widespread and the annual rate of inflation reached 900 percent.

After the military seized power in 1976, the government eliminated price controls, interest rate ceilings, credit controls, export and import taxes, and most restrictions on foreign investments in Argentina. Hence, the rate of annual inflation decreased to 45 percent by 1980, banks again began to attract savings, and unemployment declined. In terms of economic development, the Paraná-Plata and Uruguay River systems and the Atlantic Coast were given special attention.

After 1974 Argentina was involved with Uruguay in construction of the Salto Grande Dam on the Uruguay River. The Salto Grande project now generates 1890 megawatts of electricity, supplies water to irrigate 320,000 acres of land, and has fostered increased tourism between the two countries. The first rail link between Argentina and Uruguay was completed via the Salto Grande Dam in 1981, and highway connections have been completed via the dam and several bridges over the Río Uruguay.

Even greater projects are designed for the Río Paraná. The nation's largest hydroelectric projects are on the middle Paraná, along the border with Paraguay, where the Yacyretá-Apipé and Corpus dams will generate 5600 megawatts to be shared by the two countries. Atomic power plants have also been constructed but farther southward along the river. Electric power, irrigation, highway and railroad construction, navigation, and industrialization will benefit all of northeastern Argentina. Already the construction of tunnels and bridges has helped to link the Argentine Mesopotamia with the rest of the country and to stimulate its economic development.

Along the Atlantic, marine resources are increasingly being utilized. Mar del Plata, long one of Argentina's primary centers of tourism, has become its leading fishing port as well. Bahía Blanca and other cities along the route of natural gas and petroleum pipelines have become centers for the expanding petrochemical industry, and mining and smelting have increased with the expansion of the electric power supply, as illustrated by aluminum refining at Puerto Madryn and iron mining in the nearby Sierra Grande.

THE FIRST LAND LINK BETWEEN ARGENTINA AND URUGUAY IS THE FRAY BENTOS BRIDGE.

The Political Situation

One of the first great Argentine political leaders was Domingo Faustino Sarmiento, born in San Juan in 1811, a child of poor parents. At age sixteen he came upon a biography of Benjamin Franklin, and from Franklin he learned that a democracy can thrive only on the basis of an educated citizenry. During the rule of General Rosas, Sarmiento was exiled to Chile, where he became well known as a liberal writer. In 1845 he visited the United States and Europe, and became a disciple of Horace Mann. Returning to Chile, he established the first teacher-training school in Latin America, and Chile's school system became the best in Latin America.[5]

After the fall of Rosas, Sarmiento returned to Buenos Aires, where he helped to set up the new government. He was Argentina's second President, and during his presidency (1868–74) he established schools, museums, libraries, and art collections. Democracy in Argentina was well launched. After Sarmiento's death in 1888, the universities and the great newspapers of Buenos Aires carried on the liberal tradition. Yet, politics remained under the tight control of the small minority of landowners.

The established practice of constitutional government ended in 1930, when the Radical Party was tumbled from power by a military coup. From then until 1958, the government of Argentina remained in the hands of army officers, most of whom had been educated in German military schools and were inclined to take Italy or Germany as a model. Argentina proposed to become a third force in the world, neither Communist nor capitalist, and sought to lead Latin America away from dominance by the United States and Great Britain. During World War II the government strongly favored Germany and Italy, and declared war against them only in 1945 when their defeat was inevitable.

Meanwhile, an important population shift

was taking place. When the acreages of grain were reduced, tenants planted their farms with alfalfa and moved away. Having no place to go in the rural areas, they moved in large numbers to the cities, especially Buenos Aires. There was no unemployment problem, for Argentina was rapidly building new industries and undertaking programs of slum clearance in the city. To a much greater degree than in 1916, here was a worker group free from political control. The workers were not at all indoctrinated with traditional ideas of democracy, however, for their background was Italian and Spanish. It was at this point in 1943 that Perón, an obscure army colonel with great personal ambitions, became Secretary of Labor in the military government. He promoted a series of social and economic benefits for the workers and became their hero. He called the workers his *descamisados,* or "shirtless ones." For the first time in Argentine history, there was a political leader working for the poor people. He brought them together as a class, opposed to rule by the landowners and to "Yankee imperialists." Aided by his glamorous and equally ambitious wife, Eva Duarte de Perón, he rose quickly in power, until in 1945 he assumed the presidency. By that time, the army was concerned over the rise of this new political leader and tried to remove him from office. So great was the popular uprising, however, that the army had to relent. In 1946 what was probably Argentina's first uncontrolled and entirely honest election gave the hero of the *descamisados* 1,478,372 votes, compared with 1,211,666 for his rival.

Perón's policies were calculated to increase and maintain support by the workers. He decreed numerous social changes, including higher minimum wages, paid vacations, and increased social security and medical benefits. In the process, he also ran the country heavily into debt. After the death in 1952 of Evita, the people's darling, he took desperate measures to keep his followers in line, for many were already disillusioned when his economic promises failed to materialize. Perón went so far as to arrest certain Roman Catholic priests who were charged with

[5] E. Correas, "Sarmiento's Daughters: Sixty-five Who Dared," *Américas* 32, No. 1 (1980): 49–54.

efforts to undermine his government. That, plus his plan to permit a North American oil company to undertake the development of Argentina's presumed oil reserves, led army officers to organize their revolt. Peron was exiled in 1955, and a military government assumed control.

After the overthrow of Maria Estela de Perón in 1976, the military regimes sought to curb unemployment and one of the world's highest rates of inflation, as well as to maintain domestic security throughout the country. They also initiated projects to promote regional development and national integration. Yet, national unity remains an elusive goal, and international relations have brought distress. Brazil's rapid development has threatened Argentina's traditional leadership role among the Latin American nations. Boundary disputes with Chile threatened to erupt into full-scale warfare in 1978, particularly over claims to islands in the Beagle Channel south of Tierra del Fuego. In 1982 the Argentine government under General Leopoldo Galtieri launched an invasion of the Falkland Islands. These islands, controlled by Britain since 1833 but claimed by Argentina and referred to as the *Islas Malvinas*, were occupied briefly by about 9000 Argentine troops. A counterinvasion was soon launched from Great Britain and, with strategic support from the United States, inflicted a humiliating defeat on the Argentine forces. Loss of the war discredited General Galtieri, resulting in his overthrow and demonstrations against any further military rule. His successor, General Reynaldo Bignone, promised to return the country to civilian government, which was done through elections in 1983. The new President, Raúl Alfonsin, proved to be antimilitary and a strong advocate of human rights. Another consequence of the Falkland War was that Argentina turned to the Soviet Union for moral and military support during the conflict and, in exchange, became a major supplier of grain to that country.

The Falkland Islands impasse was not the only incident that threatened Argentina's status, not only among South American states, but among world powers as well. It became apparent by 1980 that Argentina was losing its fight against inflation. As Alfonsín came to office, inflation stood at nearly 500 percent. A gargantuan foreign debt and the inability to pay interest on loans caused jitters in financial markets and depressed production. Moreover, the dilemma involved major social and political problems which the new democratic government had to face.

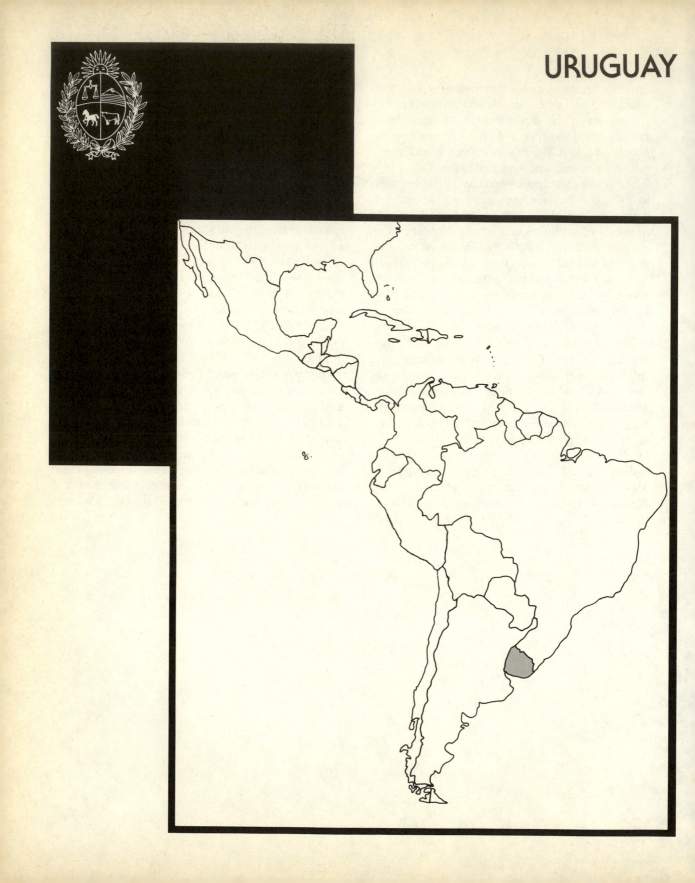

URUGUAY

28

REPÚBLICA ORIENTAL DEL URUGUAY

Land area 68,037 square miles

Population Estimate (1985) 3,000,000
Latest census (1975) 2,788,429

Capital city Montevideo 1,261,000

Percent urban 84

Birthrate per 1000 18

Death rate per 1000 9

Infant mortality rate 33.2

Percent of population under 15 years of age 27

Annual percent of increase 0.9

Percent literate 96

Percent labor force in agriculture 16

Gross national product per capita $2490

Unit of currency Uruguayan Peso

Physical Quality of Life Index (PQLI) 86

COMMERCE(expressed in percentage of values)

Exports

textiles and textile products	29	leather	14
meat	28	chemical products	4
vegetable products	15	synthetics	2

Exports to		Imports from	
Brazil	14	United States	12
Argentina	11	Nigeria	12
West Germany	9	Brazil	12
U.S.S.R.	8	Venezuela	9
United States	7	Mexico	8

Data mainly from the 1985 World Population Data Sheet of the Population Reference Bureau, the 1985 Britannica Book of the Year, and the 1985 World Almanac.

The territory now occupied by Uruguay, like that of Bolivia and Paraguay, has served throughout much of its history as a buffer zone. For several centuries it was a disputed area between the Spanish and Portuguese colonial empires in the New World. It was invaded repeatedly by both sides, as each sought to occupy and retain a strategic position at the mouth of the Río de la Plata.

Uruguay gained its independence in 1828 because of the intervention of Great Britain and an agreement between Argentina and Brazil to rec-

ognize the existence of an independent buffer state between them. Despite such delicate and difficult beginnings, this country soon became the most stable and democratic state in Latin America.

During the early twentieth century, Uruguay made rapid economic and social progress. Exports of meat, wool, leather, and grain continued to expand and enhanced the nation's prosperity, particularly during World War II and the Korean War. At the same time, social legislation transformed Uruguay from a feudal society into a

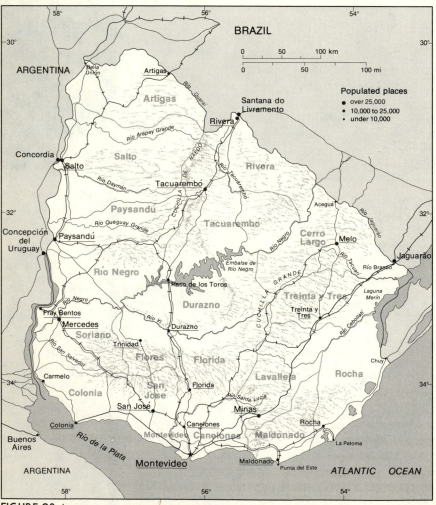

FIGURE 28-1

model "welfare state." Its citizens enjoyed the benefits of free education, free medical and hospital care, early retirement, and pensions for the elderly. Unemployment compensation was provided, the work week was reduced, and government became the nation's predominant source of jobs.

Uruaguay's "dream world" collapsed with the midcentury decline in world market prices for its commodities. Inflation became rampant, with the result that savings and investments were lost. Guerrilla activity became widespread, especially by the Tupamaro organization, followed by military takeover of the government and conflict that amounted to civil war. By 1974 the military had defeated the terrorist groups, but democracy was sacrificed for at least another decade.

The People

Uruguay, like Paraguay and Chile, possesses only one area of concentrated settlement, but this area includes the entire national territory. Uruguay is one of the few states where the effective national territory and total national territory are the same. The one cluster of people is centered around Montevideo, a city of about 1,261,000 in a total population of 3 million. This means that more than 40 percent of the people of Uruguay live in the metropolitan area of their principal city. Next in size are Paysandú and Salto, neither of which has attained a population of 100,000. The densest rural population is in the south and west, between Paysandú and Montevideo, while the remainder of the country has a scattered population with a density of less than 25 people per square mile.

Early in this century Paysandú was a major industrial community, with its manufactures based mainly on agriculture. This remains true today as Paysandú's economy focuses on the wool, sugar, beer, dairy, cement, and leather industries. Until recently, Salto served as a commercial center for the cattle-raising industry, but with construction of one of the world's largest

hydroelectric dams the city has new potential in terms of industry and tourism. Meat packing has historically been the basis of the city's industrial development and remains an important economic factor in the life of Salto today.

The composition of the Uruguayan population is similar to that of Argentina. The great majority, about 90 percent, is of European ancestry; mestizos make up about 5 percent; and blacks comprise less than 5 percent.

The Land

In many ways Uruguay is a transitional zone between the Humid Pampa of Argentina and the hilly uplands and plateaus of Brazil. A southern fringe of alluvial land borders the lower Uruguay and Plata rivers, but most of the country is hilly, with soils derived from the decomposition of underlying crystalline rocks. When the Spaniards entered what is now Uruguay, they found a landscape composed of wooded valleys with rushing streams of clear water and long, gentle, grass-covered slopes rising to distant hills. Uruguay has been called "The Purple Land," because of the faint purplish tinge given by the vistas of tall prairie grasses on smooth slopes.[1]

Uruguayan geographers divide their country into several regions, basing their divisions on the character of the surface. Along the eastern coast is a zone of lowland, composed of sandy beaches, lagoons, and wind-tossed dunes. Farther inland is a belt of hills, from southern Brazil southwestward to the southern coast of Uruguay near Montevideo. These hills form the divide between the shorter streams flowing directly to the Atlantic and the longer streams flowing westward to the Río Uruguay. Summits along the *Cuchilla Grande,* as the divide is called, reach elevations of about 1500 feet above sea level (the highest elevation in Uruguay is only 1650 feet). Westward from the Cuchilla Grande the land slopes gently

---

[1] See W. H. Hudson's description of Uruguay in the early nineteenth century in his book *The Purple Land.*

toward the Río Uruguay. The Río Uruguay itself is interrupted at several points by falls and rapids, notably at Salto. The head of navigation for ocean steamers is at Fray Bentos.

In three parts of Uruguay the granite base is covered with more recent formations: (1) Along the east coast the sandy shore deposits obscure the underlying rocks, (2) along the Uruguay-Plata shore there is a fringe of level country where the crystallines are buried under river alluvium and loess, and (3) in the central and northwestern part of the country there is an extensive cover of rock formations of later age than the granites. In the northwest, Uruguayan territory includes the southernmost part of the Paraná Plateau where flows of dark-colored lava remain as flat-lying, resistant formations bounded by sharp cliffs or cuestas.

Vegetation and Climate

Vegetation and climate are also transitional between the Argentine Humid Pampa and southern Brazil. Most of Uruguay was originally covered with a tall-grass prairie. Stream valleys were once followed by ribbons of forest, and scattered palms were, and still are, mixed with tall grass in the southeast. Today, less than 4 percent of Uruguay is forested.

The climate of Uruguay, perhaps more than that of any other part of the middle latitudes, is "temperate." Throughout the country, the average temperatures of the coldest month are near 50°F, similar to winter averages in Georgia and South Carolina. Summers, however, are cooler than in corresponding parts of North America. At Montevideo the average of the warmest month is 72°F, about the same as the average of the warmest month in Boston. Rainfall is evenly distributed throughout the year. It totals about 38 inches at Montevideo to nearly 50 inches in the north. There is a considerable irregularity in the total rainfall from year to year, but prolonged droughts are rare.

Settlement

Enthusiasts have described Uruguay, with its freedom from climatic extremes, its prevailing gentle slopes, and its abundance of clear water and nourishing grasses, as the world's finest grazing land. Yet, the physical qualities that make Uruguay a fine grazing land could not bring prosperity to the inhabitants until large international markets developed in Europe and North America, and until local people were ready to produce surpluses for export.

During the colonial period, Uruguay was remote from the centers of Portuguese and Spanish settlement. Remoteness in the case of the Portuguese was due chiefly to the great distance between Uruguay and the nearest primary settlement center at São Paulo. Yet by 1680, the Portuguese had pushed southward to the Plata shore and established a fortress at Colonia, opposite Buenos Aires. Remoteness from Spanish settlements on the Humid Pampa was due primarily to the river barrier. The Paraná-Plata is so wide, and is bordered by such a labyrinth of swamps and shifting channels, that to cross it, or even use it as a line of travel, has always been difficult. The Spaniards were more interested in their connections through Tucumán and Salta with Peru.

The Beginnings of Settlement

For nearly 200 years after the Spaniards first reached the Plata region, they made no fixed settlement in what is now Uruguay. The Banda Oriental was occupied chiefly by nomadic cattle herders, or *gauchos*. Cattle were introduced between 1611 and 1617, and were permitted to run wild and multiply. Gangs of *gauchos* followed the herds, rounding up a few animals for slaughter. Only the hides were of value unless the animals were killed near the shore, where the carcasses might be used to make tallow or salt beef. There was no attempt to claim ownership of the land or even to establish ranch headquarters. The bands of *gauchos* fought as readily for Portugal as for Spain.

The idea of landownership came slowly to Uruguay. The first contacts with the commercial world were with Brazilian traders who came to buy hides. Later, Argentine cattle buyers crossed the river from Buenos Aires to deal with the *gauchos*. These buyers eventually found it easier to employ herders to keep the cattle nearby than to follow half-wild herds over the vast interior. At first, ranch headquarters were established, and boundaries between ranches were left undefined. When more of the land was occupied, however, boundaries became important. As the zone of ranches, or *estancias*, moved northward from the Plata shore, the nomadic *gauchos* were pushed to more remote parts of the country. Landowners gradually replaced the *gauchos* with hired workers who were attached to the estates by some form of debt bondage. Even then, interior Uruguay remained a land of men, since unmarried males rather than families were preferred as the labor force.

As this type of land occupance spread northward, small villages and towns began to appear. These settlements generally were located at road junctions, and because the easiest lines of travel followed the drainage divides, or *cuchillas*, and avoided the wooded valleys, they commonly were established on the ridge-tops. At less important road junctions, perhaps only one or two retail stores would appear, surrounded by a cluster of dwellings. At the larger junctions, small towns with a considerable population of merchants would be established. Uruguay became a land of small scattered trading villages and widely spaced ranches.

The Pastoral Economy

As in Argentina, investment of British capital started Uruguay on the road to economic development. The British were the first to appreciate the potential commercial value of Uruguay's grasslands for grazing. At the time of independence, the country was already grazed by millions of cattle, but the only products exported were hides, tallow, and salt beef. Uruguay was also used for breeding mules, which were sent northward to mining communities in Brazil. In 1840 high-grade Merino sheep were introduced from Britain, and the grazing of wool sheep spread rapidly. By the middle of the century, there were an estimated 2 million sheep and 3.5 million cattle feeding on the unfenced range.

The second phase of British investment was made in 1864. The Liebig Meat Extract Company of London opened a plant at Fray Bentos on the Río Uruguay, and for the first time ranchers could sell their cattle for meat rather than hides alone. In 1868 a British company started to build a railroad to connect Montevideo with the back country.

Two inventions initiated major changes in the pastoral economy. One was barbed wire, which made possible the fencing of pastures and control of animal breeding. The other was the refrigerator ship, which made possible the shipment of frozen meat across the tropics. Both inventions came to Uruguay in the 1870s. In 1880 the first Hereford beef animals arrived from Britain, and within two decades Herefords and Shorthorns had replaced the rangy creole cattle. Meanwhile, the export of wool continued to be highly profitable. It is estimated that by 1900 there were 18.5 million head of sheep and 7 million head of cattle in Uruguay.

The first modern, large-scale refrigerated meat-packing plant, or *frigorífico*, was built in 1902 at Fray Bentos with British capital. Subsequently, three other large-scale *frigoríficos* were located at Fray Bentos. Meat exports were largely from foreign-owned plants, and a Uruguayan-owned plant was used to supply the domestic market.

Brazil is now Uruguay's largest customer for beef, taking more than half of the total exports. Most of Uruguay's meat production, however, is reserved for the domestic market. The Uruguayans probably consume more beef per capita,

A GAUCHO ROUNDS UP HIS HERD NEAR RIVERA, URUGUAY.

about 200 pounds annually, than any other people in the world. They also consume about 25 pounds of mutton annually, per person, and increasing amounts of poultry.

Land Use in Rural Uruguay

There are two major divisions of rural Uruguay. The greater part of the national territory is used for grazing on uncultivated pastures. About 90 percent of all productive land in Uruguay is used for livestock-raising, whereas crops occupy less than 10 percent. Agriculture is concentrated in the southwestern part of the country, although scattered areas of cropland are also found in the pastoral regions. Along the Río Uruguay from Paysandú southward, and along the Plata shore as far eastward as Montevideo, the land is used for crop-pasture rotation. Near the river the main crop is wheat, but farther eastward maize becomes more important. Other crops include flax, peanuts, sorghum, rice, sugarcane, soybeans, and sunflowers. The pastures of this area are used almost exclusively for cattle. Throughout this same agricultural zone are small concentrations of vineyards, from which there is a substantial production of wine.

Since World War II the government has encouraged the expansion of agriculture. An area of concentrated rice paddies has been developed in the northeast, near the Laguna Merín. Uruguay has become self-sufficient in rice production and normally produces a surplus for export. Since the 1950s sunflowers have been widely cultivated for the cooking oil their seeds provide, as have peanuts. Peanuts and sunflowers vary in acreage from year to year, and land used for these crops appears even within the pastoral zone. The government also has supported the cultivation of sugar beets in the southern part of the country, but after 1955 it also began to support the planting of sugarcane in the far northwest. Chile and Uruguay alone, among all the countries of Latin America, produce both sugar beets and sugarcane on a commercial scale. A district of orchards and vineyards, and another of intensive crop production, occur around Montevideo. There is also a zone of intensive dairy farming north and west of the city.[2]

Uruguay as a Buffer State

Independence for Uruguay resulted from influences beyond its borders. For centuries the conflict between the Spaniards and Portuguese continued without solution. When Brazil declared its independence in 1822, Uruguay was included as part of the Brazilian national territory. In 1825 the Uruguayans, with Argentine assistance, organized a resistance to the Brazilians and defeated a Brazilian force in 1827. By then it was clear that neither Argentina nor Brazil could gain complete control of this border area. Meanwhile, the British had occupied Buenos Aires in 1806 and 1807, and Montevideo in 1807. If Britain had not been involved in difficulties in Europe, the entire Plata region might have become a British dominion. In 1828 the British, whose interest in the region had not been diminished by the failure to hold the port cities, succeeded in getting both Argentina and Brazil to agree on the recognition of an independent Uruguay as a buffer state.

---

[2] Ernst Griffin, "Testing the von Thunen Theory in Uruguay," *Geographical Review* 63 (October 1973): 500–516.

Transportation

Uruguay's railroad system, like that of Argentina, was built chiefly with British capital, used British rolling stock, and burned British coal. The first line was built out of Montevideo in 1868, and by 1911 the present rail system was essentially complete. After World War II, the government used some of its sterling credits to purchase the British interests, and now the entire system is government-owned. With completion of the Salto Grande Dam across the Río Uruguay in 1982, the Uruguayan railroad system was for the first time connected with that of Argentina and other Plata River Basin countries.

About 80 percent of Uruguay's passenger and general cargo traffic moves by highway. Hence, major attention has been paid in recent years to the development of a national highway network. Present connections with Argentina include a bridge over the Río Uruguay between Fray Bentos and Puerto Unzué, a bridge between Paysandú and Colón, and a highway across the Salto Grande Dam.

Like most countries of Latin America, Uruguay operates its own airline and a system of modern telecommunications. In addition, efforts have been made to modernize the port of Montevideo to make it a primary shipping center for produce from the Plata River Basin.

Montevideo

Montevideo was not founded until 1726, when Portuguese advances forced the Spaniards to build a permanent fortress on the Uruguayan shore. They selected a site where the hilly belt of the Cuchilla Grande reaches the southern shore, providing a small sheltered harbor dominated by a low conical hill on which the fortress could be built. Almost at once Montevideo became the principal urban center of the Banda Oriental; most roads of the interior were redirected to this new port. There was not then, and there never has been since, any competitor to challenge Montevideo's supremacy as the leading center of

PUNTA DEL ESTE, A POPULAR RESORT IN URUGUAY, AT THE MOUTH OF THE RÍO DE LA PLATA.

economic, political, and social life in the country. In fact, among all the world's urban places, Montevideo is perhaps the best example of a "primate city."

Montevideo performs the functions of government, commerce, and manufacturing. Almost half of all Uruguayans live within the metropolitan area, and almost 90 percent of the nation's foreign trade passes through its port. The city serves as the major base of the South Atlantic fishing fleet and has benefited from recent expansion of the fisheries industry. Its manufacturing plants include petroleum refining, food processing, textiles, and the production of clothing and leather goods, soap, matches, and automobile tires. Montevideo is also a major resort center, attracting large numbers of tourist especially from Argentina and Brazil. Equally popular is the seaside resort of Punta Del Este, a town of about 10,000, but with a summer population of 400,000.

Uruguay as a Political Unit

Although Uruguay is the smallest nation in South America, its total area exceeds that of Denmark, Switzerland, and the Low Countries

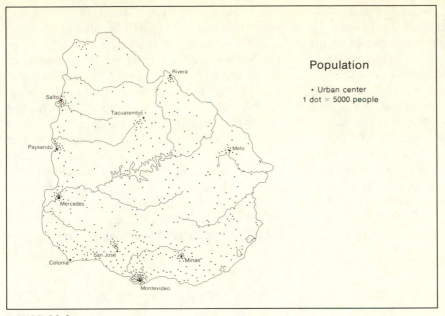

Population

• Urban center
1 dot = 5000 people

FIGURE 28-2

of Europe combined. Its population of about 3 million is more than 80 percent urban, yet, its economy is essentially rural, and national prosperity depends largely on international market prices for the products of its farms and ranches. As a buffer state, Uruguay is also immediately affected by political and economic conditions in the much larger neighboring nations.

Uruguay attained distinction among all Latin American countries through its early efforts to resolve the inequities of the traditional economic system and the injustices of political life. To a large extent these efforts resulted from the influence of José Batlle y Ordoñez, who organized the urban and working-class population under the banner of the Colorado Party. He and his successors created a welfare state that provided Uruguayans with social benefits seldom equaled elsewhere in the world. The welfare state collapsed by 1965, however, as the largely pastoral economy proved unable to support a system in which most people worked in nonproductive activities or failed to work at all.

The Economic Situation

The Uruguayan crisis was intensified in 1974 when the import price of petroleum quadrupled, while the prices paid for its primary exports, wool and meat, declined substantially. Hence, a greater degree of self-sufficiency in energy production became a high priority. Most significant was completion of the Salto Grande Dam on the Río Uruguay, a joint Argentine-Uruguayan project, which reached its full 1860-megawatt capacity in 1982. Power production of the Río Negro, long the nation's major source of hydro-electricity, was expanded by construction of the Palmar (300 megawatts) and India Muerta dams. By 1982 about 70 percent of all electricity generated in Uruguay was from water resources, and there was a steady reduction in the use of petroleum and its derivatives.

In 1981 the Uruguayan government initiated a 15-year agricultural development plan designed to provide greater self-sufficiency in food production, to stimulate industry through the processing of agricultural commodities, and to expand agricultural exports. Wool continued to be the leading export, providing more than a fourth of all export earnings, but rice, barley, citrus, and linseed were also exported. Although wheat occupies an area twice that of the second leading crop, maize, the cost of production exceeds world market prices, and Uruguay has not yet achieved self-sufficiency. Almost 90 percent of the nation's agricultural land is in family-operated farms that produce mostly for the domestic market, whereas the larger commercial farms and ranches are oriented mainly toward production for export.

Manufacturing in Uruguay has developed slowly, partly because of the lack of fuel and mineral resources and partly because of the small size of the domestic market. Such industry as does exist is based largely on the products of agriculture and ranching, such as meat packing, flour milling, the processing of oil seeds, dairying, brewing, tanning, and the manufacture of textiles. Leather products are exported, as are woolen goods produced in small factories and homes throughout much of the country. Uruguay has been actively involved in regional integration schemes, such as the Latin American Common Market and projects for development of the Paraguay-Paraná-Plata Basin, which tend to focus commercial, industrial, and transporta-

THE POSITAS BEACH IN MONTEVIDEO, URUGUAY.

tion development toward Montevideo at the mouth of the riverine system.

Waters off the Uruguayan coast abound with fish, but commercial fishing did not become a significant factor in the national economy until about 1970. Fishery exports increased from U.S. $800,000 in 1973 to more than $60 million in 1981, while tonnage during the same years increased from about 12,000 to 145,000. Related to this expansion has been increased government support for the fishing industry, the construction of new port and processing facilities.

Although small in area and population, Uruguay receives more tourists than any other country in South America. More than 90 percent are normally from Argentina, and most of these visitors arrive during the Southern Hemisphere summer, from December to April. About 1 million arrived in 1981 and occupied beach facilities extending eastward from Montevideo to Punta del Este, then northward to the Brazilian border.

Uruguay's economy suffered severely during the domestic political crisis of the 1970s. It is also strongly influenced by political and economic conditions in Argentina, and to a lesser extent by those in Brazil. By 1985, most signs pointed toward renewed prosperity within a more relaxed political environment.

The Political Situation

Uruguay became independent in 1828, but its early history was one of civil war as rival leaders vied for power. It was not until 1903, when José Batlle y Ordoñez became President, that a pattern of modern political development was formed. According to the constitution then in force, the presidential term was four years, after which the President could not legally be a candidate for reelection. In 1907 Batlle declined to continue in office, despite the fact that he had by then gained a very large public following. Uruguay has long had two major political parties, the *colorado* and *blanco*. Batlle was a *colorado*, a progressive who was vigorously opposed to dictatorships. After his first term in office he went to

Switzerland to study directly the methods of democratic government.

During his second term as President (1911–15), Batlle introduced the reforms for which he is famous. He made an accurate diagnosis of the causes of chronic disorder in his country and sought to find a political solution. He recognized that a major cause of disorder was the gap, so common in Latin America, between wealthy landowners and the workers. "It is not necessary," said Batlle, "that the rich be made poorer, only that the poor be made less poor." The reforms Batlle proposed gained many supporters not only among members of his own party, but also among the *blancos*. The constitution under which Uruguay was governed had been adapted in 1830 from that of the United States. In it the President assumed broad powers.

Batlle's influence did not end with his death in 1929. Pressure to reduce the power of the presidency became especially strong after an elected President seized control of the country between 1933 and 1938. Finally, in 1951, the *blancos* agreed to the reform originally proposed by Batlle, and in 1952 Uruguay ceased to have a President. The executive branch of the government was placed in the hands of a governing council.

The change, however, did not provide the anticipated results. With the costs of the welfare program mounting rapidly, the dominant *colorado* party, with its strength concentrated in the large industrial cities, found itself under increasing pressure from the *blancos*, who represented the majorities in the rural areas and smaller towns of the interior. In 1958 the *blancos* won the election for the first time in more than 90 years. Yet, inflation and mismanagement of certain government-owned industries aroused widespread discontent. The election of 1966 resulted in a return of the *colorados* to office. There was also a referendum regarding the form of government, as a result of which the unwieldy governing council was abandoned and Uruguay returned to a presidential system.

After 1968, the Uruguayan crisis became a

nightmare. Opposing political and social forces exploded on an unsuspecting country. An increasing foreign debt, escalating inflation, massive strikes in all sectors of the economy, conflicts between educational organizations and the government, bank robberies, and guerrilla operations spread like a forest fire. The government, under "immediate security measures," detained thousands of citizens and severely restricted civil liberties.

In 1973, amid increasing economic and political turmoil, the Uruguayan military forces seized power in a *coup d'etat.* Internal order was gradually reestablished, and by the end of 1977, Uruguayan authorities announced a plan for the country's institutional recovery. In late 1981, retired Lieutenant General Gregorio Alvarez was chosen by the armed forces to govern for a three and a half year term, to 1985, when the first elected President since the 1973 coup was chosen by democratic elections.

Uruguay, like Chile and Argentina, has known democracy, and its citizens have experienced the maximum of public benefits and personal freedom. Now, it is struggling to regain its prosperity and its traditional position as one of the world's classic examples of a democratic nation.

PORTUGUESE
SOUTH AMERICA

BRAZIL: INTRODUCTION

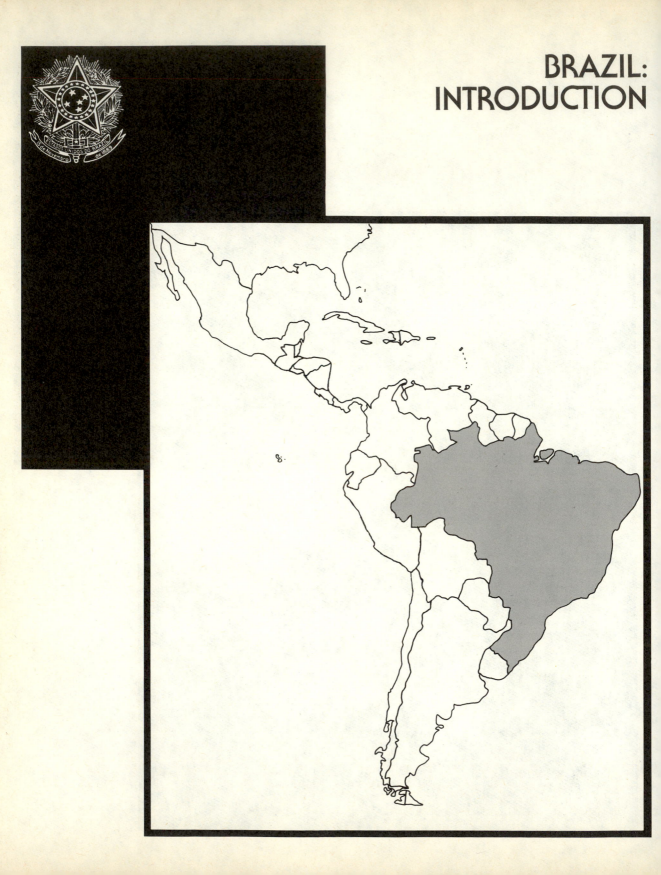

29

REPÚBLICA FEDERATIVA DO BRASIL

Land area   3,286,470 square miles

Population   Estimate (1985) 138,400,000
Latest census (1980) 119,098,992

Largest cities   Rio de Janeiro 9,014,000
São Paulo 12,589,000

Capital city   Brasília 763,254

Birthrate per 1000   31

Percent urban   68

Death rate per 1000   8

Infant mortality rate   71

Percent of population under 15 years of age   37

Annual percent of increase   2.3

Percent literate   77

Percent labor force in agriculture   36

Gross national product per capita $1890

Unit of currency   Cruzeiro

Physical Quality of Life Index (PQLI)   69

COMMERCE (expressed in percentage of values)

Exports

coffee	10.6	iron and steel	5
machinery	10	chemicals	4.5
iron ore	8.8	orange juice	2.9
animal fodder	8.7	sugar and honey	2.9
petroleum products	7.1	meat	2.6
motor vehicles	5.6		

Exports to		Imports from	
United States	20.5	Iraq	19.2
Japan	6.5	Saudi Arabia	15.2
West Germany	5.8	United States	15
Netherlands	5.6	Venezuela	5
Italy	4.9	Japan	4.6
France	4.3	West Germany	4.4

Data mainly from the 1985 World Population Data Sheet of the Population Reference Bureau, the 1985 Britannica Book of the Year, and the World Almanac.

Brazil, the world's fifth largest nation, stretches across almost half of the South American continent. It is inhabited by more than 138 million people, or about one-third of the total population of Latin America. It has large areas of thinly occupied backlands that are being developed through pioneer settlement by farmers and ranchers. It is a youthful, talented country, one in which great wealth and extreme poverty form a striking contrast. The major regions of settlement are: the Northeast, the East, São Paulo, the South, the Central-West, and the North. These also constitute official planning regions established by the Brazilian government.

Modernization and industrial growth have proceeded at a rapid pace since 1964, while the agricultural sector has developed less rapidly. Only 11 percent of the land is under cultivation, and most of the better quality agricultural lands already have been cultivated for many years. Efforts are being made to integrate the interior savannas and rain forests into the national economy, but the economic and ecological costs are high.

FIGURE 29-1

A major problem in Brazilian geography has been one of how to map this immense territory and how to evaluate its natural resources. A partial solution was provided by RADAM (Radar Amazon), an aircraft-mounted scanning radar that could operate night or day in any kind of weather. It could map soils and vegetation, as well as physical features. Helicopters dropped teams of specialists who used sophisticated techniques to supplement air photographs. The entire Amazon Basin was mapped by 1972, and all of Brazil was mapped and assessed for future development by 1975. This was a major accomplishment, yet those who have followed the story of settlement in Latin America know there are no simple answers to problems in this extraordinarily complex country.

The Land

Only a small part of the vast national territory of Brazil is too wet, too dry, or too steep to permit some kind of economic use. It is known to contain huge stores of iron ore, manganese, and other metallic minerals. On the other hand, perhaps more than any other large country in the world, and certainly more than the United States, Brazil faces problems resulting from the unfavorable geographic arrangement of its physical features. Resources are not arranged in a way to favor low-cost development. Along the greater part of the eastern coast the land faces the sea with a steep escarpment across which there are few easy routes of travel. Passage to the interior is especially difficult from the superb natural harbor at Rio de Janeiro. Meanwhile, the world's longest navigable river winds through empty forests. Little of the Brazilian territory is actually mountainous, yet the mountains are located in Brazil's core area near Rio de Janeiro and São Paulo. From the mountains, rivers radiate inland thousands of miles westward and southward to the Plata or northward to the Amazon and then eastward to the sea. The absence of a natural focus of routes scatters and isolates the clusters of population.

Surface Features

The largest plains area of Brazil is in the upper Amazon Basin, where level lands stretch eastward from the base of the Andes in Colombia, Ecuador, Peru, and Bolivia. Unlike most river lowlands, the Amazon Basin becomes narrower downstream. In the eastern part only a ribbon of floodplain carries the river through the highlands. The Atlantic Coast, especially where densely populated, is bordered by only small, discontinuous bits of lowland. There is no real coastal plain like that of eastern North America.

Most of the Brazilian territory is composed of highlands. The Brazilian highlands south of the Amazon and the Guiana highlands to the north are constructed of a basement of geologically ancient crystalline rocks. These are covered, in part, by stratified sandstones and limestones, and by sheets of diabase. Throughout the highland regions both north and south of the Amazon, three kinds of surface features occur together in a complex pattern. Where crystalline rocks are exposed, the surface is composed of gently rounded hills, deeply mantled with a fine, reddish clay soil. Above this surface, bodies of especially resistant crystalline rocks stand out boldly as low mountains. Where the covering of sandstone strata or diabase remains over the basement crystallines, the surface is tabular in form with large flattish areas along the stream divides.

The Brazilian highlands, for the most part, slope abruptly toward the Atlantic. North of Salvador there is a gentle rise from coast to interior, but from there southward to Pôrto Alegre the coast is backed by a steep, wall-like slope, the Great Escarpment, which so much resembles a range of mountains when viewed from the ocean that it is called the Serra do Mar. Along most of its course the slope is broken into a series of steps, forming parallel escarpments.

Most rivers rise in the central and southeastern part of the highlands and descend over the steep margins in falls and rapids. The Paraná system is fed by several tributaries in São Paulo, Paraná, and Santa Catarina, which rise within sight of the Atlantic Ocean, flow westward into the interior, and join the Paraná along the western border of these states. The Paraná itself drops over resistant diabase formations near the northeastern border of Paraguay, forming the Guaíra Falls (known in Brazil as the *Salto das Sete Quedas*), then downstream as far as Posadas; in Argentina, it passes through a valley cut deeply into the plateau. Similarly, the headwaters of the Rio Uruguai flow westward before turning southward along the Argentine border. The river is interrupted by rapids all the way southward to Salto in Uruguay. The entire southern part of the Brazilian highland is drained through these circuitous channels to the Río de la Plata.

Similar features are exhibited by the rivers that drain northward. The São Francisco rises north of Rio de Janeiro and flows parallel to the coast for more than a thousand miles before turning eastward along the northern border of Bahia state and descending over the Paulo Afonso Falls en route to the Atlantic. The great tributaries of the Amazon, the Tocantins-Araguaia, the Xingú, and the Tapajóz, all rise in the central area, flow northward, and descend over falls and rapids as they approach the Amazon. Only the Amazon itself is navigable far into the interior.

Climate

A considerable amount of misinformation exists regarding the temperatures of tropical countries such as Brazil. The world's highest temperatures are not found near the equator, but in the deserts at 25° or 30° of latitude. Average annual temperatures increase toward the *heat equator,* which, in South America, passes along the Caribbean and Guiana coasts through places such as Maracaibo and Georgetown. The range of temperature between coldest and warmest months, however,

decreases near the heat equator. In the equatorial regions temperatures are moderately high throughout the year but are never as high as they are during summer in the lower middle latitudes.

Human comfort depends on humidity and air movement as well as temperature. Where there is high humidity and little wind the average person tends to be uncomfortable. Such conditions occur in Rio de Janeiro during the summer months. In most parts of Brazil, rainfall is heaviest in summer, while winter tends to be dry. More than 80 inches of rain per year are received in the upper Amazon lowlands, in scattered places along the Great Escarpment, on mountain summits of the southeast, and in a small area of western Paraná state. Moisture deficiency, on the other hand, is limited to a small part of the Northeast, where the main problem is one of rainfall irregularity. There, excessive rainfall and flooding may be followed by years of extreme drought.

Natural Vegetation

The heavy rainfall of the Amazon Basin and along the coast south of Salvador is reflected in the world's largest area of rain forest, or selva. The selva is composed of evergreen and broadleaf trees, some of which achieve great size. As many as 3000 different species of trees per square mile have been identified. Overhead, branches are interlaced to form a dense canopy that blocks sunlight that might otherwise reach the forest floor. Soils under such forest, if not covered by newly deposited material, are usually poor in plant foods and in humus. Heavy rains percolating through upper layers of soil dissolve the soluble minerals, and high temperatures and humidity destroy organic material that falls to the ground.

A seasonal or semideciduous forest appears where it is less rainy or not so continuously warm. Where rainfall is nevertheless substantial and groundwater conditions are propitious, this forest is almost evergreen. Where water is less abundant, the forest is mostly deciduous and only a few species of trees retain their leaves throughout the year. The interior of Brazil, south of the evergreen forests of the Amazon, is covered with a mixture of forest and grassland. Luxuriant stands of semideciduous forest cover the wetter areas. Where the soils do not retain their moisture, the land is covered with scrubby deciduous trees and grass. Where scrub woodland predominates, there is a vegetation known as *campo cerrado* (closed grassland). In other places there are vast expanses of savanna with scattered trees called the *campo sujo* (literally, "dirty grassland"), and in other places, especially in the South, there are open grasslands (*campo limpo*, or "clean grassland") with no trees except in the valley bottoms. In the interior of the Northeast, where droughts are common, there is a thorny, deciduous, drought-resistant woodland which the Brazilians call *caatinga*.

In southern Brazil two quite different types of vegetation occur. From southern São Paulo state southward are the Araucaria or Paraná pine forests which provide most of the lumber for Brazil's construction industry. At about the same latitude, pure tall-grass prairies with dense ribbons of forest occur along the deeper river valleys. These prairies continue southward to form the predominant vegetation of Uruguay.

Mineral Resources

Few countries of the world possess mineral resources comparable to those of Brazil. The Brazilian and Guiana highlands form parts of what is geologically an ancient crystalline shield, which comprises the core of the South American continent. The world's largest reserves of iron ore are found here, forming the basis of a major iron and steel industry and a substantial export trade. Most of the iron ore mined to date has been extracted by open-pit methods in eastern Minas Gerais state, but extensive deposits have been identified recently within the Amazon Basin. In 1967 the vast reserves of the Serra dos Carajás were discovered and are now estimated to contain 20 billion tons of ore with 66 percent iron content. Manganese is mined in Minas Gerais and is also mined and exported in large quantities from Amapá territory. Other deposits of this mineral occur on the Bolivian border near Corumbá. During the eighteenth century Brazil produced about 44 percent of all the gold mined in the world. Gold is still mined in eastern Minas Gerais by underground shafts, and in many parts of the Brazilian highlands and Amazon Basin by placer methods. New gold reserves have been discovered in the Cariré area of Ceará state and in northern Pará. Brazil also possesses abundant reserves of bauxite, copper, tungsten, lead, zinc, nickel, chromium, uranium, and tin. It is the world's largest producer of industrial diamonds, semiprecious gem stones, and quartz crystals for use in radios.

The most conspicuous mineral deficit is in hydrocarbon fuels, such as coal, petroleum, and natural gas. Brazil experienced a severe economic crisis during the 1970s, when world petroleum prices skyrocketed and domestic oil production supplied only about a fourth of the country's needs. Exploration was quickly expanded from the existing oilfields of Bahia and Sergipe to other parts of the country, and especially to locations in the Atlantic Ocean offshore from the Northeast. Proven reserves, however, amount to no more than 1.5 billion barrels, while 4 billion barrels must be imported annually to supply the nation's industrial and agricultural needs.

One energy resource that Brazil does have in abundance is water power. With the advance of modern technology, huge dams have been constructed on the major rivers, and extensive transmission networks supply electricity to the growing urban industrial centers. Brazil also leads the world in the use of alcohol as a motor fuel. With the manufacture of alcohol-powered automobiles and the conversion of motors in those already built, it is estimated that more than a million vehicles will eventually run on 100 percent alcohol. At present, the alcohol is produced primarily from sugarcane, but many other crops are potential sources for the future.

The People

Perhaps nowhere on earth has there been a greater blending of different kinds of people than in Brazil. The early racial components were the native Indians, African blacks, and European Portuguese. This population has been altered greatly during the past century by the arrival of millions of immigrants from Europe and Asia. All these different elements have mixed freely, for one of the important attitudes brought over by the Portuguese was the absence of any taboo against race mixture, except among the aristocracy.

The Indians who inhabited Brazil in 1500 were chiefly of Tupi-Guaraní stock, the same linguistic group to which the Indians of Paraguay belong. They were hunters, fishers, gatherers, and shifting cultivators. They had no intertribal political organization, and they lived in small scattered groups. Their basic food was manioc, instead of maize. It is estimated that in 1500 the Indian population in all of Brazil was only about 800,000. Today, about 100,000 survive, living mostly in remote parts of the country.

The Tupi-Guaraní were quite inadequate as a source of labor. In the early years of the conquest, many died from European diseases. The women conducted agricultural labor, while the men engaged in fishing, hunting, and fighting. Intermarriage between the Portuguese men and Indian women introduced many of the physical traits into the resulting population.

The blacks, too, made an important contribution. Beginning in 1538, slaves were brought across the ocean in great numbers from Africa, especially to the Brazilian Northeast, where there was a need for field hands in the new sugar industry. The African was a good worker and possessed a knowledge of certain technological processes. The Sudanese, for example, had invented the process of iron smelting. Many of the Africans served as expert foremen on the plantations and in the mines.

The main characteristics of the Brazilians, however, came from the Portuguese who already combined a wide variety of racial and cultural elements. Like the Spaniards, the Portuguese included ingredients of Celtic, Nordic, and Mediterranean origins, and from the south they inherited Moorish and Semitic traits. When they came to the New World, their main objective was to loot the rich resources of a virgin land. They cared but little for spreading Christianity, being far more interested in opportunities for speculative profits.

Course of Settlement

In the more than 450 years since the Portuguese established colonies on the coast of South America, three products, in turn, have dominated an era. In each case there was the spectacular rise of a commercial product, a period of great profit from an expanding market, and an eventual decline because of more competitive areas of production outside Brazil. The products that have made Brazil famous are sugar, gold, and coffee. There also have been minor interludes neatly set off in time and space that have been dominated by rubber, cacao, and oranges, and, more recently, by urban construction and industry.

The early years of Portuguese colonization did not feature commercial development. Brazil was neglected until encroachment by the French and Spaniards made it imperative for the Portuguese either to establish colonies on the American coast or to relinquish their claims. The Portuguese occupation of Brazil was like the British occupation of North America in that extensive parcels of virgin land were accessible to waves of new settlers. In North America, however, the land was usually given out in family-sized holdings to newly settled immigrants, while in Brazil it was awarded in huge parcels to individuals who would organize settlements. The motive was profit.

The Portuguese carried forward their conquest of Brazil from three primary settlement centers: São Paulo, Salvador, and Recife. Founded on its present site in 1567, Rio de Janeiro was at first only a fortress and naval base, and not at all a focus for settlement of lands in the interior.

Sugar Colonies

The first Europeans to cultivate commercial crops in America were the Portuguese. For several decades before the Portuguese settled in Brazil they had been using their island possessions, such as Madeira, for the cultivation of sugarcane and by 1470 had broken the Venetian monopoly on the sale of sugar in Europe. When a lack of gold made some other source of wealth important to find, the Portuguese began to expand their sugarcane plantations on the mainland. The first were developed at São Vicente in 1532, using Indian workers who proved to be poor agriculturists except in the cultivation of tobacco. Meanwhile, plantation owners in the Northeast began to appreciate the possibility of developing a profitable economy based on sugarcane and African slaves. The problem was how to finance the purchase of slaves and sugar machinery, and how to reach a large enough market to make the investment profitable. The Dutch provided a solution. Dutch merchants financed the purchase of slaves and machinery, furnished transportation, and collected the profits from refining sugar in Antwerp and Amsterdam and selling it throughout Europe. Portugal benefited from the manufacture and sale of sugar mill equipment, and from collecting royalties. In the seventeenth century the Northeast of Brazil was the world's leading source of sugar.

From 1630 to 1654, the Dutch occupied the entire coastal area of the Northeast from the Rio São Francisco to the Amazon. Without any help from Portugal, which was involved in a struggle to gain freedom from Spain, the Portuguese in Brazil resisted the Dutch invasion, and in 1654 recaptured Recife. This was an important episode in Brazilian history, for the cooperative effort necessary to eject the Dutch built certain loyalties and traditions among the people of the Northeast. Even today this region is noted for its political solidarity.

Gold

While sugar production brought wealth to people of the Northeast, settlers in southern Brazil experienced no such prosperity. The people of São Paulo were poor, having discovered no source of wealth to exploit. From São Paulo a series of semimilitary expeditions were sent out to explore the interior of the country in search of gold. These expeditions were called *bandeiras*, and the members of the expeditions were called *bandeirantes*. Gold had already been discovered in stream gravels south of São Paulo, but southern Brazil proved to have gold only in small quantities. The *bandeirantes* found instead large numbers of Indians, many of whom had been brought together around mission stations. Intermarriage with the Indian women became common, and soon there were considerable numbers of *mamelucos*, a racial type known as mestizo in Spanish America. The *bandeirantes* traveled slowly over the vast interior of the continent, pushing the borders of Brazil far to the west and south. Searching restlessly for slaves, gold, or any other source of wealth, they grazed their animals on the savannas and stopped to plant and harvest crops along the way. They established Colonia on the shores of the Plata opposite Buenos Aires in 1680, pushed westward to the Paraguay north of Asunción, and roamed into the scrub woodland country inland from the sugar colonies of the Northeast. In 1698 they discovered rich gold-bearing gravels in central Minas Gerais, at Cuiabá in Mato Grosso in 1719, and near the former capital of Goiás in 1725. In Minas Gerais, just north of the gold fields, diamonds were discovered in 1729.

The result was a gold rush, in which not only *Paulistas* from São Paulo and Portuguese from the home country participated, but also many former sugarcane planters of the Northeast who came with their slaves. The gold period reached its peak between 1752 and 1787, and was definitely over by the beginning of the nineteenth century. During this time southern and central Minas Gerais was transformed from a wilderness into a well-populated agricultural, pastoral, and mining region partitioned among a relatively small number of landholders. Settlement of this part of Brazil led to the development of Rio de Janeiro as a port, because it become the chief outlet for the gold and the principal urban nucleus of the new region of settlement. By the beginning of the nineteenth century the best sources of gold and diamonds had been exhausted. Brazil was ready for a new form of speculative development.

Coffee and Other Commercial Crops

Brazilian agricultural history has been dominated in recent decades by the commercial production of coffee, especially in the states of São Paulo and Paraná. By 1825 there was a definite concentration of coffee in the Paraíba Valley, from which coffee planting spread westward into São Paulo state. It was coffee that supported the rise of the great city of São Paulo, which is now the leading industrial center in all of Latin America and one of the largest cities in the world.

The rubber tree (*Hevea brasiliensis*) is native to the Amazon Basin but did not become of major commercial interest until the rise of the automobile industry in the early twentieth century. An economic boom then occurred, and Brazil enjoyed a near-monopoly on the market. The boom was short-lived, however, once the development of rubber plantations in other parts of the world proved successful. Other products, such as Brazil nuts, dyes, and *maté* also supported such booms, after which Brazil was replaced by other suppliers of the world market.

Destruction of the Forest

Agricultural development in Brazil has occurred largely on land once covered by semideciduous forest. For more than four centuries the relatively small area of this forest type has supported Brazil's commercial agriculture as well as the production of food for the Brazilian people. In 1700 only the forests of the Northeast and a few spots around São Vicente and Rio de Janeiro had been cleared. Between 1700 and 1800 large areas in the vicinity of Rio de Janeiro, in the Paraíba Valley, and in southern Minas Gerais were cleared to provide food for the gold and diamond miners. By 1930 clearing of the forest had swept over São Paulo and into the southern states. It had also progressed in the Northeast. By 1950 very little virgin forest remained. Near Rio de Janeiro and São Paulo, and in the Paraíba Valley the second growth was cleared so often and at such brief intervals that today not even brush will come back on the bare land. On steep slopes soil erosion occurs at an alarming rate. Before most Brazilians were aware of the situation, much of the resource base on which the country had depended was destroyed.

Immigration

The rate of population increase in Brazil was slow until the late nineteenth century. Birth rates were relatively low and infant mortality was high because of poor hygiene and a lack of nutritious foods. With the arrival of Africans in the Northeast, this area became the most densely populated part of Brazil. As late as 1870, half of all the nation's people lived in this region.

A stream of immigrants, mostly Europeans, came to Brazil between 1822 and World War II. Most went to São Paulo state, where they worked on coffee plantations. About 34 percent were Italians, 30 percent Portuguese, 12 percent Spaniards, and 3 percent Germans. The remaining 21 percent included many other nationalities. Today, there are also about 750,000 Japanese-Brazilians, 80 percent of them Brazilian born, residing mostly in the states of São Paulo and Paraná.

Southern Brazil was first penetrated by the *bandeirantes*, who used the prairies for grazing their cattle and mules. Later, colonies of German farmers were established in Rio Grande do Sul and Santa Catarina. These small landowners were soon joined by new groups of Italians and Poles.

Cities and the Sertão

Brazil's architects lead the world in creating new forms and designs in construction. Many of Brazil's cities reflect a superbly modern appearance. Brazilian writers, however, have insisted that the "real Brazil" is not to be found in the cities, but in the back country, beyond the frontiers of concentrated settlement, in the land that Brazilians call the *sertão*.[1]

The *sertão* is not a wilderness of unexplored territory, for it has been tramped over extensively. It has been lived in, its resources have been exploited, and its landscapes have been modified over the course of more than four centuries. It forms a fringe around the margins of the effective national territory and has served as a transition zone that has withstood the forces of change. The *sertão* is essentially pastoral in nature, with small groups of people clustered around ranch headquarters or in small towns. When droughts occur in this hinterland, the unlucky victims are forced to leave in order to survive. Great masses of people migrate to the Amazon region or to the southern cities in search of employment. Many have been employed by construction companies and industrial concerns in Rio de Janeiro and São Paulo, where they earn low wages if they are unskilled. Some learn

quickly and are able to save a little money to send home to their families in the Northeast. Some settle in the *favelas*, the slum areas on the outskirts or hillsides of the cities.

Brazilians who live in the great cities have a strange fascination for the *sertão*. They have nurtured the dream of a vast wealth of resources lying dormant in the interior. The pastoral people of the *sertão*, however, are not like the city dwellers, or even like those of the agricultural areas. They are almost pure Portuguese, but with some mixture of Indian. They are essentially democratic, with no rigid class distinctions, for the ranch owners look, act, dress, and live like their workers. They are a fiercely independent people, courageous, resourceful, and superstitious. Yet, they are so widely scattered, or gathered in such small groups, that they cannot support the cost of the numerous things that bring a society forward from a pioneer life to one that can be described as modern.

Since 1960, the Brazilians have taken major steps to transform the emptiness of the backlands. First, they built a new capital city, Brasília, in the midst of the *campo cerrado*, near the headwaters of three great river systems. Then highways were built for thousands of miles through almost empty country to connect Brasília with the still more remote interior. From all over the backlands, people are attracted to the capital, and thousands of others have made clearings along the highways to grow subsistence and commercial crops.

A pessimistic expression once commonly heard in Brazil is that "This country is a land of the future — and always will be!" Yet, enormous progress has been made in recent years toward the objective stated on the Brazilian flag: "*ordem e progresso.*"

---

[1] Pronounced *sair-tówng*. The plural is *sertões*, pronounced *sair-tó-aish*.

BRAZIL:
THE NORTHEAST

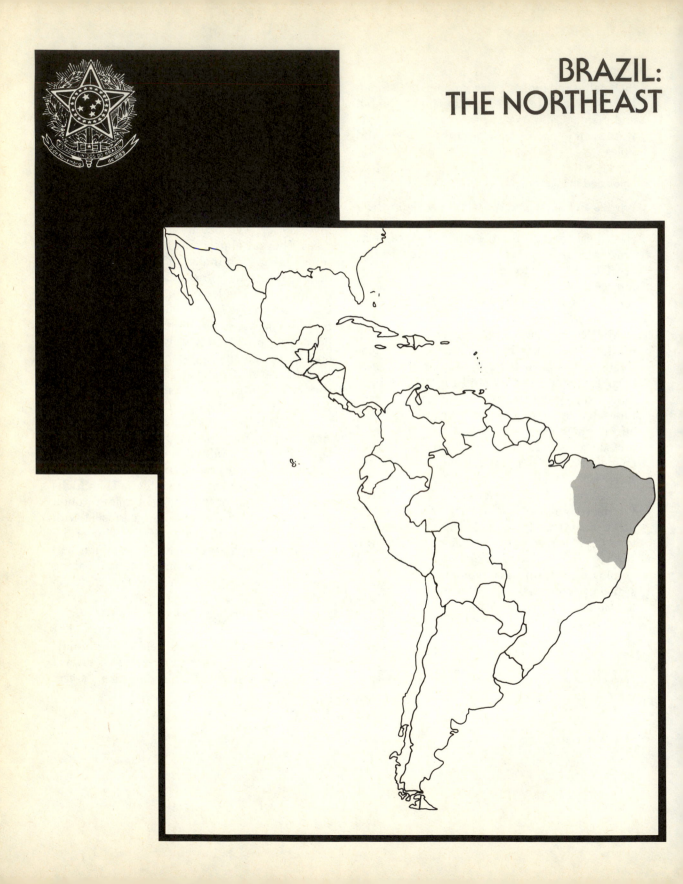

30

The Northeast has played an influential role in the growth and evolution of Brazilian life. Yet, this part of the country has resisted modernization and change. Even during prosperous years most of the people have remained consistently poor. This region includes the nine states of Maranhão, Piauí, Rio Grande do Norte, Paraíba, Pernambuco, Alagôas, Sergipe, Ceará, and Bahia. It comprises 18 percent of the nation's land area and contains about 30 percent of its total population.

The first part of Brazil to acquire spectacular wealth was the Northeast. It was here that Europeans, with the help of African slaves, first built a civilization in the tropics, and it was here that a striking contrast developed between the relative prosperity of the coastal region and the stark poverty of the *sertão*. For centuries, the principal product of the Northeast has been people, many of whom have contributed to the nation's artistic, intellectual, and political life. Others have migrated to various other parts of the country in search of greater economic opportunity. They have panned gold in Minas Gerais, tapped rubber in Amazonia, planted coffee in São Paulo, and harvested cacao in southern Bahia. Yet, the region retains a sense of unity, and those who leave do not forget their regional heritage.

The Northeast is also one of Latin America's poorest regions. It has a history of natural disasters that have added to the misery of people who are dependent on the land. Emigration has been the best solution for those unable to make a living in the backlands. Today assistance is provided to the region from many national and international organizations. Most important are basic education projects designed to improve both regional productivity and the quality of life within the northeastern states.

The Land

Two contrasting parts comprise the Northeast. First is the *mata*, a forested area corresponding to a belt of reliable rainfall, more than 40 inches annually, along the coast south of Natal. It was here that early exploration and settlement followed the discovery of a dye source, called Brazilwood. Portuguese and other Europeans developed the land, having quickly learned that as the forests were cut the fertile zone of the *mata* became ideal for the cultivation of sugarcane. The forest has been almost entirely removed, and the land it once covered is characterized by reddish clay soils. In sharp contrast to this coastal area is the zone of the *caatingas*, a land of recurring droughts and floods. Its sandy, light-colored soil provides poor farmland. Throughout the *caatingas* are islands of wetter land, especially on hilltops that were once forest-covered like the zone of the *mata*.

Surface Features

The landscapes of the Northeast include five major kinds of surface features. The first is composed of broad plains with erosion remnants that formed on ancient crystalline rocks. These plains rise gradually from sea level on the northern coast of Ceará southward to about 1500 feet in northern Bahia. The second category includes hilly uplands and low mountains whose features are massive and rounded in outline. The largest upland area is in Pernambuco and Paraíba. This is the Borborema, whose summits are about 2500 to 3000 feet above sea level. Another category of surface is the cover of sandstone strata overlying the crystalline base in Piauí and Maranhão, and in small outliers farther to the east.

Coastal *taboleiros*, or flat-topped mesa-like structures, form a fourth major surface feature.

Administrative Divisions
and their capitals

States

1. Amazonas
4. Pará
5. Maranhão
6. Piauí
7. Ceará
8. Rio Grande de Norte
9. Paraíba
10. Pernambuco
11. Alagoas
12. Sergipe
13. Bahia
14. Goiás

15. Rondônia
16. Acre
17. Mato Grosso
18. Mato Grosso do Sul
19. Espírito Santo
20. Minas Gerais
21. Rio de Janeiro
22. São Paulo
23. Paraná
24. Santa Caterina
25. Rio Grande do Sul

Territories

2. Roraima
3. Amapá

26. Distrito Federal

FIGURE 30-1

These are supported by a cover of fairly recent sedimentary strata in a narrow belt extending southward from Rio Grande do Norte state almost to Rio de Janeiro. The *taboleiros* carry a deeply leached, sandy, and infertile soil. The exposed crystalline hills are covered with a reddish clay soil that has been the primary agricultural base for more than four centuries.

The fifth category of surface features includes the sedimentary Basin of Bahia. The lowland around the *Baía de Todos os Santos* is known as the Recôncavo. Here the sedimentary strata have dropped down between two up-standing blocks of crystalline rock. This basin is of particular importance to Brazil, for not only has its clay soils supported sugarcane plantations over the centuries, but in modern times it has also been an important source of petroleum and phosphate.

The Rio São Francisco

The Brazilians have a sentimental, almost reverent attachment to the Rio São Francisco. This great river rises in central Minas Gerais state and flows 1800 miles northward through the backlands, or *sertões*. It has been compared with the Nile, but it is no Nile, for its water cannot be used to irrigate a productive floodplain. Throughout most of the dry Northeast, where water is most needed, the lay of the land is such that irrigation would be costly and could be applied to but small areas. Yet, the Rio São Francisco is far from useless.

Near the western border of Alagôas are the Paulo Afonso Falls. The crystalline rocks of this area are broken by two systems of faults. The falls occur where the water has excavated a deep trench at right angles to the course of the river at this point. Into this trench the water descends about 275 feet in a series of tremendous cascades. Current hydroelectric installations have a capacity of 535 megawatts, which is being expanded to 915 megawatts, and the electricity generated supplies seven states of the Brazilian Northeast.

In the states of Sergipe and Alagôas, in the lower São Francisco Valley, there are broad floodplains called *varzeas*. Here, an impressive type of polder project has been underway since 1975. Irrigation and drainage systems have been constructed, and extensive areas planted in rice. A resettlement program of displaced families has been implemented, and research conducted on fish production. Another program in the state of Sergipe is POLONORDESTE. This program provides credit to 8500 small farms, a colonization scheme for 1400 small farmers, and agricultural support services.

Climate and Vegetation

The profound contrast between the two parts of the Northeast is due primarily to climatic differences which, in turn, account for differences in the natural vegetation and soil. The coast south of Cape São Roque receives regular rains, brought by cold air masses as they push equatorward during the Southern Hemisphere winter. Recife, for example, receives an average annual rainfall of about 65 inches, most of it between April and July. This coastal zone of abundant rains extends inland only about 40 to 50 miles in the state of Pernambuco and is even narrower in Sergipe. On the coast northwest of Cape São Roque, São Luis de Maranhão receives an average annual rainfall of about 85 inches, most of which occurs between January and June, as in the eastern part of the Amazon Basin. These rains are brought by the indraft of warm, moist air from the equatorial North Atlantic. Between eastern Maranhão on the west and the coast of Pernambuco on the east lies a triangular-shaped zone of irregular rainfall, subject to intervals of floods and droughts. Sometimes a whole year may pass with no rainfall at all; in other years there may be excessive downpours.

The calamities of flood and drought have been the curse of the *sertão*, for both pastoralists and farmers. As population increased, so did the problem of survival. Between 1877 and 1879, Ceará had one of the most prolonged droughts in history. During this period thousands of rural people fled to the cities, and others left to gather rubber in the Amazon. Similar but less spectacular emigrations occurred in other drought years, especially 1915, 1932, 1939, 1958, and 1978. The Brazilian government has long recognized this part of the Northeast as a disaster zone and has developed programs to deal with the region's problems on both a long-term and emergency basis. The effectiveness of such programs is at times severely tested by nature.

The vegetation of the Northeast is a direct reflection of the climate. The rainy belt was originally occupied by a growth of semideciduous forest, becoming a rain forest in southern Bahia. On the higher summits of the dry area there were also patches of semideciduous forest. Where the forest has been cleared from the higher slopes, runoff is rapid and the valleys below are subject to flooding.

The broad plains of the interior are covered with caatinga. In the wetter areas the caatinga is dense, with trees as much as 30 feet high. In drier areas, the trees are low and widely spaced. After the first rains the caatinga is colorful, with green leaves and brilliant flowers. During the dry season, or in years of drought, the leaves drop from the trees, which stand bare and brown in the intensive sunshine.

In the caatinga there are many valuable plants. There are the caroá, from which fiber can be produced, and the mamona plant which yields castor oil. Among the trees are the oiticica palm, the oil of which is useful in the manufacture of paint; the carnauba palm from which a wax is extracted; and the cajú, which produces tasty fruit and the commercially valuable cashew nut. Large areas of the caatinga have been cleared for firewood and charcoal, or simply burned to make room for pasture grasses and forage crops.

Settlement

During the first decades of Portuguese settlement in Bahia, the colonists learned from the Indians how to grow tobacco and to smoke it in the form of cigars. The earliest agricultural devel-

opment in the Northeast was located on the high country west of the Recôncavo. The Indians were set to work growing tobacco, with which they were already familiar. This area has been used continuously for the production of tobacco since that time and remains one of Brazil's leading sources of the crop.

Sugar Plantations

After a few decades, the Portuguese began to realize the possibilities of profit from the commercial production of sugarcane. This Moorish crop, aleady familiar to the Portuguese through their experience with plantations in the Madeira Islands, gave excellent yields in the rainy, forested parts of the Northeast. Forest clearing and cane planting spread rapidly in parts of the Recôncavo of Bahia, near Salvador, and along the base and lower foothills of the Borborema highlands in Pernambuco, near Recife.

After 1538 in Bahia and 1574 in Recife, African slaves were imported in large numbers. Thereafter, the population of the sugar region comprised four classes: the Indians, who survived in considerable numbers only in Bahia; the African slaves; the Portuguese landowners; and the poorer people of Portuguese descent, many of whom took Indian or African wives, and most of whom settled in the towns and cities as small traders or fishermen. On the plantations, or fazendas, the wealthy landowners built substantial homes, erected churches, and set up the engenhos, or sugar mills. In contrast were the miserable slave quarters built of mud and thatch. The actual preparation of the land, planting and harvesting of the cane, and production of sugar were supervised by African foremen who were masters of sugar-producing technology. Brazil became the world's leading producer of sugar, and the cane lands were soon occupied by a relatively dense population.

The Dutch, who had played an important role in development of the sugar economy of the Northeast, were expelled from Brazil by 1654. By the end of the seventeenth century, sugar producers in Brazil faced increasing competition in the markets of Europe. During the eighteenth century, hundreds of thousands of people, including both owners and slaves, left the Northeast, heading first for the mining areas of Minas Gerais and southern Bahia and later for the coffee frontier inland from Rio de Janeiro. New owners continued to produce sugar, however, by traditionally inefficient methods. Sugar production continued after the emancipation of slaves in 1888, with the use of tenant workers.

Sugarcane in the Modern Period

Sugarcane is still grown in the Northeast, although its total production is now surpassed by that of São Paulo. In 1967 a program of economic development for the Northeast included reconstruction of the sugar industry. New machinery was installed and modern agricultural practices introduced, which included the use of fertilizers, insecticides, and improved varieties of cane. One result of the modernization was that thousands of workers were temporarily idled. Nevertheless, by 1974, Brazil had displaced Cuba as the world's largest exporter of sugar. Exports rose to 2.2 million tons, although annual domestic consumption exceeded 100 pounds per person.

Brazil is now the world's leading sugarcane producer and is second in total sugar production only to the Soviet Union, which relies on sugar beets. About a third of Brazil's sugar comes from the Northeast, which has one of the world's lowest yields per acre. A growing domestic market is the largest consumer, and the Northeast can compete with São Paulo only because of government subsidies and preferential tax measures.

The worldwide petroleum crisis of the 1970s had a particularly severe impact on countries such as Brazil which produce only a small part of their total petroleum consumption. Consequently, the Brazilian government provided strong support for a program to convert petroleum-based consumption of energy to the use of alcohol produced mainly from sugarcane. During the Iraq-Iran War in 1980, Brazilian supplies

of imported petroleum were seriously threatened; hence, the government decided to stop all sugar exports so that maximum amounts could be converted to alcohol as a motor fuel. Brazilian production of sugar is controlled by the Instituto de Açúcar e Alcool (IAA), which assigns production quotas to each mill and distillery for the crop year.

Tobacco in Bahia

Tobacco has been grown in Bahia for many centuries. The highland area just west of the Recôncavo receives abundant rainfall; a sandy, porous soil permits rapid drainage; and a nearly flat surface retards soil erosion. The first clearing and planting were done by the Indians and the sale of tobacco paid for the early imports of slaves from Africa. Tobacco from Brazil was soon available for consumption all over Europe.

The first type of tobacco exported from Brazil was "twist tobacco," made by twisting partly dried leaves into ropes that were wound around poles and fermented in the sun. This type may be used for pipe tobacco, chewing tobacco, or cigarettes. Very little is now exported, but a substantial amount is used domestically.

Cigar tobacco is produced mainly in the states of Alagôas and Bahia. Salvador, the capital of Bahia, is the center of the cigar trade. Leaf tobacco from northeastern Brazil, both the dark cigarette and cigar types, is strictly for exports. Competition from Zimbabwe and other African countries has become quite severe. Hence, the Brazilian government has sought to supplement tobacco with more attractive alternatives, such as black beans and castor oil.[1]

Other Specialized Crops

Along the Atlantic Coast between Salvador and Natal, and in small humid spots in the hilly interior, many areas rely on a single commercial crop. In Pernambuco coffee is grown on the wet, east-

---

[1] Attaché Report, U.S. Department of Agriculture, 1980, BP-0011, São Paulo, Brazil.

facing slopes of the Borborema; around Campina Grande in Paraíba there is agave; in the zone of the *mata* along the coast there are coconut plantations; and along the lower São Francisco rice is grown on miniature sections of floodplain.

The most important commercial crop in the *caatinga* is cotton. Although two-thirds of Brazil's cotton is grown in São Paulo, the Northeast remains an important producer. About 60 percent of the crop in the Northeast is "tree cotton," a variety that grows for several years without replanting. Yields are irregular and low. It is estimated that about 3 million acres are devoted to cotton in the Northeast, but it is interplanted with maize, beans, and other crops for domestic use. Of considerable secondary value is the cot-tonseed oil used in cooking, and cottonseed cake which is used as feed for cattle.

In the zone of the *caatinga* are many tiny spots of intensive agriculture along the banks of rivers that are alternately flooded and exposed by the uncontrolled rise and fall of water. These strips are called *vazantes,* and the intensive cultivation practiced on them is called *vazante agriculture.* The most important are along the north-flowing rivers of Piauí, Ceará, and Rio Grande do Norte.

Forest Products

The agricultural economy of the Northeast is supplemented by a variety of products from wild plants. At an early date, Indians collected tree

DRESSED TO PROTECT THEMSELVES FROM INSECTS AND THE SUN, THESE WOMEN PICK COTTON IN NORTHEAST BRAZIL.

cotton and used it to weave textiles. Some tree cotton is collected today to make strong cord. The caroá plant, which grows in the *caatinga*, has bayonet-like leaves that are made into a fiber used to weave hammocks. Also from the *caatinga* is the carnauba palm, the leaves of which supply a wax used in the manufacture of self-polishing floor wax, lipstick, shoe polish, carbon paper, and protective coatings for metal. With proper harvesting, the carnauba palms can remain productive as long as two centuries. Another valuable tree is the babaçú palm, which produces a great quantity of nuts and oil. Maranhão is the leading source, with lesser amounts coming from Piauí, Goiás, Ceará, and Bahia. With proper care, two harvests a year are possible, but destructive methods have destroyed millions of the trees. In Rio Grande do Norte approximately 61,000 acres are planted in cashew trees, and there are also extensive areas in Ceará. Fortaleza is a center for the processing and export of cashew nuts.

Cacao in Southern Bahia

The southern part of Bahia is not at all like the rest of the Northeast, for there is enough precipitation to support a tropical rain forest. It was here that another Brazilian boom crop developed. That crop was cacao, grown on red clay soils under dense forest cover.

Despite destructive methods and conflicts over property rights, the region between the inner margin of the *taboleiros* and the base of the Great Escarpment is Brazil's leading cacao producer. In the late 1950s, nearly 20 percent of the world's cacao came from this area. For many years Brazil was second only to Ghana, but in 1975 it slipped to third place after Ghana and Nigeria. Cacao-growing is concentrated especially around Itabuna and Ilhéus.

Economic Development

Partly because great numbers of Brazilians from the Northeast have found economic success in other parts of the country, an intense regional pride and almost spiritual attitude prevail with regard to the potential economic opportunities in the Northeast itself. Yet, the region overall remains poor.

An inadequate transportation system hampered communications in the Northeast for many generations. Even when a network of new roads was built during World War II, the great distances involved made distribution of food and materials difficult. By 1950 this network was connected with that of the Rio de Janeiro and São Paulo area, with the result that migration was facilitated southward. It is estimated that about 30 percent of the migrants settled on Brazil's emerging agricultural frontier, which followed the spread of coffee from São Paulo to Paraná. Migrants also were attracted toward the new capital of Brasília and the Belém-Brasília Highway. It was hoped that poor farmers from the Northeast would resettle in the Amazon region, but efforts to facilitate such resettlement proved costly and were opposed by the large landowners.[2]

A rapid increase in the use of trucks took place as a result of the road-building program. One market town after another became linked in a new pattern of trade. No longer were *sertanejos*, the inhabitants of the *sertão*, forced to endure a continuing isolation and deepening poverty. Concern about the prevailing poverty has prompted the government to offer tax incentives for the transfer of capital investment from other parts of Brazil and has encouraged land settlement projects and commercial fishing in previously underexploited areas.

In 1959 the Brazilian government established SUDENE in order to speed the process of development in the Northeast and to lessen the gap between levels of living in this region and in the Rio de Janeiro and São Paulo region. Another step was the construction of schools as part of an attack on widespread illiteracy. The proportion

---

[2] "INTERCOM," Washington, D.C.: Population Reference Bureau, 1979, 7:11 and 12:8.

of rural children who never attend school is no longer large, but low attendance and a high drop-out rate are continuing problems. The biggest problem of all is the enormous number of pupils who enter school every year as a result of high birth rates. The birth rate in the Northeast is 50 percent higher than that in the industrialized areas of Brazil.

Part of the SUDENE program has been devoted to the improvement of agriculture. Modern technology and more productive crops have been made available. Steep and badly eroded lands have been reforested, and more than 40 centers have been established to promote the improved breeding of cattle and hogs. A vigorous program of industrial development has also been pursued.

A search for minerals continues. Oil wells have been drilled in the Recôncavo Basin and in the bay it surrounds, while new sources of petroleum have been discovered in Sergipe, Alagôas, Piauí, and Maranhão. By 1980 offshore reserves surpassed those onshore and were located principally adjacent to Rio Grande do Norte, Sergipe, and southern Bahia. New sources of gypsum, tungsten, and iron ore also were found.

Some effort has been expended to promote tourism in the Northeast, where many miles of beautiful beaches line the Atlantic Coast. Colonial architecture, a distinctive cuisine, modern hotels, and numerous historical sites are additional attractions. Recife, the commercial capital of the Northeast, has a population of more than 1.5 million. Like Rio de Janeiro, it is famous for its celebration of Carnival just prior to the Lenten season and has become a regional center for tourism.

The Northeast, once the center of Brazilian progress and prosperity, is now an economically depressed region that tends to retard the process of modernization in the country as a whole. At the same time, it contributes much to Brazilian culture, including historical traditions, music and dance, and religious practices. It also is the source of a wide variety of agricultural, mineral, and marine resources which contribute to the nation's overall productivity.

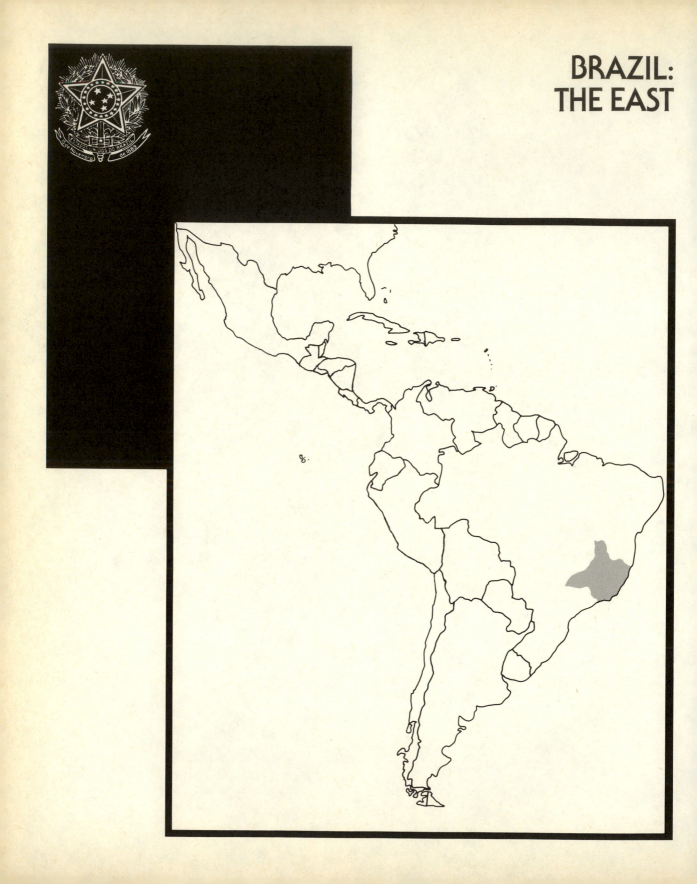

BRAZIL:
THE EAST

31

Midway between the North and South of Brazil is Guanabara Bay. Portuguese sailors entered the bay in January 1502 and named it Rio de Janeiro (River of January). The first settlement on its shores, however, was made by an expedition of French Huguenots. In subsequent years there were many skirmishes between the French and Portuguese for possession of one of the world's finest natural harbors, with the Portuguese finally gaining control in 1567. Today, the city of Rio de Janeiro is the focus and reflection of Brazil's past and present. The old capital of Brazil, Bahia, was replaced by Rio de Janeiro in 1763, which remained the seat of government for almost two centuries. A new capital was established in Brasília in 1960.

The settlement of the East gave Rio de Janeiro its start, and the roads of this region and the *sertão* beyond still lead to the shores of Guanabara Bay. Unlike the Northeast, this part of Brazil features no strong regional consciousness, for the people think of themselves as belonging to the state of Rio de Janeiro, of Minas Gerais, or of Espírito Santo. The East comprises about 8 percent of the land area of Brazil and 18 percent of its total population.

The outstanding characteristics of the region are the intricate arrangement of its surface features and the absence of a natural focus of lines of travel. The population is scattered in small, isolated units and is loosely attached to the land. Settlers who first came to the East in search of gold remained to gain a living from agricultural and pastoral activities. Repeatedly, new speculative forms of land use were introduced, but at the end of each speculative cycle, the land was returned to pasture.

The Land

The East is composed of many small natural areas, each distinct from the others. Climatic conditions and natural vegetation show wide transition zones and fewer striking contrasts, as compared with those of the Northeast.

Surface Features

For the most part, the East is a complex arrangement of crystalline hilly uplands and low mountains. There are few large level areas, except for the deltas of the Rio Paraíba and Rio Doçe, and the floodplain of the middle Paraíba. Along the coast, inland from the swampy lowlands, the Great Escarpment rises like a wall. A break in the outliers of this mountain front allows the

A DRAMATIC ENTRANCE TO RIO DE JANEIRO; "SUGAR LOAF" GUARDS THE HARBOR AT GUANABARA BAY.

sea to enter through a narrow opening. This is the spectacular entrance to Guanabara Bay, on which Rio de Janeiro is located. It is an opening guarded by the famous Sugar-Loaf, a knobby peak typical of the rainy tropics where crystalline mountains are rounded by the natural processes of weathering and erosion. Shores of the bay are generally low, swampy, and fringed with mangrove.

The massive, rounded mountains of the East stand above the general surface of crystalline hilly uplands. The highest peak in Brazil is the Pico da Bandeira (9462 feet), on the border between the states of Minas Gerais and Espírito Santo. The southern end of the Serra do Espin-

haço, in Minas Gerais, helps form the divide between the tributaries of the Rio São Francisco and 10 rivers that drain eastward to the Atlantic Ocean. The landforms here, beginning on the east coast, rise in a series of steps toward the west. These are the result of consecutive uplifts of the highlands, each starting a new erosion cycle. Successive uplifts have resulted in still lower erosion surfaces, as rivers have cut back into the higher elevations. The Serra do Espinhaço is important also for its gold and diamonds, which supported settlement in the eighteenth century, and for its vast deposits of iron ore and manganese.

Between Rio de Janeiro and São Paulo is the

valley of the Rio Paraíba, parallel to the Atlantic Coast. It is a major route of transportation between the two largest cities of Brazil, and is also a center of heavy industry and a major agricultural area. Another lowland is the Baixada Fluminense, which surrounds Guanabara Bay and includes both plains and low rounded hills shaped like half oranges. The northern edge of the Baixada Fluminense is sharply terminated by the base of the Great Escarpment. East of Petropolis this escarpment is surmounted by the Serra dos Orgãos.

Climate and Vegetation

The narrow coastal fringe of the East does not have excessively high temperatures at any time of the year, and in winter the weather can be surprisingly cool owing to the passage of cold air masses from Antarctica. Rio has an almost ideal climate, but during the warmer months, many of its residents retire to beach resorts or to Petropolis, where the altitude is 2748 feet and where the average temperature is 9° lower.

Most of eastern Brazil receives 40 to 60 inches of rainfall, but the slopes of the Great Escarpment and mountainsides near the coast have recorded more than 80 inches in scattered localities. Throughout the area, maximum rainfall occurs during the summer months, and the cool season is comparatively dry.

Over much of the region the original vegetation has been changed dramatically. It seems probable that the coastal zone in front of the Great Escarpment and some rainy areas on the inland mountains were originally covered by a dense rain forest. The southern and eastern parts supported a semideciduous forest that extended as far inland as the Serra do Espinhaço. The Serra itself, as well as the São Francisco Basin to the west, was and still is covered mostly with scrub woodland. An exception is in the South, where the use of wood for charcoal has destroyed the last of the original forest.

Settlement

At first, the highlands of the East remained less well known than many more remote parts of Brazil. When Rio de Janeiro was finally established, after the French had been dislodged in 1567, its function was that of a defense post. In fact, its site on the shores of Guanabara Bay was deliberately selected because of its inaccessibility from the interior and the consequent unlikelihood of attack from that direction.

The first explorers to reach the highlands did so through the valley of the Rio Doce from Vitória. This great natural highway to the interior was occupied by the warlike Botocudo Indians, however, who resisted European occupation. The only settlements that survived the Indian attacks were places like Vitória, on an island, and Ilhéus, on a site that could be defended from the interior.

Minas Gerais did not become important until 1698, when *bandeirantes* from São Paulo discovered gold in the stream gravels of the southern Serra do Espinhaço. Paulistas had also pushed eastward into the middle Paraíba Valley, where Taubaté became the principal urban community.

The Gold Period

After 1698, when gold was discovered throughout Minas Gerais, people rushed to settle in the East. Mining in the eighteenth century was entirely from stream gravels, and where it proved especially profitable, towns were established. One of the most important settlements was Villa Rica, now Ouro Preto. After the discovery of diamonds in 1729 near Tijuco, now Diamantina, there was another dash of settlers still farther north. A large number came from Portugal, some from São Paulo, and many from the Northeast. Those from Bahia brought their slaves with them to work in the new zone of exploitation.

The gold fever lasted about a century. During this time, the streams and hillsides were gutted and torn, and then abandoned as other parts of the world produced gold at a lower cost. Many miners returned to São Paulo with considerable capital; some planted coffee; and others returned to their beloved *sertão*.

Today, Minas Gerais produces tin, chrome, nickel, silver, and gold. Diamantina is still engaged in the mining and processing of semiprecious stones. Iron ore, however, accounts for a large part of the state's income. It is estimated that the iron ore reserves exceed 38 million tons.

Belo Horizonte and Ouro Preto

In the late nineteenth century major changes occurred in the pattern of settlement. One of these was the transfer of the state capital of Minas Gerais from the old mining town of Ouro Preto to the entirely new city of Belo Horizonte in 1896. Ouro Preto occupied a mountainous site in the Serra do Espinhaço where there was no room for expansion. The new site was literally on the edge of the *sertão* on the northern and western side of the Serra do Espinhaço, where a large area of gently sloping land was available. New government buildings and residences, wide streets, and a general pattern similar to that of Washington, D.C., give Belo Horizonte an atmosphere entirely different from that of the older cities. The new capital has grown rapidly. In 1920 there were 55,000 inhabitants; in 1950 about 350,000; and by 1980 close to 2 million. Its economy is based largely on industry, including steel, automobiles, tractors, textiles, and leather goods. In 1933 Ouro Preto was made a national monument. This historical city is a favorite of tourists, and once a year it becomes the capital again, but only for a day.

Land Use

Land-use patterns in the East have been marked by several cycles of speculative growth, followed by subsequent collapse. Among the agricultural cycles that have left their imprint on the land of the rural East are those of sugar, coffee, rice, and oranges. Considerable parts of the East, however, are not touched by any of these.

The Sugar Cycle

The cultivation of sugarcane was the first of the speculative cycles to develop in the East. This crop had been grown since early colonial times at low elevations along the coast. Many growers owned plantations around the outskirts of Rio de Janeiro, and the word *engenho* (sugar mill) is preserved in the names of many sections of the city. The expansion of cane plantations was deterred temporarily because of gold fever, but as gold mining declined, sugarcane became a crop of major importance. Yet, sugarcane also declined, first because of impoverishment of soils after years of hard use, and second because it became more profitable to fatten beef for the city market than to grow sugarcane.

The Coffee Cycle

Coffee trees had been grown around Rio de Janeiro and in the Paraíba Valley since about 1774, but coffee became commercially important only with the creation of a market in the European and North American cities. The Paraíba Valley was the first area of rapid expansion. Here, peak production was reached by the middle of the century, but around 1860 coffee plantations began to decline. Competition with the better areas of São Paulo state sealed the doom of these coffee *fazendas*.

Brazil remains the world's largest producer of coffee, but this crop no longer completely dominates the economy. Great effort has been made to diversify agricultural production and to moderate the effects of recurring natural disasters. Yet, in 1975, Brazil experienced its worst frost in 30 years, with devastating effects on a vast number of coffee plantations. It became necessary to plant 600 million new trees, many in Minas Gerais, rather than farther southward, so as to be free from similar occurrences in the future. By

1980 Brazil had made great progress in recovering its coffee production as frost and rust-resistant types of coffee were introduced.

The Rice Cycle

Another agricultural cycle appeared with the development of a rice district in the middle Paraíba Valley. Widely grown as a subsistence crop, rice is a major item in the Brazilian diet. Thus, it was only natural that a specialized area should appear to support the large urban populations of Rio de Janeiro and São Paulo, and it was fortuitous that a section of the Paraíba was physically well suited for the cultivation of this crop. Between 1918 and 1920, the Paraíba floodplain began to be used for the production of rice on a commercial scale. At first, the harvest was consumed only locally. Then, as prices for rice in São Paulo and Rio de Janeiro rose, land values rose accordingly, and a speculative boom began. Many landowners became wealthy and departed for the good life in the city, leaving their estates in the hands of tenants.

Brazil is now the largest producer of rice in the Western Hemisphere, although its yields have always been low. Moreoever, there has been a heavy dependence on upland rice grown on dry land and thus extremely vulnerable in years of drought. A government program has been developed to establish new irrigated rice areas in Rio Grande do Sul and Santa Catarina, and in selected regions of other states.

The Orange Cycle

The most recent boom crop in the East of Brazil is the orange. Its cycle began in 1930, when trees were planted in the Baixada Fluminense on steep hill slopes, and then extended rapidly along the line of the Central railroad through the Paraíba Valley and into São Paulo. During World War II, plantations were almost abandoned, but by 1968 Brazil surpassed the United States as an exporter and a decade later sold more frozen orange juice to the United States than to any other country. Today, almost 90 percent of the orange juice traded on the world market is of Brazilian origin. The major producing area is in Minas Gerais near the town of Frutal. Brazil does not export fresh oranges to the United States, but the export of concentrate to Florida is a large part of this growing industry. American orange juice processors mix the Brazilian concentrate with Florida juice, and the products are indistinguishable.

Use of Land for Pasture

In most speculative agricultural developments, there is always a possibility that the land will be returned to pasture. Cattle have provided a steady source of income through the centuries, and a large proportion of southern Minas Gerais and Rio de Janeiro states remain devoted exclusively to grazing. The traditional system of land rotation includes scattered, temporary fields planted to field crops, such as maize, rice, beans, manioc, and bananas. Close to large cities, however, as in the Baixada Fluminense and Paraíba Valley, there is substantial pressure on the land to produce food. This condition allows increasingly shorter rest periods for the soil. As a result, the soil has become so leached and eroded that even tree growth cannot return. The carrying capacity of the land for cattle has also declined.

The demand of city dwellers for food, basically rice and meat, increases steadily, and so do food prices. Cattle formerly brought from open ranges of the *sertões* to fattening pastures, and then sent by rail to be slaughtered in the city, are now slaughtered in the area of production. The meat is then shipped in refrigerated trucks to the consuming area, especially to Belo Horizonte. Livestock has been improved by the introduction of Zebu, Charolais, Holstein, Jersey, and Guernsey breeding stock. Along with the pastoral activities there has been an increased development of dairying, which produces cheese, butter, and powdered milk.

Fishing

Traditionally, the little fishing that occurred in the East was conducted by fishermen from coastal villages who went out to sea in small colorful sailboats called *jangadas*. These raft-like boats were primitive, and the catch was never large, but the real problem was that the market for fish was small and restricted to the immediate shore.

The Brazilian ocean waters are rich in lobsters, shrimp, and many other edible species of crustaceans and fish. Modernization of the fishing industry is still sorely needed. Foreign countries, especially Japan, are supplying modern fishing boats, equipment, and spare parts, a move that it is hoped will stimulate fishing and fish processing in the East and in the remainder of Brazil as well.

Minerals and Industries in the East

Brazil's East is endowed with a notable store of minerals. Despite the decline of placer mining soon after the settlement of Minas Gerais, a wealth of gold, diamonds, and semiprecious stones remains. In addition, there is an abundance of minerals that are vitally important to modern large-scale industries. From remote

THE ITABIRA IRON ORE MINE IN MINAS GERAIS WHERE 150 CARS ARE HAULED BY 5 ELECTRIC DIESEL LOCOMOTIVES.

areas of the *sertão* come small but important supplies of such minerals as zirconium, chromium, molybdenum, nickel, tungsten, titanium, industrial diamonds, beryl, mica, and quartz crystals for electrical use. Among the renowned semiprecious stones are topaz, tourmaline, amethyst, and aquamarine.

One source of gold has supported a stable mining community for 150 years. This is the famous Morro Velho mine which was opened at Nova Lima, near Ouro Prêto in Minas Gerais state, by a British company in 1834. Morro Velho has the distinction of being the second deepest gold mine in the world, after the famous Boksburg mine of South Africa. Its shafts have followed a rich gold-bearing vein for more than 8000 feet, extending well below sea level at the mine's deepest excavation.

Iron and Steel

The largest reserves of iron ore in the world are found in Brazil. A major deposit is in the "Iron Quadrilateral," in the state of Minas Gerais, where the ore has an iron content of 66 percent. A century before gold was discovered in Minas Gerais, Brazilians were aware of the huge deposits of iron ore, but no value could be placed on this resource until modern methods of large-scale mining and a demand for things made of steel developed.

At the time of World War I, Brazilian ore bodies had hardly been touched. When the quantity and quality of these ores were first described in 1910, there was a scramble for mining concessions. The ore bodies were quickly divided among British, North American, French, Ger-

THE VOLTA REDONDA STEEL MILL.

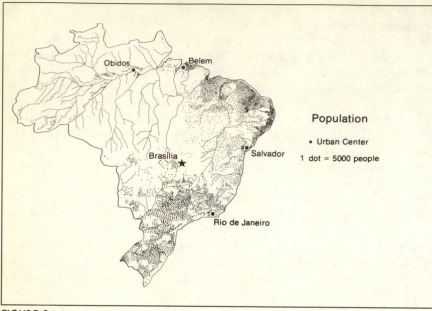

Population

• Urban Center

1 dot = 5000 people

FIGURE 31-1

man, and Brazilian investors. The largest ore body was at Itabira, on the eastern side of the Serra do Espinhaço. Another large deposit was found at Lafaiete, near the main line of the Central Railroad. In 1919 the North American businessman Percival Farquhar planned for the large-scale mining and export of iron ore from Itabira.[1] He employed the best engineers, had sufficient capital to finance the project, and was assured of an adequate market in the Ruhr of Germany. In Brazil, however, he ran into stubborn opposition. The Brazilians felt that local resources should be used to enrich Brazil. As a result, South America's first large-scale mining and export of iron ore took place in Venezuela. For many years, the steel industry was dependent on charcoal as a fuel. This meant a rapid destruction of the forest. Brazil did build a modern steel plant during World War II, at Volta Redonda in the middle Paraíba Valley. North American technicians were assigned the task of building the plant and training Brazilian workers. Volta Redonda began production in 1946, and since then 15 other steel mills have been constructed. Numerous small plants have also been built in Minas Gerais where pig iron and steel are produced. By the mid-1960s, Brazilian steel production reached 1.5 million tons; by 1980 the figure had risen to 8.3 million.

An ample supply of iron ore, manganese, limestone, and water has supported the iron and steel industry. Further stimulus has been provided by those industries that rely on steel, such as automobile manufacture and shipbuilding, both of which have grown at an astonishing rate.

Rio de Janeiro

Rio de Janeiro is one of the world's most strikingly beautiful cities. It was originally established where the towering ribs of the coastal mountains border the western side of Guanabara Bay. From there it spread to several adjacent narrow lowlands between projecting rocky ridges. It has

[1] See the masterful study of this important person by Charles A. Gauld, *The Last Titan, Percival Farquhar: American Entrepreneur in Latin America,* Hispanic American reports, Special Issue, Stanford University, 1964.

never stopped growing, as it extended both northward behind the coastal mountains and southward along the crescent-shaped white sand beaches of the Atlantic Ocean. High above, on Corcovado Peak, stands a statue of *Christ the Redeemer* with outstretched arms as if to extend its blessings on the city.

Copacabana, once clearly the favorite beach of Rio de Janeiro, now competes with eight other beaches that are reached from the downtown area through a series of 13 tunnels leading to the beach communities of Ipanema, Gávea, Leblon, and Barra da Tijuca. Broad avenues speed traffic through the bustling business district, past modern homes and apartments, and along the waterfront.

The city has extensive docks that border Guanabara Bay, but the mangrove swamps once infested with fever-carrying mosquitoes have long since been cleared away. The downtown area has the highrise structures common to most large cities, as well as the traditional public buildings, hotels, stores, and restaurants. A network of roads surrounds the urban area and provides spectacular views of the city and its natural setting.

Rio de Janeiro also has its slums, the *favelas*. The nearly 10 million people who live in metropolitan Rio include many who are crowded into flimsy shacks clinging to hillside slopes, some overlooking the business district or the beach at Copacabana. These people have little opportunity to appreciate the magnificence of the landscape of which they are a part. Yet, those who have migrated to Rio consider themselves far better off and experience a vastly superior quality of life than when they lived in the backlands.

Becoming adjusted to the *Carioca* way of life is not difficult.[2] Many young men and women devote themselves to pleasure, and this is easy to do in the establishments that cater to youthful activities on the beach, in the restaurants and cafes, and in the many discotheques. The age of the chaperone is dead. Women are free to seek employment in various fields, but the number of jobs available in the modern sector is limited and the competition keen. Most jobs go to men, although improved employment prospects exist for the better educated single women. With the advent of transnational corporations and export-oriented industries, employment opportunities also have developed where there is a dependence on low-paid female labor.[3]

Much of the wealth of Brazil was concentrated in Rio de Janeiro while it served as the national capital. The leaders of the country congregated here. When the capital was moved to Brasília, and efforts were made to move people into the backlands, it seemed that Rio de Janeiro might decline in importance, but this has not happened. The city has continued to grow and now claims about 10 million people in its metropolitan area. It is a well-loved city, of which the Brazilians are justly proud, and it continues to attract travelers from all over the world.

---

[2] *Carioca* is a term applied to the people and features of Rio de Janeiro.

[3] K. Newland, "Women, Men, and the division of Labor," *Worldwatch Paper* 37. No. 13 (1980).

BRAZIL: SÃO PAULO

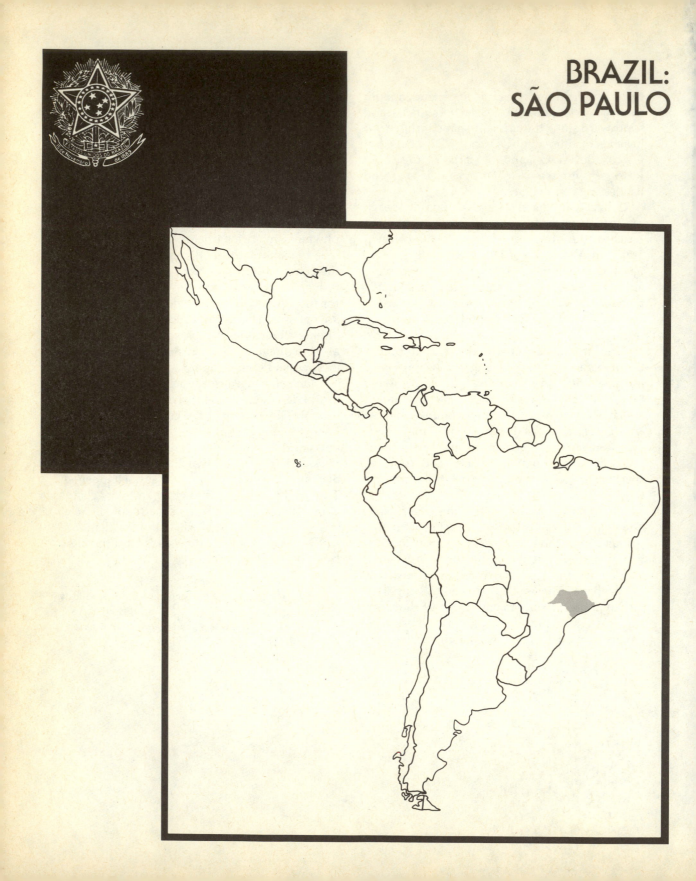

32

Two hundred miles southwest of Rio de Janeiro is the thriving, dynamic city of São Paulo, with a metropolitan area that includes about 13 million people. The activity of this burgeoning population makes a profound impression on the traveler and on the 600,000 migrants who pour into the city annually from the Northeast and other parts of Brazil in search of employment. It is the fastest growing city in the world, with a projected metropolitan-area population of 20 million by 1990, and it is the largest of all Brazilian urban communities.

The state of São Paulo occupies only 3 percent of Brazil's total area, but it is inhabited by about 20 percent of its population. The story of São Paulo is one of vigorous economic activity; of speculative wealth from products such as sugar, gold, and coffee; and of a persistent search for maximum profits from minimal investments. Its most important event was the spread of coffee cultivation across the state and the resultant wave of speculative exploitation. The origin of this phenomenon dates back to the early nineteenth century, when "coffee shoppes" had become a fad in England and when coffee was being advertised as the cure for a long list of diseases. The new beverage gained swiftly in popularity. As with the spread of sugarcane three centuries earlier, Brazil began to produce at the beginning of a rising market and went on to capture a major share of the international trade.

Early in the present century, Brazil began to be plagued by an overproduction of coffee, but by 1966 the government took effective steps to reduce production by uprooting millions of trees. Now, an International Coffee Agreement sets quotas for coffee-producing countries of the world.

In addition to coffee, São Paulo state has been a leader in the development of commercial crops such a sugarcane, cotton, oranges, and bananas. Furthermore, a spectacular development of manufacturing, especially since World War II, has been concentrated in and around the city of São Paulo. As a result, this city has achieved a level of economic production unequaled in any other part of the country.

The Land

The physiography of São Paulo state is more convenient for its inhabitants than is that of the East. There is the well-known combination of low mountains, crystalline uplands, and tablelands of stratified rock, but the organization of these elements is fairly simple. Climatic conditions along the coast are similar to those that occur in the rainy tropics, but in the highlands much lower temperatures are experienced. In the southern part of the state, there is a zone of transition between areas suited to tropical plants and areas where such plants cannot grow because of low temperatures and frost.

Surface Features

Inland from Santos, the Great Escarpment rises to elevations of nearly 2900 feet and forms an unbroken slope from the edge of the upland to the sea. This is the Serra do Mar. At the foot of the Escarpment is a narrow alluvial lowland that nature has fashioned into long curving beaches of white sand. Santos is built at the easternmost end of this lowland fringe.

The upland, west of the Great Escarpment, is drained by tributaries of the Rio Paraná. The valleys are wide and generally swampy along their bottoms, whereas the interfluves rise in broad sweeping curves to summit elevations of about 2600 feet above the sea. Between the gently sloping hill country of the São Paulo Basin and the Paraíba Valley there is a sharp drop. The Central Railroad which connects São Paulo with Rio de Janeiro descends about 550 feet in 15 miles as it leaves the São Paulo Basin.

Only the southern and eastern parts of São Paulo state lie within the area of crystalline rocks. The oldest sedimentary layers, which rest on the crystallines, are subject to stream erosion, with the result that where the layers outcrop, they have been excavated to form an *inner lowland.* The inner lowland is terminated to the north and west by the steep face of a cuesta that is capped by almost vertical cliffs of diabase. In western São Paulo, most of the Paraná Plateau is composed of sandstones resting on the diabase. Wherever the rivers encounter diabase in their channels they form rapids, as is the case along much of the course of the Rio Paraná.

Soils

In São Paulo state, soils are closely related to the underlying rock. On the crystalline upland the reddish clay soil, known as *massapê,* is similar to that found throughout eastern Brazil where crystallines are exposed under a semideciduous forest cover. Among the soils of the Paraná Plateau, the *terra roxa,* formed on outcrops of diabase, is best known. This is a deep porous soil containing considerable humus, which can be easily recognized by its dark reddish-magenta color. When wet it is so slippery and sticky that travel over it is very difficult, whereas in dry weather it gives off a powdery red dust that stains everything it touches. On the outcrops of sandstone, a light-colored, sandy soil is formed, known as *terre arenosa.*

These soils cannot be described as good or bad, fertile or infertile. The *terra roxa,* for example, is excellent for coffee trees, since its porosity permits roots to penetrate far into the ground. Its chemical composition, however, is such that for cotton it is an inferior soil, because the plant tends to form branches and leaves instead of fiber. Many of the soils of São Paulo erode rapidly, especially when they are used without fertilizer for the production of cotton. If the *terra roxa* is allowed to dry at the surface, it is apt to erode during the next heavy rain.

Climate and Vegetation

This entire region is one of abundant moisture. Cloudiness and rainfall along the coast are substantial, and the zone of the Escarpment is one of the rainiest places in Brazil. The belt of heavy rains, however, is narrow. São Paulo city, less than 35 miles from Santos, receives an average annual rainfall of 56 inches, and most of the state receives between 50 and 60 inches.

There is an important temperature boundary as well as rainfall boundary in southern São Paulo and northern Paraná, where the northern limit of frosts appears. Frosts occur at times in the higher mountains of the East, in Minas Gerais, but south of the latitude of Sorocaba they come frequently enough to make the planting of tropical crops somewhat hazardous. In the boundary area, frosts occur only in the higher valleys and on south-facing slopes.

Climatic features are reflected in the pattern of natural vegetation. On the rainy slopes of the escarpment and along the coast there is a dense tropical rain forest composed of evergreen broadleaf species. In the cloud zone of the upper Escarpment trees are moss covered, and the ground is soaked with moisture. On the other hand, forests and grasslands are intermingled on the highlands. Semideciduous forests once grew abundantly on the slopes of the mountains and over the crystalline uplands, whereas denser forests marked the outcrops of diabase. The Rio Paraná forms a sharp vegetation boundary, as immediately west of it are woodland savannas which the Brazilians call *campo cerrado*. The São Paulo Basin was originally grass covered, whereas the inner lowland was occupied by savanna with galeria forest along the streams. South of Sorocaba, the savannas were transformed into pure grass prairies, which are known as *campo limpo*.

Settlement Before the Coffee Period

An Indian mission, founded by a Jesuit priest on the site of the city of São Paulo in 1554, became the first permanent settlement in the highlands of Brazil. The mission was located on lower terraces which overlooked the swampy valley of the Rio Tieté, in the midst of tall grass and scattered thickets of scrub trees. The Portuguese *bandeirantes* infiltrated this frontier with comparative ease. They pushed the Jesuits south, married Indian women, and often sold the Indian men into slavery.

Settlers who came to São Paulo, and São Vicente as well, were not wealthy. Anyone in Portugal with private means or with any standing at court went to Salvador or Recife. The less important came to São Paulo. They were a hardy, adventurous people and, coming mostly from the south of Portugal, inherited no small amount of Moorish restlessness. They could not afford African slaves, and the Indians were poor workers, nor was the land in the São Paulo Basin productive of speculative crops. So, they began to look for other sources of wealth. Some benefited when gold was discovered in Minas Gerais and Mato Grosso, but gold did not bring wealth to São Paulo; coffee did.

The Spread of Coffee across São Paulo

Between 1885 and 1900 the São Paulo region was transformed from an outlying part of the East to a new and independent region of settlement that focused on the city of São Paulo and the port of Santos. Three developments accounted for the transition: an expansion of the North American and European markets for coffee; the spread of coffee over the state of São Paulo; and the rapid immigration into São Paulo of millions of Europeans. A forerunner to this was the establishment of some 400 German peasants who became tenants, or *colonos*, in 1847.

Immigration

The immigrants and their descendants settled mostly in the Paraíba Valley, and in the mountains and hilly uplands around Campinas, Sorocaba, and São Paulo. The rapid increase in population during the eighteenth century was the result of a subsidized immigration that made the economy prosper. Italians and Portuguese came in large numbers, followed by Germans, Spaniards, and Japanese. By 1900, São Paulo had become the state with the highest percentage of foreign-born. Today, the Japanese in São Paulo are more numerous than any other such community outside of Japan. They have been especially successful in agriculture, although they have also gone into professional fields. Although they constitute only 0.5 percent of Brazil's population, they produce 60 percent of the nation's soybeans, 71 percent of its potatoes, and 94 percent of its tea.[1] Other immigrant workers included Syrians, Lebanese, and Arabs. Emigrants from other parts of Brazil also came to this region, with the states of Bahia and Minas Gerais contributing the largest number of immigrants.

Wages paid to *colonos* were adequate only for subsistence and were not a real share of the profits. However, many of the immigrants became entrepreneurs, a tradition that was passed on to second and third generations, and the Paulista *colono* had the advantage of free rent and near self-sufficiency in food production. This allowed him to save enough to eventually buy small parcels of land.[2]

Distribution of the Coffee Fazendas

When coffee plantations were developed in São Paulo state, they followed the lines of existing roads. As the *fazendeiros* began to recognize the soil and surface features best suited to coffee, however, a pattern evolved that increasingly corresponded to the underlying land.

The great wave of coffee planting in the period between 1885 and 1900 not only followed the routes out of Campinas, but also moved westward, encountering the diabase cuesta near Botucatú. Where forests once vaguely reflected the distribution of *terra roxa* soils, coffee began increasingly to do so, but with greater precision.

The spread of coffee plantations did not stop at the western edge of the *terra roxa*. The lure of virgin lands beyond was too strong. On the interfluves between the various streams draining to the Paraná, coffee plantations were pushed rapidly westward and northwestward, occupying the poorer sandy lands that gave high yields when they were first cleared. Fingers of settlement extended along the railroad lines to Barretos, Rio Prêto, and the Mato Grosso beyond, and also as far southward as the Rio Paraná.

[1] *Américas*, 37, No. 1 (1985): 51.
[2] T. W. Merrick, and D. H. Graham, *Population and Economic Development in Brazil: 1800 to the Present*, (Baltimore and London: Johns Hopkins University Press, 1979), pp. 114–115.

Other Uses of Land in the Coffee Region

Before 1875 coffee was grown on only part of the cultivated land, and the remainder was used for the traditional Brazilian combination of rice, beans, and maize. Occasionally, there was a little sugarcane and cotton. Almost all of the new plantations were devoted to coffee, and this land was mostly on the ridges. Whenever there was diabase, there was *terra roxa* and wherever there was *terra roxa*, there was coffee. Certain districts specialized in other crops. The lowland around Sorocaba was devoted to cotton, fruit orchards surrounded Limeira, and around the outskirts of São Paulo there was truck gardening. Throughout the state the area in pasture regularly exceeded that used for crops.

Brazil gained entry to the international orange juice market in 1962, when Florida suffered a serious freeze in its orange groves. The center of this industry is concentrated within a belt extending from Campinas in the south to São José do Rio Prêto and Colombia in the north. About 5000 commercial farms are devoted to citrus, from which approximately 80 percent of the production is processed into frozen concentrate and nearly all of this is exported. From small beginnings at a plant in Bebedouro, output grew from 21.6 million boxes to 150 million boxes within a 16-year period. Important byproducts are citrus pulp pellets used for cattle feed, essential oils from the peel of the citrus, and stripper oils used as solvents in soaps and perfumes.

Sugarcane Cultivation   Sugarcane has been one of the leading crops in São Paulo state since the earliest period of settlement. It has gained even greater significance in recent years, however, as a source of alcohol to substitute for diesel fuel and gasoline. By 1980 a mixture of 20 percent alcohol and 80 percent gasoline was a common fuel for Brazilian automobiles, and many fully alcohol-powered cars were in operation. São Paulo, the principal sugar-producing state in Brazil, provides two-thirds of the nation's supply of alcohol. Less than 4 percent of the country's agricul-

A SUGAR CANE MILL AT PIRACICABA WHICH PRODUCES ALCOHOL FOR LIQUOR AND FOR CARS.

tural area is used for sugarcane and other alcohol components. In 1978, as sugarcane cultivation was expanding, Brazil imported basic foodstuffs at a cost of U.S. $2 billion. On the other hand, thousands of new jobs have been created in agriculture and industry by the nation's alcohol program, and by 1980 the use of alcohol to substitute for petroleum resulted in savings of $1.3 billion annually in much-needed foreign exchange.

Transportation

One of Brazil's major problems has been an inadequate transportation system, even in the state of São Paulo. Part of the difficulty has been that of climbing the Great Escarpment. The British became pioneers in the development of railroad systems when they built a line from Santos to São Paulo in 1867. The trains were lifted or lowered over the Serra do Mar by means of cables and stationary steam engines, burning British coal. Nearly 80 years later, in 1946 the interior of São Paulo was served by three main railroad zones, but these were not well managed or maintained.

The Brazilian railroads began to decline in importance during World War II, being unable to compete with road and air transportation. Hence, in 1957 the government consolidated most of the 22 individual railroads into one com-

FIGURE 32-1

pany. In 1973 Brazil, stunned by the world oil crisis, undertook a costly program of modernization that would in the long run rejuvenate the system through electrification and save millions of dollars on oil imports.

The Decline of Coffee

The speculative wealth generated by the spread of coffee supported the rise of São Paulo city and gave the state a commanding position in Brazilian politics. Early in the twentieth century, however, there were signs of serious financial problems.

In 1899 Brazil produced 9 million bags of coffee, and in the next three years production nearly doubled. In 1902 there were 530 million trees of bearing age and 135 million more trees about to come into production. The government immediately banned new plantings. However, by 1906, when the whole world consumed 12 million bags a year, Brazil's crop was 20 million bags. Brazilian planters continued to expand their coffee acreage until in 1930 they had a crop of more than 28 million bags. By then the world was in deep Depression. Coffee which had sold at 24.8 cents a pound in 1929 brought 7.6 cents in 1931. The financial structure collapsed, and with it there was a revolution that brought Getúlio Vargas to power. The first Brazilian republic thus came to an end.

São Paulo still is a major producer of coffee and along with Paraná will continue to supply about 75 percent of Brazil's crop. However, the world coffee market is notably unstable, despite implementation of the International Coffee Agreement. Yet another problem is illegal coffee trade. It has been reported that between 500,000 and 1 million bags of coffee are smuggled across the Brazilian border into Paraguay each year, constituting a significant loss of revenue to the Brazilian treasury.

Settlement in Northern Paraná

The northern part of Paraná state is a transition zone between land that has frost every winter and land that is frost-free. By 1930 the coffee frontier began to spread across the Rio Parana-panema into northern Paraná. In advance of the zone of settlement were large properties occupied by scattered groups of Indians. A British land company combined several of these properties and proceeded to survey the land. Clearings were made, roads and railroads were laid out, and towns were located at intervals suitable to the proposed economy. The company permitted increasingly large acreages to be planted to coffee until in 1953 widespread frosts wiped out many of the Paraná plantations. Since then other waves of coffee plantings have occurred, and there have continued to be periodic killing frosts.

Agricultural policies have changed radically in Brazil during recent years, especially in the state of São Paulo. Lands have been made more productive through the use of chemical and organic fertilizers, terraces and other soil conservation measures have slowed erosion and rebuilt soils, new breeds of animals have been fattened with less feed, and new varieties of plants have in-creased yields per acre. An example of the new plants is a variety of cotton classified as IAC-17 that yields up to 30 percent more than other types. Also notable is a hybrid corn that has doubled yields during the past three decades. Mechanization, however, has lagged. In São Paulo only about 10 percent of the cotton is picked by machines, which can do the work of 200 pickers. It is estimated that 50 percent of the planting and 75 percent of the cultivation are by mule.

São Paulo City

When the railroad first came to São Paulo, it was a city of only 26,000 inhabitants and contained about 4000 houses. It differed little from many other small towns that were founded in the colonial era. Its streets were narrow, irregular, and mostly unpaved. Campinas to the northwest had grown rapidly before 1883 because it, rather than São Paulo, was the focus of two major rail lines. With the spread of coffee the major focus of routes shifted from Campinas to São Paulo, and the growth of the city was spectacular. New industries by 1905 included textile mills, shoe factories, and others using local raw materials. Cotton textile manufacturing alone employed 39,000 workers.

Manufacturing

Today, São Paulo is a city of perpetual motion. Its automobile industry manufactures 2800 cars daily. More than 1.5 million vehicles circulate in heavy traffic, contributing to the air and noise pollution. American auto makers such as General Motors, Ford, and Chrysler are located in the city, as are 75 percent of the manufacturers producing items such as rubber goods, machinery, and electrical supplies. The city also produces refrigerators, television sets, cosmetics, textiles, shoes, furniture, plastics, clothing, locomotives, and steel. Well over 3 million people work at jobs in 65,000 business establishments or industrial concerns.

THE CITY OF SÃO PAULO.

A new industry is the manufacture of armaments. Aided by an ambitious advertising campaign, Brazil's arms sales reached nearly $500 million in 1980. The industry was born more than a decade ago when the military government was struggling to crush several guerrilla groups. When the guerrillas were defeated, the government realized that some of its specialized equipment was unavailable elsewhere in the world, except in the Soviet Union. Items now manufactured are light tanks, armored cars, helicopters, amphibious trucks, and military training aircraft. The center for this manufacturing is Santo-Amaro, 7 miles from the city of São Paulo. The industry is small in comparison to that of the United States, the Soviet Union, France, or Great Britain. However, it has become a significant exporter, especially to Iraq, where an estimated 5000 Brazilians work on a myriad of construction projects.

Urban Pattern

The city of São Paulo is not at all like Rio de Janeiro. It is certainly a less attractive city, as "modernization" has made it appear similar to many other large world metropolises; yet, its buildings include some architectural masterpieces. The famous *Triangulo* is the site of the central business district. As late as 1949 there were only three skyscrapers; now the entire city is a maze of highrise glass, steel, and concrete buildings. For many miles around the city there are the usual residential suburbs, whereas along the railroads within and outside the city there are crowded industrial districts. Farther from the urban center, there are satellite industrial towns. Invading the zone of old truck farms around the periphery of the metropolitan area are new residential areas.

São Paulo is as modern as any of the world's great cities. Its life is a mixture of vitality and optimism, combined with frustrating social problems caused by rapid and unplanned urban growth. It is a challenging city, with a relatively high crime rate and an incredible number of daily traffic accidents. Yet, São Paulo also has a great variety of cultural attractions, including restaurants of every nationality, theaters, sports clubs, recreation centers, night clubs, boutiques, art galleries, and museums. In large part, São Paulo reflects the driving nature of modern Brazil.

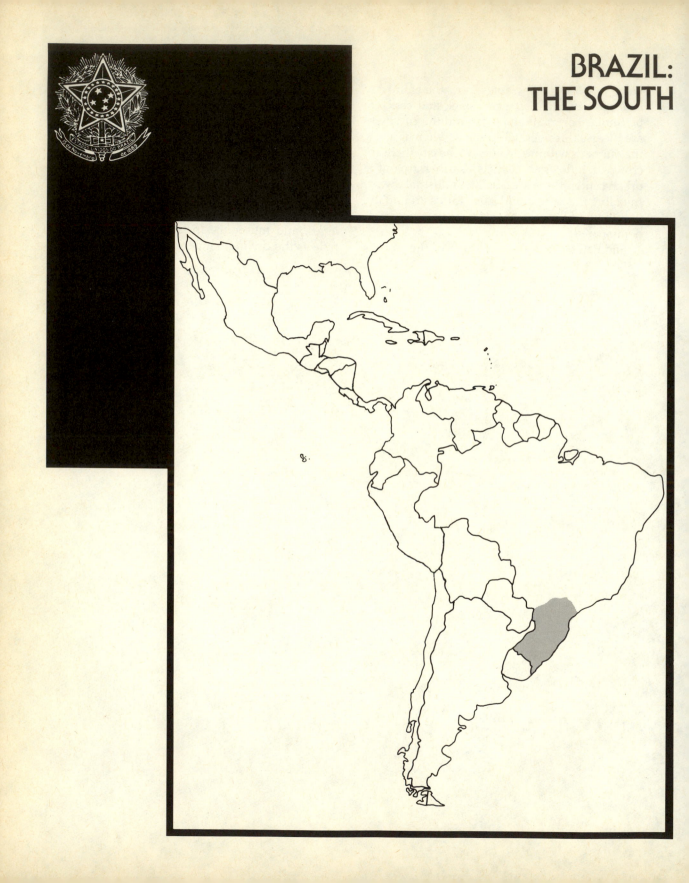

BRAZIL:
THE SOUTH

33

The South of Brazil includes three states: Rio Grande do Sul, Santa Catarina, and Paraná. These states comprise 7 percent of the total national territory and are occupied by about 18 percent of the total population. They differ markedly from São Paulo state, and from the East and Northeast. When pioneers settled along new roads in the South of Brazil, they adopted the traditional Brazilian system of shifting cultivation. If there were no established nuclei, settlers made temporary clearings by the old slash and burn technique. There, they cultivated maize, rice, beans, potatoes, and manioc, and raised hogs to be sold in the market. After a few years, they abandoned the clearings and moved on to other parts of the region. Destruction of the remaining forests proceeded rapidly.

The Land

Surface features in the northern part of this region are similar to those of São Paulo. There are the coastal zone and the Great Escarpment, the crystalline hilly uplands, the inner lowlands, the east-facing cuestas, and the tablelands of the Paraná Plateau. The Great Escarpment continues almost to Pôrto Alegre. In Paraná it forms one single slope, whereas in Santa Catarina it is broken by faults into a wide zone of blocks and is threaded by torrential streams.

The crystalline zones bend westward in Paraná, so that Curitiba is located in the midst of a wide zone of crystalline hills. In Santa Catarina the diabase swings eastward until the cuesta stands on the crest of the Great Escarpment. This surface feature is called the Serra Geral. Markedly different are the surface features of the southern half of Rio Grande do Sul. There, the diabase cuesta turns sharply to the west, crossing the state in an east-west direction, overlooking the valley of the Rio Jacuí. To the south of the cuesta are flat-topped mesas that extend southward into Uruguay.

Only between São Francisco do Sul and Tubarão is the southern Brazilian coast rocky and precipitous. From southern São Paulo a swampy, flat country extends southward to Paranaguá and São Francisco, while Itajaí and Florianópolis are located in hill country. South of Tubarão, the hilly land of the interior is fringed by a wide coastal zone of alternating sandbars and lagoons. Largest of the lagoons is the Lagôa dos Patos.

In northern Rio Grande do Sul, Santa Catarina, and Paraná the diabase is exposed at the surface over the territory west of the cuesta. This forms one of the world's largest lava plateaus. Deep canyons have been cut into the resistant diabase by the Rio Paraná and its tributaries, and at the heads of these canyons are spectacular waterfalls. The Guaíra Falls on the Paraná, known in Brazil as the *Salto das Sete Quedas,* are located at the northeast corner of Paraguay. Even better known are the Iguaçu Falls on the border of Brazil and Argentina.

Many different mineral resources exist in the varying geological structures. There are high-quality iron ores in Santa Catarina and Paraná. In the crystalline areas of Rio Grande do Sul and Paraná there are copper deposits, and coal is found in all three states. The largest coal deposit is located in southern Santa Catarina near Criciúma, where 75 percent of Brazil's coal is mined. A smaller amount comes from the São Jerônimo area in Rio Grande do Sul.

Climate and Vegetation

Contrasts in climate and vegetation occur along the southern border of São Paulo, which marks approximately the northern limit of frosts. Frosts never occur along the coast, nor in the deep Paraná Valley. Rainfall is abundant in every month of the year throughout the South. Temperatures

IGUAÇU FALLS ON THE BORDER OF BRAZIL AND ARGENTINA.

during the warmest month of Pôrto Alegre average 77° F, and those of the coldest month average 57° F.

The inland forest is a mixture of pine and deciduous broadleaf trees, but much of Rio Grande do Sul is covered by tall grass prairie. A dense rain forest which begins in southern Bahia extends southward almost to Pôrto Alegre, where it is replaced by lighter semideciduous trees that continue in a band along the south-facing slope of the highland across Rio Grande do Sul.

Population Clusters

Each of the three states of the South has a distinct core of concentrated settlement. In Paraná the main area of settlement is on the upland in an area of crystalline hills, whereas the coastal zone is well settled only in the vicinity of Paranaguá. The main urban center, Curitiba, lies inland from the Great Escarpment. In Rio Grande do Sul and Santa Catarina, the population clusters occupy valleys and lowlands, whereas the highlands are sparsely settled.

Rio Grande do Sul

The largest population cluster in the South is that of Rio Grande do Sul. In 1822 the Brazilian emperor realized that he had to thwart the northern expansion of the Spaniards. Consequently, in 1824 a group of German laborers, craftsmen, and peasants were brought to the new colony of São Leopoldo, in the previously ignored forest lands north of Pôrto Alegre. During the next 35 years, the government brought in more than 200,000 Germans and settled them on small farms in this region. Most of the colonies were arranged in a semicircular form along the terraces north of the Jacuí Valley and on the lower slopes of the cuesta. The forest was soon cleared, and typical German crops, such as potatoes and rye, were planted. Later, maize was added as feed for hogs.

At first, the hardships were many. All-weather roads were lacking, colonies were lost in the wilderness without a chance to market their products, and there was no place to buy necessities. Colonies that solved the transportation problems were successful; others were not. The sad plight of isolated pioneers in Rio Grande do Sul was reported in Germany in 1859, and for a time further emigration to the region was prohibited. Gradually, the problem was solved, in part by connections with river ports along the Rio Taquarí or other tributaries of the Jacuí which gave access to Pôrto Alegre, and in part by the construction of a railroad westward from that city. Land values were, and still are, determined more by proximity to a line of transportation than by the quality of the land itself.

Italian pioneers arrived in Rio Grande do Sul between 1870 and 1890. They settled along the crest of the diabase cuesta, in the belt of semideciduous forest. Unlike the settlers of São Paulo, the Northeast, or East, whose homes were temporary, camplike structures, the Italians occupied small farms and built substantial homes. In addition, Italian settlements could always be distinguished by the presence of vineyards.

The German colonies expanded rapidly and by World War II numbered about 600,000 people. The Germans worked hard to establish themselves; yet, many eventually sold their properties at prices far greater than the original value, or left their property to descendants. From São Leopoldo they moved westward, founding a string of German settlements that extended beyond Santa Maria. Then, along with other Europeans, they entered the forests along the Rio Uruguai in the northwestern part of the state.

Zones of Settlement   In the rural area tributary to Pôrto Alegre, there are five contrasting zones of settlement. The first of these is the oldest. It is the pastoral zone south of the Rio Jacuí, extending to the borders of Uruguay and Argentina. West of the Lagôa dos Patos are small farms that produce rice, beans, and maize, but on the open prairies there are vast cattle ranches used exclusively for pasture. This is the land of the *gaucho*. The products of its pastoral economy are hides, wool, and dried beef. The chief processing and exporting center is Pelotas, where the largest meat-packing plants are located.

A second zone is the densely populated flood-plain of the Rio Jacuí, and that of its northern tributary, the Rio Taquarí. Most of the inhabitants are Luso-Brazilians, or Brazilians of Portuguese origin, and the main product is rice, a low-yielding crop that the government intends to upgrade by establishing new irrigation areas. Rice cultivation has increased steadily throughout Brazil, as a result of a growing domestic demand, and Rio Grande do Sul is the leading producer.

The third zone consists of German settlements on the terraces and cuesta slopes to the north of Pôrto Alegre which feature well-kept homes and a distinctive type of architecture. Maize, hogs, rye, and potatoes are produced, and around Santa Cruz is one of Brazil's main tobacco-growing districts.

A zone of Italian colonies high above the valley of the Jacuí is the fourth zone. Sixty percent of the grapes, and a large part of the wines produced in Brazil, come from this area. Centers of production include Caxias do Sul, Garibaldi, and Bento Gonçalves. The wines are not exported in significant quantity but, rather, supply a growing domestic market. Occupation of the fifth zone, a hilly surface of the Paraná Plateau, in northern Rio Grande do Sul, has occurred only since 1940. Generous government grants have made this area a major producer of wheat.

The Soybean Revolution   Soybeans became an important crop in the northwest of Rio Grande do Sul during the late 1960s, and the city of Passo Fundo became the trade center for the region. From 1968 to 1977, soybean cultivation was enormously successful, with a thirteen-fold increase in tonnage produced.[1] The expansion of

soybean cultivation has continued, to the extent that soybeans may eventually replace coffee as the nation's principal agricultural export; Brazil has already become a major challenger to the United States in the world's soybean market.

Pôrto Alegre   During the past 30 years, substantial economic development has been achieved in Rio Grande do Sul, especially in the flourishing capital city of Pôrto Alegre. Leather tanning has long been one of the city's major activities, as German settlers made a home industry of hides purchased from their *gaucho* neighbors. Other industries are represented by wineries, breweries, textile mills, and shops that produce jewelry from semiprecious agates, as well as metals and transportation equipment. Industrialized suburbs extend to the north.

Pôrto Alegre is built on a ridge of hills, near the junction of five waterways that connect the city to its hinterland. Railroads link Pôrto Alegre with Argentine lines at Uruguaiana and with the Uruguayan lines at Sant'Anna do Livramento. A suburban mass transportation system is being constructed from the central business district of Pôrto Allegre 17 miles northward to the town of Sapucaia. The system, which will accommodate 48,000 passengers per hour in each direction, is designed to benefit low-income residents of the metropolitan area and to reduce travel time, fuel costs, and noise and air pollution. Pôrto Alegre is estimated to have a population of 1.1 million, but almost 2 million people inhabit the metropolitan area.

Another urban community, settled by Italian immigrants, is Caxias do Sul. It is located 60 miles north of Pôrto Alegre and is the center of a district devoted to intensive viticulture. This city produces some of the finest wines in Latin America and is also noted for the manufacture of metalware. Outlying parts of this region are used largely for cattle and hog production.

Coal   The South has two main coal mining districts. Most important is the district around São Jerônimo, Rio Grande do Sul, in the open *campos*

---

[1] David R. Hicks, *Agricultural Land Use and Related Innovation and Government Assistance in Rio Grande do Sul, Brazil* (Ph.D. dissertation, Michigan State University, East Lansing, Michigan, 1980).

south of the Rio Jacuí. This coal is of a low grade and has been of little significance in the Brazilian economy. Impurities are such that special equipment is required when the coal is used for heat or gas. About 90 percent of Brazil's total proven coal reserves are located in Rio Grande do Sul. The second coal district is in Santa Catarina, near Criciúma. This coal is of better quality and is used for metallurgical purposes, particularly in the manufacture of iron and steel.

Santa Catarina

North of Rio Grande do Sul is the state of Santa Catarina. The first settlers were pioneers from São Paulo who arrived in the 1660s. German mercenaries settled above the crest of the Serra Geral in the early 1820s but were forced to abandon their homes because of frequent Indian attacks. They fled to coastal settlements around Florianopolis, and even after the Indian menace was eliminated, the area remained virtually unoccupied.

Dr. Herman Blumenau, a German surgeon, realized the potential of this region for settlement. The landscape reminded him of the Rhine Valley, and he persuaded 17 Germans from Pomerania who were seeking liberty to come to this "new land." They arrived in 1850, and by 1870 more than 6000 others followed. The town of Blumenau was established upstream from the port of Itajaí, and Joinville was settled just inland from São Francisco do Sul.

Itajaí has become an important boatbuilding and fishing center, whereas Joinville is a busy industrial center. Fields outside Joinville are carefully cultivated, and cattle are important to the local economy. Joinville has a number of textile mills, distilleries and breweries, furniture factories, and shipyards. It also ships rice, maize, tobacco, arrowroot, and dairy products through the port of São Francisco. Itajaí exports rosewood, mahogany, and cedar, and duplicates Joinville in the export of rice, sugar, and dairy products.

Paraná

The settlement of Paraná was distinctly different from that of the other southern states. There were fewer Germans, for example, and most of those came from Santa Catarina. The earliest pioneers were Italians. The people today are mostly of Slavic origin, including Ukrainians, Poles, and Russians.

European colonization in Paraná is much more recent than that farther south. Good roads were built from Curitiba throughout the new pioneer area. Then, with state aid, Poles and Italians were successfully settled on the land. By 1885 there was no room for new colonists, so later arrivals had to proceed into the lowland to the west or remain in Curitiba as merchants. After 1890, many new colonies were planned and settled, with various degrees of success.

Paraná is the most densely populated state in southern Brazil, and more than 80 percent of its nearly 10 million people have been settled during the past 40 years. The economy is no longer based solely on coffee but has shifted to diversified agriculture.

Characteristics of the South

The three southern states form one of the most productive, progressive regions of Brazil. Until recently, this region has been one of European immigration and agricultural colonization. Now the land is largely occupied, and frontier development has moved westward into Paraguay and northwestward to Mato Grosso, Rondônia, and beyond.

The settlement of Paraná in modern times has been primarily the result of immigration from São Paulo and is associated with the cultivation of coffee. Until recently, this state has led all others in coffee production, but devastating frosts have reduced its coffee output. Urbanization and industrial development have been based primarily on the processing of agricultural commodities and the exploitation of forest resources, especially the so-called Paraná pine which has long been the nation's principal source of lumber.

Santa Catarina is the least developed state of the South, partly because of its relatively small size, rugged terrain, and lack of natural resources. It is characterized by small farms, European settlement, and lack of integration with the rest of the republic. The eastern part of the state is the most urbanized and prosperous, especially along the Atlantic Coast.

Rio Grande do Sul is one of the leading states in Brazil in almost every respect. It is prominent in agricultural production, especially in soybeans, wheat, maize, rice, potatoes, beans, grapes, and onions. Cattle- and sheep-raising are likewise important. The capital and primary urban center, Pôrto Alegre, is one of Brazil's largest cities and a major commercial and industrial center. Since the Rio Jacuí is not navigable for the largest ocean-going vessels, however, heavy manufacturing has increasingly concentrated nearer the coast, as at Pelotas and Rio Grande. There, chemical plants, petroleum refineries, grain elevators, and metallurgical industries have been created to serve interior markets or the export trade. Politically, too, Rio Grande do Sul plays an important role in national affairs. Like Virginia in the United States, Rio Grande do Sul state is noted as the "birthplace of presidents" and other national leaders.

The South of Brazil, along with São Paulo state, contributes in a major way to the economic well-being of the country and to its cultural heritage. Yet, it is little traveled and little known by Brazilians of other regions, and even less by the world at large.

BRAZIL:
THE CENTRAL-WEST

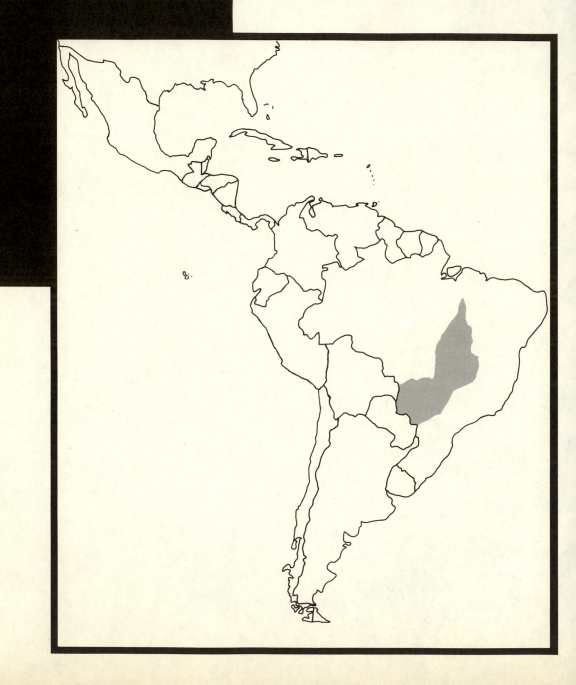

34

Although the Central-West contains 22 percent of the total area of Brazil, it is occupied by only 4 percent of the people. It is administratively divided into the Federal District and the states of Goías and Mato Grosso do Sul. For centuries the region was neglected, but during the past 25 years it has been Brazil's fastest growing region. New roads and airports have brought waves of settlers from southern Brazil and from the Northeast.

Here again is the *sertão*. It is toward this great, comparatively empty interior of Brazil that many of the country's leaders have sought to direct mass migration, a *marcha para oeste*. It is here that an entirely new city, Brasília, became the national capital in 1960. Within 10 years, as migrant workers streamed from the Northeast and as public officials were transferred from Rio de Janeiro, the new capital grew to more than 500,000. Today, the metropolitan area is estimated to be home for more than 760,000 people.

The Land

The Central-West has received less attention than many other parts of Brazil; hence, much of the region seems relatively unexplored. Nevertheless, it was well known to the *bandeirantes*, itinerant miners, and pioneer settlers.

Surface Features

The vast, rolling surface of the Brazilian Planalto Central has been exposed to the leveling action of running water for millions of years and is considered one of the oldest such surfaces yet identified.[1] Resistant quartzites to the east stand out as low mountains, and the headwaters of the Paraná, São Francisco, and Tocantins-Araguaia drain in several directions from the still undissected remnants of this ancient crystalline structure. On this erosion surface the mantle of soil has lost most of its soluble minerals, and the finer rock particles have been carried down by percolating water, leaving a coarse, porous soil, low in fertility for shallow-rooted plants. Under the surface soil is a layer of cemented iron

---

[1] L. C., King, "A Geomorphologia do Brasil Oriental," *Revista Brasileira de Geografia* 18 (1956): 147–265; see also P. E. James, "The Geomorphology of Eastern Brazil, As Interpreted by Lester C. King," *Geographical Review* 49 (1959): 240–46.

oxide, 10 feet or more in thickness. This lateritic formation is know as *cangá*. To break it requires dynamite and bulldozers.

In Mato Grosso do Sul the high surface descends almost imperceptibly into a tabular upland that continues westward. Far to the west, along the eastern side of the Paraguay Valley, diabase and sandstone layers support a steep, west-facing scarp, from beneath which the crystalline rocks emerge.

Climate and Vegetation

In the Central-West region the winter months, July to September, are cool and dry, whereas summer, from October to April, is hot and rainy. The passage of cold air masses from the South brings heavy frontal rainfall. The vegetation reflects these general climatic conditions, but in greater detail mirrors the underlying conditions of soil and water. The geographer Leo Waibel thought that the vegetation had been affected by long-continued burnings by the Indian inhabitants.[2] Others believe that the intricate intermingling of savanna and scrub woodland may have been modified in detail by fire, but that the vegetation as a whole reflects the moisture and temperature conditions of a tropical wet and dry climate.

Several types of vegetation have been significant in forecasting the value of an area for agricultural settlement. The *mata da primeira classe* (first-class forest) is a dense semideciduous forest reflecting an abundance of water, with trees reaching 100 feet in height. The *mata de segunda classe* (second-class forest) is composed of trees no more than 60 feet in height, with 30 percent losing their leaves in the dry season. The *cerradão* is a dense scrub woodland, with trees only 30 to 50 feet in height, yet close enough to form a complete cover over the ground. The *campo cerrado*, which covers 75 percent of the total area of the upland surface, is composed of savanna grass-lands with scattered scrub trees, or grasslands with no trees. On the wet lowlands of the Paraguay Valley the vegetation is a savanna with scattered palms. The Brazilians call this area the *Pantanal.*

Until the early 1970s the *campo cerrado* was used primarily for grazing, but a federal agency concluded that conditions exist for the development of agriculture. Specialists agree that the Central-West could some day become one of the country's most important farming districts.

Some attempts have been made to settle colonies of Europeans in this kind of country, but these have not prospered. Japanese farmers have been successful in truck farming near the larger cities, but efforts to grow cotton in another area were a complete failure. These scrub lands have a low natural fertility but do have a good soil structure and sufficient rainfall. Modern methods of farming and government support have already resulted in an expansion of areas sown to crops.[3] The crops being produced are maize, soybeans, and sorghum. Most conspicuous, however, is an enormous expansion in the cultivation of upland rice.

Settlement of the Sertões

More than three centuries ago, the *sertões* beyond the São Paulo frontier were penetrated by the *bandeirantes*. Moving in groups, they were accompanied by soldiers, slaves, and livestock. They traveled great distances through forests, across highlands, and along rivers. In 1682 they found gold in Goiás, and in many scattered places between Goiás and Cuiabá they found both gold and diamonds in small quantities. However, by the nineteenth century the principal activity in the *sertões* for pioneers and their descendants was the grazing of cattle in southern

---

[2] L. Waibel, "Vegetation and Land Use in the Planalto Central of Brazil," *Geographical Review* 38 (1948): 529-554.

[3] S. M. Cunningham, "Recent Developments in the Centre-West Region," *Bank of London and South America Review* 14 (1980): 2, 45.

Goiás and in Mato Grosso do Sul. The northern tropical forests of both states were occupied by Indians.

Routes from all the *sertões* of the west and north came to focus on the town of Uberaba, whence a well-traveled course could be followed southward to Campinas and São Paulo. Between 1910 and 1940, railroads were extended to the far west and south, but most of the lines were single-track and costly to operate. Soon the railroads were subject to competition from trucks, as highways were extended far into the interior even beyond the railroad lines.

Most recently, both Mato Grosso do Sul and Goiás have benefited from the improvement of infrastructure. New highways and access roads have been built, and telephone and electric lines have been expanded. An adequate water supply is also being provided to most of the people.[4] A truck can now travel from Goiás to São Paulo in several days, and an airplane can arrive in a few hours; yet, the economic infrastructure is still in an early state of development. Livestock and agriculture remain the predominant feature of the regional economy.

Settlement in Mato Grosso do Sul

Cattle-raising on open range has been customary in most of Mato Grosso do Sul. Estates averaged 5000 acres in size, but each employed only a minimum number of herders. Animals grazed on the *Pantanal* during the dry part of the year and ascended to high pasture in the rainy season. Campo Grande became a leading trade center of the region when it was connected by railroad to the east. It is now a shipping center for livestock, hides, and skins, and agricultural products. It became a state capital in 1979, when Mato Grosso do Sul was separated politically from the formerly vast state of Mato Grosso.

Near the border of Paraguay, the forested land was quickly occupied by pioneer farmers after World War II. The forests were felled, and coffee was planted in the clearings. The farmers then planted rice, beans, maize and manioc among the young trees. By the time the coffee trees reached bearing age, the railroad had extended southward from Campo Grande to Ponta Porã on the border of Paraguay.

After the Paraguayan War, the Paraguay River was opened to international traffic and Corumbá, at the head of navigation, increased in importance. Since World War II huge deposits of manganese and iron ore have been mined at Morro do Urucum, about 15 miles away. Corumbá has metal works, blast furnaces, lumber mills, and a meat-packing plant, all of which contribute to the country's exports.

Goiânia

The Brazilians have shifted their state capitals on three occasions, creating new cities where no city existed before. The earliest was when Teresina became the capital of Piauí in 1852. The second was the shift of the capital of Minas Gerais from Ouro Preto to Belo Horizonte in 1898. Another was the shift of the capital of Goiás state from the old gold mining town of Goiás to Goiânia in 1937. Goiás was established in 1682 when gold was discovered in the stream gravels along one of the headwaters of the Rio Araguaia, at an elevation of only 1900 feet. Gioânia, 75 miles to the southeast of Goiás, was high enough to be free of malarial mosquitoes. Goiânia is now a fast growing city, with more than 360,000 inhabitants. It is a city of tall buildings, but it also has its museums, parks and gardens, and is noted for nearby deposits of diamonds, nickel, and quartz crystal.

Brasília

After Brazil declared its independence from Portugal in 1822, the statesman José Bonifácio suggested that a new capital be established on the Planalto Central near the geographical center of

[4] S. M. Cunningham, "Recent Developments in the Centre-West Region," *Bank of London and South America Review* 14, No. 2 (1980): 44–52.

the national territory and that it be named Brasília. In every constitution Brazil adopted after 1889, there was a section authorizing the establishment of a Federal District and a new capital city in the interior. Brazilians knew that this would become a reality, because it was part of the mystique of the *sertão*. The project was delayed, however, by the remoteness of the interior and the magnitude of investment required. It was President Juscelino Kubitschek's determination that finally translated words into action in the late 1950s.

In 1957 the Brazilian Congress approved the proposed construction of a new capital. The government was officially transferred to Brasília on April 21, 1960, and on that date the previous Federal District became the state of Guanabara. Brasília has become one of the most spectacular accomplishments in Brazilian history and a source of deep pride for most Brazilians.

The new capital site was in the midst of *campo cerrado*, 80 miles northeast of Anápolis. A contest was held for the design of the city, and the winner was an engineer, Lúcio Costa, whose plan of Brasília resembled an airplane with residential districts along the "wings" and administrative offices of the government on the "fuselage." The famous Brazilian architect Oscar Niemeyer designed all major public buildings. The commercial district was located at the intersection of the "wings" and "fuselage," whereas along the monumental axis were the cathedral and a museum to house important documents.

Along the residential axes there are separate groups of apartment buildings, each group arranged around its own recreation area containing

BRASILIA'S CONGRESS BUILDING.

a school, theater, shopping center, and other facilities. Along the side of the northern residential axis are the spacious grounds of the *Universidade do Brasília.* The city is bordered to the east by Lake Paranoá, created by a dam that backs waters toward the southwest and northwest. At the easternmost point of land, almost surrounded by water, is the *Palácio da Alvorada,* or Palace of Dawn, which is the home of the President. Around the shores of the lakes are lots reserved for private homes, clubs, hotels, restaurants, and the embassies of foreign nations. The embassies present a striking variety of structure, as some nations have employed world famous architects to design their buildings.

The planners of Brasília sought to prevent the formation of shantytowns, or *favelas,* within the city. They have been largely successful, but numerous worker communities have developed adjacent to Brasília, among which Taguatinga is larger than the capital city itself. Meanwhile, people continue to move to Brasília faster than homes can be built. *Favelas,* therefore, cannot be entirely eliminated. Around Brasília a zone of market gardens has developed. There, small farmers mostly of Japanese ancestry, who have migrated from São Paulo, are engaged in supplying the city with fresh fruit and vegetables. Brasília itself is planned to serve only one function, that of national capital. There are consequently few manufacturing establishments or commercial enterprises beyond those needed to serve the needs of the city.

Brasília's Connections From Brasília there are flights to other parts of the country, including previously remote areas of the North and Northeast. Paved roads link the capital with all regions of the country, and "feeder" road connections have been improved. A paved highway from Brasília leads westward to Anápolis, Goiânia, and Cuiabá, and then through the selva of Rondônia and Acre to connect with the Marginal Highway of Peru. The Belém-Brasília Highway leads northward to connect with the Trans-American Highway and is followed by an almost continuous stream of heavily laden trucks. This has encouraged many persons to settle along the road in a previously unoccupied part of Brazil. Other paved roads extend southward from the Federal District to Belo Horizonte and Rio de Janeiro, and to São Paulo via Goiânia. A railroad extension to Brasília from Belo Horizonte was completed in 1968, but is little used. The Central-West region is rapidly being integrated with the rest of the Brazilian nation. Consequently, the pioneer fringe has moved westward to Rondônia and to Acre territory and northward to the states and territories of Brazilian Amazônia.

BRAZIL: THE NORTH

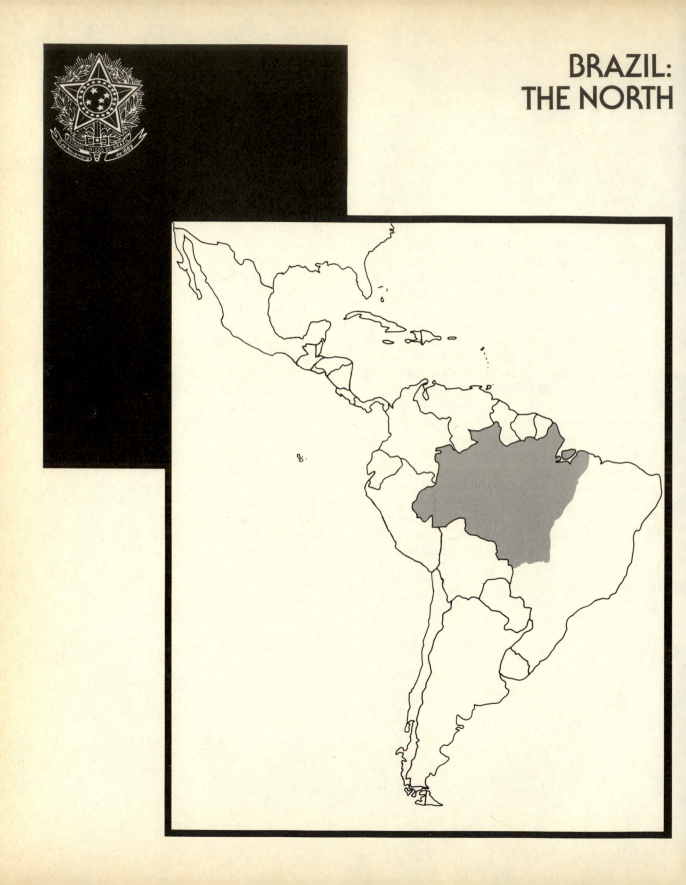

35

The North of Brazil includes the states of Amazônas, Acre, Mato Grosso, and Rondônia, plus the territories of Amapá and Roraima. Included in this region is the Amazon Basin, a vast, thinly populated area with immense stretches of tropical rain forest and a complex pattern of rivers. Containing 58 percent of Brazil's total land mass, the drainage basin of the Amazon River is the world's largest. The Amazon itself is by far the world's largest river and is second only to the Nile in length. The North, like the Northeast, the East, and São Paulo, has gone through periods of speculative development, most notably when it was the world's major source of rubber. It continues to be an area of exploitation, pioneer settlement, and unknown long-term destiny.

The Land

The Amazon region has been surrounded by an aura of myth, mystery, and misinformation. Perhaps this is because of preconceptions concerning the effect of hot, tropical, steamy climates on people of European origin, or perhaps because of sometimes exaggerated tales of Brazilian frontier life, or because of rumors of land to be had for the taking and the expectation of discovering great natural resources. Actually, the Amazon region does contain great natural wealth, but any rational plan for regional development must involve careful consideration of a variety of ecological factors.

Surface Features

Only a small portion of the Amazon region can be described as a plain. Above the junction with the Rio Negro and the Rio Madeira, the plain widens out like a spatula, until a distance of some 800 miles separates the highlands to the north and to the south. This is the part of the basin lying just east of the Andes, drained by the Purús, the Juruá, the Javary, and the main stream.[1] Most of the

---

[1] The main course of the Amazon is given different names in different sections: The Peruvians call it the Río Marañon; from the Brazilian border eastward as far as the junction of the Rio Negro, the Brazilians call the main stream the Solimões; and from the Rio Negro to the sea they call it the Amazônas. The English name, Amazon, is used in this book to refer to the entire course of the main stream as far westward as the Pongo de Manseriche in Peru where the Río Marañón emerges from the Andes.

surface of this large area is underlain by unconsolidated gravels, sands, clays, and silt. About 90 percent is above the level of the highest floods, because the floodplain of the main stream is mostly less than 50 miles wide.

Because of gradual submergence of the land where the Amazon empties into the sea, the mouth of the river is embayed. There is no delta, although the yellow, silt-laden waters discolor the ocean as far as 200 miles offshore. The floodplain is bordered by sharp bluffs that stand 150 to 200 feet above the swamps along the river. As the main stream meanders across the lowland, its channel shifts frequently, leaving oxbow lakes and swamps.

The crystalline hilly uplands of the Guiana and Brazilian highlands nearly join near the mouth of the Amazon. They are surmounted by a few massive ranges or conical-shaped mountain remnants, as well as by sandstone tabular uplands or plateaus. On the northern side of the floodplain, northeast of Santarém, the edge of younger strata forms a *taboleiro* that rises to 1150 feet above sea level.

The Amazon River System

The Amazon River is fed by more than 1000 tributaries and drains more water to the sea than the three next largest rivers of the world combined. This constitutes 11 percent of all the water drained from the continents to the oceans. Furthermore, the water is remarkably pure. The chemical purity of the water in the Rio Negro, the major Amazon tributary from the north, is said to be equivalent to that of distilled water.

Ocean-going ships reach far into the interior of South America because of the great depths of the river. The Amazon tributaries, on the other hand, where they cross areas of crystalline rocks are interrupted by falls and rapids. West of the Madeira they are navigable for riverboats far upstream into Acre state, but east of the Madeira they are all interrupted by rapids within 200 miles of the main stream.

Considerable attention has been focused on

the Amazon Basin, for it is here that more than 50 percent of the remaining tropical forests of the world are located. Major development projects require the clearance of this segment of the ecological environment, whereas ecologists favor agricultural systems, such as shifting cultivation, that would leave but a fleeting impression on the landscape.[2] In Amazônia the dominant procedure has been to convert the newly deforested land into pasture, a system that some agronomists and planners find unacceptable.

Soils

The myth persists that tropical soils are fertile and productive. Actually, forests of the rainy tropics are luxuriant because of the warm, moist climate; the soil itself tends to be poor. On tropical plains where the soil is exposed throughout the year to the percolation of water under conditions of high temperature the soluble minerals are leached out, leaving only insoluble iron and aluminum compounds at the surface. The finer soil particles are carried down, leaving the surface horizon coarser than before. Organic matter falling on the ground is quickly destroyed so that relatively little of it is mixed with the mineral content to form humus. Only on the river floodplains, where new layers of silt are deposited with each flood, are there any substantial areas of fertile soil.

If the rain forest is destroyed, the source of many raw materials, such as resins, gums, fibers, timber, fruits, and nuts, will be gone forever. It has been predicted that some day Amazônia could look like Madagascar, where only 12,000 out of 228,000 square miles of original forest remain.[3] Human vision and ingenuity can defeat such a dire prediction, of course, but constructive leadership will be required. To the individual

---

[2] M. Hiraoka, "The Development of Amazônia," *Geographical Review* 72, No. 1 (1972): 94–98.

[3] W. M. Denevan, "Development and the Imminent Demise of the Amazon Rain Forest," *The Professional Geographer* 25, No. 2 (1973): 130–135.

RIVER TRANSPORTATION IN AMAZONAS, BRAZIL.

settlers, control of the forest is the single greatest challenge. If this work is relaxed the unrelenting cover of growing things creeps back again to hide from view all traces of human action.

Climate

Another common misconception concerning the Amazon region is that temperatures are unbearably high. In reality, they are lower than those that occur in summer in the Mississippi Valley. Temperatures in the Amazon region are indeed high, but not excessive. At Manaus, for example, the highest temperature ever recorded is 101.5°F, and the lowest is 69.4°. The highest temperature recorded in New York is 106°F. Hu-

midity, however, can be high enough in some areas to create discomfort. Elsewhere, the steady movement of easterly trade winds from the ocean not only brings great amounts of moisture onto the land, but also relieves the sensation of excessive heat. Rainfall throughout the North is abundant, even in the drier part of the year.

Settlement

In the early colonial period, the cities of Salvador and Recife were bastions of wealth; settlement near the mouth of the Amazon, however, was discouraged because of the dense forests and annual floods. An expedition sent out from São Luis de Maranhão to secure this northern bit of Portuguese territory from Dutch and English invasion founded the city of Belém in 1615.

It was the missionaries who first successfully penetrated the Amazon and who, unintentionally, brought misfortune to the native people. Epidemics felled the Indians, who were gathered around the mission stations. Those who survived were the victims of slave traders.

The Rubber Period

Rubber did not become a product of major importance until Charles Goodyear discovered the vulcanizing process in 1839. This process made it possible to keep rubber from becoming sticky in hot weather or brittle when cold. Especially important was the subsequent manufacture of various mechanical and electrical devices in which rubber was an essential ingredient, such as electric insulation and automobile tires.

A new world market suddenly appeared, and Brazil was in the fortunate situation of possessing a monopoly on the raw material necessary to supply this market. The main problem was a scarcity of labor. There was a wild rush to buy land in the rubber forests and a hasty recruitment of laborers. A large percentage of the Brazilians scattered over this wide territory came originally from the *sertão* of the Northeast. Especially numerous were workers from the drought-stricken regions of Ceará, from which as many as 20,000 per year poured into the Amazon.

The rubber gatherers were actually trapped in the forest. The owner of a tract of land recruited workers in Belém or Manaus, loaned them funds to buy essential equipment and food, and then deposited them and their families on a riverbank where they built primitive shelters that served as homes. From each isolated camp the workers carved paths through the forest leading to perhaps 200 rubber trees. The latex tapped from the trees was brought to camp and formed into solid rubber balls by smoking over a slow fire. Then, at intervals the owner would show up to collect the

product and leave supplies for which the worker was always in debt.

Collapse of the Rubber Business

The boom was great while it lasted. Towns were founded at the head of navigation, roads were built allowing oxcarts to pass through the forest, and railroads were projected. The chief concentration of people and wealth was in the cities of Manaus and Belém, where money was squandered on elaborate mansions. Few could have foreseen the collapse of this fabulous rubber economy.

In 1876 an Englishman named Henry Wickham collected some *Hevea* rubber seeds in the Tapajós Valley and smuggled them to the Kew Gardens in London. Transplanted to the Botanical Gardens of Colombo, Ceylon, and then to Malaya and Sumatra in 1896, they provided a new source of natural rubber. Wild trees in the Amazon yielded only 3 pounds per tree per year, whereas the Malayan varieties produced 10 to 17 pounds. Moreover, near the rubber-growing region of Malaya and Sumatra were the densely populated lands of Java and India, where great numbers of workers could be recruited. In less than 20 years British Malaya and Dutch Sumatra produced 93 percent of the world's rubber supply, whereas Brazil's rubber economy collapsed and many small towns in the interior were gradually engulfed by the advancing forest.

In 1927 the Ford Motor Company purchased land in the Amazon Basin and established plantations to supply its needs for rubber. At great expense, the company overcame the technical difficulties involved in plantation rubber production, but owing to a scarcity of workers the project was finally abandoned. The plantations were sold to the Brazilian government and continue to operate on a small scale.

Present Centers of Settlement

Cuiabá, the capital of Mato Grosso state, is a former gold-mining center in the geographic center of South America. Today it is know for the

LIKE BUSY ANTS, MINERS SWARM OVER A MINING SITE IN SERRA PELADA, BRAZIL.

this mineral wealth and now exports not only manganese ore, but also small amounts of gold, rubber, cattle, and maize.

Largest of the Amazon Basin cities is Belém, near the mouth of the great river in a position to control the trade of the vast Amazonian region. It is not only the capital of Pará state, but is also a busy port handling all products that come from the Amazon, such as fish, fruit, flowers, minerals, and cattle. Ocean-going ships and fishing vessels are constantly loading and unloading their varied cargo at the city's port. Marajó Island, opposite Belém, has long been used for the grazing of cattle. Belém derives a part of its prosperity as a trade center from this island, which comprises a total area larger than that of Switzerland.

Santarém, once a small missionary settlement surrounded by a shifting cultivation of cacao, sugarcane, and other tropical commodities, has become a bustling port of about 150,000 inhabitants. The city was settled in 1865 by Southerners from Tennessee and South Carolina who fled the Confederate states when slavery was abolished. Today, it serves as a trading center for the mining, rubber, lumber, and jute industries of the region.

Manaus, the capital of Amazônas state, is presently a city of about 500,000 people. Because it is a free port, it has become a dynamic center of commerce, specializing in the export of jute, lumber, rubber, black pepper, and Brazil nuts. Its many shops are stocked to overflowing with all kinds of foreign goods such as cameras, watches, air conditioners, and television sets. Its highrises stand side by side with fabulous mansions of the rubber boom days. Supplying this city with food and other necessities is an expensive proposition, as ocean-going cargo ships must navigate 1000 miles up the Amazon to reach the city. Exotic Manaus has become a significant tourist attraction, most visitors arriving by plane. There is much to see, including an animal reserve with crocodiles, turtles, snakes, monkeys, tapirs, and jaguars that can be viewed from floating pavilions. Still standing in the center of the city is the world famous opera house, *Teatro do Amazônas,*

harvesting of rubber and palm nuts. It is also the home of a surprisingly large community of people from the Middle East. This modern city of more than 100,000 is growing and sharing in benefits from the Amazon region's federally funded projects.

A significant area of settlement has been created from the mining of manganese in Amapá. Manganese ores were discovered in 1944 in the dense forests north of the mouth of the Amazon. A 120-mile railroad was built to connect this ore body with the port of Macapá, and since 1959 Brazil has been producing more than a million tons of manganese per year, mostly from this federal territory. Its capital, the once sleepy little town of Macapá, has been a direct beneficiary of

THE TEATRO AMAZONAS, THE GRAND OPERA HOUSE OF MANAUS, WAS BUILT IN 1896 AND IS STILL IN USE.

built during the rubber boom at a cost of $10 million and now restored to its original glittering condition.

Rural Settlement

Above all, the North of Brazil is a land of migration. People come and go occupying this vast frontier area and moving on when opportunities appear greater elsewhere. In recent years an influx of 5500 settlers per month has poured into Rondônia, which in 1981 became Brazil's twenty-third state. Other settlers have moved on to Acre in search of wealth from mining, grazing, or agriculture.

The story of Japanese settlement in the Amazon region is an important part of Brazilian history. The first Japanese colonists arrived in 1925, when the state of Pará gave them 124,000 acres of land along the Rio Acará about 50 miles south of Belém. They cleared the forest and planted vegetables that were familiar in Japan, but the people of Belém who were used to a diet of rice, beans, and *farinha* (manioc flour) refused to try strange foods such as turnips, tomatoes, and cabbage. Attempts to grow cacao were also unsuccessful, owing to poor yields. Then, in 1933, a Japanese immigrant brought in some 20 sprouts of *piper nigrum* (black pepper), a plant that required ample rainfall and high humidity. The venture was a great success. Pepper became the most valuable product of Pará and gave the state a new source of revenue. The Japanese also were successful in growing jute on the Amazon floodplain and in supplying vegetables for Manaus.

In the 1960s it was thought that only about

10,000 Indians inhabited the great empty forests of Amazônia. Today, it is estimated that 200,000 live around the gold mining centers of Crepuri, Rio Amana, and Al Patro Signo alone. Flocking into the Crepuri area are an estimated 25,000 prospectors annually, seeking gold and competing for living space on Indian land. Although the Brazilian government legislates humane Indian policies and tries to integrate the Indians into a modern society, primitive tribes and their cultures are gradually disappearing.

Programs of Economic Development

In 1966 the federal government of Brazil formulated a program to hasten economic development in the North. The responsible agency was SUDAM, similar to SUDENE which has coordinated development in the Northeast. A new tax program allowed individuals or corporations to make only half of their scheduled income tax payments for investments in the North. The result was a flow of new capital into a little-developed part of the country. As new roads were built, trucks loaded with shoes, clothing, canned goods, barbed wire, machinery, and books went north. Trucks going south were loaded with jute, pepper, minerals, and fruit. Fish from the Amazon appeared on the menus of restaurants in Brasília. Loads of timber from the Amazon forests were sent to factories in São Paulo. With new roads, goods can now be exchanged between São Paulo and Belém within a week rather than within two months.

The most ambitious effort to develop transportation in the North of Brazil was the construction of the Trans-Amazonian Highway. It now extends from João Pessoa and Recife, on the Atlantic Coast, to Cruzeiro do Sul on the western border with Peru, a distance of 2000 miles. Yet, this is but one of the highways constructed or planned to crisscross the Amazon Basin and help integrate this vast region into the rest of the Brazilian nation.

The greatest immediate potential for develop-

ment of the North appears to lie with forestry. There can be no doubt that the Amazon forests include a great variety of useful timber. There is wood as light as the balsa used to make rafts, or so dense that the best ax can make little impression on it. There are woods of great strength for beams or bridge timbers, and others that are elastic. There are woods that resist marine borers or are immune to the attacks of termites. There are woods with figured grain in different colors that can be used for veneer in furniture or paneling. Also within the selvas, in addition to rubber, are plants from which a great variety of products can be made: chicle, dyes, tannin, resins, gums, waxes, *guaraná* (a soft drink), edible nuts, vegetable ivory nuts (for buttons), essential oils, ipecac, rotenone, and many others.

The Jarí Project

In the 1950s, when a North American billionaire, Daniel K. Ludwig, envisioned a worldwide paper shortage he began a search for the perfect pulpwood tree. He found it in the fast-growing, disease-resistant *Gmelina arborea*. In 1967 he paid nearly $6 million for a piece of jungle and tropical rain forest about 200 miles upstream from the mouth of the Amazon. Lying astride the Jarí River, a tributary of the Amazon, this unproductive tract, about the size of Connecticut, was thinly populated by small farmers, nutgatherers, and fishermen. Ludwig began to clear the forest and replant it not only wth *Gmelina* but also with other fast-growing, disease-resistant trees, such as Caribbean pine and eucalyptus. A power plant and pulp mill were assembled in less than a year by a shipyard in Japan. Since the mill and power plant were too large to pass through the Panama Canal, the giant structures were towed 15,500 miles by way of the Philippines, through the Indian Ocean, across the Atlantic, and into the Amazon River, a journey that required three months.

Thirteen years later, Jarí Florestal e Agropecuaria Ltda. was the largest development enterprise ever attempted by a private individual and included the world's largest tree farm. Flood-plains were converted into huge ricefields. An open-pit kaolin mine was developed, and a plant was built to process the fine clay used in making glossy paper, cosmetics, ceramics, and "Kaopectate."

It is estimated that about 40,000 people were involved in the Jarí project, 7000 of whom lived at Monte Dourado, a few miles north of the mill site. All modern facilities and services were provided to assure that everyone involved could enjoy a comfortable lifestyle in the warm and humid tropics.

Ecologists and others soon voiced concern over destruction of the "natural forest," although scientists working on the project maintained that all possible precautions were being taken to protect the fragile ecosystem.[4]

Given the relentless opposition of environmentalists, changing world economic conditions, and growing political pressures from within Brazil, Ludwig finally sold the entire project to Brazilians in 1982. Like the projects of Henry Ford many years earlier, it lost millions of dollars and demonstrated that both natural and human constraints must be overcome before even well-financed development efforts can be successful in the Amazonian wilderness.

In 1978 a Landsat space survey found that 1.5 percent of the Brazilian rain forest had been cut or burned, not an alarming amount in itself but a 168-percent increase over that reported in a survey completed three years earlier. If the trend were to continue, the forest would be depleted within a few decades, and the delicate relation-ship between animals and plants would be destroyed.

The Greater Carajas Project

Largest of all development schemes is the Greater Carajas Project, begun in 1973. It provides for the construction of a dam on the Tocantins River to generate 4 million kilowatts of electricity, which will aid in the exploitation of enormous iron ore reserves estimated at up to 50 billion tons. Bauxite reserves along the Trombetas river, a tributary of the Tocantins, will also be developed. A railroad will transport iron ore and aluminum from this site about 350 miles from Belém to the Atlantic Coast for export. In the long run mining may prove more valuable for Amazonian development than forestry or agriculture and less destructive to the natural environment.

Tourism

There are few places left on earth that cannot be penetrated by the aggressive tourist. Even the backlands of Brazil have become a tourist "attraction." The Amazon River system has 1000 known tributaries, and travelers are being urged to "explore" some of these previously inaccessible rivers, some of which are more than 1000 miles long. The Brazilian government has made Bananal Island, in northern Goiás, into a national park. It, along with land on either side along the Rio Araguaia, has been named a national hunting and fishing preserve, where tourists can catch a wide variety of fish. Largest is the *pirarucú*, the world's largest fresh-water variety, which can be six feet long and weigh 500 pounds. The time seems to have arrived when no part of Brazil remains outside the effective national territory or free of external concerns.

---

[4] D. T. Halperin, "The Jarí Project: Large Scale Land and Labor Utilization in the Amazon," *Geographical Survey* 9, No. 1 (1980): 13–21.

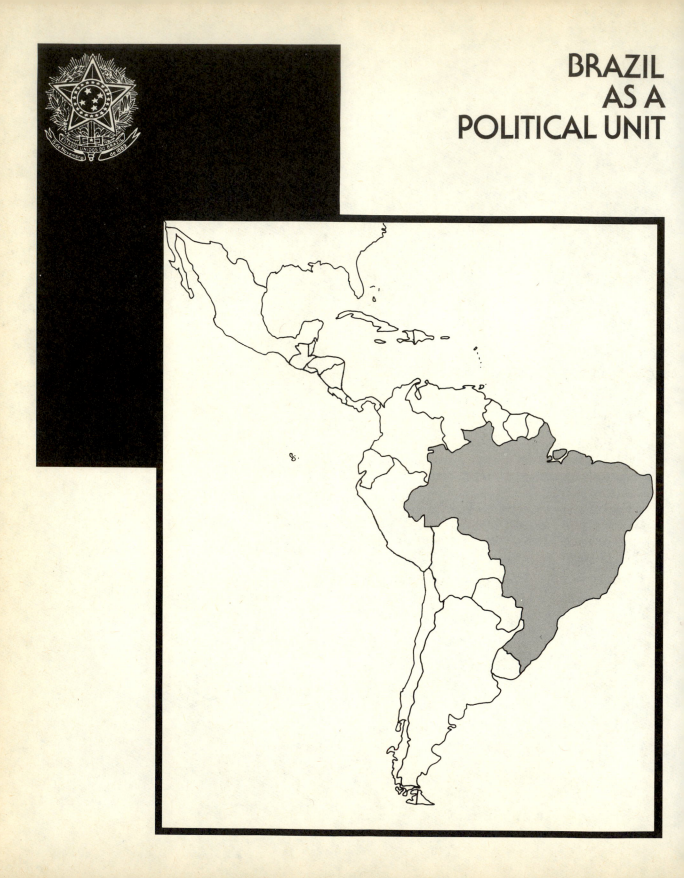

BRAZIL
AS A
POLITICAL UNIT

36

Brazil is a country of vast proportions, containing more than 40 percent of the total land area of Latin America. It is exceeded in size only by the Soviet Union, Canada, China, and the United States. With such magnitude, it is not surprising that Brazil is also a country of great complexity. It is, for example, impossible to describe the typical Brazilian because there are many kinds of Brazilians, a wide variety of people who contribute to the artistic, political, economic, and intellectual life of the nation.

Brazil is also one of the world's foremost "developing countries." As previously noted, Brazilians themselves have commonly joked that "Brazil is a country of the future, and always will be!" However, such pessimism was sharply diminished after 1964, when a succession of military governments achieved both political stability and economic development. In less than two decades, Brazil was converted from a country characterized by primitive agriculture and a large expanse of unoccupied territory to a significant industrial power with a large supply of hydroelectric energy.

Many unsolved social, economic, and political problems remain. The worldwide realities of inflation, recession, and terrorism are manifest in Brazil as in most other developing countries. The increasingly sharp contrasts between rural and urban life, between cities and the frontier, and between the various geographic regions are ever present. The relentless stream of rural people into the cities makes the search for solutions even more difficult.

Starting in 1920, São Paulo state had a greater percentage of industrial labor than any other state in Brazil and a major share of the total national value of manufactured goods. This rapid advance attracted an increasing number of workers to São Paulo but limited the rate of gain in industrial wages. Wages did not keep pace with increasing economic production. Attempts to correct the imbalance between São Paulo and the rest of the country have become a major element of government policy. In the 1950s development agencies were created to supervise the several economic regions. These encountered difficulties in the 1960s due to a lack of governmental support and to political problems. Actually, the gap between São Paulo and the East as compared with the less developed regions widened.

Population

Brazil's population growth rate has slowed in recent years but is still relatively high. In 1985 it was estimated that the total population of the country exceeded 138 million and that it would pass 251 million by the year 2020. The rate of increase is 2.3 percent, not the highest in Latin America, yet adequate cause for concern. The rapid rate of population increase is a direct result of programs for the improvement of health. As early as 1930, there was an attack on malaria and yellow fever. Tremendous progress was made in World War II and in later years as the use of insecticides, antibiotics, X-rays, and the laser beam helped to eliminate or control many fatal diseases. The death rate has been reduced to 8 per thousand, while the birth rate stands at 31 per thousand. No longer do most babies die within the first month after birth. Since 37 percent of the population is under 15 years of age, a continued high birth rate is almost inevitable in the years ahead.

Immigration is no longer the factor in population growth that it was in the late nineteenth century. At that time it reached a peak exceeding 1 million people annually. This influx of people made Brazil as culturally varied as the United States and more ethnically mixed. Particularly notable are the African cultural influence in the Northeast and the European influence in São Paulo and the South. Immigration has been controlled since the 1930s, under a quota system that applies to all countries except Portugal. Yet, Brazil received many refugees from Europe following World War II and remained receptive to Japanese immigration as well. Now a law has been enacted that will strictly control the entry of foreigners. This is to assure that the limited number of new jobs available each year will be available to a seemingly unlimited number of young Brazilians entering the nation's workforce.

The Agricultural Economy

Two quite different, but overlapping, economic systems support the Brazilian people. In one, there are workers who produce things for sale and use the money to buy goods and services. In the other are those who produce primarily for their own consumption. The latter constitute about 38 percent of the total Brazilian population and include subsistence farmers who sell a minute surplus of farm products so as to buy certain necessities other than food.

For as long as most people can remember, commercial coffee production has dominated the Brazilian economy. Before World War I, Brazil produced more than 75 percent of the world's coffee. This crop has lost its preeminence, however, as Brazil has embarked on a wide-ranging program of agricultural expansion. Although coffee is still one of Brazil's valuable exports, other agricultural commodities such as soybeans, sugar, and orange juice also have become major factors in Brazil's export trade.

Cotton, which at one time was Brazil's second-ranked export, never became a Brazilian monopoly. Most of the cotton came from the Northeast, which for a short time produced 96 percent of its crop for export. There was a boom in cotton production during the early 1970s in southern Goiás, but after two years of crop failure, cotton cultivation declined. Whereas small amounts are grown in several other states, the two main areas of production are in São Paulo, west and northwest of São Paulo city, and to the northeast.

For the first time in history, Brazil in 1980 exported more manufactured goods than primary products, and at the same time it was the world's second-largest food exporter, after the United States. There is, however, a growing internal demand for food supplies, which in future years may reduce the volume of agricultural commodities available for export. Investment in agriculture, furthermore, offers relatively quick returns. Brazil possesses unusual advantages among the world's few major food exporting nations in that it can readily produce both tropical and mid-latitude crops.

The Brazilian government has committed itself to a program that will lead to higher agricultural productivity. Most notable has been the expansion of soybean cultivation in the South.

PAULO AFONSO HYDROELECTRIC PLANT ON THE SAO FRANCISCO RIVER OF NORTHEAST BRAZIL.

Specialized experiment stations have been established throughout the country to expand not only soybean production but also production of livestock and crops such as maize, cotton, cassava, sugarcane, cacao, citrus, and rice.

Electric Power

Industrial concerns cannot prosper without some inexpensive source of power. In the absence of abundant coal deposits, hydroelectricity is the obvious answer, because Brazil has had a vast amount of unused water power. The early start of São Paulo as a manufacturing center was due partly to power developments on the Rio Tietê and at Cubatão. Rio de Janeiro, on the other hand, had an inadequate supply of electricity from a plant on the Rio Paraíba.

More than two decades ago Brazil initiated an ambitious program to increase the supply of electric energy; there are now many hydroelectric plants in operation. These include a plant at the Paulo Afonso Falls on the Rio São Francisco, the Tres Marias plant on the upper São Francisco, the Furnas Dam on the Rio Grande in Minas Gerais, a plant on the Rio Doce at Mascarenas, and the Urubupungá complex on the Paraná River. The Ilha Solteira plant is expected to produce twice as much energy as the Aswan Dam on the River Nile.

Turbines began operating in 1983 at the world's largest hydroelectric complex, Itaipú. This is a joint venture with Paraguay. The size and amount of machinery, equipment, and manpower involved in the project have not been equaled elsewhere. Twenty-eight thousand workers, 600 gigantic machines, and a volume of concrete equivalent to that needed to build a city of 4 million inhabitants were employed as the work progressed. Itaipú will supply one-fifth of Brazil's total power requirement. By the year 2000 Brazil also plans to have 63 atomic-powered generating plants. The country will then be well advanced in solving its energy needs.

Iron Ore, Steel, and Other Industries

For many years the Brazilian government resisted the involvement of foreigners in the development of its mineral resources, such as iron ore. In 1966, however, it decreed that privately owned ore-shipping ports could be built. Subsequently, both government and private interests have been involved in making Brazil the largest exporter of iron ore in Latin America.

Brazil has had a steel industry since 1817, but the development of modern large-scale steel mills had to await a number of technological developments that did not become available until the 1930s. The lack of coking coal seemed to be an insuperable obstacle because of the high cost of imports. Consequently, the early mills all made use of charcoal, and produced steel of exceptional quality. Even now many Brazilian mills use charcoal as their fuel.

The nationally owned steel mill, at Volta Redonda in the Paraíba Valley, now known as the Gilherme Guinle Steel Plant, was built during World War II and began production in 1946. Since then 15 other steel mills have been built. About 20 million tons of steel ingots are produced annually. Additional mills are to be con-

structed with the help of British, Japanese, German, and Italian financial and technical interests.

Industries that use steel have been established at an increasing rate since 1960. For example, Brazil has become a leading shipbuilding nation, with six major shipyards. Four are located on Guanabara Bay in Rio de Janeiro, one is at nearby Angra dos Reis, and the other is at the southern port of Rio Grande. These shipyards construct small fishing boats and river craft as well as refrigerated bulk carriers and petroleum tankers.

The manufacture of automobiles is now the fifth most important industry in Brazil. Two-thirds of the industry is owned by foreign car makers. Among the companies are: Mercedes Benz, Chrysler, Volkswagen, General Motors, Alfa Romeo, Ford, Toyota, and Fiat. Brazil has become an exporter of airplanes, subway cars, and armored vehicles as well.

Regional Balance

One of Brazil's most persistent problems is the regional imbalance of economic development. In the Northeast, the Central-West, and the North, income per capita is dangerously low in comparison with the productivity and income of São Paulo. In most modern industrial nations there is a core area from which a large part of the gross national product is derived, but the problem in Brazil arises from too great a concentration in one state. The question is how to enhance the development of other regions in Brazil without damaging the leadership role of São Paulo. The effects of massive injections of capital into various sectors of the economy are visible throughout the country, but the cost is high. In 1980, Brazil's foreign debt of nearly $50 billion was the largest of any non-oil exporting developing country, with interest rates running as high as $1.5 billion a year. Brazil's imports outpace its exports because it must buy 80 percent of its oil abroad until its various energy programs are fully established. It must borrow additional funds to refinance existing loans and borrow even more for capital investment. Thus, there is a constant gamble between the ability to borrow and the capacity to repay.

The Political Situation

Brazil became independent from Portugal in 1822. Its first emperor, Dom Pedro I, who is said to have been despotic, abdicated in favor of his five-year-old son in 1840. Until 1889, Dom Pedro II was the monarch, and proved to be an enlightened and popular statesman. It was the emancipation of slaves in 1888 which brought an end to the institution of monarchy and the landed aristocracy on which the monarchy was based.

In 1894, after a few years of military government, there began a succession of civilian presidents. By this time modern political alignments had made their appearance. São Paulo state was already the strongest economically, but Minas Gerais had the largest population, and the states of the Northeast, acting as a unit for political purposes, often held the balance of power. For many years the presidency went alternately to a candidate from São Paulo and then to one from Minas Gerais. Some of the presidents were capable and honest administrators, but there were also examples of corruption and confusion. From 1925 to 1927 a Communist leader, Luiz Carlos Prestes, led rebellious soldiers through the backlands of Brazil trying to arouse the spirit of revolt among those in the *sertões* and among the serf-like workers on the coffee plantations and cattle ranches.

From 1930 until 1945 Getúlio Vargas was in control of Brazil, setting up what he called the *Estado Novo*. Although the Brazilian people tolerated their dictator, they would not be regimented in the Fascist pattern. In 1945 the army removed Vargas and called for new elections and the end of the New State. Elections were held and the ballots honestly counted. After an interval,

Vargas himself was elected President in 1951 and held office until his suicide in 1954.

Between 1956 and 1964 Brazil had a succession of civilian presidents who continued to resist foreign investment. Juscelino Kubitschek was the President who was chiefly responsible for getting the construction of Brasília underway and for moving the capital from Rio de Janeiro. He was followed by two presidents whose economic policies have been described as irresponsible. Despite runaway inflation, the government took no effective steps to bring it under control. In fact, such steps were highly unpopular, especially among those who gained from the rapid increase in land values. These presidents also brought Brazil closer to the Communist orbit by promoting closer trade relations with the Soviet Union and by patronizing the Communist leaders of Cuba. In 1964 the army intervened, exiled the President in power, João Goulart, and took control of the government. The next five presidents were military leaders, army officers who did not seek personal gain but were devoted to what they considered to be an economically sound program.

In 1977, under President Ernesto Geisel's administration, there was a cooling of Brazilian-U.S. relations when the United States criticized the human rights situation and opposed a Brazilian treaty with West Germany for the development of atomic energy. The Church, too, became involved in the human rights issue and appealed to the people to gain support for the values of freedom, justice, and dignity.

Succeeding Geisel was João Baptista da Figueiredo who came to office in 1979. Then in 1985, after two decades of military dictatorship, Tancredo Neves was elected as Brazil's first civilian President since the 1964 military coup. Brazil had clearly entered a new era.

Among all of the nations of Latin America, Brazil clearly has the greatest potential for the development of its own natural resources and human capacity. It is also the Latin American nation that appears most likely to have a significant impact on future world affairs.

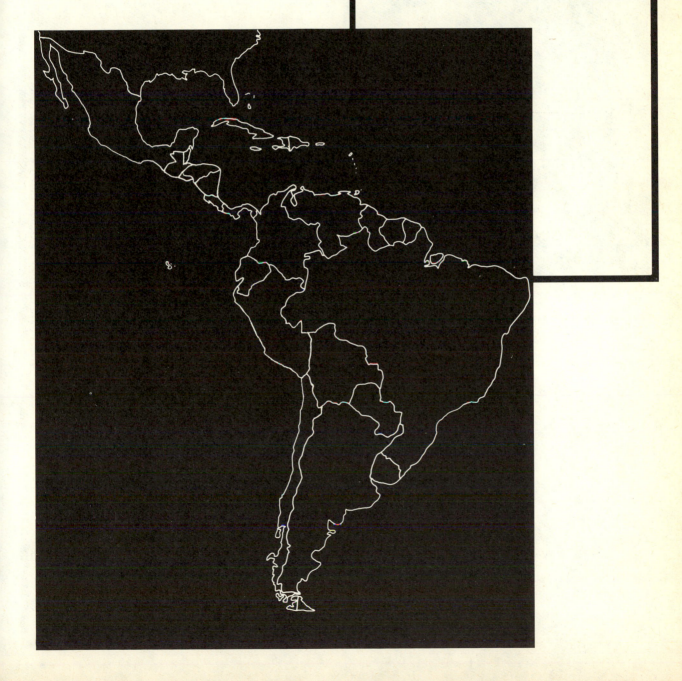

CONCLUSION
AND
APPENDICES

PROBLEMS AND PROSPECTS

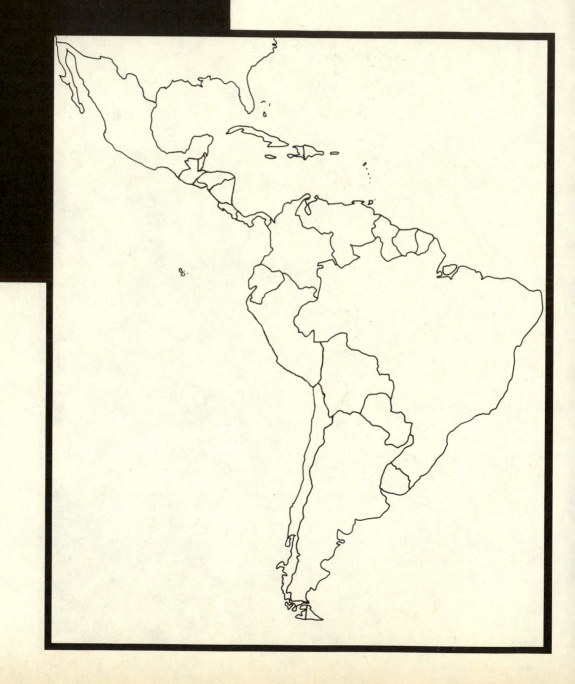

37

Each edition of this book has emphasized the great differences that exist within and between the various countries of Latin America. However, similarities among these countries are equally striking and cover a wide range of phenomena. A high birth rate and high unemployment are common in many areas. Rapid urbanization and a high mortality rate characterize both Middle and South America. Dependence on a single crop or mineral for foreign exchange is a trait that has typified most of the countries throughout their existence. Historically, there has been a concentration of population around each nation's capital city, with considerable empty country beyond. The predominant religion in most countries is Roman Catholic, and the predominant language is Latin-based. Most of the countries are struggling to achieve economic progress and social reform in the face of unstable political conditions. This is a culture region where the industrial and democratic revolutions have drastically transformed preexisting ways of living, thus making it part of a worldwide process of cultural innovation.[1]

Continuing Concerns

No one has yet learned to predict the future with any significant degree of accuracy. Yet an examination of the past, along with a consideration of current problems and trends, suggests some themes that may be of continuing concern within Latin America during the years ahead. Among these are: (1) population growth, (2) economic development, (3) external indebtedness, (4) foreign trade and regional integration, (5) territorial conflicts, and (6) political stability.

Population Growth

It is important to consider cultural change in Latin America within an appropriate context. The world's greatest population explosions are occurring in Africa and Latin America. An almost impossible task, yet enduring problem, exists in trying to keep the rate of economic expansion ahead of population growth. Three parts of the population problem in Latin America must be emphasized. First is the extraordinary rate of increase; second is the increasing concentration

---

[1] This perspective is developed in P. E. James and K. E. Webb, *One World Divided,* 3d ed. (New York: John Wiley & Sons, 1980).

of people in urban areas; and third is the effort being made, in certain countries, to relocate people and industries away from areas of excessively concentrated settlement.

It is estimated that by the year 2000, the total population of Latin America will exceed 550 million, which is more than the present population of the Soviet Union and United States combined. By 1995 the workforce of Latin America will number 179 million and could have a powerful influence on development, if gainfully employed, and also if unemployed. Despite the burgeoning population, Latin America has been transformed from a backward rural region of the 1940s to an economically aggressive urban region in the 1980s. It is also the most highly industrialized region in the developing world. Women as well as men have entered the labor force in massive numbers as most Latin American economies have become integrated into the worldwide political and economic system.[2]

All over Latin America the population of cities is increasing much faster than the total population. People tend to move away from the thinly settled areas and concentrate increasingly in areas already well populated. The population pattern of Latin America features an urban core in the midst of each separate cluster of people and in most cases there is only one regional trade center. Because the clusters generally have remained separate, the service areas seldom overlap. In addition, most countries have a primate city that is at least several times larger than any other city in the nation. Exceptions include Brazil, where the cities of São Paulo and Rio de Janeiro are of comparable size, and Ecuador, where Quito and Guayaquil are of roughly equal magnitude.

The concentration of people in cities is one of the distinguishing characteristics of an industrial society. The proportion of people employed in agriculture and the proportion employed in cities may be used as rough indices of economic devel-

opment. In Latin America, however, definite peculiarities exist. In the urbanized societies of Western Europe, Anglo-America, and Australia and New Zealand, employment in manufacturing was the initial reason for urban growth. Industries appeared first, and this led to a multiplication of service employment. In Latin American cities, service occupations appeared before there was a rapid increase of jobs in the industrial sector.

In 1960 slightly less than half of all Latin Americans, 100 million out of 201 million, lived in cities. Today, 226 million out of 397 million, or 65 percent, reside in urban areas. Most notable in South America is the example of Peru, which in 1981 experienced an urban increase of 5.1 percent. With such urbanization have come traffic congestion and pollution, severe housing shortages, inadequate public services, and political stress. One response has been to promote decentralization and regional development, which would at least diminish the flow of migration to the central cities. Such efforts have not been conspicuously successful to date but are worthy of continued emphasis.

Economic Development

Since World War II, throughout Latin America there has been a "rising tide of expectations." During the war, Latin America served as a major supplier of food and raw materials to Anglo-America and Western Europe. Imports of industrial goods were restricted, however, because these were needed in the all-out military effort. As a consequence, financial resources accumulated in Latin America, and domestic industries were established to compensate for the decline in manufactured goods from abroad.

With an intensive demand for economic development has come a concomitant disregard, until recently, for preservation of the natural environment. Examples abound. In Mexico City industrialization and traffic congestion have increased to the extent that this city, more than almost any other in the world, suffers the health

---

[2] V. P. Vaky, "Hemispheric Relations: Everything Is Part of Everything Else," *Foreign Affairs* 59, No. 3 (1981): 617–647.

hazards of severe environmental pollution. The Lake of Valencia, once the center of Venezuela's principal agricultural region, is now the receptacle of industrial discharge from factories all around its perimeter. Having no external drainage, this beautiful fresh-water lake soon may become an open cesspool unless prompt remedial action is taken. In Brazil, development has been given clear priority over conservation. Widespread international concern has been aroused over destruction of the Amazonian forests and its Indian tribes, as modern highways, agriculture, and industries penetrate the once virgin region.[3] In southern Brazil, near Cubatão, heavy industry is so concentrated that air and water pollution has become a threat to human existence. Hence, the area around Cubatão has become known as the "Valley of Death." A challenge for the future is to achieve economic growth within a context of long-term environmental preservation.

External Indebtedness

Nations, like individuals, can benefit from loans to achieve long-range economic objectives or to survive short-term financial crises. The ability to benefit from borrowing, however, depends on how wisely the money is invested, the terms of the loan, and a variety of other factors. Most important is a steady and dependable source of income, which for an individual usually means remunerative employment and for a nation revolves around a healthy balance of trade. More specifically, income from the export of goods is generally a critical factor if a country is to pay its debts in a prompt and orderly manner.

The rapid rise of population in Latin America, together with increased demands per capita, has placed severe pressure on the economic system. Where domestic demand exceeds the production of goods and services, an imbalance of trade and growing external indebtedness are among the

likely results. If the economic and political systems cannot meet perceived needs, there may be a clamor for revolutionary change.

After 1981, the nations of Latin America collectively faced the most severe economic crisis since the Great Depression of the 1930s. Resulting in part from the preceding and concurrent energy crisis, this condition has featured increased unemployment, reduced production, a high rate of domestic inflation, and severe balance of payment deficits. The deficits reached a record figure of $39 billion for the region in 1981. At the same time, the growing costs of interest payments and amortization on external debts absorbed a high proportion of the region's financial capacity and limited the possibilities of overall economic gain.

The total external debt of the Latin American nations rose to $336 billion by the end of 1983, compared with $75 billion in 1975. More than 85 percent of the debt was owed by the seven largest and most developed nations: Brazil, Mexico, Argentina, Venezuela, Chile, Colombia, and Peru. The proportion of foreign exchange earnings required to service the collective external debt of the Latin American nations rose from 27 percent in 1975 to 59 percent in 1982. Brazil, the region's leading country in size, population, and potential, achieved the dubious distinction of having the world's greatest external indebtedness: some $92 billion in 1983. This represented a record 72 percent of Brazil's total export earnings during the same year.[4] Other countries of the region faced parallel problems, and it was apparent that responsible action would be required both by the debtor governments and by the private lending institutions in the more developed nations of the world.

Foreign Trade and Regional Integration

Just as World War II had a global impact in terms of destruction, so also did the bold actions taken

---

[3]Emilio F. Moran, *Developing the Amazon: The Social and Ecological Consequences of Government-Directed Colonization Along Brazil's Transamazonian Highway* (Bloomington: Indiana University Press, 1981).

[4] Inter-American Development Bank, *Economic and Social Progress in Latin America: Natural Resources* Washington, D.C., 1983).

to achieve recovery and rehabilitation. Certain aspects of the reconstruction period were especially noted by leaders of nations in the less developed parts of the world. First, through the Marshall Plan, the United States provided massive financial and material aid to the devastated nations of Western Europe. Second, the European Economic Community (EEC) was formed, whereby the formerly separate and competitive nations of Europe began to work together, through a Common Market approach. Soon Western Europe not only recovered from wartime destruction and deprivation, but also gained a degree of prosperity exceeding anything it previously had known.

The leaders of Latin America observed the example of Western Europe and felt it could be applied equally well to their poverty-stricken region. They consequently made persistent appeals to the United States for a Marshall Plan-type of aid. The Alliance for Progress, proclaimed by President John F. Kennedy in 1961 and formulated by representatives of all the American nations (except Cuba) in a meeting at Punta del Este, Uruguay, provided the response. With $80 billion from the Latin American nations themselves through reform of their tax laws and $10 billion from the United States, supplemented by another $10 billion in private investments from the United States. Western Europe, and Japan, the major problems of Latin America were to be addressed within a period of 10 years. Among the goals were to eliminate illiteracy, improve health conditions, provide decent housing, achieve land reform, and strengthen democratic institutions throughout the region. Private enterprise was to be encouraged, along with social development, and support was to be given to economic integration.

The Alliance for Progress was welcomed enthusiastically in Latin America, and substantial progress actually was achieved. However, it was soon realized that the level of funding was insufficient and, more important, that a 10-year effort to help get Western Europe "back on its feet" was totally inadequate as a model to resolve the basic problems that had deterred economic progress in Latin America for several centuries.

Even the world's most advanced nations have experienced serious economic problems during the early 1980s. Hence, foreign aid is increasingly channeled in the form of loans through private banks and with less than generous terms of interest and repayment. Given current realities, most Latin American leaders understand that economic development must be largely self-generated and self-sustained. Consequently, they have placed great emphasis on import substitution, agricultural self-sufficiency, diversification of export commodities and markets, and the export of manufactured goods as opposed to raw materials. The plea of "trade, not aid" is presented forcefully by Latin Americans to gain tariff and quota concessions for the export of their goods to Western Europe, the United States, and Japan.

The unified approach to regional development had an impressive beginning in Latin America with formation of the Central American Common Market (CACM) in 1960. The five small nations of the Central American isthmus (Guatemala, El Salvador, Honduras, Nicaragua, and Costa Rica) prospered dramatically until this harmonious union was disrupted by the so-called Soccer War between Honduras and El Salvador in 1969.

Almost simultaneous with creation of the CACM was that of the more ambitious, though less successful, Latin American Free Trade Association (LAFTA). It included Mexico, Venezuela, Colombia, Ecuador, Peru, Bolivia, Chile, Paraguay, Argentina, Uruguay, and Brazil. In 1969, an Andean Group (Colombia, Ecuador, Peru, Bolivia, and Chile) was formed within LAFTA, in the belief that these countries by acting in concert could better compete with the larger nations: Mexico, Argentina, and Brazil. Venezuela became a Pact member in 1973; Chile withdrew in 1976. LAFTA, in turn, was restructured and renamed by the Treaty of Montevideo in 1980 as the Latin American Integration Association (LAIA), while retaining the original 11-nation membership. This treaty recognized differing levels of economic development among the member countries by establishing three separate categories of tariff rates.

The countries of the Paraguay-Paraná-Plata Basin, namely, Argentina, Uruguay, Brazil, Bolivia, and Paraguay, signed a treaty in 1969 to achieve a physical integration of this region. Special attention has been focused on the construction of dams, bridges, and tunnels to facilitate power development and to improve the regional transportation system. Although significant progress has been made, effective integration is hindered by the conflicting interests of Argentina and Brazil over use of the Río Paraná and by the imbalance of economic and political power among the participating nations.

Yet another grouping of nations was formed in 1968 as the Caribbean Free Trade Area (CARIFTA) following dissolution of the West Indies Federation in 1962. CARIFTA, in turn, was restructured in 1973 as the Caribbean Community (CARICOM), which includes Barbados, Guyana, Jamaica, Trinidad and Tobago, Belize, Dominica, Grenada, Montserrat, Saint Lucia, Saint Vincent, Antigua, and Saint Kitts-Nevis.

Territorial Conflicts

Considering the number of Latin American countries and the length of their international boundaries, prominent territorial conflicts are relatively few. Those that do exist, however, are persistent. Among these are Guatemala's territorial claim to all of Belize, a claim that is recognized only by Guatemala among all of the world's nations. El Salvador and Honduras have engaged in open warfare over disputed boundary territory as part of the 1969 "Soccer War," and no final settlement has yet been achieved.

In South America, two extensive land areas are subject to dispute. Venezuela's designs on the Essequibo region of Guyana, if implemented, would deprive Guyana of more than half of its total national territory. The disputed land is thinly populated but offers potential for forestry and mineral exploitation. Ecuador has already lost half of its national territory, mostly in the Amazon Basin, during a 1941 war with Peru. Although it had little choice but to sign the Protocol of Rio de Janeiro in 1942, which ended the war, Ecuador officially rejected the Protocol in 1960 and has retained a considerable degree of resentment toward Peru. In addition to the territorial question, the southern boundary of Ecuador remains to be defined, as reflected in a brief outbreak of fighting between Ecuador and Peru in 1981.

Several additional conflicts remain. One relates to Bolivia's demand for an outlet to the sea, which was lost as an outcome of the war of the Pacific, 1879–83. Another pertains to Argentina's claim to the Falkland or Malvinas Islands, over which a disastrous war was fought with Great Britain in 1982. Both disputes have been the subject of negotiation over a period of many decades; neither appears likely to be resolved permanently by the use of force.

Of particular interest to most Latin American countries is control of adjacent seas. In 1948, three years after President Harry S Truman proclaimed the jurisdiction of the United States over the seabed resources of the continental shelf, Chile, Peru, and Ecuador claimed maritime zones extending 200 miles from their coasts, embracing the seabed and the water column above. The variety and number of claims that ensued made it clear that a Law of the Sea Conference must be held. The first meeting was convened in 1974, when 76 countries claimed 12 to 200 miles of territorial seas. Approximately 40 percent of the world's oceans lies within 200 miles from shore, and each coastal state has been quick to anticipate potential gains. A case in point occurred in 1979 when Aves Island assumed a central role in a brief but bloody "Lobster War" between Miami-based Cuban lobstermen and the Bahamian government.

Aves Island is a mere dot of land in the eastern Caribbean, more than 200 miles north of the Venezuelan mainland and about 100 miles west of the Windward Islands. Oil-rich Venezuela claims all economic rights, including fishing, in a vast area surrounding the tiny island. Venezuela's gain means a substantial loss of ocean space to islands such as Puerto Rico, Guadeloupe, Dominica, and Montserrat. The 200-mile rule that would give rights to many countries would result in the overlapping of rights with one another. The rule awards more territory to the southern Caribbean countries and less to the northern islands, where fishing is by nature the poorest. Despite more than a decade of discussion, wrangling, and compromise, the proposed 350-article Sea Law Treaty has not been implemented. At stake is much more than fishing rights. Also involved is the freedom of navigation and overflight, protection of the marine environment, conservation and management of fisheries resources, recovery of minerals from the sea, and exploration for petroleum and natural gas.[5]

Political Stability

Nothing is more important than political stability for the achievement of economic growth and modernization. Stability may be achieved through military dictatorship, as it has been in Latin America many times during the past. However, such stability is usually followed by a period of recession and chaos.

Among the larger nations of Latin America, only a few have been able to maintain political stability within a democratic system for any extended period of time. Most notable is Mexico, which has moved far through basic reforms achieved initially by its Revolution, 1910–15,

and its resulting constitution of 1917. Other examples are Venezuela and Colombia, where stability and democratic government have prevailed since about 1959.

Four other countries of South America have begun a slow, yet deliberate, change in political foundations. Long-entrenched military regimes in Argentina, Brazil, Chile, and Uruguay are attempting to open the way for a lasting democracy, and at the same time are trying to avoid a return to periods of turbulence and instability. These four countries have a total population of about 178 million. Hence, whatever political moves they make are likely to influence developments throughout the continent.

The countries of Latin America are not easily governed, because social inequalities are deeply rooted and have fostered a tradition of rebellion. Within the universities of the region, student involvement in political issues is intensive and is considered no less normal than involvement in preparation for one's professional career. Stability throughout much of the region is tenuous at best, yet there is an irresistible demand for improvement in all aspects of life, including government.

The people of Latin America are for the most part young, dynamic, and impatient. They insist on being recognized and respected. They are growing in numbers and in influence internationally. Unfortunately, they and their countries are generally poorly understood by the outside world. Most needed for all Americans is enlightened leadership founded on a basic understanding of the people and the conditions in which they live. A spirit of Pan Americanism is needed to help achieve Latin American unity and hemispheric solidarity.

---

[5] S. L. Richardson, "Power, Mobility and the Law of the Sea," *Foreign Affairs* 58, No. 4 (1980): 902–919.

REFERENCES

The following references provide further information of a geographic nature on Latin America. Additional references are available from a variety of bibliographic sources. Of major importance is the *Handbook of Latin American Studies*, edited at the Hispanic Foundation, Library of Congress, and published annually since 1935 (University Presses of Florida). This handbook contains an annotated list of publications concerning all aspects of Latin American studies. The American Geographical Society publishes its *Current Geographical Publications*, which is the accession list of the Society's library.

For data concerning geographic features and places, see *Webster's New Geographical Dictionary*, G. & C. Merriam Company, Publishers, 1980. The most recent statistical data regarding population, production, and trade can be found in the current *Britannica Book of the Year*. A summary of data has been published annually since 1955 by the Committee on Latin American Studies, University of California at Los Angeles, under the title *Statistical Abstract of Latin America*. Also useful are the news releases and monographs of the Inter-American Development Bank, the World Bank, and the Organization of American States. The Area Handbook Series compiled by the American University and published by the U.S. Government Printing Office, Washington, D.C., includes individual volumes on all the major countries of Latin America and is an indispensable source of information.

Other government publications are available on specific topics. Examples are *Foreign Agriculture*, published by the U.S. Department of Agriculture, and the *Mineral Trade Notes* and *Minerals Yearbook* of the U.S. Bureau of Mines. Current events may be followed in the *Times of the Americas*, the only English-language newspaper in the United States devoted exclusively to news from Latin America. Another newspaper with excellent coverage of Latin America is *The Miami Herald*. A catalogue of doctoral dissertations dealing with research on Latin America is available through University Microfilms International, of Ann Arbor, Michigan, whereas useful articles on Latin America appear in periodicals such as the *Journal of the Developing Areas, Journal of Inter-American Studies, Journal of Inter-American Economic Affairs,* and *Journal of Latin American Studies.*

Two organizations that promote geographic research of Latin America are the Conference of Latin Americanist Geographers (CLAG), based at Ball State

University, Muncie, Indiana, and the Pan American Institute of Geography and History (PAIGH), with headquarters in Mexico City. Both maintain an active publication program.

Latin America — General

(Including South America and Middle America)

Avery, W. P., 1983, "The Politics of Crisis and Cooperation in the Andean Group," *Journal of Developing Areas,* 17:2, 155–184.

Barraclough, S., 1973, *Agrarian Structure in Latin America,* D. C. Heath, Lexington, Mass.

Blakemore, H., and Smith, C. T., 1983, *Latin America: Geographical Perspectives,* 2d ed., Methuen, London.

Boehm, R. G., and Visser, S. (eds), 1984, *Latin America: Case Studies,* Kendall/Hunt, Dubuque, Ia.

Caviedes, C., 1981, "Natural Hazards in Latin America: A Survey and Discussion," *Proceedings of the Conference of Latin Americanist Geographers,* Muncie, Ind., 8:280–294.

Choucri, N., 1982, *Energy and Development in Latin America,* D. C. Heath, Lexington , Mass.

Cole, J. P., 1975, *Latin America: An Economic and Social Geography,* Butterworths, London.

Collier, D. (ed.), 1980, *The New Authoritarianism in Latin America,* Princeton University Press, Princeton, N.J.

Craig, A., and Lagos, R. A., 1981, "Geomorphologic Research in Western South America: The Surprising Seventies," *Proceedings of the Conference of Latin Americanist Geographers,* Muncie, Ind., 8:301–311.

Davidson, W. V., and Parsons, J. J. (eds.), 1980, *Historical Geography of Latin America,* Louisiana State University, Baton Rouge.

deLaubenfels, D. J., 1970, "The Vegetation Formations of Latin America: A Bibliography," *Revista Geográfica,* 72:97–138.

Dickenson, J.P., 1977, "A Decade of British Geographical Work on Latin America: Introduction and Bibliography," *Revista Geográfica,* 85:213–227.

Dickinson, J. C., III, 1972, "Alternatives to Monoculture in the Humid Tropics of Latin America," *Professional Geographer,* 24:3, 217–222.

Enders, T. O., and Mattion, R. P., 1984, *Latin America: The Crisis of Debt and Growth,* Brookings Institution, Washington D.C.

Farley, R., 1972, *The Economics of Latin America: Development Problems in Perspective,* Harper and Row, New York.

Ferris, E. G., 1982, "National Support for the Andean Pact," *Journal of the Developing Areas,* 16:2, 249–270.

Fox, R. M., 1975, *Urban Population Growth Trends in Latin America,* Inter-American Development Bank, Washington, D.C.

Gilbert, A., 1974, *Latin American Development: A Geographical Perspective,* Penguin Books, Middlesex, U.K.

Gilbert, A., Hardoy, J., and Ramírez, R., 1982, *Urbanization in Contemporary Latin America,* Wiley, New York.

Gregersen, H. M., and Contreras, A., 1975, *U.S. Investment in the Forest-based Sector in Latin America: Problems and Potentials,* Johns Hopkins University Press, Baltimore.

Griffin, E. C., and Ford, L., 1980, "A Model of Latin American City Structure," *Geographical Review,* 70:4, 397–422.

Griffin, E. C., and Minkel, C. W., 1971, "A Bibliography of Theses and Dissertations on Latin America by U.S. Geographers, 1960–1970," *Revista Geográfica,* 74:115–134.

Hardoy, J. E. (ed.), 1975, *Urbanization in Latin America: Approaches and Issues,* Anchor Press, New York.

Harrison, K., et al., 1975, *Improving Food Marketing Systems in Developing Countries: Experiences from Latin America,* Latin American Studies Center, Michigan State University.

Hill, A. D. (ed.), 1973, *Latin American Development Issues,* CLAG Publications, Inc., East Lansing, Mich.

Hunter, J. M., and Foley, J. W., 1975, *Economic Problems of Latin America*, Houghton Mifflin, Boston.

Inter-American Development Bank, 1983, *Economic and Social Progress in Latin America: Natural Resources*, Washington, D.C.

Inter-American Development Bank, 1984, *External Debt and Development in Latin America: Background and Prospects*, Washington, D.C.

Knight, T. J., 1979, *Latin America Comes of Age*, Scarecrow Press, Metuchen, N.J.

Krause, W., and Jud, G. D., 1973, *International Tourism and Latin American Development*, Bureau of Business Research, University of Texas, Austin.

Kritz, M. M., and Gurak, D. T. (eds.), 1979, "International Migration in Latin America," Special Issue, *International Migration Review*, 13:3.

"Latin America, 1984", *Current History*, 83:490, 49–82, Philadelphia.

McCoy, T. L. (ed.), 1974, *The Dynamics of Population Policy in Latin America*, Ballinger, Cambridge, Mass.

McGaughey, S. E., and Gregersen, H. M. (eds.), 1983, *Forest-based Development in Latin America*, Inter-American Development Bank, Washington, D.C.

Maos, J. O., 1984, *The Spatial Organization of New Land Settlement in Latin America*, Westview Press, Boulder, Colo.

Martinson, T. L., and Elbow, G. S. (eds.), 1981, *Geographic Research on Latin America: Benchmark 1980*, Conference of Latin Americanist Geographers, Ball State University, Muncie, Ind.

Martinson, T. L., Lentnek, B., and Carmin, R. L. (eds.), 1971, *Geographic Research on Latin America: Benchmark, 1970*, Proceedings of the Conference of Latin Americanist Geographers, Ball State University, Muncie, Ind.

Morris, A., 1981, *Latin America: Economic Development and Regional Differentiation*, Hutchinson, London.

Morse, R. M., 1975, "The Development of Urban Systems in the Americas in the Nineteenth Century," *Journal of Interamerican Studies*, 17:1, 4–26.

Mulchansingh, V., 1971 "The Location of Oil Refining in Latin America and the Caribbean," *Revista Geográfica*, 75:85–126.

Mytelka, L. K., 1979, *Regional Development in a Global Economy: The Multinational Corporation, Technology and Andean Integration*, Yale University Press, New Haven.

Nelson, M., 1973, *The Development of Tropical Lands: Policy Issues in Latin America*, Johns Hopkins University Press, Baltimore.

Odell, P. R., and Preston, D. A., 1973, *Economics and Societies in Latin America: A Geographical Interpretation*, Wiley, London.

Olson, G. W., 1981, "Progress in Use of Soil Resource Inventories in Latin America," *Proceedings of the Conference of Latin Americanist Geographers*, Muncie, Ind., 8:295–300.

Richardson, H. W., and Richardson, M., 1975, "The Relevance of Growth Center Strategies to Latin America," *Economic Geography*, 51:163–178.

Robinson, D. J. (ed.), 1979, "Social Fabric and Spatial Structure in Colonial Latin America," *Dell Plain Latin American Studies*, Syracuse University, Syracuse, N.Y.

Robinson, D. J., 1980, *Studying Latin America: Essays in Honor of Preston E. James*, Department of Geography, Syracuse University, Syracuse, N.Y.

Sable, M. H., 1971, *Latin American Urbanization: A Guide to the Literature, Organizations and Personnel*, Scarecrow Press, Metuchen, N.J.

Sánchez-Albornoz, N., 1974, *The Population of Latin America: A History*, University of California Press, Berkeley.

Sauer, C. O., 1966, *The Early Spanish Main*, University of California Press, Berkeley.

Silvert, K. H., 1966, *The Conflict Society, Reaction and Revolution in Latin America*, American Universities Field Staff, New York.

Smith, T. L., 1967, *The Process of Rural Develop-*

ment in Latin America, University of Florida Press, Gainesville.

Stepan, A., 1980, "The U.S. and Latin America: 1979," *Foreign Affairs,* 58:3, 659–692.

Switzer, K. A., 1973, "The Andean Group: A Reappraisal," *Inter-American Economic Affairs,* 26:4, 69–81.

Taeuber, L. B., 1974, *General Censuses and Vital Statistics in the Americas,* U.S. Government Printing Office, Washington, D.C.

Tambs, L. A., 1970, "Latin American Geopolitics: A Basic Bibliography," *Revista Geográfica,* 73:71–105.

Thomas, R. N. (ed.), 1973, *Population Dynamics of Latin America,* CLAG Publications, East Lansing, Mich.

Thomas, R. N., and Hunter, J. M. (eds.), 1980, *Internal Migration Systems in the Developing World: With Special Reference to Latin America,* G. K. Hall, Boston.

Tosi, J. A., and Voertman, R. F., 1964, "Some Environmental Factors in the Economic Development of the Tropics," *Economic Geography,* 40:189–205.

United Nations, 1981, *Economic Survey of Latin America, 1979,* United Nations, New York.

Veliz, C., 1980, *The Centralist Tradition of Latin America,* Princeton University Press, Princeton, N.J.

West, R. C., and Augelli, J. P., *Middle America: Its Lands and Peoples,* 2d ed., Prentice-Hall, Englewood Cliffs, N.J.

Wiarda, H. J. (ed.), 1980, *The Continuing Struggle for Democracy in Latin America,* Westview Press, Boulder, Colo.

Wilkie, J. W. (ed.), 1980, *Statistical Abstract of Latin America,* Latin American Center Publications, University of California, Los Angeles.

World Bank, 1978, *Land Reform in Latin America: Bolivia, Chile, Mexico, Peru and Venezuela,* Washington, D.C.

Young, R. C., 1970, "The Plantation Economy and Industrial Development in Latin America," *Economic Development and Cultural Change,* 18:3, 342–357

Mexico

Allen, E. A, 1981, "The State and the Region: Some Comparative Developmental Experiences from Mexico and Brazil," *Revista Geográfica,* 93:61–77.

Arreola, D. D., 1982, "Nineteenth-Century Townscapes of Eastern Mexico," *Geographical Review,* 72:1, 1–19.

Balán, J., Browning, H. L., and Jelin, E., 1973, *Men in a Developing Society: Geographic and Social Mobility in Monterrey, Mexico,* Institute of Latin American Studies, University of Texas, Austin.

Ball, J. M., 1971, *Migration and the Rural Municipio in Mexico,* Georgia State University, Atlanta.

Bank of London and South America, 1975, "Mexico: The State of Nuevo León," *BOLSA Review,* 9:9, 494–498.

Barkenbus, J. N., 1974, "The Trans-Peninsular Highway: A New Era for Baja California," *Journal of Interamerican Studies,* 16:3, 259–273.

Bullard, F. J., 1968, *Mexico's Natural Gas: The Beginning of an Industry,* Bureau of Business Research, University of Texas, Austin.

Calvert, P., 1973, *Mexico,* Praeger, New York.

Corlos, M. L., 1974, *Politics and Development in Rural Mexico: A Study of Socio-Economic Modernization,* Praeger, New York.

Coale, A. J., 1979, "Population Growth and Economic Development: The Case of Mexico," *Foreign Affairs,* 56:2, 415–429.

Collier, G. A., 1975, *Fields of the Tzotzil: The Ecological Bases of Tradition in Highland Chiapas,* University of Texas Press, Austin.

Cornelius, W. A., 1975, *Politics and the Migrant Poor in Mexico City,* Stanford University Press, Stanford, Calif.

DeWalt, B. R., 1979, *Modernization in a Mexican Ejido: A Study in Economic Adaptation,* Cambridge University Press, New York.

Doolittle, W. E., 1983, "Agricultural Expansion in a Marginal Area of Mexico," *Geographical Review,* 73:301–313.

Fagen, R. R., 1977, "The Realities of U.S.-Mexican Relations," *Foreign Affairs,* 55:4, 685–700.

Fox, D. J., 1972, "Patterns of Morbidity and Mortality in Mexico City," *Geographical Review,* 62: 151–185.

Freebairn, D. K., 1983, "Agricultural Interactions Between Mexico and the United States," *Journal of Interamerican Studies,* 25, 3: 215–298.

Freivalds, J., 1975, "Mexico: The Agricultural and Livestock Sectors," *Bank of London and South America Review,* 9:8, 440–447.

Gerhard, P., 1975, "Continuity and Change in Morelos, Mexico," *Geographical Review,* 65: 335–352.

Ginneken, W. V., 1980, *Socio-Economic Groups and Income Distribution in Mexico: A Study Prepared for the ILO World Employment Programme,* St. Martin's Press, New York.

Gonzáles, A., 1966, "Problems of Agricultural Development in a Pioneer Region of Southwestern Mexico," *Revista Geográfica,* 64:29–52.

Grayson, G. W., 1980, "Oil and Policies in Mexico," *Current History,* 78:454, 53–56.

Griffin, E. C., and Ford, L. R., 1976, "Tijuana: Landscape of a Culture Hybrid," *Geographical Review,* 66: 435–447.

Gunn, R. L., 1976, "A Land-Use Study of Commercial Agriculture in Chiapas, Mexico," Ph.D. Dissertation, Michigan State University, East Lansing.

Harnapp, V. R., 1979, "Landsat as a Land Evaluation Tool in Gulf Coast Mexico," *Revista Geográfica.* 90:151–156.

Henderson, D. A., 1965, "Arid Lands Under Agrarian Reform in Northwest Mexico," *Economic Geography,* 41: 300–312.

Lentnek, B., Charnews, M., and Cotter, J. V., 1978, "Commercial Factors in the Development of Regional Urban Systems: A Mexican Case Study," *Economic Geography,* 54: 291–308.

Levy, P., 1980, *Mexico: An Economic Survey,* Economics Department, Citibank, New York; México, D.F.

Looney, R. E., 1978, *Mexico's Economy: A Policy Analysis with Forecasts to 1990,* Westview Press, Boulder, Colo.

MacGregor, M. T., and Valverde, C., 1975, "Evolution of the Urban Population in the Arid Zones of Mexico, 1900–1970," *Geographical Review,* 65: 215–228.

Nelson, C., 1971, *The Waiting Village: Social Change in Rural Mexico,* Little, Brown & Co. Boston.

Nugent, J. B., and Tarawneh, F. A., 1983, "The Anatomy of Changes in Income Distribution and Poverty among Mexico's Economically Active Population between 1950 and 1970," *Journal of the Developing Areas,* 17:2, 197–226.

Pfeifer, G., 1966, "The Basin of Puebla—Tlaxcala in Mexico," *Revista Geográfica,* 85–107.

Platter, S., 1980, "Economic Development and Occupational Change in a Developing Area of Mexico," *Journal of the Developing Areas,* 14:4, 469–482.

Pyle, J., 1970, "Market Locations in Mexico City," *Revista Geográfica,* 73: 59–69.

Sanders, T. G., 1979, "Mexico in 1978," *American University Field Staff Reports,* 31.

Sauer, C. O., 1932, "The Road to Cíbola," *Ibero-Americana, 3,* University of California Press, Berkeley.

Sawatsky, H. L., 1971, *They Sought a Country: Mennonite Colonization in Mexico,* University of California Press, Berkeley.

Scott, I., 1982, *Urban and Spatial Development in Mexico,* Johns Hopkins University Press, Baltimore.

Silvers, A., and Crosson, P., 1980, "Rural Development and Urban-bound Migration in Mexico," *Resources for the Future*, Washington, D.C.

Smith, S. K., 1981, "Determinants of Female Labor Force Participation and Family Size in Mexico City," *Economic Development and Cultural Change*, 20:1, 129–152.

Snyder, D. E., 1966, "Urbanization and Population Growth in Mexico," *Revista Geográfica*, 73–84.

Tata, R. J., 1973, "A Space Potential Interpretation of Transportation Efficiency in Northeast Mexico," *Revista Geográfica*, 78: 35–53.

Venezian E. L., and Gamble, W. K., 1969, *The Agricultural Development of Mexico: Its Structure and Growth Since 1950*, Praeger, New York.

Wagner, P. L., 1962, "Natural and Artificial Zonation in a Vegetation Cover: Chiapas, Mexico," *Geographical Review*, 52: 253–274.

Wagner, P. L., 1963, "Indian Economic Life in Chiapas," *Economic Geography*, 39: 156–164.

Walton, M. K., 1977, "The Evolution and Localization of Mezcala and Tequila in Mexico," *Revista Geográfica*, 85: 113–129.

West, R. C., and Parsons, J. J., 1941, "The Topia Road: A Trans-Sierran Trail of Colonial Mexico," *Geographical Review*, 34: 574–586.

Williams, B. J., 1972, "Tepetate in the Valley of Mexico," *Annals of the Association of American Geographers*, 62: 618–626.

Williams, E. J., 1979, *The Rebirth of the Mexican Petroleum Industry: Development Directions and Policy Implications*, D.C. Heath, Lexington, Mass.

Central America

Arbingash, S. A., et al., 1979, *Atlas of Central America*, Bureau of Business Research, University of Texas, Austin.

Crowley, W., and Griffin, E. C., 1983, "Political Upheaval in Central America," *Focus*, 34:1.

Epsy, H. C., and Creamer, L., Jr., 1970, *Another World: Central America*, Viking Press, New York.

Feinberg, R. E. (ed.), 1982, *Central America: International Dimensions of the Crisis*, Holmes and Meier, New York.

Gibson, J. R., 1970, *A Demographic Analysis of Urbanization: Evolution of a System of Cities in Honduras, El Salvador, and Costa Rica*, Latin American Studies Program, Cornell University, Ithaca, N.Y.

McCleeland, D. H., 1972, *The Central American Common Market: Economic Policies, Economic Growth, and Choices for the Future*, Praeger, New York.

May, J. M., and McLellan, D. L., 1972, *The Ecology of Malnutrition in Mexico and Central America*, Hafner, New York.

Miller, E. E., 1982, "Small Farms in Central America," *Revista Geográfica*, 38–46.

Minkel, C. W., 1967, "Programs of Agricultural Colonization and Settlement in Central America," *Revista Geográfica*, 19–53.

Orantes, I. C., 1972, *Regional Integration in Central America*, D. C. Heath, Lexington, Mass.

Portig, W. H., 1965, "Central American Rainfall," *Geographical Review*, 55: 68–90.

World Bank, 1978, *A Model of Agricultural Production and Trade in Central America*, Washington, D.C.

Guatemala

Biechler, M. J., "The Regionalization of Coffee Culture in Guatemala," *Revista Geográfica*, 77:33–55.

Dickinson, J. C., 1974, "Fisheries of Lake Izabal, Guatemala," *Geographical Review*, 64:385–409.

Early, J. D., 1982, *The Demographic Structure and Evolution of a Peasant System: The Guatemalan Population,* University Presses of Florida, Gainesville.

Fletcher, L. B., et al., 1970, *Guatemala's Economic Development: The Role of Agriculture,* Iowa State University Press, Ames.

Grieb, K. J., 1979, *Guatemalan Caudillo,* Ohio University Press, Athens.
Griffith, W. J., 1965, *Empires in the Wilderness: Foreign Colonization and Development in Guatemala, 1837–1844,* University of North Carolina Press, Chapel Hill.

Horst, O. H., 1967, "The Specter of Death in a Guatemalan Highland Community," *Geographical Review,* 57:151–167.
Hoy, D. R., 1970, "A Review of Development Planning in Guatemala," *Journal of Inter-American Studies,* 12:2, 217–228.

Johnson, D., 1981, "Agricultural Location Using Production Data: The Example of Beans in Guatemala," *Revista Geográfica,* 94:117–121.
Jones, J. P., III, 1982, "The Periodic and Daily Indian Market System of Southwest Guatemala," *Revista Geográfica,* 96:47–53.

Melville, T., and Melville, M., 1971, *Guatemala: The Politics of Landownership,* Free Press, New York.

Thomas, R. N., 1977, "The Migration System of Guatemala City: Spatial Inputs," *Revista Geográfica,* 75:73–84.

Veblen, T. T., 1977, "Native Population Decline in Totonicapán, Guatemala," *Annals of the Association of American Geographers,* 67:484–499.
Veblen, T. T., 1978, "Forest Preservation in the Western Highlands of Guatemala," *Geographical Review,* 68:417–434.

World Bank, 1978, *Guatemala: Economic and Social Position and Prospects,* Washington, D.C.

Belize

Ashcraft, N., 1972, "Belize, the Awakening Land," *National Geographic Magazine.*
Ashcraft, N., 1973, *Colonialism and Underdevelopment: Processes of Political Economic Change in British Honduras,* Teachers College Press, New York.

Bolland, O. N., 1977, *The Formation of a Colonial Society: Belize, from Conquest to Crown Colony,* Johns Hopkins University Press, Baltimore.

Frost, M. D., 1981, "Patterns of Human Influence on Landscape and Wildlife: Selected Case Studies in Belize," *Revista Geográfica,* 94:89–100.
Grant, C. H., 1976, *The Making of Modern Belize: Politics, Society, and British Colonialism in Central America,* Cambridge University Press, New York.

Kearns, K. C., 1973, "Belmopan: Perspective on a New Capital," *Geographical Review,* 63:147–169.

Menon, P. K., 1979, "The Anglo-Guatemalan Territorial Dispute over the Colony of Belize (British Honduras)," *Journal of Latin American Studies,* 11, 2, 343–371.
Minkel, C. W., and Alderman, R. H., 1970, *A Bibliography of British Honduras, 1900–1970,* Latin American Studies Center, Michigan State University, East Lansing.

Sawatsky, H. L., 1969, *Mennonite Settlement in British Honduras,* Geography Department, University of California, Berkeley.
Setzekorn, W. D., 1975, *Formerly British Honduras: A Profile of the New Nation of Belize,* Dumbarton Press, Newark, Calif.

El Salvador

Browning, D., 1971, *El Salvador: Landscape and Society,* Clarendon Press, Oxford.
Browning D., 1983, "Agrarian Reform in El Salvador," *Journal of Latin American Studies,* 15:2, 399–426.

Daugherty, H. E., 1969, "Man-Induced Ecologic Change in El Salvador," Ph.D. Dissertation, University of California, Los Angeles.

Davis, L. H., and Weisenborn, D. E., 1981, "Small Farmer Market Development: The El Salvador Experience," *Journal of the Developing Areas*, 15:3, 407–416.

Durham, W. H., 1979, "Scarcity and Survival in Central America: Ecological Origins of the Soccer War," Stanford University Press, Stanford, Calif.

Karush, G. E., 1978, "Plantations, Population, and Poverty: The Roots of the Demographic Crisis in El Salvador," *Studies in Comparative International Development*, 8:3, 59–75.

Satterthwaite, R., 1971, "Campesino Agriculture and Hacienda Modernization in Coastal El Salvador, 1949 to 1969," Ph.D. Dissertation, University of Wisconsin, Madison.

White, A., 1973, *El Salvador*, Praeger, New York.

World Bank, 1979, *El Salvador: Demographic Issues and Prospects*, Washington, D.C.

Honduras

Checchi, V., and Assoc., 1959, *Honduras: A Problem in Economic Development*, Twentieth Century Fund, New York.

Croner, C. M., 1972, "Spatial Characteristics of Internal Migration to San Pedro Sula, Honduras," Ph.D. Dissertation, Michigan State University, East Lansing.

Davidson, W. V., 1972, "Historical Geography of the Bay Islands, Honduras: Anglo-Hispanic Conflict in the Western Caribbean," Ph.D. Dissertation, University of Wisconsin, Milwaukee.

Durham, W. H., 1979, *Scarcity and Survival in Central America: Ecological Origins of the Soccer War*, Stanford University Press, Stanford, Calif.

Ruhl, J. M., 1984, "Agrarian Structure and Political Stability in Honduras," *Journal of Inter-American Studies*, 26:1, 33–68.

Shirey, R. I., 1970, "An Analysis of the Location of Manufacturing: Tegucigalpa and San Pedro Sula, Honduras," Ph.D. Dissertation, University of Tennessee, Knoxville.

Thompson, N. R., III, 1973, "The Economic Geography of the Mining Industry of Honduras, Central America," Ph.D. Dissertation, University of Tennessee, Knoxville.

Wheeler, J. O., and Thomas, R. N., 1973, "Urban Transportation in Developing Economies: Work Trips in Tegucigalpa, Honduras," *Professional Geographer*, 25:2.

Nicaragua

American University, 1983, *Area Handbook for Nicaragua*, Washington, D.C.

Denevan, W. M., 1961, "The Upland Pine Forests of Nicaragua: A Study in Cultural Plant Geography," *University of California Publications in Geography*, 12:251–320.

Incer, J., 1973, *Geografía Ilustrada de Nicaragua*, Librería y Editorial Recalde, Managua.

Nietschmann, B., 1973, *Between Land and Water: The Subsistence Ecology of the Miskito Indians, Eastern Nicaragua*, Seminar Press, New York.

Parsons, J. J., 1955, "The Miskito Pine Savanna of Nicaragua and Honduras," *Annals of the Association of American Geographers*, 45:36–63.

Patten, G. P., 1971, "Dairying in Nicaragua," *Annals of the Association of American Geographers*, 61:303–315.

U.S. Department of State, 1983, "Nicaragua," Pub. 7772.

Walker, T., 1981, *Nicaragua: The Land of Sandino*, Westview Press, Boulder, Colo.

Costa Rica

Bergoeing, J. P., 1979, "Aspectos Geomorfológicos del Litoral Pacífico Central de Costa Rica," *Revista Geográfica*, 90:133–150.

Britt, K., 1981, "Costa Rica Steers the Middle Course," *National Geographic Magazine,* 160:1, 32–57.

Carvajal, A. G., 1983, "Les Migrations Intérieurs à Costa Rica: Une Approche Regionale au Probleme," *Revista Geográfica,* 98:91–114.

Ellefsen, R., Rodríguez, H., and Raburn, R., 1979, "Using Remote Sensing to Monitor Urban Growth: A Costa Rican Example," *Revista Geográfica,* 90:115–132.

Golley, F. B., Olien, M. D., and Hoy, D. R., 1971, "Cognized Environments of San Carlos Valley Settlers," *Revista Geográfica,* 74:33–50.

Hall, C., 1984, "Regional Inequalities in Well-being in Costa Rica," *Geographical Review,* 74:48–62.

Henshall, J., 1972, "Golden Caribbean Costa Rica," *Geographical Magazine,* 45:3, 192–198.

Klijzing, F.K.H., and Taylor, H. W., 1982, "Spatial Order in the Demographic Transition: The Costa Rican Case," *Revista Geográfica,* 96:54–59.

Nunley, R. E., 1960, *The Distribution of Population in Costa Rica,* National Academy of Sciences — National Research Council, Washington, D.C.

Parsons, J. J., 1963, "Agricultural Colonization in Costa Rica," *Geographical Review,* 53:451–454.

Revista Geográfica, 1977 and 1978, Special volume on Costa Rica, 86/87.

Seligson, M. A. 1979, "The Impact of Agrarian Reform: A Study of Costa Rica," *Journal of the Developing Areas,* 13:2m 161–174.

Spielmann, H. P., 1972, "La Expansión Ganadera en Costa Rica: Problemas de Desarrollo Agropecuaria," *Revista Geográfica,* 77:57–84.

Stouse, P.A.D., 1970, "Instability of Tropical Agriculture: The Atlantic Lowlands of Costa Rica," *Economic Geography,* 46:78–97.

Panama

Fox, D., 1964, "Prospects for the Panama Canal," *Tijdschrift voor Economishche en Sociale Geografie,* 55:86–101.

Fuson, R. H., 1964, "House Types in Central Panama," *Annals of the Association of American Geograhers,* 54:190–208.

Mayer, H. M., 1973, "Some Geographic Aspects of Technological Change in Maritime Transportation," *Economic Geography,* 49:2, 145–155.

Merrill, W. C., et al., 1975, *Panama's Economic Development: The Role of Agriculture,* Iowa State University Press, Ames.

Nyrod, R. F. (ed.), 1980, *Panama: A Country Study,* American University, Washington, D.C.

Organization of American States, 1978, *República de Panamá: Projecto de Desarrollo Integrado de la Región Oriental de Panamá—Darién,* Washington, D.C.

The Antilles

Auty, R., 1976, "Caribbean Sugar Factory Size and Survival," *Annals of the Association of American Geographers,* 66:76–88.

Belisle, F. J., 1983, "Tourism and Food Production in the Caribbean," *Annals of Tourism Research,* 10:497–513.

Blume, H., 1967, "Types of Agricultural Regions and Land Tenure in the West Indies," *Revista Geográfica,* 67:7–20.

Clarke, G. C., 1971, "Population Problems in the Caribbean," *Revista Geográfica,* 75:32–48.

Demas, W. G., 1978, "The Caribbean and the New International Economic Order," *Journal of Interamerican Studies,* 10:3, 229–263.

Doerr, A. H., and Hoy, D. R., 1957, "Karst Landscapes of Cuba, Puerto Rico, and Jamaica," *The Scientific Monthly,* 85:178–187.

Eyre, A., 1971, *A New Geography of the Caribbean*, 4th ed., George Philip, London.

Hoy, D. R., and Fisher, J. S., 1974, "Primary Production and the Measurement of Agricultural Potential in the Caribbean, *Revista Geográfica*, 80:71–87.

Lowenthal, A. F., 1982, "The Caribbean," *The Wilson Quarterly*, 4:113–141.

Lowenthal, D., 1972, *West Indian Societies*, Oxford University Press, New York.

Mitchell, H., 1972, *Caribbean Patterns*, 2d ed., Wiley, New York.

Pearcy, G. E., 1965, *The West Indian Scene*, Princeton University Press, Princeton, N.J.

Perkins, D., 1966, *The United States and the Caribbean*, Harvard University Press, Cambridge, Mass.

Richardson, B. C., 1972, "The Agricultural Dilemma of the Post-Plantation Caribbean," *Inter-American Economic Affairs*, 26:1, 59–70.

Waddell, D.A.G., 1967, *The West Indies and the Guianas*, Prentice-Hall, Englewood Cliffs, N.J.

World Bank, 1978, *The Commonwealth Caribbean*, Johns Hopkins University Press, Baltimore.

Cuba

Acosta, M., and Hardoy, J. E., 1973, *Urban Reform in Revolutionary Cuba* (Trans. M. Bochner), Antilles Research Program, Yale University Press, New Haven, Conn.

Domínguez, J. I. (ed.), 1982, *Cuba: Internal and International Affairs*, Sage, Beverly Hills, Calif.

Dyer, D. R., 1967, "Cuban Sugar Regions," *Revista Geográfica*, 67:21–30.

González, A., 1971, "The Population of Cuba," *Caribbean Studies*, 11:74–84.

Guerra y Sánchez, R., 1964, *Sugar and Society in the Caribbean: An Economic History of Cuban Agriculture*, Yale University Press, New Haven,Conn.

Mesa-Lago, C., 1971 (ed.), *Revolutionary Change in Cuba*, Pittsburgh University Press, Pittsburgh, Pa.

Mesa-Lago, C., 1981, *The Economy of Socialist Cuba*, University of New Mexico Press, Albuquerque.

Nelson, L., 1970, "Cuban Population Estimates, 1953-1970," *Journal of Interamerican Studies*, 12:3, 367–378.

Pérez, L., 1982, "Iron Mining and Socio-Demographic Change in Eastern Cuba, 1886–1940," *Journal of Latin American Studies*, 14:2, 381–406.

Pérez, L. A., Jr., 1982, "The Collapse of the Cuban Planter Class, 1868–1968," *Inter-American Economic Affairs*, 36:3, 3–22.

Ritter, A. R., 1974, *The Economic Development of Revolutionary Cuba: Strategy and Performance*, Praeger, New York.

Hispaniola

Bodini, H., 1984, *Viability of Small-Farmer Communities in the Enriquillo-Cul de Sac Depression of Hispaniola*, Ph.D. Dissertation, University of Tennessee, Knoxville.

Logan, R. W., 1968, *Haiti and the Dominican Republic*, Oxford University Press, New York.

Pagán, D. P., 1979, *Bibliografía General de la Isla de Santo Domingo*, Universidad Central del Este, San Pedro de Macoris, D.R.

Palmer, C. E., 1976, *Land Use and Landscape Change Along the Dominican-Haitian Borderlands*, Ph.D. Dissertation, University of Florida, Gainesville.

Haiti

Anglade, G., 1981, *Atlas Critique d'Haiti*, Les Presses de l'Université de Québec.

Lundahl, M., 1979, *Peasants and Poverty: A Study of Haiti*, St. Martin's, New York.

Marshall, D. I., 1979, *"The Haitian Problem": Ille-*

gal Migration to the Bahamas, Institute of Social and Economic Research, University of the West Indies, Kingston, Jamaica.

Murray, G. F., 1977, *The Evolution of Haitian Peasant Land Tenure: A Case Study in Agrarian Adaptation to Population Growth*, Ph.D. Dissertation, Colombia University, New York.

Organization of American States, 1972, *Haiti: Mission d'Assistance Technique Intégrée*, Washington, D.C.

Rothberg, R. I., and Clague, C. K., 1971, *Haiti: The Politics of Squalor*, Houghton Mifflin, Boston.

Rubin, V., and Schaedel, R. P. (eds.), 1975, *The Haitian Potential: Research and Resources of Haiti*, Teachers College Press, Columbia University, New York.

Tata, R. J., 1982, *Haiti: Land of Poverty*, University Press of America, Washinton, D.C.

Wood, H. A., 1963, *Northern Haiti: Land, Land Use, and Settlement*, University of Toronto Press, Toronto.

Dominican Republic

Antonini, G. A., and Boswell, T. D., 1979, ''El Uso de Censos en la Regionalización Agro-socioeconómica de un País Latinoamericano,'' *Revista Geográfica*, 90:9–31.

Antonini, G. A., and York, M. A., 1979, ''Integrated Rural Development and the Role of the University of the Caribbean: The Case of Plan Sierra, Dominican Republic,'' *Revista Geográfica*, 90:97–113.

Aquino, C. A., 1980, ''La Estrategia de los Asentamientos Humanos en la República Dominicana,'' *Revista Geográfica*, 91/92:41–68.

Augelli, J. P., 1962, ''Agricultural Colonization in the Dominican Republic,'' *Economic Geography*, 38:15–27.

Clausner, M. D., 1973, *Rural Santo Domingo: Settled, Unsettled, and Resettled*, Temple University Press, Philadelphia.

Fuente, S. de la, 1976, *Geografía Dominicana*, Editorial Colegial Quisqueyana, Santo Domingo, D.R.

Girault, C., 1978, La Comercialización del Cafe en la República Dominicana, *Revista Geográfica*, 88:9–60.

Organization of American States, 1967, *Reconocimiento y Evaluación de los Recursos Naturales de la República Dominicana*, Washington, D.C.

Symanski, R., and Burley, N., 1973, ''The Jewish Colony of Sosua,'' *Annals of the Association of American Geographers*, 63:366–378.

World Bank, 1978, *Dominican Republic: Its Main Economic Development Problems*, Washington, D.C.

Puerto Rico

Boswell, T. D., 1977, ''Distance and Migration Selectivity in Puerto Rico Prior to Economic Development,'' *Revista Geográfica*, 85:189–203.

Kimber, C. T., 1973, ''Spatial Patterns in the Dooryard Gardens of Puerto Rico,'' *Geographical Review*, 63:6–26.

Monk, J. J., and Alexander, C. S., 1979, ''Modernization and Rural Population Movements: Western Puerto Rico,'' *Journal of Interamerican Studies*, 21:4, 523–550.

Niddrie, D. L., 1965, ''The Problems of Population Growth in Puerto Rico,'' *Journal of Tropical Geography*, 20:26–33.

Picó, R., 1974, *The Geography of Puerto Rico*, Aldine, Chicago.

Tata, R. J., 1981, ''Cambios en la Conducta Económica de Puerto Rico,'' *Revista Geográfica*, 94:101–115.

Jamaica

Ayub, M. A., 1980, *Made in Jamaica*, Johns Hopkins University Press, Baltimore.

Bounds, J. H., 1979, "Restoration of Mined Land to Farming in Jamaica," *Revista Geográfica,* 80:105–110.

Eyre, L. A., 1972, *Geographic Aspects of Population Dynamics in Jamaica,* Florida Atlantic University Press, Boca Raton.

Eyre, L. A., 1972, "The Shantytowns of Montego Bay, Jamaica," *Geographical Review,* 62:394–413.

Eyre, L. A., 1984, "Political Violence and Urban Geography in Kingston, Jamaica," *Geographical Review,* 74:24–37.

Floyd, B., 1970, "Agricultural Innovations in Jamaica: The Yallahs Valley Land Authority," *Economic Geography,* 46:63–77.

Floyd, B., 1979, *Jamaica: An Island Microcosm,* St. Martin's Press, New York.

Lewis, V. A., 1983, "The Small State Alone: Jamaican Foreign Policy, 1977—1980, *"Journal of Interamerican Studies and World Affairs,* 25:2, 139–170.

Mulchansingh, V. C., 1977, "Extra-Kingston Location of Industry," *Revista Geográfica,* 85:149–187.

Naughton, P. W., 1982, "Some Aspects of Rainfall for Jamaica, West Indies," *Revista Geográfica,* 96:60–64.

Norton, A., and Symanski, R., 1975, "The Internal Marketing System of Jamaica," *Geographical Review,* 65:461–475.

Porter, A.R.D., Jackson, T. A., and Robinson, E., 1982, *Minerals and Rocks of Jamaica,* Jamaica Publishing House, Kingston.

Stone,C., 1982, "Socialism and Agricultural Policies in Jamaica in the 1970's," *Inter-American Economic Affairs,* 35:4, 3–29.

Trinidad and Tobago

Augelli, J. P., and Taylor, H. W., 1960, "Race and Population Patterns in Trinidad," *Annals of the Association of American Geographers,* 50:123–138.

Granger, O. E., 1983, "The Hydroclimatonomy of a Developing Tropical Island: A Water Resources Perspective," *Annals of the Association of American Geographers,* 73:183–205.

Harewood, J., 1963, "Population Growth in Trinidad and Tobago in the Twentieth Century," *Social and Economic Studies,* 12:1–26.

Mulchansingh, V. C., 1983, "Use of Natural Gas in Trinidad and the Point Lisas Pole de Croissance," *Caribbean Geography,* 1:2, 133–139.

Richardson, B. C., 1975, "Livelihood in Rural Trinidad in 1900," *Annals of the Association of American Geographers,* 65:240–251.

The Lesser Antilles

Adams, J. E., 1978, "From Landsmen to Seamen: The Making of a West Indian Fishing Community," *Revista Geográfica,* 88:151–166.

Agor, W. H., 1980, "Private Sector Investment in the Development of the Bahamas: Lessons Learned from Successful Program Implementation, *Inter-American Economic Affairs,* 34:1, 83–96.

Albuquerque, K. de, and McElroy, J., 1983, "Agricultural Resurgence in the United States Virgin Islands," *Caribbean Geography,* 1:2, 121–132.

Benchley, P., 1982, "Boom Times and Buccaneering: The Bahamas," *National Geographic Magazine,* 162:3, 364–395.

Berleant-Schiller, R., 1977, "The Social and Economic Role of Cattle in Barbuda," *Geographical Review,* 67:299–309.

Bounds, J. H., 1972, "Industrialization of the Bahamas," *Revista Geográfica,* 77:95–113.

Bounds, J. H., 1978, "The Bahamas Tourism Industry: Past, Present, and Future," *Revista Geográfica,* 88:177–219.

Bounds, J. H., 1981, "Land Use on New Providence Island, Bahamas, 1960–1979," *Revista Geográfica,* 94:123–153.

Brierley, J. S., 1974, *Small Farming in Grenada,*

West Indies, Department of Geography, University of Manitoba, Winnipeg, Canada.

Carson, M. A., and Tam, S. W., 1977, "The Land Conservation Conundrum of Eastern Barbados," *Annals of the Association of American Geographers*, 67:185–203.

Day, M., 1983, "Doline Morphology and Development in Barbados," *Annals of the Association of American Geographers*, 73:206–219.

Fentem, A. D., 1960, *Commercial Geography of Dominica*, Indiana University, Bloomington.

Fentem, A. D., 1961, *Commercial Geography of Antigua*, Indiana University, Bloomington.

Harris, D. R., 1965, "Plants, Animals, and Man in the Outer Leeward Islands, West Indies: An Ecological Study of Antigua, Barbuda, and Anguilla," *University of California Publications in Geography*, 18:1-184.

Henshall, J. D., and King, L. J., 1966, "Some Structural Characteristics of Peasant Agriculture in Barbados," *Economic Geography*, 42:74–84.

Hoy, D. R., 1962, "Changing Agricultural Land Use on Guadeloupe, French West Indies," *Annals of the Association of American Geographers*, 52:441–454.

Husbands, W., 1983, "The Genesis of Tourism in Barbados: Further Notes on the Welcoming Society," *Caribbean Geography*, 1:2, 107–120.

Kingsbury, R. C., 1960, *Commercial Geography of the British Virgin Islands*, Indiana University, Bloomington.

Long, F., 1983, "Industrialization and the Role of Industrial Development Corporations in a Caribbean Economy: A Study of Barbados 1960–1980," *Inter-American Economic Affairs*, 37:3, 33–56.

McQuillan, D. A., 1984, "Accessibility and the Development of Export Agriculture in Dominica," *Caribbean Geography*, 1:3, 149–163.

Marshall, D. T., 1979, *The Haitian Problem: Illegal Migration to the Bahamas*, University of the West Indies, Kingston, Jamaica.

Oxtoby, F. E., 1970, "The Role of Political Factors in the Virgin Islands Watch Industry," *Geographical Review*, 60:463–474.

Potter, R. B., 1983, "Urban Development, Planning and Demographic Change 1970–1980, in Barbados," *Caribbean Geography*, 1:1, 3–12.

Potter, R. B., 1984, "Mental Maps and Spatial Variations in Residential Desirability: A Barbados Case Study," *Caribbean Geography*, 1:3, 186–197.

The Guianas

Gritzner, C. F., Jr., 1964, "French Guiana Penal Colony: Its Role in Colonial Development," *Journal of Geography*, 63:314–319.

Hope, K. R., 1979, *Development Policy in Guyana: Planning, Finance, and Administration*, Westview Press, Boulder Colo.

Lowenthal, D., 1960, "Population Contrasts in the Guianas," *Geographical Review*, 50: 41–58.

Mandle, J. R., 1973, *The Plantation Economy: Population and Economic Change in Guyana, 1838–1960*, Temple University Press, Philadelphia, Pa.

Nystrom, J. W., 1942, *Surinam: A Geographic Study*, The Netherlands Information Bureau, New York.

Richardson, B. C., 1974, "Distance Regularities in Guyanese Rice Cultivation," *Journal of the Developing Areas*, 8:2, 235–256.

Sackey, J. A., 1979, "Dependence, Underdevelopment and Socialist-Oriented Transformation in Guyana," *Inter-American Economic Affairs*, 33:1, 29–50.

Vining, J. W., 1977, "Presettlement Planning in Guyana," *Geographical Review*, 67:4, 469–480.

Venezuela

Betancourt, J. F., 1978, "Estimating Interstate Internal Migration from 'Place-of-Birth' Data," *Revista Geográfica*, 88:61–78.

Betancourt, R., 1978, *Venezuela's Oil*, Allen and Unwin, Boston.

Bilbas, R. R., 1982, "To Sow or Not to Sow: A Historiographical Essay on the Venezuelan Agrarian Question, 1973–1980," *Revista de la Historia de América*, 94:133–150.

Blank, D. E., 1980, "Oil and Democracy in Venezuela," *Current History*, 78:454, 71–75.

Dufresne, L. S., 1981, "Procesos de Modernización de la Agricultura en los Andes Venezolanos," *Revista Geográfica*, 93:37–60.

Ellner, S., 1980, "Party Dynamics and the Outbreak of Guerrilla Warfare in Venezuela," *Inter-American Economic Affairs*, 34:2, 3–24.

Harris, D., 1971, "The Ecology of Swidden Cultivation in the Upper Orinoco Rain Forest, Venezuela," *Geographical Review*, 61:475–495.

Hendelman, H., 1979, *Scarcity Amidst Plenty in Oil-Rich Venezuela*, Indiana University Press, Bloomington.

Hyde, D. M., 1980, "The Mineral Industry of Venezuela," *Bureau of Mines Yearbook*, Washington, D.C.

Jones, R. C., 1978, "Myth Maps and Migration in Venezuela," *Economic Geography*, 54:75–91.

Jones, R. C., 1982, "Regional Income Inequalities and Government Investment in Venezuela," *Journal of the Developing Areas*, 16:3, 373–389.

Karlsson, W., 1975, *Manufacturing in Venezuela: Studies on Development and Location*, Almqvist and Wiksell, Stockholm.

Lombardi, J. V., 1982, *Venezuela: The Search for Order, the Dream of Progress*, Oxford University Press, New York.

McBeth, B. S., 1983, *Juan Vicente Gómez and the Oil Companies in Venezuela, 1908–1935,* Cambridge University Press, New York.

Sequera de Segnini, I., 1978, *Dinámica de la Agricultura y su Expresión en Venezuela*, Ariel-Seix Barral, Caracas.

Colombia

Aschmann, H., 1960, "Indian Pastoralists of the Guajira Peninsula," *Annals of the Association of American Geographers*, 50:408–18.

Belisle, F. J., and Hoy, D. R., 1980, "The Perceived Impact of Tourism by Residents: A Case Study in Santa Marta, Colombia," *Annals of Tourism Research*, 7:1, 83–101.

Berry, R. A., and Soligo, R. (eds.), 1979, *Economic Policy and Income Distribution in Colombia*, Westview Press, Boulder, Colo.

Blasier, C., 1966, "Power and Social Change in Colombia: The Cauca Valley," *Journal of Inter-American Studies*, 8:386–410.

Brunnschweiler, D., 1972, *The Llanos Frontier of Colombia: Environment and Changing Land Use in Meta*, Latin American Studies Center, Michigan State University, East Lansing.

Edwards, W. M., 1981, "Ten Issues in Carrying Out Land Reform in Colombia," *Inter-American Economic Affairs*, 35:1, 55–68.

Gade, D. W., 1967, "The Guinea Pig in Andean Culture," *Geographical Review*, 57:213–224.

Griffin, E. C., 1974, "The Changing Role of the Río Magdalena in Colombia's Economic Growth," *Geographical Survey*, 3:1, 14–24.

Horna, H., 1982, "Transportation Modernization and Entrepreneurship in Nineteenth Century Colombia," *Journal of Latin American Studies*, 14:1, 33–53.

Kline, H. F., 1981, "The Coal of 'El Cerrejón': An Historical Analysis of Major Colombian Policy Decisions and MNC Activities," *Inter-American Economic Affairs*, 35:3, 69–88.

McGreevey, W. P., 1974, "Urban Growth in Colombia," *Journal of Inter-American Studies,* 16:4, 387–408.

Nelson, R. R., Schultz, T. P., and Slighton, R. L., 1971, *Structural Change in a Developing Economy: Colombia's Problems and Prospects,* Princeton University Press, Princeton, N.J.

O'Dea Gauhan, T., 1977, "Housing and the Urban Poor: The Case of Bogotá, Colombia," *Journal of Inter-American Studies,* 19:1, 99–124.

Ortiz, S. R., 1973, *Uncertainties in Peasant Farming: A Colombian Case,* Humanities Press, New York.

Parsons, J. J., 1948, "Antioqueño Colonization in Western Colombia," *Ibero-Americana* 32, University of California Press, Berkeley.

Parsons, J. J., and Bowen, W. A., 1966, "Ancient Ridged Fields of the San Jorge River Floodplain," *Geographical Review,* 56:317–343.

Shlemon, R. J., and Phelps, L. B., 1971, "Dredge-Tailing Agriculture on the Río Nechi, Colombia," *Geographical Review,* 61:396–414.

Sloan, J. W., 1979, "Regionalism, Political Parties, and Public Policy in Colombia," *Inter-American Economic Affairs,* 33:3, 25–46.

Smith, T. L., 1967, *Colombia: Social Structure and the Process of Development,* University of Florida Press, Gainesville.

Wood, W. A., 1970, "Recent Glacier Fluctuations in the Sierra Nevada de Santa Marta, Colombia," *Geographical Review,* 60:374–392.

Williams, L. S., and Griffin, E. C., 1978, "Rural and Small Town Depopulation in Colombia," *Geographical Review,* 31:13–30.

Ecuador

Agosin, M. R., 1979, "An Analysis of Ecuador's Industrial Development Law," *Journal of the Developing Areas,* 13:3, 263–273.

Basile, D. G., 1974, *Tillers of the Andes,* Department of Geography, University of North Carolina, Chapel Hill.

Bromley, R. J., 1977, *Development Planning in Ecuador,* Centre for Development Studies, University of Wales, Swansea.

Burt, A. L., et al., 1960, "Santo Domingo de los Colorados: A New Pioneer Zone in Ecuador," *Economic Geography,* 36:221–230.

Hiraoka, M., and Yamamoto, S., 1980, "Agricultural Development in the Upper Amazon of Ecuador," *Geographical Review,* 70:4, 423–445.

Hurtado, O., 1981, *Political Power in Ecuador,* University of New Mexico Press, Albuquerque.

Kasza, G. J., 1981, "Regional Conflict in Ecuador: Quito and Guayaquil," *Inter-American Economic Affairs,* 35:2, 3–41.

Martz, J. D., 1980, "The Quest for Popular Democracy in Ecuador," *Current History,* 78:454, 66–70.

Preston, D. A., Taveras, G., and Preston, R. A., 1981, "Emigración Rural y Desarrollo Agrícola en la Sierra Ecuatoriana," *Revista Geográfica,* 93:7–35.

Quintanilla, V. G., "Fitogeografía de las Islas Galápagos: Observaciones Preliminares en la Isla de San Cristóbal," *Revista Geográfica,* 98:58–78.

Redclift, M. R., 1978, *Agrarian Reform and Peasant Organization on the Ecuadorian Coast,* Athlone Press, London.

Revista Geográfica, 1976, Special number dedicated to Ecuador, 84.

Smith, V., 1975, "Marketing Agricultural Commodities in Pichincha Province, Ecuador," *Geographical Review,* 65:3, 353–363.

Stewart, N. R., Belote, J., and Belote, L., 1976, "Transhumance in the Central Andes," *Annals of the Association of American Geographers,* 66:3, 377–397.

U.S. Department of State., 1980, "Ecuador-Peru," *International Boundary Study,* Washington, D.C.

Wood, H. A., 1972, "Spontaneous Agricultural Colonization in Ecuador," *Annals of the Association of American Geographers,* 62:599–617.

Peru

Alberts, T. (ed.), 1983, *Agrarian Reform and Rural Poverty: A Case Study of Peru,* Westview Press, Boulder, Colo.

Amiran, D.H.K., 1970, "El Desierto de Sechura, Peru: Problems of Agricultural Use of Deserts," *Revista Geográfica,* 72:7–12.

Barker, M. L., 1980, "National Parks, Conservation, and Agrarian Reform in Peru," *Geographical Review,* 70:1, 2–18.

Caviedas, N. C., 1975, "El Niño 1972: Its Climatic, Ecological, Human and Economic Implications," *Geographical Review,* 65:4, 493–509.

Cole, J. P., and Mather, P. M., 1972, "Peru Prominence Level Factor Analysis," *Revista Geográfica,* 77:7–32.

Coutu, A. J., and King, R. A., 1969, *The Agricultural Development of Peru,* Praeger, New York.

Davies, T. M., Jr., 1974, *Indian Integration in Peru: A Half Century of Experience, 1900–1948,* University of Nebraska Press, Lincoln.

Denevan, W. M., 1971, "Campo Subsistence in the Gran Pajonal, Eastern Peru," *Geographical Review,* 61:496–518.

Dickinson, J. C., 1969, "The Eucalypt in the Sierra of Southern Peru," *Annals of the Association of American Geographers,* 59:294–307.

Gade, D., and Escobar, M., 1982, "Village Settlement and the Colonial Legacy in Peru," *Geographical Review,* 72:430–449.

Goodsell, C. T., 1975, "The Multinational Corporation as Political Actor in a Changing Environment: Peru," *Inter-American Economic Affairs,* 29:3, 3–21.

Gordon, D. R., 1982, "The Andean Auto Program and Peruvian Development: A Preliminary Assessment," *Journal of the Developing Areas,* 16:2, 233–246.

Gorman, S. M., 1982, "Geopolitics and Peruvian Foreign Policy," *Inter-American Economic Affairs,* 36:2, 65–88.

Gorman, S. M. (ed.), 1982, *Post-Revolutionary Peru: The Politics of Transportaton,* Westview Press, Boulder, Colo.

Handelman, H., 1975, *Struggle in the Andes: Peasant Political Mobilization in Peru,* Institute of Latin American Studies, University of Texas, Austin.

Jameson, K. P., 1979, "Designed to Fail: Twenty-five Years of Industrial Decentralization Policy in Peru," *Journal of the Developing Areas,* 14:1, 55–70.

Kelly, K., 1965, "Land-Use Regions in the Central and Northern Portions of the Inca Empire," *Annals of the Association of American Geographers,* 55:327–338.

Kus, J. S., 1972, "Selected Aspects of Irrigated Agriculture in the Chimú Heartland, Peru," Ph.D. Dissertation, University of California, Los Angeles.

Mikesell, R. E., 1975, *Foreign Investment in Copper Mining: Case Studies of Mines in Peru and Papua New Guinea,* Johns Hopkins University Press, Baltimore.

Miller, R., 1982, "The Coastal Elite and Peruvian Politics," *Journal of Latin American Studies,* 14:1, 97–120.

Morisset, J., 1973, "The Department of Puno as a Territory to Be Developed in Southern Peru: Geographic Views on the Problematics of Integration," *Revista Geográfica,* 79:11–40.

Moseley, M., 1978, "An Empirical Approach to Prehistoric Agrarian Collapse: The Case of the Moche Valley, Peru," *Social and Technological Management in Dry Lands,* Westview Press, Boulder, Colo.

Orlove, B. S., 1977, *Alpacas, Sheep, and Men: The Wool Export Economy and Regional Society in Southern Peru,* Academic Press, New York.

Parsons, J. R., and Psuty, N. P., 1975, "Sunken Fields and Prehistoric Subsistence on the Peruvian Coast," *American Antiquity,* 40:3, 259–281.

Robinson, D. A., 1971, *Peru in Four Dimensions,* Blain Ethridge, Detroit.

Skeldon, R., 1977, "The Evolution of Migration Patterns During Urbanization in Peru," *Geographial Review,* 65:4, 493–509.

Snyder, D. E., 1967, "The 'Carretera Marginal de la Selva,' A Geographical Review and Appraisal," *Revista Geográfica,* 67:87–100.

Stewart, N. R., 1965, "Migration and Settlement in the Peruvian Montaña: The Apurimac Valley," *Geographical Review,* 55:143–157.

Stewart, N. R., Belote, J., and Belote, L., 1976, "Transhumance in the Central Andes," *Annals of the Association of American Geographers,* 66: 377–397.

Stidd, C. K., 1976, "Tradewinds and Soybeans," *Oceans,* 9:4, 30–33.

White, S., 1978, "Cedar and Mahogany Logging in Eastern Peru," *Geographical Review,* 68:394–416.

Bolivia

Buechler, R. M., 1981, *The Mining Society of Potosí, 1776–1810,* University Microfilms International, Ann Arbor, Mich.

Eckstein, S., and Hagopian, F., 1983, "The Limits of Industrialization in the Less Developed World: Bolivia," *Economic Development and Cultural Change,* 32:2, 63–95.

Edelmann, A. T., 1967, "Colonization in Bolivia," *Inter-American Economic Affairs,* 20:39–54.

Fifer, J. V., 1967, "Bolivia's Pioneer Fringe," *Geographical Review,* 57:1–23.

Fifer, J. V., 1972, *Bolivia: Land, Location, and Politics Since 1825,* Cambridge University Press, New York.

Fifer, J. V., 1972, "The Search for a Series of Small Successes: Frontiers of Settlement in Eastern Bolivia," *Journal of Latin America Studies,* 14:2, 407–432.

Gade, D. W., 1970, "Spatial Displacement of Latin American Seats of Government: From Sucre to La Paz as the National Capital of Bolivia," *Revista Geográfica,* 73:43–57.

Glassner, M. I., 1970, "The Río Lauca: Dispute Over an International river," *Geographical Review,* 60:192–207.

Hiraoka, M., 1980, "Settlement and Development of the Upper Amazon: The East Bolivian Example," *Journal of the Developing Areas,* 14:3, 327–347.

Klein, H. S., 1965, "The Creation of the Patiño Tin Empire," *Inter-American Economic Affairs,* 19:3–23.

Morales, W. Q., 1980, "Bolivia Moves Toward Democracy," *Current History,* 78:454, 76–79.

Nash, J., 1979, *We Eat the Mines and the Mines Eat Us: Dependency and Exploitation in Bolivian Tin Mines,* Columbia University Press, New York.

Preston, D. A., 1970, "Freeholding Communities and Rural Development: The Case of Bolivia," *Revista Geográfica,* 73:29–41.

St. John, R. B., 1977, "Hacia el Mar: Bolivia's Quest for a Pacific Port," *Inter-American Economic Affairs,* 31:3, 41–73.

Sheriff, F., 1979, "Cartografía Climática de la Región Andina Boliviana," *Revista Geográfica,* 89:45–68.

South, R. B., 1977, "Coca in Bolivia," *Geographical Review,* 67:22–33.

Stouse, P.A.D., Jr., 1971, "Regional Specialization in Developing Areas: The Altiplano of Bolivia," *Revista Geográfica,* 74:51–70.

Thomas, R. N., and Wittick, R. I., 1981, "Migrant Flows to La Paz, Bolivia, as Related to the Internal Structure of the City: A Methodological Treatment," *Revista Geográfica*, 91:41–51.

Weil, C., 1983, "Migration Among Landholdings by Bolivian Campesinos," *Geographical Review*, 73:182–197.

Weil, C., and Weil, J., 1983, "Government, Competition, and Business in the Bolivian Chapare: A Case Study of Amazonian Occupation," *Inter-American Economic Affairs*, 36:4, 29–62.

Zuvekas, C., Jr., 1979, "Measuring Rural Underemployment in Bolivia," *Inter-American Economic Affairs*, 32:4, 65–83.

Chile

Bahre, C. J., 1979, *Destruction of the Natural Vegetation of North Central Chile*, University of California Press, Berkeley.

Caviedes, C. L., 1971, "Perturbations During the Predominance of Anticyclonic Summer Weather in Central Chile," *Revista Geográfica* 74:71–81.

Caviedes, C. L., 1984, *The Southern Core Realities of the Authoritarian State in South America*, Rowman and Allanheld, Totowa, N.J.

Glassner, M. I., 1969, "Feeding a Desert City: Antofagasta, Chile," *Economic Geography*, 45:339–348.

Glassner, M. I., 1970, "The Río Lauca: Dispute Over an International River," *Geographical Review*, 60:192–207.

MacPhail, D., 1973, "The Geomorphology of the Río Teno Lahar," *Geographical Review*, 63:517–532.

Morales, M., and Labra, P., 1980, "Condicionantes Naturales, Metropolización y Problemas de Planificación del Gran Santiago, Chile," *Revista Geográfica*, 91/92:179–221.

Peña, O., 1982, "Comentario Sobre las Clasificaciones Climáticas en Uso y Proposiciones de Revisión en el Caso Chileno," *Revista Geográfica* 96:65–90.

Porteous, J. D., 1972, "Urban Transplantation in Chile," *Geographical Review*, 62:455–478.

Quintanilla, V. G., 1979, "Los Perfiles Fitogeográficos del Semiarido Chileno," *Revista Geográfica*, 89:69–97.

Rasheed, K.B.S., 1977, "Depopulation of the Oases in Northern Chile," *Revista Geográfica*, 74:101–113.

Salinas, R. M., 1980, "Uso del Suelo y Estructura Urbana de Valparaiso," *Revista Geográfica*, 91/92:153–178.

Stallings, B., 1978, *Class Conflict and Economic Development in Chile, 1958–1973*, Stanford University Press, Stanford, Calif.

Paraguay

Arnold, A. F., 1971, *Foundations of an Agricultural Policy in Paraguay*, Praeger, New York.

Bank of London and South American Review, 1971, "Paraguay: A Change in Development Policy," 5:50.

Gillespie, F., 1983, "Comprehending the Slow Pace of Urbanization in Paraguay Between 1950 and 1972," *Economic Development and Cultural Change*, 31:2, 355–374.

Gobierno del Paraguay—Secretaría General de la O.E.A., 1983, *Desarrollo Regional Integrado del Chaco Paraguayo*, Gobierno de la República del Paraguay, Asunción.

Gorham, J. R., (ed.), 1973, *Paraguay: Ecological Essays*, Academy of Arts and Sciences of the Americas, Miami, Fla.

Grow, M., 1981, *The Good Neighbor Policy and Authoritarianism in Paraguay*, Regents Press of Kansas, Lawrence.

Lewis, P. H., 1980, *Paraguay Under Stroessner*, University of North Carolina Press, Chapel Hill, N.C.

McDonald, R. H., 1981, "The Emerging New Politics in Paraguay," *Inter-American Economic Affairs*, 35:1, 25–44.

Mitchell, G. H., and Barron, T. E., 1974, *A Development View of the Paraguayan Beef Industry*, New Mexico State University, Las Cruces.

Nickson, R. A., 1981, "Brazilian Colonization of the Eastern Border Region of Paraguay," *Journal of Latin American Studies*, 13:111–131.

Stewart, N. R., 1967, *Japanese Colonization in Eastern Paraguay*, National Academy of Sciences—National Research Council, Washington, D.C.

Warren, H. G., 1978, *Paraguay and the Triple Alliance: The Postwar Decade, 1869–1978*, Institute of Latin American Studies, University of Texas, Austin.

Williams, J. H., 1982, "Paraguay's Unchanging Chaco," *Américas*, 34:4, 14–19.

Williams J. H., 1983, "Stroessner's Paraguay," *Current History*, 82:481, 66–68.

World Bank, 1978, *Paraguay: Regional Development in Eastern Paraguay*, Washington, D.C.

World Bank, 1979, *Paraguay: Economic Memorandum*, Washington, D.C.

Young, G., 1982, "Paraguay," *National Geographic Magazine*, 162:2, 240–269.

Argentina

Ackerman, W. F., 1975, "Development Strategy for Cuyo, Argentina," *Annals of the Association of American Geographers*, 65:36–47.

Bank of London and South American Review, 1979, "Argentina," 13:464–470.

Bowen, N., 1975, "The End of British Economic Hegemony in Argentina: Messer-Smith and the Eady-Miranda Agreement," *Inter-American Economic Affairs*, 28:4, 3–24.

Bruniard E. D., and Bolsi, A. S., 1974, "La Región Funcional de Resistencia (Provincia del Chaco, República Argentina), *Revista Geográfica*, 81:7–46.

Carty, W. P., 1981, "Ring Around the City," *Américas*, 33:2, 3–5.

Denis, P., 1980, "Espacio Agrario y Centros Urbanos de la Provincia de Córdoba: Conceptualización Geográfica y Problematica Actual de las Relaciones Ciudad-Campo a Nivel Regional," *Revista Geográfica*, 91/92:101–139.

Eidt, R. C., 1968, "Japanese Agricultural Colonization: A New Attempt at Lane Opening in Argentina," *Economic Geography*, 44:1–20.

Eidt, R. C., 1977, *Pioneer Settlement in Northeast Argentina*, University of Wisconsin Press, Madison.

Federación Argentina de la Industria Gráfica y Afines, 1977, *Argentina: A Latin American Country*, Buenos Aires.

Ferrer, A., 1980, "The Argentine Economy, 1976–1979," *Journal of Inter-American Studies*, 22:2, 131–162.

Goebel, J., 1982, *The Struggle for the Falkland Islands: A Study in Legal and Diplomatic History*, Yale University Press, New Haven, Conn.

Hansis, R. A., 1977, "Land Tenure Hazards, and the Economy: Viticulture in the Mendoza Oasis, Argentina," *Economic Geography*, 53:368–371.

Jones, D. M., 1975, "Shifting Patterns of Sugarcane Production in Northwest Argentina," Ph.D. Dissertation, Michigan State University, East Lansing.

Kirchner, J. A., 1980, *Sugar and Seasonal Labor Migration: The Case of Tucumán, Argentina*, Department of Geography, University of Chicago.

Miller, E. E., 1979, "The Frontier and the Development of Argentine Culture," *Revista Geográfica*, 90:183–198.

Morris, A. S., 1969, "Development of the Irrigation Economy of Mendoza, Argentina,"

Annals of the Association of American Geographers, 59:97–115.

Morris, A. S., 1972, "Dairying in Argentina," *Revista Geográfica,* 76:103–120.

Morris, A. S., 1977, "The Failure of Small Farmer Settlements in Buenos Aires Province, Argentina," *Revista Geográfica,* 85:63–77.

Reina, R. E., 1973, *Paraná: Social Boundaries in an Argentine City,* Institute of Latin American Studies, University of Texas, Austin.

Revista Geográfica, 1982, Special number dedicated to Argentina, 12 articles (all in Spanish).

Rock, D. (ed.), 1975, *Argentina in the Twentieth Century,* University of Pittsburgh Press, Pittsburgh, Pa.

Sargent, C. S., Jr., 1971, "Elements of Urban Plat Development: Greater Buenos Aires, Argentina," *Revista Geográfica,* 74:7–32.

Smith, P. H., 1980, "Argentina: The Uncertain Warriors," *Current History,* 78:454, 62–65, 85–86.

Solberg, C. E., 1982, "Peopling the Prairies and the Pampas," *Journal of Inter-American Studies,* 24:2, 131–161.

Stern, F., 1978, "Between Repression and Reform: A Strange Impression of Argentina and Brazil," *Foreign Affairs,* 56:4, 800–818.

Sternberg, R., 1972, "Occupance of the Humid Pampa 1856–1914," *Revista Geográfica,* 76:61–102.

Tata, R. J., 1977, "Population Geography of Argentina," *Revista Geográfica,* 85:79–95.

Whiteford, S., 1981, *Workers from the North: Plantations, Bolivian Labor, and the City in Northwest Argentina,* University of Texas Press, Austin.

Williams, G., 1979, "Welsh Settlers and Native Americans in Patagonia," *Journal of Latin America Studies,* 11:1, 41–66.

Winsberg, M. D., 1968, *Modern Cattle Breeds in Argentina,* Center for Latin American Studies, University of Kansas, Lawrence.

Winsberg, M. D., 1970, "Una Regionalización Estadística de la Agricultura en la Pampa Argentina," *Revista Geográfica,* 72:46–60.

Wynia, D. P., 1981, "Illusion and Reality in Argentina," *Current History,* 80:463, 62–65.

Uruguay

Alisky, M., 1969, *Uruguay: A Contemporary Survey,* Praeger, New York.

Daly, H. E., 1965, "Uruguayan Economy: Its Basic Nature and Current Problems," *Journal of Inter-American Studies,* 8:316–330.

Fletcher, L. B., and Merrill, W. C., 1970, *Uruguay's Agricultural Sector: Priorities for Policies, Investment Programs and Projects,* Inter-American Development Bank, Washington, D.C.

Griffin, E., 1972, "The Agricultural Land Use Regions of Uruguay," *Revista Geográfica,* 76:121–151.

Griffin, E., 1973, "Testing the Von Thünen Theory in Uruguay," *Geographical Review,* 63:500–516.

Griffin, E., 1974, "Causal Factors Influencing Agricultural Land Use Patterns in Uruguay," *Revista Geográfica,* 80:13–33.

Kleinpenning, J.M.G., 1981, "Uruguay: The Rise and Fall of a Welfare State Seen Against a Background of Dependency Theory," *Revista Geográfica,* 93:101–117.

McDonald, R. H., 1972, "Electoral Politics and Urguayan Political Decay," *Inter-American Economic Affairs,* 26:1, 25–45.

McDonald, R. H., 1975, "The Rise of Military Politics in Uruguay," *Inter-American Economic Affairs,* 28:4, 25–43.

Prozecanski, A. C., 1977, "Authoritarian Uruguay," *Current History,* 72:424, 73–75, 85–86.

Redding, D. C., 1967, "The Economic Decline of Uruguay," *Inter-American Economic Affairs,* 20:55–72.

Vanger, M. A., 1980, *The Model Country. José Batlle y Ordóñez of Uruguay, 1907–1915,* University Press of New England, Hanover, N.H.

World Bank, 1979, *Uruguay: Economic Memorandum,* Washington, D.C.

Brazil

Baer, W., 1979, *The Brazilian Economy: Its Growth and Development,* Grid Publishing, Columbus, Ohio.

Baumann Neves, R., 1982, "The Expansion of Manufactured Exports," *Bank of London and South American Review,* 16:2, 64–76.

Bergman, R. W., 1980, *Amazon Economics,* University Microfilms International, Ann Arbor, Mich.

Brazilian Embassy, London, 1976, *Amazonia,* Belmont Press, Northampton, U.K.

Brazilian Embassy, London, 1976, *Brazil: A Geography,* Belmont Press, Northampton, U.K.

Bunker, S. G., 1982, "The Cost of Modernity: Inappropraite Bureaucracy, Inequality, and Development Program Failure in the Brazilian Amazon," *Journal of the Developing Areas,* 16:573–596.

Cole, M. M., 1960, "Cerrado, Caatinga, and Pantanal: The Distribution and Origin of Savanna Vegetation of Brazil," *Geographical Journal,* 126:168–179.

Colson, R. F., 1981, "The Proalcool Programme: A Response to the Energy Crisis," *Bank of London and South American Review,* 15:2, 60–70.

Correia de Andrade, M., 1980, *The Land and People of Northeast Brazil,* University of New Mexico Press, Albuquerque.

Cunningham, S. M., 1980, "Recent Developments in The Centre-West Region," *Bank of London and South American Review,* 14:2, 44–51.

Dawsey, C. B., III, 1979, "Income and Residential Location in Piraçicaba, São Paulo, Brazil," *Revista Geográfica,* 89:185–189.

Dawsey, C. B., III, 1983, "Push Factors and Pre-1970 Migration to Southwest Paraná, Brazil," *Revista Geográfica,* 98:54–57.

Dean, W., 1966, "The Planter as Entrepreneur: The Case of São Paulo," *Hispanic American Historical Review,* 46:138–152.

Denevan, W. D., 1970, "The Aboriginal Population of Western Amazonia in Relation to Habitat and Subsistence," *Revista Geográfica,* 72:61–86.

Dickenson, J. P., 1978, *Brazil: Studies in Industrial Geography,* Westview Press, Boulder, Colo.

Dozier, C. L., 1956, "Northern Paraná, Brazil: An Example of Organized Regional Development," *Geographical Review,* 46:318–333.

Elias de Castro, I., 1983, "Housing Projects: Widening the Controversy Surrounding the Removal of 'Favelas'," *Revista Geográfica,* 97:56–69.

Epstein, D. G., 1973, *Brasília: Plan and Reality,* University of California Press, Berkeley.

Evenson, N., 1973, *Two Brazilian Capitals: Architecture and Urbanism in Rio de Janeiro and Brasília,* Yale University Press, New Haven.

Faissol, S. (ed.), 1975, *Urbanização e Regonalização: Relações com o Desenvolvimento Econômico,* Instituto Brasileiro de Geografia e Estatística, Rio de Janeiro.

Fishlow, A., 1978/1979, "Flying Down to Rio: Perspectives on U.S.-Brazil Relations," *Foreign Affairs,* 57:2, 387–404.

Forman, S., 1975, *The Brazilian Peasantry,* Columbia University Press, New York.

Galey, J., 1979, "Industrialist in the Wilderness: Henry Ford's Amazon Venture," *Journal of Inter-American Studies,* 21:2, 261–289.

Galloway, J. H., 1968, "The Sugar Industry of Pernambuco During the Nineteenth Century," *Annals of the Association of American Geographers,* 58:285–303.

Gauld, C. A., 1964, "The Last Titan, Percival Farquar: American Entrepreneur in Latin America," *Hispanic American Reports,* Special Issue.

Gauthier, H. L., 1975, "Migration Theory and the Brazilian Experience," *Revista Geográfica*, 82:51–61.

Goodland, R.J.A., and Irwin, H. S., 1975, *Amazon Jungle: Green Hell or Red Desert?*, Elsener Scientific Publishing, New York.

Haller, A. O., 1982, "A Socio-economic Regionalization of Brazil," *Geographical Review*, 72:4, 450–464.

Henshall, J. D., and Momsen, R. P., Jr., 1976, *A Geography of Brazilian Development*, Westview Press, Boulder, Colo.

Hicks, R. H., 1980, "Agricultural Land Use and Related Innovation and Government Assistance in Rio Grande do Sul, Brazil," Ph.D. Dissertation, Michigan State University, East Lansing.

James, P. E., 1952, "Observations on the Physical Geography of Northeast Brazil," *Annals of the Association of American Geographers*, 42:153–176.

James, P. E., 1953, "Patterns of Land Use in Northeast Brazil," *Annals of the Association of American Geographers*, 43:98–126.

Johnson, D., 1979, "Agroclimatological Location of Maize and Grain Sorghum in Northeast Brazil," *Revista Geográfica*, 89:37–43.

Johnson, F. I., 1982, "Sugar in Brazil: Policy and Production," *Journal of Developing Areas*, 17:2, 243–256.

Katzman, M. T., 1977, *Cities and Frontiers in Brazil: Regional Dimensions of Economic Development*, Harvard University Press, Cambridge, Mass.

Kemp, G., 1978, "Scarcity and Strategy," *Foreign Affairs*, 56:2, 398–411.

Kerr, J. A., 1975, "Cattle-Raising in Brazil," *Revista Geográfica*, 83:95–108.

Knight, P. T., 1971, *Brazilian Agricultural Technology and Trade: A Study of Five Commodities*, Praeger, New York.

Kohlhepp, G., 1981, "Ócupação e Valorização Econômica da Amazonia," *Revista Geográfica*, 94:67–88.

Lecocq Muller, N. 1983, "Demographic Growth and Urban Expansion in the Metropolitan Area of São Paulo," *Revista Geográfica*, 97:29–30.

Levine, R. M., 1980, "Brazil: Democracy Without Adjectives," *Current History*, 78:454, 49–52, 82–83.

Love, J. L., 1980, *São Paulo in the Brazilian Federation, 1889–1937*, Stanford University Press, Stanford, Calif.

McDowell, E., 1980, "Japanese in Brazil," *Américas*, 32:5, 33–38, Washington, D.C.

Mahar, D. J., 1976, "Fiscal Incentives for Regional Development: A Case Study of the Western Amazon Basin," *Journal of Inter-American Study*, 18:3, 357–378.

Malloy, J. M., 1979, *The Politics of Social Security in Brazil*, University of Pittsburgh Press, Pittsburgh, Pa.

Martine, G., 1979, "Adaptation of Migrants or Survival of the Fittest?: A Brazilian Case," *Journal of the Developing Areas*, 14:1, 23–41.

Merrick, T. W., and Graham, D. H., 1979, *Population and Economic Development in Brazil: 1800 to the Present*, Johns Hopkins University Press, Baltimore.

Momsen, R. P., 1979, "Projeto Radam: A Better Look at the Brazilian Tropics," *Geo Journal*, 3:1, 3–14, Wiesbaden, Federal Republic of Germany.

Momsen, R. P., Jr., 1963, "Routes Over the Serra do Mar," *Revista Geográfica*, 32:5–167.

Moran, E. F., 1981, *Developing the Amazon: The Social and Ecological Consequences of Government-Directed Colonization Along Brazil's Transamazonian Highway*, Indiana University Press, Bloomington.

Mougeot, L. J. A., 1982, "Ascenção Socio-econômica e Retenção Migratoria Durante o Desenvolvimento da Fronteira na Região Norte do Brazil," *Revista Geográfica*, 96:107–130.

Needell, J. D., 1983, "Rio de Janeiro at the Turn of the Century: Modernizaton and the Parisian Ideal," *Journal of Inter-American Studies*, 25:1, 83-103.

Nimer, E., 1964, "Circulação, Atmosférica do Nordeste e Suas Consequencias — o Fenomeno, das Secas," *Revista Brasileira de Geografia*, 26:147–157.

Oltman, R. E., et al., 1964, *Amazon River Investigations, Reconnaissance Measurements of July 1963*, U.S. Geological Survey, Washington, D.C.

Paiva, R. M., Schattan, S., and Trench de Freitas, C. F., 1973, *Brazil's Agricultural Sector: Economic Behavior, Problems and Possibilities*, XV International Conference of Agricultural Economists, São Paulo.

Poppino, R. E., 1973, *Brazil: The Land and the People*, Latin American Histories, Oxford University Press, London.

Pyle, G. F., 1970, "Some Geographical Aspects Basic to Understanding Brazilian Urbanization," *Revista Geográfica*, 73:5–27.

Robock, S. H., 1963, *Brazil's Developing Northeast: A Study of Regional Planning and Foreign Aid*, Brookings Institution, Washington, D.C.

Roett, R. (ed.), 1976, *Brazil in the Seventies*, American Enterprise Institute for Public Policy Research, Washington, D.C.

Sahota, G. S., and Rocca, C. A., 1981, "Process of Production and Distribution in Brazilian Agriculture," *Economic Development and Cultural Change*, 29:4, 683–721.

Sanchez, P. A., et al., 1982, "Amazon Basin Soils: Management for Continuous Crop Production," *Science*, 216:821–827.

Santos, M., 1963, "La Culture du Cacao dans l'État de Bahia," *Les Cahiers d'Outre-Mer*, 16:366–378.

Saunders, J. (ed.), 1971, *Modern Brazil: New Patterns and Development*, University of Florida Press, Gainesville.

Selcher, W. A., 1978, *Brazil's Multilateral Relations: Between First and Third Worlds*, Westview Press, Boulder, Colo.

Semple, R. K., Gauthier, H. L., and Youngmann, C. E., 1972, "Growth Poles in São Paulo, Brazil," *Annals of the Association of American Geographers*, 62:591–598.

Siqueira, I., and Ferreira de Souza, A., 1983, "The Evolution of Metropolitan Spaces in Brazil," *Revista Geográfica*, 97:10–28.

Slade, J., III, 1979, "Nordic Farmers and Latin Cattle Barons Faced by German Settlers in Southern Brazil," *Revista de Historia de América*, 87:127–139.

Smith, N., 1980, "Anthrosols and Human Carrying Capacity in Amazonia," *Annals of the Association of American Geographers*, 70:4, 553–566.

Smith, N.J.H., 1978, "Agricultural Productivity Along Brazil's Transamazon Highway," *Agro-Ecosystems*, 4:415–432.

Smith, N.J.H., 1981, *Man, Fishes, and the Amazon*, Columbia University Press, New York.

Smith, N.J.H., 1982, *Rainforest Corridors: The Transamazon Colonization Scheme*, University of California, Berkeley.

Smith, P. S., 1983, "Reaping the Whirlwind: Brazil's Energy Crisis in Historical Perspective," *Inter-American Economic Affairs*, 37:1, 3–20.

Smith, T. L., 1972, *Brazil: People and Institutions*, 4th ed., Louisiana State University Press, Baton Rouge.

Sternberg, H. O'R., 1975, "The Amazon River of Brazil," Franz Steiner Varlag GMBH, Wiesbaden, West Germany.

Stevens, R. L., and Brandão, P. R., 1961, "Diversification of the Economy of the Cacao Coast of Bahia (Brazil)," *Economic Geography*, 37:231–253.

Taylor, H. W., 1973, "São Paulo's Hollow Frontier," *Revista Geográfica*, 79:149–166.

Thomas, V., 1981, *Pollution Control in São Paulo, Brazil: Costs, Benefits and Effects on Industrial Location*, World Bank, Washington, D.C.

Trends and Perspectives of the Brazilian Economy, 1980, 26, Banco Lor Brasileiro, S.A.-Chase, New York.

Webb, K. E., 1959, "Origins and Development of a Food Economy in Central Minas Gerais,"

Annals of the Association of American Geographers, 49:409–419.

Webb, K. E., 1974, *The Changing Face of Northeast Brazil,* Columbia University Press, New York.

Wesche, R., "Planned Rainforest Family Farming on Brazil's Transamazonic Highway," *Revista Geográfica,* 81:105–114.

World Bank, 1979, *Brazil: Human Resources Special Report,* Washington, D.C.

B

GUIDES TO PRONUNCIATION

Spanish

Syllabication

A word has as many syllables as it has single vowels (Bu-ca-ra-man'-ga) or vowels and diphthongs, such as *ue* or *ai*, (Bue'-nos Ai'-res). A single consonant between vowels forms a syllable with the following vowel (Li'-ma). Two consonants between vowels are separated (San-tia'-go), except when the second consonant is *l* or *r*, in which case both consonants form a syllable with the following vowel (Chi-cla'-yo, Su'-cre). **ch, ll,** and **rr** are considered single letters in Spanish and cannot be separated.

Accent or Stress

In general, words ending in a consonant, except *n* or *s*, are stressed on the last syllable (La Li-ber-tad'). Words ending in a vowel, or in *n* or *s*, are stressed on the next to last syllable (Gra-na'-da, Ma-ni-za'-les). Exceptions to this rule are shown by a written accent on the vowel of the stressed syllable [Que-ré-ta-ro, Tu-cu-mán, Bo-lí-var).

Vowels

a is pronounced as in f*a*ther (H*a*-b*a*'-n*a*). **e** is pronounced as *a* in f*a*te (San Jo-s*é*). **i** is pronounced as in mach*i*ne (N*i*-ca-ra'-gua). **o** is pronounced as in n*o*te (C*o*-lón). **u** is pronounced as in fl*u*te (Pe-r*ú*). **u** is silent after **q** (*Qu*e-re-ta-ro); **it is** also silent in the combinations **gue** and **gui,** in which case it makes the *g* hard, as in *g*o (San Mi-*gu*el'). A diphthong is formed if the *u* in *gue* bears a diaeresis (Ca-ma-*güey*', pronounced Ca-ma-*gwāĭ*). **y** is also a consonant (see under consonants).

Diphthongs

Consist of a strong vowel (*a, e,* or *o*) and a weak vowel (*i* [*y*] *or u*), or of the two weak vowels, and are stressed on the strong vowel. Common diphthongs are: **ua** and **ue,** the *u* approximating the sound of *u* in q*u*ality (G*ua*n-to, pronounced Gwan'-to; B*ue*-na-ven-tú-ra, pronounced B*wā*-na-ven-tú-ra); **ai** (or **ay**) pronounced like *ai* in *ai*sle (B*ue*'-nos *Ai*'-res); **ei** (or **ey**) pronounced like *ey* in th*ey* (N*ei*'-va; **oi** (or **oy**) pronounced like *oy* in b*oy* (To-ron-t*oy*'); and **ia** in which the two syllables are slightly slurred in pronouncing them (San-t*ia*-go). Two strong vowels do not form a diphthong (Ca-l*la*'-o).

Consonants

d, f, l, m, n, p, and t

These are pronounced as in English. **b** and **v** are similar to a combination of *b* and *v* in English. **c** before *e* or *i* is pronounced as *s* in similar (Va-len'-*c*ia); otherwise as in *cac*tus (Ca-ra'-*c*as). **ch** is pronounced as in *ch*ur*ch* (*Ch*ia'-pas). **g** is pronounced as in *g*o (Bo-*g*o-tá), but **g,** before *e* or *i*, and **j** are similar to the German *ch* (like English *h* forcibly hissed), as in Car-ta-*ge*'-na, *Ja*-lis'-co. (See also the explanation of **gu** under vowels and diphthongs.) **h** is always silent (*H*on-du-ras, pronounced On-du'-ras, and *H*ua-nu-co, pronounced Wa-nu'-co). **k** is not a Spanish letter; it is found only in foreign words, in which it has the same sound as in English. **ll** in Central America and in parts of Mexico is pronounced as *y* in *y*es (Ciu-dad Tru-ji-*llo*, pronounced Sēudad Truhē-*y*o); in Spain and in some parts of Spanish America it has the sound of *lli* in mi*lli*on (Vi-*lla*-ri-ca [Paraguay], pronounced Ve*ly*are'ca); in southern South America and parts of Mexico it has the sound of *zh* in a*z*ure (A-ve-*lla*-neda, pronounced Avā*zh*anaā'da). **ñ** is pronounced like *ny* in can-*y*on (Na-ri-*ñ*o, pronounced Narē-*ny*o). **q** is always followed by *u* and is pronounced as in li*qu*or (*Qu*e-ré-ta-ro). **r** is slightly trilled on the tip of the tongue; initial **r** is pronounced with more vibration, and **rr** is pronounced like initial *r*. **s** is pronounced as in similar. **x,** in Mexico, when between vowels, is pronounced like Spanish *j* (O-a-*x*a'-ca); otherwise like *s* in similar (Ta*x*-co, pronounced Tas'co); in other parts of Spanish America it is generally pronounced as in ta*x*. **y** as a consonant is pronounced as in *y*et, but with

more force (Yu-ca-tán). **z** is always pronounced as *s* in similar.

Portuguese

Syllabication

A word has as many syllables as it has single vowels (Pa-ra-ná) or vowels and diphthongs (São Pau'-lo). A single consonant between vowels forms a syllable with the following vowel (A-ma-zo'-nas). Two consonants between vowels are separated (San'-tos).

Vowels

a is pronounced as in f*a*ther (P*a*-rá). **e** is pronounced either as in b*e*t (P*e*r-nam-bu'-co) or as *a* in f*a*te (C*e*-a-rá). **i** varies from m*i*lk (Es-p*i*-ri-to San'-to) to mach*i*ne (San'-ta Ca-ta-r*i*'na). **o** varies from m*o*ral to c*o*ld, and, when final, is pronounced like *u* in fl*u*te (Cam-po F*o*r-m*o-s*o, pronounced Cam' pu Formō' su). **u** is pronounced as in fl*u*te (Per-nam-b*u*'-co). It is silent in the combination **gue** and **gui,** in which case it makes the *g* hard as in *g*o; the combination **gua** forms a diphthong as in Spanish (*Gua*-ra-tingue-tá, pronounced *Gwa*ratingātá). For the pronunciation of **u** after **q**, see **q** below.

Diphthongs

Consist of a strong vowel (*a, e,* or *o*) and a weak vowel (*i* or *u*), as in Mi'-nas Ge-r*ais*', or of the two weak vowels, as in J*uiz*' de Fo'-ra. Exception to this rule are shown by a written accent (Pa-ra-í-ba). Two strong vowels do not form a diphthong (A-la-gô'-as).

Nasalization

The tilde always nasalizes the vowel it covers, silencing the following vowel and resulting in a sound approximating *owng* (São Pau-lo, pronounced, approximately, Sowng Pow'-loo). Final *m* nasalizes the preceding vowel (Be-lém

[Bāleng]; Jardim [Zharding']), as does final *ns* (To-can-tins [Tocantings']).

Consonants

b, d, f, k, l, p, t, and **v** are pronounced as in English. **c,** before the vowels *a* or *o* and before all consonants, is pronounced as in *ca*ctus (Cam'-pos); **c,** before the vowels *e* or *i,* and **ç** are pronounced as in cement (Ce-a-rá, Al-co-bá-ça). **ch** is pronounced as *sh* in *sh*awl (C*h*a-pa'da). **g,** before *e* or *i,* is pronounced as *zh* in a*z*ure (Mi'nas Ge-rais'); otherwise it is pronounced as in *g*o (Por'-to A-le'-*g*re). **h,** following *l* is pronounced as *lli* in mi*lli*on (Il-*h*e-os, pronounced Il*y*ā'ozh); following *n,* it is pronounced as *ny* in ca*ny*on (U-be-ra-bin-*h*a, pronounced Uberabin'*y*a); otherwise it is silent. **j** is pronounced as *zh* in a*z*ure (São João, pronounced Sawng Z*h*ōawang'). **m** is pronounced as in *m*other (Re-*m*e'-dios], except when it ends a word, in which case it loses its identity, combining with the preceding vowel to form a nasal (Be-lé*m*). **n,** before hard *g,* is pronounced as in si*ng* (Guarati*n*guetá); **n,** before final *s,* nasalizes the preceding vowel (Tocan-tins); otherwise it is pronounced as in *n*ame (Dia-man-ti'*n*a). **q** is always followed by *u;* **qu** before *e* or *i* is pronounced as in li*qu*or (Pe-*que*'-no); before *a* or *o* it forms a diphthong and is pronounced *kw* (Puer-ta Je-ri-co-a-*qua*-ra, pronounced Pwer'ta Zhericōakwara). **r** is pronounced in the throat and slightly trilled. **s,** between vowels, is pronounced as in ro*s*e (Cam'po Formo'*s*o). When final, or when preceding *b, v, d, g, l, m, n,* or *r,* it is pronounced, in Portugal and in Rio de Janeiro as *zh* a a*z*ure (Mi-na*s* Ge-rais [Mē'na*zh* Zharizh']); in the outlying districts of Brazil it is pronounced as *s* in *s*imilar in these instances. In all other cases it is pronounced as *s* in *s*imilar. **x,** between vowels, is pronounced as *z* in E*z*ekiel (Fa-*xi*'-ma [Fa*z*ēma]; otherwise, in Portugal and Rio de Janeiro, it is pronounced as *sh* in *sh*awl (Xin-gú [*Sh*ingu']), and in the outlying districts of Brazil as *s* in *s*imilar (Xingú [*S*ingu']). Recent regulations have excluded the use of the letter **y. z,** before a vowel, is pronounced as in E*z*ekiel (San'-ta Lu-*z*i'-a); otherwise, in Portugal and Rio de Janeiro, it is pronounced as *sh* in *sh*awl or *zh* in a*z*ure (Santa Cruz [San'ta Cru*sh*']), and in the outlying districts of Brazil it is pronounced as *s* in *s*imilar (Santa Cruz [San'ta Cru*s*']).

Photo Credits

INDEX